Brief Applied Calculus

Frank C. Wilson

Green River Community College

HOUGHTON MIFFLIN COMPANY

Boston   New York

Publisher: Richard Stratton
Senior Sponsoring Editor: Molly Taylor
Editorial Associate: Andrew Lipsett
Project Editor: Shelley Dickerson
Editorial Assistant: Katherine Roz
Project Manager: James Lonergan
Senior Marketing Manager: Jennifer Jones
Marketing Associate: Mary Legere

Cover photographs: © Michael Rosenfeld/Getty Images; © Royalty-Free/Corbis; © Dan Herrick/ZUMA/Corbis; © Creatas Images/Jupiterimages

Photo Credits: **Chapter 1**: p. 1, © Lou Dematteis/Reuther/CORBIS; p. 43, © Richard-Cummins/CORBIS; **Chapter 2**: p. 53, © Frederic J. Brown/Getty Images; p.102, © Reuters/CORBIS;p.154, Bureau of Labor Statistics; **Chapter 3**: p. 155, © Alex Fezer/CORBIS; p. 166, © Tony Duffy/Getty Images; **Chapter 4**: p. 216, © Royalty Free/CORBIS; p.233, © Taxi/Getty Images; **Chapter 5**: p. 261, © Ariel Skelley/CORBIS; p. 326, © Stone/Getty Images; **Chapter 6**: p. 338, © Karen Kasmauski/CORBIS; p. 355, © Reuters/CORBIS; **Chapter 7**: p. 406, © Royalty Free/CORBIS; p. 442, © Arthur Morris/CORBIS; p. 449, © Lawson Wood/CORBIS; p. 449, © Jeffrey L. Rotman/CORBIS; p.449, ©Stuart Westmoreland/CORBIS; p. 456, © David Fleetham/CORBIS; **Chapter 8**: p. 459,© Royalty Free/CORBIS; p. 477, Author; p. 477, Author; p. 477, Author; p. 477, Author.

Printed in the U.S.A.
Library of Congress Control Number: 2006937100

Instructor's exam copy:
ISBN 13: 978-0-618-91001-4
ISBN 10: 0-618-91001-8

For orders, use student text ISBNs:
ISBN 13: 978-0-618-61105-8
ISBN 10: 0-618-61105-3

1 2 3 4 5 6 7 8 9 – VH – 11 10 09 08 07

Contents

Preface

To the Student

Have you ever asked, "When am I ever going to use this?" or "Why do I care?" after learning a new mathematical procedure? Many students have. This book seeks to answer those questions by teaching mathematical skills which, when applied, can improve your quality of life. Whether calculating how long it will take to pay off a car loan or predicting how much tuition will be next year, examples and exercises in this book are based on interesting and engaging real-life data. Consequently, when you find a mathematical solution, you will learn something new about the world in which we live. To make real-life data analysis even more meaningful, you will become skilled in collecting and analyzing data from your own life through the *Make It Real* projects. The analytical skills learned through these projects will remain with you long after you have left this course.

This book is written in a casual, reader-friendly style. Although key mathematical terms and concepts are appropriately addressed, the focus is conceptual understanding not mathematical jargon. If you are pursuing a degree in business, social science, or a similar field, this book is written specifically for you.

You will find that your understanding of key concepts will be enhanced through the use of a graphing calculator. The TI-83 Plus (or the TI-83 or TI-84) is the ideal calculator for this course. Because learning how to use the calculator is a challenge for many students, Technology Tips are integrated throughout the text allowing you to learn a new calculator technique when it is needed. These tips detail how to graph a function, solve an equation, find the maximum value of a function, and so on. Rather than giving a broad overview of a procedure, the tips take you through the actual keystrokes on the TI-83 Plus and show you calculator screenshots so that you can verify that you're doing each step correctly.

I have learned much about teaching from my students and deeply value their input. Likewise, I'm interested in hearing from you. Let me know how this book works for you and feel free to share any feedback you may have on how to improve this text. See the "contact us" form on the text's website: **college.hmco.com/pic/wilsonBAC**. Or, from the text's website, click on my name to find my biography, which includes a link to my website.

To the Instructor

Thank you for selecting this book. I believe you will find its approach refreshing and its content interesting to you and your students. It is written specifically for students pursuing business, social science, or a related field. As you know, many of these students do not enjoy mathematics and are taking this course only because it is required for their major. (Many business schools require a finite math—applied calculus—statistics sequence.) Several features are included in this text to make the course content more accessible to these students including:

- an informal writing style that emphasizes conceptual understanding without becoming bogged down in mathematical jargon

- examples and exercises throughout the text based on interesting and engaging real-life data. Over 500 real-life applications, featuring over 70 businesses, products, and associations, help to make math real for your students! (See Index of Applications and Index of Businesses, Products, and Associations inside the front and back covers at the very beginning and end of the book.)
- *Make It Real* projects which allow students to collect and analyze data relevant to their personal lives
- detailed *Technology Tips* (including screenshots) which teach students how to use the graphing calculator as a tool to analyze real-life data
- a bare-bones approach to course content. Topics covered in detail include those that are most relevant to everyday people

I believe that candid feedback from colleagues is helpful in enhancing teaching effectiveness. Please feel free to contact me with any recommendations, comments, or other feedback that you feel will enhance the effectiveness of future editions of this text. See the "contact us" form on the text's website: **college.hmco.com/pic/wilsonBAC.** Or, from the text's website, click on my name to find my biography, which includes a link to my website.

Disclaimer

In this book, I have incorporated real-world data from the financial markets to the medical field. In each case, I have done my best to present the data accurately and interpret the data realistically. However, I do not claim to be an expert in financial, medical, and other similar fields. My interpretations of real-world data and my associated conclusions may not adequately consider all relevant factors. Therefore, readers are encouraged to seek professional advice from experts in the appropriate fields before making decisions related to the topics addressed herein.

Despite the usefulness of mathematical models as representations of real-world data sets, most mathematical models have a certain level of error. It is common for model results to differ from raw data set values. Consequently, conclusions drawn from a mathematical model may differ (sometimes dramatically) from conclusions drawn by looking at raw data sets. Readers are encouraged to interpret model results with this understanding.

Acknowledgments

This textbook would not have been possible without the contributions of many colleagues. I greatly appreciate all of the people who contributed time and talent to bring this book to fruition.

The feedback from the following reviewers was invaluable and helped to shape the final form of the text: Bill Ardis, *Collin County Community College, TX;* James J. Ball, *Indiana State University, IN;* Michael L. Berry, *West Virginia Wesleyan College, WV;* Marcelle Bessman, *Jacksonville University, FL;* Mike Bosch, *Iowa Lakes Community College, IA;* Emily Bronstein, *Prince George's Community College, MD;* Dean S. Burbank, *Gulf Coast Community College, FL;* Andra Buxkemper, *Bunn College, TX;* Roxanne Byrne, *University of Colorado—Denver, CO;* Scott A. Clary, *Florida Institute of Technology, FL;* David Collingwood, *University of Washington, WA;* Mark A. Crawford Jr., *Western Michigan University, MI;* Khaled Dib, *University of Minnesota Duluth, MN;* Lance D. Drager, *Texas Tech University, TX;* Klara Grodzinsky, *Georgia Institute of Technology, GA;* Lucy L. Hanks, *Virginia Polytechnic Institute and State Uni-*

versity, VA; Jean B. Harper, *State Univiversity of N.Y. – College at Fredonia, NY;* Kevin M. Jenerette, *Coastal Carolina University, SC;* Cynthia Kaus, *Metropolitan State University, MN;* Michael LaValle, *Rochester Community and Technical College, MN;* Roger D. Lee, *Salt Lake Community College, UT;* Lia Liu, *University of Illinois Chicago, IL;* Alan Mabry, *Unitersity of Texas at El Paso;* Quincy Magby, *Arizona Western College, AZ;* Mary M. Marco, *Bucks County Community College, PA;* Nicholas Martin, *Shepherd College, WV;* William C. McClure, *Orange Coast College, CA;* James McGlothin, *Lower Columbia College, WA;* Victoria Neagoe, *Goldey Beacom College;* David W. Nelson, *Green River Community College;* Ralph W. Oberste-Vorth, *Marshall University, WV;* Armando I. Perez, *Laredo Community College;* Cyril Petras, *Lord Fairfax Community College, VA;* Mihaela Poplicher, *University of Cincinnati;* John E. Porter, *Murray State University, KY;* David W. Roach, *Murray State University, KY;* R. A. Rock, *Daniel Webster College, NH;* Arthur Rosenthal, *Salem State College, MA;* Kimmo I. Rosenthal, *Union College, NY;* Sharon Mayhew Saxton, *Cascadia Community College, WA;* Edwin Shapiro, *University of San Francisco, CA;* Denise Szecsei, *Stetson University;* Abolhassan S. Taghavy, *Richard J. Daley College, IL;* Muhammad Usman, *University of Cincinnati, OH;* Jorge R. Viramontes Olivas, *University of Texas at El Paso, TX;* Beverly Vredevelt, *Spokane Falls Community College, WA;* Michael L. Wright, *Cossatot Community College, AR.*

I extend special thanks to Cindy Harvey, who provided detailed recommendations on how to improve an early draft of this text, to Helen Medley, who reviewed the final draft text for accuracy, and to Paul Lorczak, who conducted the accuracy review for the typeset manuscript. Their suggestions greatly enhanced the quality of the text.

My appreciation goes to Rob Jonas, who wrote most of the solutions to the text.

I appreciate the exceptional work of my student, Jon Austin, who developed some of the mathematical models used in the text.

I'm thankful for Erica Carlson, who assisted with the art manuscript and for Uli Gersiek, who created most of the artwork in the text.

I also thank Texas Instruments for providing a way to enhance student learning by creating the TI-83 Plus calculator. Without it, the Technology Tips would not have been possible.

On a personal note, I could not have written this text without the tireless support of my wife, Shelley Wilson. She worked overtime caring for our home and children for the three years I worked on the text. Despite this challenge, she never faltered in offering me encouragement, support, and love. For this, I am eternally grateful and indebted to her.

Frank C. Wilson

Real Applications! Real Data! Real Learning!

Chapter 3

The Derivative

It is impossible to determine how quickly a person is running from a single photograph, since speed is calculated as a change in distance over a change in time. Nevertheless, we may estimate a person's speed at a particular instant in time by determining the distance traveled over a small interval of time (e.g., one second). A runner's speed may be classified as a rate of change in distance. A key component of calculus is the study of rates of change.

3.1 Average Rate of Change
- Calculate the average rate of change of a function over an interval

3.2 Limits and Instantaneous Rates of Change
- Estimate the instantaneous rate of change of a function at a point
- Use derivative notation and terminology to describe instantaneous rates of change

3.3 The Derivative as a Slope: Graphical Methods
- Find the equation of the tangent line of a curve at a given point
- Numerically approximate derivatives from a table of data
- Graphically interpret average and instantaneous rates of change
- Use derivative notation and terminology to describe instantaneous rates of change

3.4 The Derivative as a Function: Algebraic Method
- Use the limit definition of the derivative to find the derivative of a function
- Use derivative notation and terminology to describe instantaneous rates of change

3.5 Interpreting the Derivative
- Interpret the meaning of the derivative in the context of a word problem

155

Chapter Opener

Each chapter opens with a quick introduction to a key concept presented within a real-life context, accompanied by a related photo.

A detailed overview of the sections provides a clear picture of the topics to be presented.

Getting Started

Each section opens with *Getting Started,* real-life applications or mathematical scenarios, which shows the relevance of the section content to a student's everyday life. Section objectives are also provided.

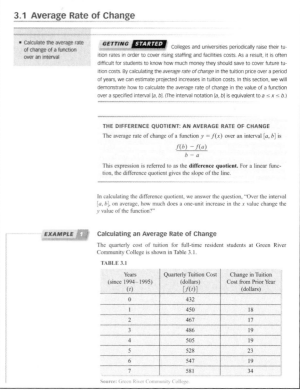

156   CHAPTER 3   The Derivative

3.1 Average Rate of Change

- Calculate the average rate of change of a function over an interval

GETTING STARTED   Colleges and universities periodically raise their tuition rates in order to cover rising staffing and facilities costs. As a result, it is often difficult for students to know how much money they should save to cover future tuition costs. By calculating the *average rate of change* in the tuition price over a period of years, we can estimate projected increases in tuition costs. In this section, we will demonstrate how to calculate the average rate of change in the value of a function over a specified interval [a, b]. (The interval notation [a, b] is equivalent to $a \le x \le b$.)

THE DIFFERENCE QUOTIENT: AN AVERAGE RATE OF CHANGE

The average rate of change of a function $y = f(x)$ over an interval [a, b] is

$$\frac{f(b) - f(a)}{b - a}$$

This expression is referred to as the **difference quotient.** For a linear function, the difference quotient gives the slope of the line.

In calculating the difference quotient, we answer the question, "Over the interval [a, b], on average, how much does a one-unit increase in the x value change the y value of the function?"

EXAMPLE 1   Calculating an Average Rate of Change

The quarterly cost of tuition for full-time resident students at Green River Community College is shown in Table 3.1.

TABLE 3.1

Years (since 1994–1995) (t)	Quarterly Tuition Cost (dollars) [$f(t)$]	Change in Tuition Cost from Prior Year (dollars)
0	432	
1	450	18
2	467	17
3	486	19
4	505	19
5	528	23
6	547	19
7	581	34

Source: Green River Community College.

Real Applications! Real Data! Real Learning!

Technology Tips with Keystrokes

The understanding of core concepts is often enhanced through the use of a graphing calculator. Technology Tips, incorporated throughout the text, guide students through new techniques on the calculator such as graphing a function, solving an equation, and finding the value of a function.

Many tips show actual calculator keystrokes, often with multiple, sequential screenshots, teaching students how to use graphing calculators as a tool to analyze real-life data.

When the use of graphing technology is advised, a graphing calculator icon will appear ![icon].

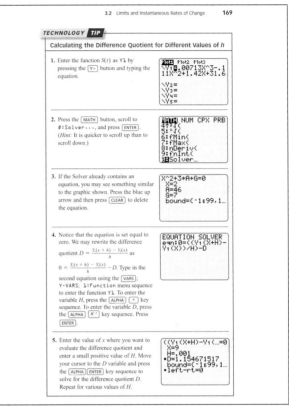

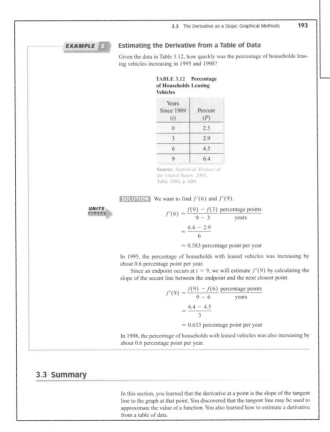

Easy to Read

Incorporating plain language, an informal writing style, and navigational aids make this text easy to read without compromising the integrity of the math. Conceptual understanding is emphasized.

Examples

Real-life applications, examples, and data help engage students—even those who have never enjoyed mathematics. Citing sources—especially those of interest to business and social sciences students—help to "make it real".

Accompanying many examples is a ruler icon **UNITS** reminding students to pay close attention to unit analysis.

Summary

Concise end-of-section summaries help to keep the focus on conceptual understanding and the big picture.

Real Applications! Real Data! Real Learning!

According to the model, at what rate is the value of the cassette tapes shipped changing when the number of cassette tapes shipped is 250 million? What was the estimated shipment value when 251 million cassette tapes were shipped? (Use a tangent-line approximation.)

18. 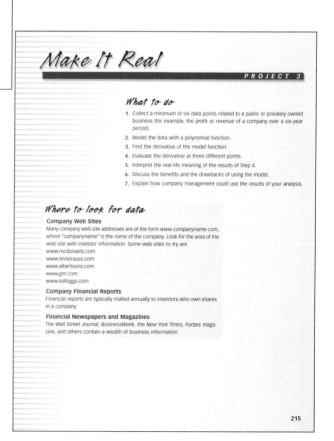 **Decreasing Popularity of the Cassette Tape** Since the advent of the compact disk and mp3 player, cassette tape sales have dropped off dramatically. In fact, in 2004 there were 5.2 million cassettes shipped compared to 442.2 million in 1990. (Source: *Statistical Abstract of the United States, 2006*, Table 1131.) Use the model in Exercise 17 to calculate the rate of change in cassette tape shipment value in 1990 and in 2004.

19. **College Attendance** Based on data from 2000–2002 and Census Bureau projections for 2003–2013, *private* college enrollment may be modeled by

$$P(x) = 0.340x - 457 \text{ thousand students}$$

where x is the number of students (in thousands) enrolled in *public* colleges. (Source: Modeled from *Statistical Abstract of the United States, 2006*, Table 204.) Calculate the rate of change in *private* college enrollment when there are 12,752 thousand *public* college students enrolled. Use a tangent line approximation to estimate the number of students enrolled in *private* colleges when there are 12,753 thousand *public* college students.

20. **Radio Broadcasting Wages** Based on data from 1992 to 1998, the average wage of a worker in the radio broadcasting industry may be modeled by

$$W(t) = 164.3t^2 + 778.6t + 23,500$$

dollars, where t is the number of years since 1992. (Source: Modeled from *Statistical Abstract of the United States, 2001*, Table 1123, p. 703.)

How quickly were wages of radio broadcasting employees increasing at the end of 1998? What was the average salary in 1999? (Use a tangent-line approximation.)

In Exercises 21–30, estimate the specified derivative by using the data in the table. Then interpret the result.

21. **Television Broadcasting Wages**

Average Annual Wage per Worker: Television Broadcasting

Year Since 1992 (t)	Annual Wage (dollars) [$W(t)$]
0	41,400
1	42,200
2	43,700
3	47,200
4	51,100
5	51,000
6	54,600

Source: *Statistical Abstract of the United States, 2001*, Table 1123, p.703.

Estimate $W'(5)$.

22. Using the table in Exercise 21, estimate $W'(6)$.

23. **Daily Newspapers**

Different Daily Newspapers

Years Since 1970 (t)	Newspapers [$N(t)$]
0	1,748
5	1,756
10	1,745
15	1,676
20	1,611
25	1,533
30	1,480

Source: *Statistical Abstract of the United States, 2001*, Table 1130, p. 706.

Estimate $N'(25)$.

24. Use the table in Exercise 23 to estimate $N'(30)$.

Make It Real Projects

These end-of-chapter projects ask students to collect and analyze data from their own experiences and interests. By providing choices in the selection, students are better able to process these concepts and connect math to their own lives.

Exercises

Over 500 real life applications featuring over 70 business, products, and associations help to make math real for your students! (See *Index of Applications* and *Index of Businesses, Products, and Associations* inside the front and back covers at the very beginning and very end of the book.) Using real-world data from real companies such as Starbucks and Wal-Mart, and interesting topics such as debit cards and student loans, the exercises bring a current and immediate motivation for learning mathematical concepts.

A globe icon [icon] indicates the use of real-life data in a problem.

A globe icon accompanied by an M [icon] identifies problems featuring the modeling of real-life data.

Make It Real

PROJECT 3

What to do

1. Collect a minimum of six data points related to a public or privately owned business (for example, the profit or revenue of a company over a six-year period).
2. Model the data with a polynomial function.
3. Find the derivative of the model function.
4. Evaluate the derivative at three different points.
5. Interpret the real-life meaning of the results of Step 4.
6. Discuss the benefits and the drawbacks of using the model.
7. Explain how company management could use the results of your analysis.

Where to look for data

Company Web Sites

Many company web site addresses are of the form *www.companyname.com*, where "*companyname*" is the name of the company. Look for the area of the web site with investor information. Some web sites to try are
www.mcdonalds.com
www.levistrauss.com
www.albertsons.com
www.gm.com
www.kelloggs.com

Company Financial Reports

Financial reports are typically mailed annually to investors who own shares in a company.

Financial Newspapers and Magazines

The *Wall Street Journal, BusinessWeek*, the *New York Times, Forbes* magazine, and others contain a wealth of business information.

Additional Resources—Get the Most out of Your Textbook!

Supplements for the Instructor

Online Instructor's Solutions Manual   Written by Frank Wilson, this manual offers step-by-step solutions for all text exercises.

HM Testing™ (Powered by Diploma®) *"Testing the way you want it"*   HM Testing offers all the tools needed to create, author, deliver, and customize multiple types of tests—including authoring and editing algorithmic questions.

 *Online Teaching Center*

Supplements for the Student

Student Solutions Manual   Written by Frank Wilson, this manual offers step-by-step solutions for all odd-numbered text exercises.

Excel Guide   The guide provides an introduction to Excel, as well as step-by-step examples designed to teach students how to use Excel to solve selected types of problems found in this course.

Online Study Center   Includes *Online Graphing Calculator Guide,* ACE quizzes, a graphing calculator simulator, and more.

Instructional DVDs   Hosted by Dana Mosely, these text-specific DVDs cover selected sections of the text and provide explanations of key concepts, examples, exercises, and applications in a lecture-based format.

Eduspace® (powered by Blackboard™)   Eduspace is a web-based learning system that provides instructors with powerful course management tools and students with text-specific content to support all of their online teaching and learning needs. Eduspace makes it easy to deliver all or part of a course online. Resources such as algorithmic automatically-graded homework exercises, tutorials, instructional video clips, an online multimedia eBook, live online tutoring with SMARTHINKING™, and additional study materials all come ready-to-use. Instructors can choose to use the content as is, modify it, or even add their own.

Visit *www.eduspace.com* for more information.

SMARTHINKING™ Live, online tutoring   SMARTHINKING provides an easy-to-use and effective online, text-specific tutoring service. A dynamic **Whiteboard** and **Graphing Calculator function** enable students and e-structors to collaborate easily.

Visit **smarthinking.college.hmco.com** for more information.

Online Course Content for Blackboard®, WebCT®, and eCollege®   Deliver program or text-specific Houghton Mifflin content online using your institution's local course management system. Houghton Mifflin offers homework, tutorials, videos, and other resources formatted for Blackboard®, WebCT®, eCollege®, and other course management systems. Add to an existing online course or create a new one by selecting from a wide range of powerful learning and instructional materials.

For more information, visit **college.hmco.com/pic/wilsonBAC**
or contact your Houghton Mifflin sales representative.

Chapter 1

Functions and Linear Models

Mathematical functions are a powerful tool used to model real-world phenomena. Whether simple or complex, functions give us a way to forecast expected results. Remarkably, anything that has a constant rate of change may be accurately modeled with a linear function. For example, the cost of filling your car's gas tank is a linear function of the number of gallons purchased.

1.1 Functions
- Distinguish between functions and nonfunctions in tables, graphs, and words
- Use function notation
- Graph functions using technology
- Determine the domain of a function

1.2 Linear Functions
- Calculate and interpret the meaning of the slope of a linear function
- Interpret the physical and graphical meaning of x- and y-intercepts
- Formulate the equation of a line given two points
- Recognize the slope-intercept, point-slope, and standard forms of a line

1.3 Linear Models
- Use technology to model linear and near-linear data
- Use a linear equation to describe the relationship between directly proportional quantities
- Recognize and model naturally occurring linear, near-linear, and piecewise linear relationships

1.1 Functions

- Distinguish between functions and nonfunctions in tables, graphs, and words
- Use function notation
- Graph functions using technology
- Determine the domain of a function

GETTING STARTED   Our society is a complex system of relationships between people, places, and things. Many of these relationships are interconnected. In mathematics, we often model the relationship between two or more interdependent quantities by using a **function**. In this section, we will show how to distinguish between functions and nonfunctions and will practice using function notation. We will also demonstrate how to use technology to draw a function graph, and discuss how to find the domain of a function.

DEFINITION: FUNCTION

A function is a rule that associates each input with *exactly one* output.

Often the rule is represented in a table of data with the inputs on the left-hand side and the outputs on the right-hand side. For example, the amount of money we pay to fill up our gas tank is a function of the number of gallons pumped (see Table 1.1).

TABLE 1.1

Gallons Pumped	Total Fuel Cost
10	$15.99
15	$23.99
20	$31.98

In this case, the input to the function is *gallons pumped* and the output of the function is *total fuel cost*. Fuel cost is a function of the number of gallons pumped because each input has exactly one output.

Similarly, the weekly wage of a service station employee is a function of the number of hours worked. The *number of hours worked* is the input to the function, and the *weekly wage* is the output of the function. Since weekly wage is a function of the number of hours worked, an employee who works 40 hours expects to be paid the same wage each time she works that amount of time.

EXAMPLE 1

Determining If a Table of Data Represents a Function

A car dealer tracks the number of blue cars in each of three shipments and records the data in a table (see Table 1.2). Is the number of blue cars a function of the number of cars in the shipment?

TABLE 1.2

Number of Cars in the Shipment	Number of Blue Cars
22	6
24	7
24	5

same input { ... } different outputs

SOLUTION According to the definition of a function, each input must have exactly one output. The input value 24 has two different outputs: 5 and 7. Since the input 24 has more than one output, the number of blue cars is *not* a function of the number of cars in the shipment.

Function Notation

When we encounter functions in real life, they are often expressed in words. To make functions easier to work with, we typically use symbolic notation to represent the relationship between the input and the output. Let's return to the fuel cost table introduced previously (Table 1.3).

TABLE 1.3

Gallons Pumped	Total Fuel Cost
10	$15.99
15	$23.99
20	$31.98

Observe that the fuel cost is equal to $1.599 times the number of gallons pumped. We represent this symbolically as $C(g) = 1.599g$, where $C(g)$ represents the total fuel cost when g gallons are pumped. [$C(g)$ is read "C of g"]. The letter C is used to represent the name of the rule, and the letter g in the parentheses indicates that the rule works with different values of g (see Figure 1.1).

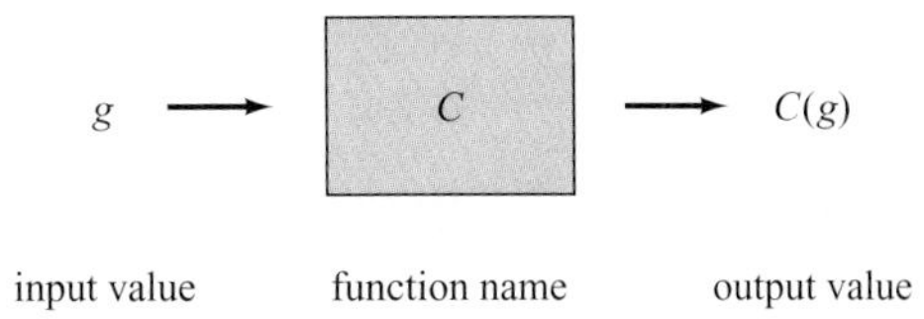

$$g \longrightarrow \boxed{C} \longrightarrow C(g)$$

input value   function name   output value

FIGURE 1.1

We call the output variable of a function the **dependent variable** because the value of the output variable *depends* upon the value of the input variable. The input variable is called the **independent variable.** (One way to remember the meaning of the terms is to observe that both *input* and *independent* begin with *in.*) From the table, we see that

$$C(10) = 15.99$$
$$C(15) = 23.99$$
$$C(20) = 31.98$$

For this function, the independent variable took on the values 10, 15, and 20, and the dependent variable assumed the values 15.99, 23.99, and 31.98.

EXAMPLE **2**   **Determining a Linear Model from a Verbal Description**

An electronics store employee earns $8.50 per hour. Write an equation for the employee's earnings as a function of the hours worked. Then calculate the amount of money the employee earns (in dollars) by working 30 hours.

SOLUTION Since the employee earns $8.50 for each hour worked, the employee's total earnings are equal to $8.50 times the total number of hours worked. That is,

$$E(h) = 8.50h$$

where E is the employee's earnings (in dollars) and h is the number of hours worked. To calculate the amount of money earned by working 30 hours, we evaluate this function at $h = 30$.

$$E(30) = 8.5(30)$$
$$= 255$$

The employee earns $255 for 30 hours of work.

Function notation is extremely versatile. Suppose we are given the function $f(x) = x^2 - 2x + 1$. We may evaluate the function using either numerical values or nonnumerical values. For example,

$$f(2) = (2)^2 - 2(2) + 1 \qquad f(\triangle) = (\triangle)^2 - 2(\triangle) + 1$$
$$= 4 - 4 + 1$$
$$= 1$$

$$f(a + 2) = (a + 2)^2 - 2(a + 2) + 1$$
$$= (a^2 + 4a + 4) - 2a - 4 + 1$$
$$= a^2 + 2a + 1$$

In each case, we replaced the value of x in the function $f(x) = x^2 - 2x + 1$ with the quantity in the parentheses. Whether the independent variable value was 2, $\triangle$, or $a + 2$, the process was the same.

EXAMPLE **3**   **Evaluating a Function Using Function Notation**

Evaluate the function $s(t) = t^3 + 4t$ at $t = 3$, $t = \triangle$, and $t = a^2$.

SOLUTION

$$s(3) = (3)^3 + 4(3) \qquad s(\triangle) = (\triangle)^3 + 4(\triangle)$$
$$= 27 + 12$$
$$= 39$$

$$s(a^2) = (a^2)^3 + 4(a^2)$$
$$= a^6 + 4a^2$$

Graphs of Functions

Functions are represented visually by plotting points on a Cartesian coordinate system (see Figure 1.2). The horizontal axis shows the value of the independent variable (in this case, x), and the vertical axis shows the value of the dependent variable (in this case, y).

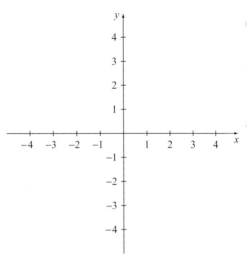

FIGURE 1.2

When using the coordinate system, y is typically used in place of the function notation $f(x)$. That is, $y = f(x)$. This is true even if the function has a name other than f.

The point of intersection of the horizontal and vertical axes is referred to as the **origin** and is represented by the ordered pair $(0, 0)$. To graph an ordered pair (a, b), we move from the origin $|a|$ units horizontally and $|b|$ units vertically and draw a point. If $a > 0$, we move to the right. If $a < 0$, we move to the left. Similarly, if $b > 0$, we move up, and if $b < 0$, we move down. For example, consider the table of values with its associated interpretation in Table 1.4 and the graph in Figure 1.3.

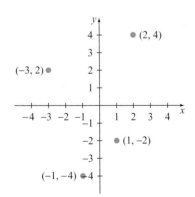

FIGURE 1.3

TABLE 1.4

x	y	Horizontal	Vertical
-3	2	left 3	up 2
-1	-4	left 1	down 4
1	-2	right 1	down 2
2	4	right 2	up 4

When we are given the equation of a function, we can generate a table of values and then plot the corresponding points. Once we have drawn a sufficient number of points to be able to determine the basic shape of the graph, we typically connect the points with a smooth curve. For example, the function $y = x^3 - 9x$ has the table of values and graph shown in Figure 1.4.

x	y
−4	−28
−3	0
−2	10
−1	8
0	0
1	−8
2	−10
3	0
4	28

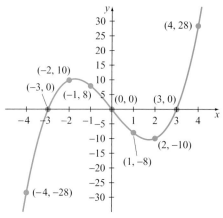

FIGURE 1.4

Estimating Function Values from a Graph

Estimate $f(-3)$ and $f(2)$ using the graph of $f(x) = x^3 - 16x$ shown in Figure 1.5.

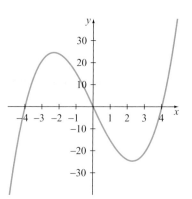

FIGURE 1.5

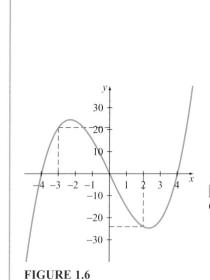

FIGURE 1.6

SOLUTION It appears from Figure 1.6 that $f(-3) \approx 20$ and $f(2) \approx -25$. Calculating these values with the algebraic equation, we see that

$$f(-3) = (-3)^3 - 16(-3) \qquad\qquad f(2) = (2)^3 - 16(2)$$
$$= -27 + 48 \qquad \text{and} \qquad\qquad = 8 - 32$$
$$= 21 \qquad\qquad\qquad\qquad\qquad = -24$$

One drawback of using a graph to determine the values of a function is that it is difficult to be precise. For this reason, algebraic methods are typically preferred when precision is important.

Not all data sets represent functions. If any value of the independent variable is associated with more than one value of the dependent variable, the table of data and its associated graph will not represent a function. For example, consider $y = \pm\sqrt{2x}$ (see Figure 1.7).

x	y
0	0
0.5	−1
0.5	1
2	−2
2	2
4.5	−3
4.5	3
8	−4
8	4

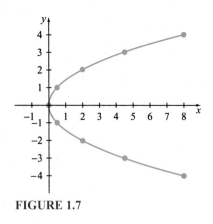

FIGURE 1.7

Each positive value of x is associated with two different values of y. We can easily determine this from the graph by observing that a vertical line drawn through any positive value of x will cross the graph twice. This observation leads us to the Vertical Line Test.

VERTICAL LINE TEST

If every vertical line drawn on a graph intersects the graph in at most one place, then the graph is the graph of a function. Otherwise, the graph is not the graph of a function.

EXAMPLE 5

Determining If a Graph Represents a Function

The graph of $y = x^2$ is shown in Figure 1.8. Does the graph represent a function?

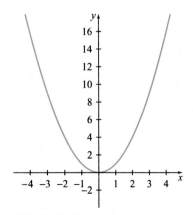

FIGURE 1.8

TECHNOLOGY **TIP**

Graphing a Function

1. Bring up the graphing list by pressing the $\boxed{\text{Y=}}$ button.

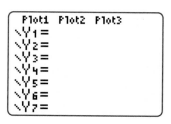

2. Type in the function using the $\boxed{\text{X,T,}\theta\text{,n}}$ button for the variable and the $\boxed{\text{^}}$ button to place an expression in an exponent. Make sure you use parentheses as needed.

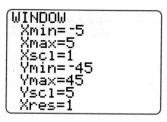

3. Specify the size of the viewing window by pressing the $\boxed{\text{WINDOW}}$ button and editing the parameters. The Xmin is the minimum x value, Xmax is the maximum x value, Ymin is the minimum y value, and Ymax is the maximum y value. The Xscl and Yscl are used to specify the spacing of the tick marks on the graph.

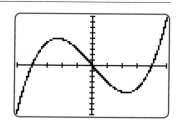

4. Draw the graph by pressing the $\boxed{\text{GRAPH}}$ button.

Domain and Range

We are often interested in the set of all possible values of the independent variable and the set of all possible values of the dependent variable.

DOMAIN AND RANGE

The set of all possible values of the independent variable of a function is called the **domain.** The set of all possible values of the dependent variable of a function is called the **range.**

It is easy to remember the meaning of the terms if we observe that *input, independent variable,* and *domain* all contain the word *in.*

Consider a common kitchen blender. If we put an orange into a blender and turn the blender on, the orange is transformed into orange juice. We say that *orange* is in the domain of the blender function and *orange juice* is in the range of the blender function. On the other hand, if we put a rock into the blender and turn it on, the blender will self-destruct. We say that *rock* is not in the domain of the blender function, since blenders are unable to process rocks.

Finding the domain of a function frequently involves solving an equation or an inequality. Recall that *solving an equation* means finding the value of the variable that makes the equation a true statement.

The domain of most frequently used mathematical functions is the set of all real numbers. However, there are three common situations in which the domain of a function is restricted to a subset of the real numbers. They are:

1. A zero in the denominator
2. A negative value under a square root symbol (radical)
3. The context of a word problem

Let's look at an example for each of the situations.

EXAMPLE 8

Determining the Domain of a Function

What is the domain of $g(x) = \dfrac{3x - 1}{2x + 6}$?

SOLUTION   We know that the value in the denominator must be nonzero, since division by zero is undefined. That is, $2x + 6 \neq 0$. To find the value of x that must be excluded from the domain, we must solve the following equation:

$$2x + 6 = 0$$
$$2x + 6 - 6 = 0 - 6 \qquad \text{Subtract 6 from both sides}$$
$$2x = -6$$
$$\frac{2x}{2} = \frac{-6}{2} \qquad \text{Divide both sides by 2}$$
$$x = -3$$

The domain of the function is all real numbers except -3. We may rewrite the function equation with the domain restriction as follows:

$$g(x) = \frac{3x - 1}{2x + 6}, \qquad x \neq -3$$

EXAMPLE 9

Determining the Domain of a Function

What is the domain of $y = \sqrt{x - 3}$?

SOLUTION   We know that the value underneath the radical must be nonnegative, since the square root of a negative number is undefined in the real number system. Therefore, $x - 3 \geq 0$. Solving for x, we get

$$x - 3 \geq 0$$
$$x - 3 + 3 \geq 0 + 3 \qquad \text{Add 3 to both sides}$$
$$x \geq 3$$

The domain of the function is all real numbers greater than or equal to 3.

EXAMPLE 10

Determining the Domain of a Function

The revenue R from the sale of x gallons of gasoline is given by the equation

$$R(x) = 1.379x$$

What is the domain of the function?

SOLUTION   In the context of this problem, it doesn't make sense to talk about selling a negative number of gallons of gas. So $x \geq 0$. The domain of the function is all nonnegative real numbers.

1.1 Summary

In this section, you learned to distinguish between functions and nonfunctions, and you practiced using function notation. You learned how to use technology to graph a function, and you discovered how to find the domain of a function. Mastering each of these techniques will help you understand the subsequent concepts covered in this chapter.

1.1 Exercises

In Exercises 1–6, determine whether the output is a function of the input.

1. $W(a) =$ your weight in pounds when you were a years old.

2. $S(n) =$ your score on test number n in a finite math course. (Assume you could only take the test once.)

3.

Time of Day	Temperature (°F)
11:00 a.m.	68
1:00 p.m.	73
3:00 p.m.	75

4.

Time of Day	Temperature (°F)
10:00 a.m.	66
12:00 p.m.	74
2:00 p.m.	74

5.

Fish Caught	Salmon in Catch
169	24
182	32
182	47

6.

Fish Caught	Salmon in Catch
252	74
276	74
301	92

In Exercises 7–13, calculate the value of the function at the designated input and interpret the result.

7. $C(x) = 39.95x$ at $x = 4$, where $C(x)$ is the cost of buying x pairs of shoes.

8. $C(x) = 29.95x + 200$ at $x = 300$, where $C(x)$ is the cost of making x pairs of shoes.

9. $H(t) = -16t^2 + 120$ at $t = 2$, where $H(t)$ is the height of a cliff diver above the water t seconds after he jumped from a 120-foot cliff.

10. 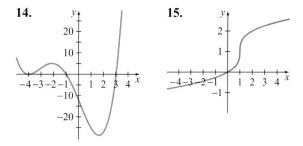 **Revenue** Find $R(t)$ at $t = 4$, where $R(t)$ is the quarterly revenue for an international tortilla producer t quarters after December 1999 as shown below.

Quarter	Revenue (in millions of dollars)
1	460.0
2	452.0
3	474.0
4	483.8
5	466.0

Source: Gruma, S.A. de C.V.

11. **Earnings per Share** Find $E(t)$ at $t = 4$, where $E(t)$ is the quarterly earnings per share for an international tortilla producer t quarters after December 1999.

Quarter	Earnings per Share (in dollars)
1	0.13
2	−0.11
3	0.15
4	0.06
5	0.03
6	0.06

Source: Gruma, S.A. de C.V.

12. **Stock Price** Find $P(t)$ at $t = 3$, where $P(t)$ is the stock price of an e-learning company at the end of the day and t is the number of days after November 4, 2001.

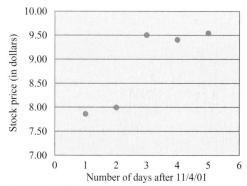

Source: Digital Think Corporation.

13. **Stock Price** Find $P(t)$ at $t = 4$, where $P(t)$ is the stock price of a computer company at the end of the day and t is the number of days after November 4, 2001.

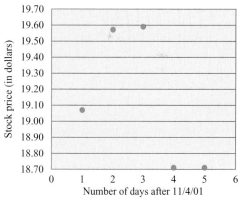

Source: Apple Computer Corporation.

In Exercises 14–19, determine whether the graphs represent functions by applying the Vertical Line Test.

14.

15.

16.

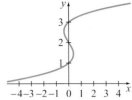

17.

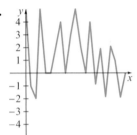

18.

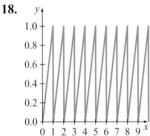

19.

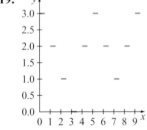

In Exercises 20–23, graph the function on your graphing calculator, using the specified viewing window. Note that $a \leq x \leq b$ means x min $= a$ and x max $= b$. Similarly, $c \leq y \leq d$ means y min $= c$ and y max $= d$.

 20. $y = x^2 - 5x; -3 \leq x \leq 7, -8 \leq y \leq 8$

 21. $y = -x^2 - 4; -3 \leq x \leq 3, -10 \leq y \leq 1$

 22. $y = -x + 2; -3 \leq x \leq 3, -2 \leq y \leq 6$

 23. $y = -2x^2 + 1; -2 \leq x \leq 2, -3 \leq y \leq 2$

In Exercises 24–27, use the graph to estimate the value of the function at the indicated x value. Then calculate the exact value algebraically.

24. $y = 0.25(x + 1)^2(x - 3); x = 4$

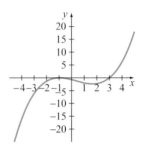

25. $y = 5|x| - x^2; x = 1$

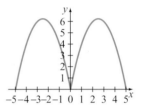

26. $y = \dfrac{|x + 2|}{x^2 + 1}; x = 3$

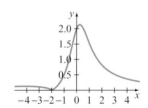

27. $y = -\dfrac{4(x^2 - 4)}{x^3 + 2x^2 + 3x + 6}; x = -2$

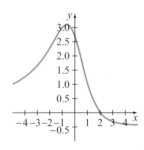

In Exercises 28–41, determine the domain of the function.

28. $f(x) = 5x^3 - 5x$ **29.** $f(p) = p^2 - 2p + 1$

30. $C(r) = \dfrac{r^2 - 1}{r + 1}$ **31.** $h(r) = \dfrac{r^2 - 1}{r^2 + 1}$

32. $S(t) = \dfrac{t-3}{4t+1}$ **33.** $g(t) = \dfrac{3t-3}{4t-4}$

34. $H(x) = \sqrt{x-5}$

35. $f(a) = a^3 - \sqrt{a+1}$

36. $f(x) = \dfrac{\sqrt{x+5}}{x+2}$ **37.** $f(x) = \dfrac{\sqrt{2x+6}}{x^2+3}$

38. $C(n) =$ the cost of buying n apples

39. $P(n) =$ the profit earned from the sale of n bags of candy

40. $W(n) =$ the average birth weight of someone born in year n

41. $H(n) =$ the average height of someone born in year n

Exercises 42–46 are intended to challenge your understanding of functions and graphs.

 42. Graph the function $f(x) = \dfrac{x-1}{x^2-1}$ on your calculator. What do you think is the domain of f?

43. Determine the domain of $f(x) = \dfrac{x-1}{x^2-1}$ algebraically and compare your answer to the solution of Exercise 42.

44. Are all lines functions? Explain.

45. Does the table of data represent a function? Explain.

x	y
0	5
1	6
1	6
3	0
4	2

46. For a given date and location, is air temperature a function of the time of day? Explain.

1.2 Linear Functions

- Calculate and interpret the meaning of the slope of a linear function
- Interpret the physical and graphical meaning of x- and y-intercepts
- Formulate the equation of a line given two points
- Recognize the slope-intercept, point-slope, and standard forms of a line

GETTING STARTED   The cost of filling our car's gas tank, the number of calories we consume by eating a few bags of fruit snacks, and the amount of sales tax we pay when we buy new clothes are examples of linear functions. In this section, we will show how to calculate the slope of a linear function. We will discuss how to interpret the physical and graphical meaning of slope, x-intercept, and y-intercept. We will also show three different ways to write a linear equation and demonstrate how to find the equation of a linear function from a data set.

DEFINITION: GRAPH OF A LINEAR FUNCTION

The line passing through any two points (x_1, y_1) and (x_2, y_2) with $x_1 \neq x_2$ is referred to as the **graph of a linear function.**

Linear functions are characterized by a constant rate of change. That is, increasing a domain value by one unit will always change the corresponding range value by a constant amount. The converse is also true. Any table of data with a constant rate of change represents a linear function. The constant rate of change is referred to as the **slope** of the linear function.

DEFINITION: SLOPE

The slope of a linear function is the change in the output that occurs when the input is increased by one unit. The slope m may be calculated by dividing the difference of two outputs by the difference in the corresponding inputs. That is,

$$m = \frac{y_2 - y_1}{x_2 - x_1}$$

where (x_1, y_1) and (x_2, y_2) are data points of the linear function.

The number of calories C in n bags of fruit snacks is shown in Table 1.5.

TABLE 1.5

Bags (n)	Calories (C)
1	100
2	200
3	300

Source: Fruit Smiles 9-oz. box label

Notice that the calorie count increases by 100 calories for each additional bag. Since the dependent variable C is changing at a constant rate (100 calories per bag), the data may be modeled by a linear function. In this case, the linear function is $C = 100n$.

To calculate the slope of the function, we may use any two data points. Using the data points $(x_1, y_1) = (1, 100)$ and $(x_2, y_2) = (3, 300)$, the slope is

UNITS

$$m = \frac{300 - 100 \text{ calories}}{3 - 1 \text{ bags}}$$

$$= \frac{200}{2} \frac{\text{calories}}{\text{bags}}$$

$$= 100 \text{ calories per bag}$$

A one-bag increase in the input results in a 100-calorie increase in the output.

Will we get the same result if we use different points? Let's check using $(1, 100)$ and $(2, 200)$ (see Figure 1.13).

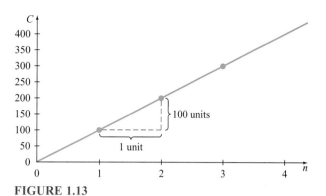

FIGURE 1.13

UNITS

$$m = \frac{100 - 200 \text{ calories}}{1 - 2 \text{ bags}}$$

$$= \frac{-100}{-1} \frac{\text{calories}}{\text{bag}}$$

$$= 100 \text{ calories per bag}$$

The result is the same! With linear functions, we may use any two points to calculate the slope.

Not all lines are functions. Vertical lines fail to pass the Vertical Line Test, so they are not functions. As shown in Example 1, vertical lines have an undefined slope.

EXAMPLE 1 **Determining the Slope of a Line from Two Points on the Line**

What is the slope of the line going through (2, 4) and (2, 8) (Figure 1.14)?

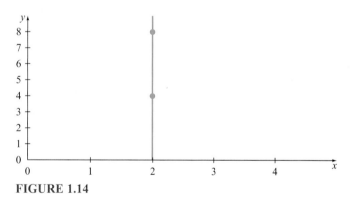

FIGURE 1.14

SOLUTION

$$m = \frac{8 - 4}{2 - 2}$$

$$= \frac{4}{0}$$

$$= \text{undefined}$$

Since division by zero is not defined, the line has an undefined slope. Vertical lines are the only lines that have an undefined slope.

EXAMPLE **2**   **Determining the Slope of a Line from Two Points on the Line**

What is the slope of the line going through (2, 4) and (5, 4) in Figure 1.15?

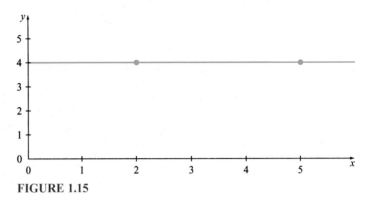

FIGURE 1.15

SOLUTION

$$m = \frac{4 - 4}{5 - 2}$$

$$= \frac{0}{3}$$

$$= 0$$

Any line with a zero slope is a horizontal line.

The absolute value of the slope is referred to as the **magnitude** of the slope. In a general sense, slope is a measure of steepness: The greater the magnitude of the slope, the greater the steepness of the line. The graph of a line with a negative slope falls as the independent variable increases. The graph of a line with a positive slope rises as the independent variable increases.

EXAMPLE **3**   **Determining the Sign and Magnitude of a Line's Slope from a Graph**

Determine from the graph in Figure 1.16 which lines have a negative slope and which lines have a positive slope. Then identify the line whose slope has the greatest magnitude.

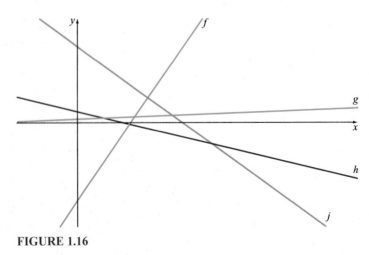

FIGURE 1.16

SOLUTION Since lines h and j fall as the value of x increases, they have negative slopes. Since lines f and g rise as the value of x increases, they have positive slopes.

The steepest line is the line whose slope has the greatest magnitude. Although both f and j are steep, line f is steeper. Therefore, f has the slope with the greatest magnitude.

Intercepts

In discussing linear functions, it is often useful to talk about where the line crosses the x and y axes. Knowing these **intercepts** helps us to determine the equation of the line.

DEFINITION: y-INTERCEPT

The y-intercept is the point on the graph where the function intersects the y axis. It occurs when the value of the independent variable is 0. It is formally written as an ordered pair $(0, b)$, but b itself is often called the y-intercept.

EXAMPLE 4

Finding the y-intercept of the Graph of a Linear Function

What is the y-intercept of the linear function $y = 3x + 5$?

SOLUTION At the y-intercept, the x coordinate is 0.

$$y = 3(0) + 5 \quad \text{Substitute 0 for } x$$
$$= 5$$

So the y-intercept of the function is $(0, 5)$.

DEFINITION: x-INTERCEPT

The x-intercept is the point on the graph where the function intersects the x axis. It occurs when the value of the dependent variable is 0. It is formally written as an ordered pair $(a, 0)$, but a itself is often called the x-intercept.

EXAMPLE 5

Finding the x-intercept of the Graph of a Linear Function

What is the x-intercept of the linear function $y = 3x + 5$?

SOLUTION At the x-intercept, the y coordinate is 0.

$$y = 3x + 5$$
$$0 = 3x + 5 \quad \text{Substitute 0 for } y$$
$$-3x = 5$$
$$x = -\frac{5}{3}$$

So the x-intercept of the function is $\left(-\frac{5}{3}, 0\right)$.

EXAMPLE 6   **Determining the Slope and Intercepts of the Graph of a Linear Function**

Determine the slope, the y-intercept, and the x-intercept of the linear function from its graph (Figure 1.17).

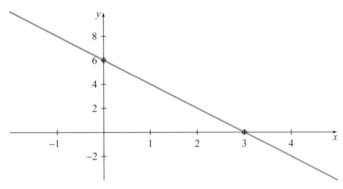

FIGURE 1.17

SOLUTION   Since the graph of the function crosses the y axis at $(0, 6)$, the y-intercept is $(0, 6)$. Since the graph crosses the x axis at $(3, 0)$, the x-intercept is $(3, 0)$.

Since the line is falling as x increases, the slope will be negative. The slope of the line is

$$m = \frac{y_2 - y_1}{x_2 - x_1}$$

$$= \frac{6 - 0}{0 - 3}$$

$$= \frac{6}{-3}$$

$$= -2$$

Linear Equations

The graph of any line may be represented by a linear equation. Vertical and horizontal lines have the simplest linear equations.

EQUATIONS OF VERTICAL AND HORIZONTAL LINES

■ The equation of a vertical line passing through a point (a, b) is $x = a$.
■ The equation of a horizontal line passing through a point (a, b) is $y = b$.

If we know the slope and y-intercept of a linear function, we are able to easily determine the slope-intercept form of the linear equation.

SLOPE-INTERCEPT FORM OF A LINE

A linear function with slope m and y-intercept $(0, b)$ has the equation

$$y = mx + b$$

EXAMPLE 7 ## Determining the Slope-Intercept Form of a Line

What is the slope-intercept form of the linear function with slope 5 and y-intercept $(0, 4)$?

SOLUTION Since $m = 5$ and $b = 4$,

$$y = 5x + 4$$

is the slope-intercept form of the line.

EXAMPLE 8 ## Finding a Linear Model from a Verbal Description

For breakfast, we decide to eat an apple containing 5.7 grams of dietary fiber and a number of servings of Cheerios™, each containing 3 grams of dietary fiber. (**Source:** Cheerios box label.) Write the equation for our dietary fiber intake as a function of number of servings of cereal.

SOLUTION We know that if we don't eat any cereal, we will consume 5.7 grams of dietary fiber (from the apple). So the y-intercept of the fiber function is $(0, 5.7)$. Since each serving of Cheerios contains the same amount of fiber, we know that the function is linear. The slope of the function is 3 grams per serving. So the equation of the fiber function is

$$F(n) = 3n + 5.7$$

where $F(n)$ is the amount of dietary fiber (in grams) and n is the number of servings of cereal.

To check our work, we directly calculate the number of grams of dietary fiber in a breakfast containing two servings of cereal and one apple.

$$\text{Fiber} = 3 + 3 + 5.7 = 11.7 \text{ grams}$$

Using the fiber function formula, we get

$$F(2) = 3(2) + 5.7$$
$$= 11.7 \text{ grams}$$

The results are the same, so we are confident that our formula is correct.

EXAMPLE 9 ## Determining a Linear Model from a Table of Data

The amount of sales tax paid on a clothing purchase in Seattle is a function of the sales price of the clothes, as shown in Table 1.6.

TABLE 1.6

Sales Price (p)	Tax (T)
$20.00	$1.72
$30.00	$2.58
$40.00	$3.44

Source: www.cityofseattle.net.

If the function is linear, write the equation for the sales tax as a function of the sales price.

SOLUTION We must first determine if the function is linear. Since each $10 increase in sales price increases the sales tax by a constant $0.86, we conclude that the function is linear. The slope of the function is

UNITS

$$m = \frac{2.58 - 1.72 \text{ dollars}}{30 - 20 \text{ dollars}}$$

$$= \frac{0.86 \text{ dollar}}{10 \text{ dollars}}$$

$$= 0.086 \text{ tax dollar per sales price dollar}$$

(In other words, for each dollar increase in the sales price, the sales tax increases by 8.6¢.) Since a sale of $0 results in $0 sales tax, we know that the y-intercept is (0, 0). The equation of the function is

$$T = 0.086p$$

To check our work, we substitute the point (40, 3.44) into the equation.

$$3.44 = 0.086(40) \qquad p = 40 \text{ and } T = 3.44$$

$$3.44 = 3.44$$

The statement is true, so we are confident that our equation is correct.

EXAMPLE 10 **Finding a Linear Model from the Graph of a Linear Function**

The graph in Figure 1.18 shows the balance of a checking account as a function of the number of ATM withdrawals from the account.

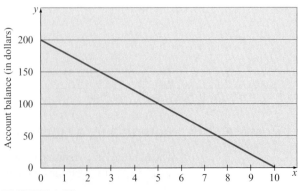

FIGURE 1.18

(a) Identify the *x*-intercept and *y*-intercept and interpret their physical meaning.

(b) Write the linear equation for the function.

SOLUTION

(a) The *x*-intercept is (10, 0). When there have been 10 withdrawals, the account balance is $0. The *y*-intercept is (0, 200). When there have been no withdrawals, the account balance is $200.

(b) The slope of the function is

$$m = \frac{200 - 0 \text{ dollars}}{0 - 10 \text{ withdrawals}}$$

$$= -\$20 \text{ per withdrawal}$$

So the slope-intercept form of the linear equation is

$$B = -20w + 200$$

where *B* is the account balance and *w* is the number of withdrawals.

Finding the Equation of a Line

As demonstrated in Example 10, to find the slope-intercept form of a line from two points, we need to calculate the slope and the *y*-intercept of the function. To do this, we proceed as follows:

1. Calculate the slope.
2. Write the function in slope-intercept form, substituting the slope for *m*.
3. Select one of the points and substitute the output value for *y* and the input value for *x*.
4. Solve for *b*.
5. Write the function in slope-intercept form, substituting the *y*-intercept for *b*.

EXAMPLE 11   **Finding the Slope-Intercept Form of a Line from Two Points**

Find the equation of the line passing through the points (3, 5) and (7, 1).

SOLUTION

$$m = \frac{5 - 1}{3 - 7}$$

$$= \frac{4}{-4}$$

$$= -1$$

So the slope is −1. Substituting in the slope and the point (3, 5), we get

$$y = -1 \cdot x + b$$

$$5 = -1(3) + b$$

$$b = 8$$

The *y*-intercept is (0, 8).

The slope-intercept form of the line is $y = -1 \cdot x + 8$ or $y = -x + 8$.

Other Forms of Linear Equations

There are two additional forms of linear equations that are commonly used: standard form and point-slope form.

STANDARD FORM OF A LINE

A linear equation may be written as

$$ax + by = c$$

where a, b, and c are real numbers. If $a = 0$, the graph of the equation is a horizontal line. If $b = 0$, the graph of the equation is a vertical line. In the equation, a and b cannot both be zero.

The standard form of a linear equation is extremely useful when working with systems of equations or solving linear programming problems. We will discuss the standard form of a line further when we introduce these topics.

The point-slope form of a line is especially useful when we know a line's slope and the coordinates of a point on the line.

POINT-SLOPE FORM OF A LINE

A linear function written as

$$y - y_1 = m(x - x_1)$$

has slope m and passes through the point (x_1, y_1).

EXAMPLE 12   **Finding a Linear Model from a Verbal Description**

Based on data from 1980 to 1999, per capita consumption of milk as a beverage has been decreasing by approximately 0.219 gallon per year. In 1997, the per capita consumption of milk as a beverage was 24.0 gallons. (**Source:** *Statistical Abstract of the United States, 2001*, Table 202, p 129.) Find an equation for the per capita milk consumption as a function of years since 1980.

SOLUTION The slope of the line is $m = -0.219$. The point $(17, 24.0)$ lies on the line, since 1997 is 17 years after 1980. The point-slope form of the line is

$$y - 24.0 = -0.219(t - 17)$$

If preferred, the equation may be rewritten in slope-intercept form,

$$y = -0.219(t - 17) + 24.0$$
$$y = -0.219t + 3.723 + 24.0$$
$$y = -0.219t + 27.723$$

or standard form,

$$0.219t + y = 27.723$$

Graphing Linear Functions

To graph a linear function, we first generate a table of values by substituting different values of x into the equation and calculating the corresponding value of y. For example, if we are given the linear equation $y = 4x - 8$, we may choose to evaluate the function at $x = -1$, $x = 0$, $x = 1$, $x = 2$, and $x = 3$, as shown in Table 1.7.

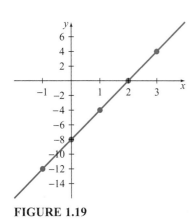

FIGURE 1.19

TABLE 1.7

x	y
-1	-12
0	-8
1	-4
2	0
3	4

We then plot the points and connect them with a straight line, as shown in Figure 1.19. (Although we plotted multiple points, only two points are necessary to determine the line.)

EXAMPLE 13

Graphing a Linear Function

Graph the function $y = 2x - 3$.

SOLUTION From the equation, we see that the y-intercept is $(0, -3)$. We need to find only one more point. Evaluating the function at $x = 2$ yields $y = 1$, so $(2, 1)$ is a point on the line. We plot each point and connect the points with a straight line, as shown in Figure 1.20.

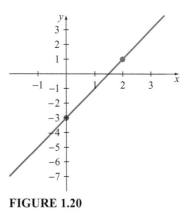

FIGURE 1.20

Recall that the standard form of a line is $ax + by = c$. Graphing a line from standard form is remarkably easy. The y-intercept of the graph occurs when the x value is equal to zero. Similarly, the x-intercept of the graph occurs when the

y value is equal to zero. Using these facts, we can quickly determine the x- and y-intercepts of the function. To find the x-intercept, we set $y = 0$.

$$ax + by = c$$
$$ax + b(0) = c$$
$$ax = c$$
$$x = \frac{c}{a}$$

Notice that the x coordinate of the x-intercept is the constant term divided by the coefficient on the x term.

To find the y-intercept, we set $x = 0$.

$$ax + by = c$$
$$a(0) + by = c$$
$$by = c$$
$$y = \frac{c}{b}$$

Notice that the y coordinate of the y-intercept is the constant term divided by the coefficient on the y term. Using this procedure to find intercepts will allow us to graph linear equations in standard form quickly by hand.

EXAMPLE 14

Graphing a Linear Function

Graph the linear function $2x + y = 4$.

SOLUTION   The x-intercept is found by dividing the constant term by the coefficient on the x term.

$$x = \frac{4}{2}$$
$$= 2$$

The point $(2, 0)$ is the x-intercept.

The y-intercept is found by dividing the constant term by the coefficient on the y term.

$$y = \frac{4}{1}$$
$$= 4$$

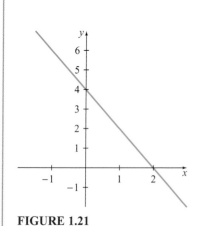

FIGURE 1.21

The point $(0, 4)$ is the y-intercept. We graph the x- and y-intercepts and then draw the line through the intercepts, as shown in Figure 1.21.

1.2 Summary

In this section, you learned how to calculate the slope of a linear function and how to interpret the physical and graphical meaning of slope, x-intercept, and y-intercept. You also discovered how to calculate the equation of a linear function from a set of data and learned three different ways to write a linear equation.

1.2 Exercises

In Exercises 1–6, calculate the slope of the linear function passing through the points.

1. (2, 5) and (4, 3)   **2.** (−3, 4) and (0, −2)

3. (1.2, 3.4) and (2.7, 3.1)

4. (7, 11) and (9, 2)   **5.** (2, 2) and (5, 2)

6. (4, 3) and (4, 7)

In Exercises 7–12, find the x-intercept and y-intercept of the linear function.

7. $y = 5x + 10$ **8.** $y = -3x + 9$

9. $y = 2x + 11$ **10.** $y = \frac{1}{2}x - 2$

11. $3x - y = 4$ **12.** $4x - 2y = 5$

In Exercises 13–18, write the equation of the linear function passing through the points in slope-intercept form, point-slope form, and standard form.

13. (2, 5) and (4, 3)

14. (−3, 4) and (0, −2)

15. (1.2, 3.4) and (2.7, 3.1)

16. (7, 11) and (9, 2)

17. (−2, 2) and (5, 2)

18. (−3.1, 4.5) and (2.1, −3.4)

In Exercises 19–25, graph the line.

19. $y = 4x - 2$ **20.** $x - y = 3$

21. $y - 4 = 0.5(x - 2)$ **22.** $y = -5x + 10$

23. $2x - 3y = 5$ **24.** $y - 9 = -3(x + 2)$

25. $y = -\frac{2}{3}x + \frac{4}{3}$

In Exercises 26–30, use the graph to determine the equation of the line.

26.

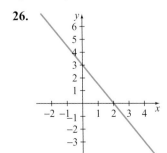

27.

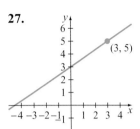

28.

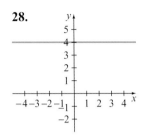

29.

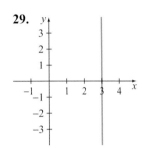

30.

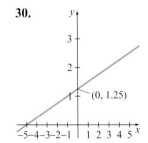

In Exercises 31–35, determine if the table of data represents a linear function. If so, calculate the slope and interpret its real-life significance.

31. **Average Personal Income**

Year	U.S. Average Personal Income (in terms of year 2000 dollars)
1989	18,593
1999	28,525

Source: www.census.gov.

32. **Internet Access**

Year	Number of People with Internet Access at Home or at Work (in thousands)
1997	46,305
1998	62,273
1999	83,677
2000	112,949

Source: www.census.gov.

33. **Take-Home Pay**

Months (since Sept. 2001)	Take-Home Pay (dollars)
0	3,167.30
1	4,350.31

Source: Employee pay stubs.

34. **Museum Admission Cost**

People in Group	Total Cost of Admission for a Group Visiting the Experience Music Project
3	$59.85
4	$79.80
5	$99.75
6	$119.70

Source: www.emplive.com.

35. **Solid Waste Disposal**

Clean Wood (pounds)	Cost to Dispose of Clean Wood at Enumclaw Transfer Station
500	$18.75
700	$26.25
900	$33.75
1,000	$37.50

Source: www.dnr.metrokc.gov.

In Exercises 36–38, find the linear equation that models the data and answer the specified questions.

36. **Nutrition** The Recommended Daily Allowance for dietary fiber is 25 grams for a person on a 2,000-calorie-per-day diet. A $\frac{3}{4}$-cup serving of Post® Fruity Pebbles® contains 0.2 gram of dietary fiber. (Source: Package labeling.) A 1-cup serving of 2 percent milk fortified with vitamin A doesn't contain any dietary fiber. A large banana $\left(8 \text{ to } 8\frac{7}{8} \text{ inches long}\right)$

contains 3.3 grams of dietary fiber. (Source: www.nutri-facts.com.) How many servings of Fruity Pebbles with milk would you have to eat along with a large banana to consume 8 grams of dietary fiber? (Round up to the nearest number of servings.)

37. **Nutrition** The Recommended Daily Allowance for dietary fiber is 25 grams for a person on a 2,000-calorie-per-day diet. A $\frac{3}{4}$-cup serving of General Mills Wheaties® contains 2.1 grams of dietary fiber. (Source: Package labeling.) A 1-cup serving of 2 percent milk fortified with vitamin A doesn't contain any dietary fiber. A large $\left(8 \text{ to } 8\frac{7}{8} \text{ inches long}\right)$ banana contains 3.3 grams of dietary fiber. (Source: www.nutri-facts.com.) How many servings of Wheaties with milk would you have to eat along with a large banana to consume 8 grams of dietary fiber? (Round up to the nearest number of servings.)

38. **Nutrition** The Recommended Daily Allowance for dietary fiber is 30 grams for a person on a 2,500-calorie-per-day diet. A large apple $\left(3\frac{1}{4} \text{ inches in diameter}\right)$ contains 5.7 grams of dietary fiber. A large banana (8 to $\left(8\frac{7}{8} \text{ inches long}\right)$ contains 3.3 grams of dietary fiber. (Source: www.nutri-facts.com.) How many apples would you have to eat along with a banana to consume 30 grams of dietary fiber?

Exercises 39–51 are intended to challenge your understanding of linear equations and their graphs.

39. What is the equation of the line that passes through $(4, 7)$ and $(4, -1)$?

40. Do the two equations represent the same line? Explain.

$$3x + 4y = 12; \quad y + 3 = -0.75(x - 8)$$

41. Explain why the equation of a vertical line cannot be written in slope-intercept or point-slope form.

42. **Alaska Population** The U.S. Census Bureau projects that the population of Alaska will grow at an approximate rate of 9,321 people per year between 1995 and 2025. It estimates that 791,000 people will be living in Alaska in 2025. (Source: www.census.gov.) What is the projected population of Alaska in 2035?

43. Are all lines functions? Explain.

44. Two lines are said to be parallel if they do not intersect. If $f(x) = mx + b$ and $g(x) = nx + c$ are parallel lines, what conclusions can you draw about b, c, m, and n?

45. Two lines are perpendicular if the slope of one line is the negative reciprocal of the slope of the other line. Find the equation of the line perpendicular to $y = 4$ that passes through the point $(3, -2)$.

46. **M** **World Population** Based on data from 1980 to 2001, the population of the world may be modeled by $P = (81.3t + 4460)$ million people, where t is the number of years since 1980. (**Source:** Modeled from *Statistical Abstract of the United States, 2001*, Table 1324, p. 829.) According to the model, how quickly will the world population be increasing in 2010?

47. The equation of a nonvertical line is given by $ax + by = c$. What are the slope, x-intercept, and y-intercept of the line?

48. Is the function shown in the table linear? Explain.

x	y
2	11
5	17
8	23
10	29
14	35

49. In order for the function shown in the table to be linear, what must be the value of b?

x	y
2	7
5	13
8	19
10	b
14	31

50. In order for the function shown in the table to be linear, what must be the value of a?

x	y
2	6
5	12
8	18
a	26
14	30

51. What is the point of intersection of the lines $x = a$ and $y = b$?

1.3 Linear Models

- Use technology to model linear and near-linear data
- Use a linear equation to describe the relationship between directly proportional quantities
- Recognize and model naturally occurring linear, near-linear, and piecewise linear relationships

GETTING STARTED The amount of money deducted from a paycheck for Social Security, the annual income of a sales representative, and the profit from a club car wash may be modeled by linear functions. In this section, we will look at a wide variety of information encountered in everyday life and learn how to determine if the data are linear, near-linear, or piecewise linear. We will also demonstrate how to use technology to determine the linear equation that best fits a data set. Additionally, we will show how to find linear equations to model real-life phenomena.

Direct Proportions

Two quantities are said to be **directly proportional** if the ratio of the output to the input is a constant. Quantities that are directly proportional to each other may be modeled by a linear equation of the form

$$y = kx$$

The constant $k = \dfrac{y}{x}$ and is called the **constant of proportionality.**

EXAMPLE 1   **Determining If a Data Set Is Directly Proportional**

For the data in Table 1.8, determine whether the employee's weekly income is directly proportional to the number of hours worked. If so, find a linear function that models the data.

TABLE 1.8

Hours Worked (h)	Income (I)
35.0	$210.00
38.5	$231.00
42.3	$253.80

SOLUTION   Calculate the ratio $\dfrac{I}{h}$ for each ordered pair.

UNITS

$$\frac{\$210}{35 \text{ hours}} = \$6/\text{hr}$$

$$\frac{\$231}{38.5 \text{ hours}} = \$6/\text{hr}$$

$$\frac{\$253.80}{42.3 \text{ hours}} = \$6/\text{hr}$$

Since $\dfrac{I}{h} = 6$ for each ordered pair, the weekly income is directly proportional to the hours worked with a constant of proportionality $k = 6$. The linear function is

$$I = 6h$$

EXAMPLE 2   **Determining If a Data Set Is Directly Proportional**

Determine whether the cost of filling a gas tank as given in Table 1.9 is directly proportional to the number of gallons put into the tank. If so, find a linear function that models the data.

Number of Gallons (g)	Cost (C)
3.50	$5.25
6.28	$9.41
17.34	$25.99

TECHNOLOGY **TIP**

Drawing a Scatter Plot

1. Bring up the Statistics Plot menu by pressing the 2nd button, then the Y= button.

2. Open Plot1 by pressing ENTER.

3. Turn on Plot1 by moving the cursor to On and pressing ENTER. Confirm that the other menu entries are as shown.

4. Graph the scatter plot by pressing ZOOM and scrolling to 9:ZoomStat. Press ENTER. This will graph the entire scatter plot along with any functions in the Graphing List. The ZoomStat feature automatically adjusts the viewing window so that all of the data points are visible.

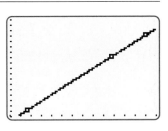

The graph shows the data points along with the linear model. We see that the linear model fits the data well.

EXAMPLE **4** **Using the Line of Best Fit to Forecast Sales**

The annual sales of McDonalds Corporation are shown in Table 1.14.

TABLE 1.14

Years Since 1990 (*t*)	Franchised Sales (*S*) (millions of dollars)
1	12,959
2	14,474
3	15,756
4	17,146
5	19,123
6	19,969
7	20,863
8	22,330
9	23,830
10	24,463
11	24,838

Source: www.mcdonalds.com.

(a) Draw a scatter plot of the data. Will a linear function fit the data well?
(b) Find the equation of the linear model that best fits the data.
(c) Graph the linear model simultaneously with the scatter plot. How well does the model predict the data?
(d) What does the slope of the function tell us about the company's sales?

SOLUTION

(a) Figure 1.23 shows a scatter plot of the data.

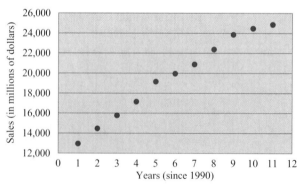

FIGURE 1.23

The sales revenue increases every year, but the amount by which it increases varies from year to year. Since the rate of increase is not constant, a linear function won't fit the data perfectly. However, the data do appear to be near-linear, so a linear model may fit the data well.

(b) Using linear regression, we determine that the linear model for sales is $S = 1233.5t + 12{,}213$ million dollars where t is the number of years since the end of 1990.

(c)

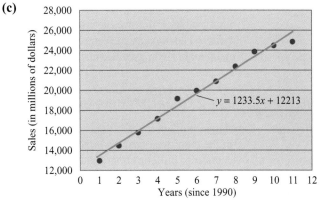

FIGURE 1.24

As shown in Figure 1.24 and Table 1.15, the linear model fits the data fairly well.

TABLE 1.15

Years Since 1990 (t)	Franchised Sales (S) (millions of dollars)	Estimated Sales (E) (millions of dollars)	Error ($S - E$) (millions of dollars)	% Error $\left(\dfrac{S - E}{S}\right)$
1	12,959	13,447	−488	−3.76%
2	14,474	14,680	−206	−1.42%
3	15,756	15,914	−158	−1.00%
4	17,146	17,147	−1	−0.01%
5	19,123	18,381	743	3.88%
6	19,969	19,614	355	1.78%
7	20,863	20,848	16	0.07%
8	22,330	22,081	249	1.12%
9	23,830	23,315	516	2.16%
10	24,463	24,548	−85	−0.35%
11	24,838	25,782	−944	−3.80%

Although the actual sales and the estimated sales from the model differed by as much as 944 million dollars, the model projections were within 3.88 percent of the actual values.

(d) The slope of the sales function is 1233.5 million dollars per year. According to the model, we expect that, on average, the sales will increase by about 1233.5 million dollars per year. We may use this information to help us determine whether to buy stock in the company. We should be aware that past performance is no guarantee of future results. Before making a stock purchase, we would want to look at other indicators, such as earnings per share, recent news, analyst ratings, and so on.

Finding Linear Functions to Model Real-Life Phenomena

Not all of the data we encounter in the real world are laid out nicely in an input-output table. Often we have to ferret out the information needed to formulate a model. The next two examples will illustrate real-life situations where using mathematics can give us financial leverage.

EXAMPLE 5   **Using Linear Models to Make Business Decisions**

Many companies offer their salespeople a base salary plus a commission based on the employee's sales. In November 2001, a Nebraska firm posted an ad on an electronic bulletin board advertising a sales representative position paying a $36,000 base salary plus commission. (**Source:** www.dice.com.) Based on her past sales experience, a saleswoman estimates that she can generate $600,000 in sales annually. To maintain her standard of living, she needs to earn a total of $63,000 annually.

(a) What commission rate (as a percentage of sales) must she earn to maintain her standard of living?

(b) Write the equation for her annual income as a function of her annual sales dollar volume.

(c) If she wants to increase her annual income to $75,000, by how much will she have to increase her annual sales?

(d) If she wanted to increase her annual income to $75,000 without increasing her sales volume, what commission rate must she receive?

(e) How could the answers to the previous questions help her during salary negotiations?

SOLUTION

(a) We have

$$I = ms + b$$

where I is the annual income (in thousands of dollars) and s is the annual sales (in thousands of dollars). If she doesn't sell anything (assuming she doesn't get fired), she will still earn $36,000, so $(0, 36)$ is the y-intercept. Thus

$$I = ms + 36$$

When $s = 600$, $I = 63$, so

$$63 = m \cdot 600 + 36$$
$$27 = 600m$$
$$m = 0.045$$

She must earn a 4.5 percent commission to maintain her standard of living.

(b) The equation for her annual income is $I = 0.045s + 36$ thousand dollars, where s is the sales volume (in thousands of dollars).

(c) Setting $I = 75$ and solving for s, we get

$$75 = 0.045s + 36$$
$$39 = 0.045s$$
$$s = 866.667$$

She must increase her annual sales from $600,000 to $866,667 to increase her annual income to $75,000.

(d) Letting m be the variable and setting I equal to 75, we get

$$75 = m \cdot 600 + 36$$
$$39 = 600m$$
$$m = 0.065$$

So to increase her income to $75,000 without increasing sales, she must earn a 6.5 percent commission.

(e) We know that her minimum acceptable commission rate is 4.5 percent. To increase her salary without having to increase her workload, she will have to convince the employer to increase her commission rate to 6.5 percent. If the employer is unwilling to raise the commission rate, she will have to bring in an additional $266,667 in sales to earn $75,000. This information can give her leverage in salary negotiations.

EXAMPLE 6

Using Linear Models to Make Business Decisions

In 2001, BeautiControl Cosmetics sold a new consultant a demonstration package for $119.82. (**Source:** BeautiControl Cosmetics consultant.) The consultant's director told her that the average revenue from each in-home party is $250. The consultant's cost for the cosmetics is 50 percent of the sales price.

(a) Find the linear model for the consultant's cost as a function of the dollar volume of the cosmetics sold.

(b) Find the linear model for the consultant's revenue as a function of the dollar volume of cosmetics sold.

(c) Find the linear model for the consultant's profit as a function of the dollar volume of cosmetics sold.

(d) How many dollars of cosmetics must the consultant sell before she begins to make a profit?

(e) How many in-home parties must the consultant hold in order to begin making a profit?

SOLUTION

(a) Let x be the sales volume in dollars. Then the cost function C is

$$C(x) = 0.5x + 119.82$$

The cost of the demonstration package, $119.82, is the fixed cost. The consultant's variable cost, $0.5x$ dollars, is dependent upon her sales dollar volume.

(b) The revenue function R is

$$R(x) = x$$

The consultant's revenue is equal to her sales dollar volume.

(c) The profit function is the difference between the revenue and cost functions.

$$P(x) = R(x) - C(x)$$
$$= x - (0.5x + 119.82)$$
$$= 0.5x - 119.82$$

(d) We need to find the break-even point (the sales dollar volume that makes the profit equal zero). Setting the profit equal to zero, we get

$$0 = 0.5x - 119.82$$
$$119.82 = 0.5x$$
$$x = 239.64$$

So to break even, she must sell $239.64 worth of cosmetics.

(e) Dividing the sales dollar volume needed to break even by the amount of revenue per party, we get

UNITS

$$\frac{239.64 \text{ dollars}}{250.00 \frac{\text{dollars}}{\text{party}}} = 0.95856 \text{ parties}$$

It doesn't make sense to talk about fractions of a party, so we round up. The consultant should begin to make a profit at her first party.

Piecewise Linear Models

A function of the form

$$f(x) = \begin{cases} ax + b & x \le c \\ dx + g & x > c \end{cases}$$

is called a **piecewise linear function.** The rule used to calculate $f(x)$ depends upon the value of x. In this case, if x is less than or equal to c, the rule is $f(x) = ax + b$. If x is greater than c, the rule is $f(x) = dx + g$.

A common piecewise linear function is the absolute value function $f(x) = |x|$. This function makes negative numbers positive and leaves positive numbers unchanged. In practical terms, it "erases" the negative sign of any negative number. The absolute value function is formally defined as a piecewise linear function.

$$f(x) = |x|$$
$$= \begin{cases} x & x \ge 0 \\ -x & x < 0 \end{cases}$$

What is $|-5|$? Since $-5 < 0$, we apply the second rule of the piecewise function.

$$|-5| = -(-5)$$
$$= 5$$

Piecewise linear functions are very common in a variety of real-life situations, such as the ones given in Examples 7 and 8.

EXAMPLE 7

Using a Piecewise Linear Model to Forecast Monthly Payments

An orthodontist and the parents of a girl scheduled to receive braces agreed upon the following payment plan: a $550 down payment when treatment begins and $100 per month until the treatment is paid in full. The total cost for the treatment is $1250, and the treatment is expected to take 18 months to complete. (**Source:** Author's personal payment plan.)

(a) How many months after treatment begins will the treatment be paid in full?

(b) Write the equation for the balance due as a function of the number of months since treatment began.

SOLUTION

(a) The patient's parents pay $550 when treatment begins, so $700 remains for the balance to be paid in full. Since $100 is paid each month, the treatment will be paid in full 7 months later.

(b)

$$B(t) = \begin{cases} -100t + 700 & 0 \le t \le 7 \\ 0 & 7 < t \le 18 \end{cases}$$

where $B(t)$ is the balance of the account t months after treatment begins. Since payments are made monthly, the function is defined for whole number values of t.

   We initially use a linear function for $B(t)$, since the account balance starts at $700 and is reduced by a constant $100 per month. However, after 7 months, the account balance has reached $0. If we continued to use the linear function, the balance would end up being a negative number, which doesn't make sense in the context of the problem. Therefore, no further payments are made after the seventh month, and the account balance remains at $0 from the end of the seventh month through the eighteenth month.

EXAMPLE 8

Using a Piecewise Linear Model to Make Spending Decisions

Experience Music Project, an interactive music museum in Seattle, charges adult visitors a $19.95 admission fee. Groups of 15 or more may enter for $14.50 per person. (**Source:** www.emplive.com.)

(a) Draw a scatter plot for group cost as a function of group size.
(b) Write the total cost of admission as a function of the number of people in a group. Confirm that the function and the scatter plot are in agreement.
(c) Calculate the cost of admitting a group of 14 people and the cost of admitting a group of 15 people.
(d) Determine the largest group we could bring in and still remain below the cost of a 14-person group.

SOLUTION

(a)

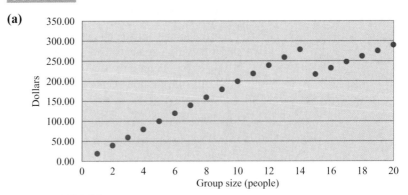

FIGURE 1.25

(b)

$$C(x) = \begin{cases} 19.95x & x < 15 \\ 14.5x & x \geq 15 \end{cases}$$

where x is the number of people in the group and C is the total cost in dollars.

(c) $C(14) = 19.95(14) = \$279.30$ is the cost for a group of 14 people.

$C(15) = 14.5(15) = \$217.50$ is the cost for a group of 15 people.

(d) From the scatter plot, it looks as if the maximum number of people we could bring in and still remain below the 14-person group price is 19 people. We can confirm the result algebraically.

$$279.30 = 14.5x$$

$$x \approx 19.26$$

Since it doesn't make sense to talk about 0.26 of a person, we round down to 19.

$$C(19) = 14.5(19)$$

$$= 275.50$$

The cost for a 19-person group is $275.50. This is slightly less than the $279.30 cost for a 14-person group.

Determining Which Modeling Method to Use

Although linear regression is a powerful tool for creating linear models, many linear models can be constructed algebraically, as shown in Examples 5 through 8. It is good practice to first attempt to find the linear model algebraically. A common pitfall of beginning students is to use the calculator to find the linear model when it is actually easier to find the model algebraically.

Whenever the data set contains exactly two points, we should find the linear model algebraically. In addition, there are other key phrases that alert us to the fact that a linear model may be constructed algebraically. Table 1.16 details how to interpret the mathematical meaning of some commonly occurring phrases.

TABLE 1.16

Phrase	Mathematical Meaning
constant rate of change	slope
increasing at a rate of 20 people per year	The slope of the linear function is $20\dfrac{\text{people}}{\text{year}}$.
tickets cost $37 per person	The slope is $37\dfrac{\text{dollars}}{\text{person}}$.
sales increase by 100 units for every $1 decrease in price	The slope is $$m = \dfrac{100 \text{ units}}{-1 \text{ dollar}}$$ $$= -100 \text{ units per dollar.}$$
initial price of $1.25	The y-intercept is 1.25.
There are 350 chairs today. The number of chairs is increasing by 10 chairs per month.	The y-intercept is 350 and the slope is 10.
The price is $2.25 and is decreasing at a constant rate of $0.02 per day.	The y-intercept is 2.25 and the slope is -0.02.

By applying the techniques demonstrated in the table, you will increase your comfort level in working with real-world problems.

1.3 Summary

In this section, you looked at a wide variety of information encountered in everyday life, and you learned how to model linear, near linear, and piecewise linear data sets. You also learned how to use technology to determine the line of best fit to model real-life phenomena.

1.3 Exercises

In Exercises 1–5, do the following:

(a) Draw the scatter plot.
(b) Find the equation of the line of best fit.
(c) Interpret the meaning of the slope and the y-intercept of the model.
(d) Explain why you do or do not believe that the model would be a useful tool for businesses and/or consumers.

1.  Share Price

Month in 2000	Harbor Capital Appreciation Fund Share Price on Last Day of Month
10	$48.16
11	$41.50
12	$35.58

Source: www.quicken.com.

2. **Share Price**

Month in 2001	CREF Stock Retirement Annuity Share Price on Last Day of Month
3	$164.5760
6	$173.1588
9	$146.4102

Source: TIAA-CREF statement.

 3. **University Enrollment**

Washington State Public University Enrollment	
Years Since 1990 (t)	Students (S)
0	81,401
1	81,882
2	83,052
3	84,713
4	85,523
5	86,080
6	87,309
7	89,365
8	90,189
9	91,543
10	92,821

Source: Washington State Higher Education Coordinating Board.

 4. **Per Capita Personal Income**

Per Capita Personal Income—Utah	
Years Since 1993 (t)	Personal Income (P) (dollars)
0	16,830
1	17,638
2	18,508
3	19,514
4	20,613
5	21,594
6	22,305
7	23,436

Source: Bureau of Economic Analysis (www.bea.gov).

5. **Per Capita Income Ranking**

Years since 1995	North Carolina Per Capita Income Ranking (out of 50 states)
0	42
1	34
2	32
3	28
4	28

Source: http://www.census.gov/statab.

In Exercises 6–16, formulate the linear or piecewise linear model for the real-life scenario. If a linear model represents a directly proportional relationship between the independent and dependent variables, so state.

6. **Personal Income** In November 2001, Hall Kinion advertised an Account Executive position in Savannah, Georgia, for a salesperson with experience selling Internet services. The job advertised a $35K base salary and $24K to $48K in commissions in the first year. (**Source:** www.dice.com.)

(a) Assuming a 20 percent commission rate, how many dollars in sales would you have to generate to earn $64K annually?

(b) If you were in the sales field, why would this type of information be valuable to you?

7. **Solid Waste Disposal** King County charges $82.50 per ton to dispose of garbage at area transfer stations. A minimum disposal fee of $13.72 per entry is charged for vehicles disposing of trash. In addition to the disposal fee, a moderate risk waste (MRW) fee of $2.61 per ton is assessed. A minimum MRW fee of $1 is charged on all transactions. Tax is charged on the combined price of the disposal and moderate risk waste fees. The tax rate is 3.6 percent. The weight of the trash is rounded to the nearest 20 pounds before the fees are calculated. (**Source:** www.dnr.metrokc.gov.)

(a) Write the equation for disposal cost as a function of the number of pounds of refuse.

(b) How much will it cost to drop off 230 pounds of trash?

(c) How much will it cost to drop off 513 pounds of trash?

(d) If you ran a construction company, how could you use the results of this exercise?

8. **Cellular Phone Plan** In 2002, Qwest offered a 150-minute cellular plan including free long distance and free roaming to its Seattle-area customers for $29.99 per month. Additional time cost $0.35 per minute or portion of a minute. (**Source:** Qwest customer monthly statement.)

(a) Write the cellular phone cost equation as a function of the number of minutes used.

(b) How could you use the cost equation if you were enrolled in the plan?

9. **Cellular Phone Plan** Sprint PCS offered a service plan to its Seattle-area customers for $29.99 per month that included 200 anytime minutes. The plan also included free long distance. Additional minutes or portions of a minute were $0.40 each. (**Source:** www1.sprintpcs.com.)

(a) Determine the cost equation for the Sprint PCS plan as a function of the number of anytime minutes used.

(b) Is the Sprint plan or the Qwest plan (from Exercise 8) a better deal for a consumer who uses 300 anytime minutes monthly? Explain.

10. **Solid Waste Disposal** The Enumclaw Transfer/Recycling Station accepts clean wood (stumps, branches, etc.) in addition to household garbage. The station charges $75 per ton with a minimum fee of $12.75 per entry for vehicles disposing of clean wood. (**Source:** www.dnr.metrokc.gov.)

(a) How many pounds of clean wood could be dropped off at the station without exceeding the $12.75 minimum fee?

(b) Write the equation for the disposal cost as a function of the number of pounds of clean wood.

(c) If you ran a landscaping business, how could you use the results of parts (a) and (b)?

11. **Used Car Value** In 2001, the average retail price of a 2000 Toyota Land Cruiser 4-Wheel Drive was $43,650. The average retail price of a 1995 Toyota Land Cruiser 4-Wheel Drive was $21,125. (**Source:** www.nadaguides.com.)

(a) Find a linear model for the value of a Toyota Land Cruiser in 2001 as a function of its production year.

(b) Use your linear model to predict the value of a 1992 Toyota Land Cruiser in 2001.

(c) A 1992 Toyota Land Cruiser had an average retail price of $14,325 in 2001. How good was your linear model at predicting the value of the vehicle?

12. **Nutrition** The Recommended Daily Allowance (RDA) for fat for a person on a 2,000-calorie-per-day diet is less than 65 grams. A McDonalds Big Mac® sandwich contains 34 grams of fat. A large order of French fries contains 26 grams of fat. (**Source:** www.mcdonalds.com.)

(a) Write the equation for fat grams consumed as a function of large orders of French fries.

(b) Write the equation for fat grams consumed as a function of Big Macs.

(c) How many large orders of French fries can you eat without exceeding the RDA for fat?

(d) How many Big Macs can you eat without exceeding the RDA for fat?

(e) How many combination meals (Big Mac and French fries) can you eat without exceeding the RDA for fat?

13. 🌐 **Nutrition**   The Recommended Daily Allowance (RDA) for fat for a person on a 2,500-calorie-per-day diet is less than 80 grams. A McSalad Shaker Chef Salad® contains 8 grams of fat. A package (44.4 ml) of ranch salad dressing contains 18 grams of fat. (**Source:** www.mcdonalds.com.)

   (a) Write the equation for fat grams as a function of Chef Salads.
   (b) How many salads can you eat and remain below the RDA for fat?
   (c) If you use one package of ranch salad dressing, how many salads can you eat and remain below the RDA for fat?
   (d) The RDA for fat for a person on a 2000-calorie diet is 65 grams. If a person on a 2,000-calorie diet uses one package of ranch salad dressing, how many salads can that person eat and remain below the RDA for fat?

14. 🌐 **Sales Tax**   The Picture People photography studio in Tacoma, Washington, charges $40 for one of its promotional photo packages. The total purchase price for the promotional package, including tax, is $43.20. (**Source:** Picture People sales receipt.)

   (a) Write the equation for Tacoma sales tax as a function of the pretax purchase price.
   (b) A family has $60 cash to pay for pictures. What is the maximum pretax purchase price that the family can afford?

15. 🌐 **Airline Ticket Cost**   The total cost of an airline ticket includes the published fare, a federal flight segment tax, a federal security fee, and an airport Passenger Facility Charge (PFC). The flight segment tax is $3 on each segment of the itinerary. (A segment is a takeoff and landing.) The PFC varies from airport to airport; it ranges from $3 to $4.50, and it is imposed by an airport on enplaning passengers. When a passenger is departing from or connecting at any airport, that airport's PFC will apply in addition to the fare. Likewise, a $2.50 security fee is imposed on all enplaning passengers.

   A Southwest Airlines refundable airline ticket from Seattle to Phoenix is advertised at $222. Write the maximum total cost of a Southwest (**Source:** www.southwest.com.) airline ticket from Seattle to Phoenix as a function of the number of segments. (Assume that a passenger travels exactly one segment per airplane. That is, if the plane stops, the passenger changes planes.)

16. 🌐 **Airline Ticket Cost**   The total cost of an airline ticket includes the published fare, a federal flight segment tax, a federal security fee, and an airport Passenger Facility Charge (PFC). The flight segment tax is $3 on each segment of the itinerary. (A segment is a takeoff and landing.) The PFC varies from airport to airport; it ranges from $3 to $4.50, and it is imposed by an airport on enplaning passengers. When a passenger is departing from or connecting at any airport, that airport's PFC will apply in addition to the fare. Likewise, a $2.50 security fee is imposed on all enplaning passengers.

   A Southwest Airlines refundable round-trip airline ticket from Seattle to Phoenix is advertised at $444. (**Source:** www.southwest.com.) Write the maximum total cost of a round-trip airline ticket from Seattle to Phoenix as a function of the number of segments.

Exercises 17–25 are intended to challenge your understanding of linear models.

17. The standard form of a linear function in three dimensions is given by $ax + by + cz = d$, where x, y, and z are variables and a, b, c, and d are constants. A business determines that the **revenue** generated by selling x cups of coffee, y bagels, and z muffins is given by

$$R = ax + by + cz$$

What is the practical meaning of the constants a, b, and c?

18. A business determines that the **cost** of producing x cups of coffee, y bagels, and z muffins is given by

$$C = ax + by + cz + d$$

What is the practical meaning of the constants a, b, c, and d?

19. The equation of the line of best fit for the data in the table is $y = 0$.

x	y
0	−1
1	1
2	0
3	0
4	1
5	−1
6	0

This model passes through three of the seven points shown in the table. Is this model a good fit for the data? Justify your answer.

20. Are production costs directly proportional to the number of items produced for most businesses? Explain.

21. Does the fact that a linear model passes through the point $(0, 0)$ imply that the dependent variable of the original data is directly proportional to the independent variable of the original data? Explain.

22. A linear model has a correlation coefficient of $r = -1$. Does the fact that the linear model passes through the point $(0, 0)$ imply that the output of the original data is directly proportional to the input of the original data? Explain.

23. Use your calculator to find the line of best fit for the data in the table.

x	y
0	1
1	1
2	1
3	1

Why do you think the correlation coefficient is undefined? Does this mean that the model doesn't fit the data? Justify your answer.

24. Find the piecewise linear function that best fits the data in the table.

x	y
0	2
1	4
2	6
3	7
4	8
5	9
6	10

25. Find the piecewise linear function that best fits the data in the table.

x	y
−2.0	14
1.0	8
2.5	5
4.0	2
8.0	4
10.0	5

Chapter 1 Review Exercises

Section 1.1   *In Exercises 1–3, calculate the value of the function at the designated input and interpret the result.*

1. $C(x) = 49.95x$ at $x = 2$, where C is the cost (in dollars) of buying x pairs of shoes.

2. $C = 9.95x + 1200$ at $x = 400$, where C is the cost of making x pairs of shoes.

3. $H = -16t^2 + 100$ at $t = 2$, where H is the height of a cliff diver above the water t seconds after he jumped from a 100-foot cliff.

In Exercises 4–7, determine the domain of the function.

4. $f(x) = 4x^3 - 10x$

5. $f(p) = p^2 - 9p + 15$

6. $C = \dfrac{r^2 - 1}{2r + 1}$ **7.** $C = \dfrac{r^2 + 1}{r^2 - 1}$

In Exercises 8–9, interpret the meaning of the indicated point on the graph.

8. 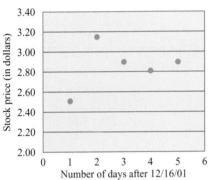 **Stock Price** Find P at $t = 4$, where P is the stock price of an e-learning company at the end of the day and t is the number of days after December 16, 2001.

Source: Click2Learn Corporation.

9. 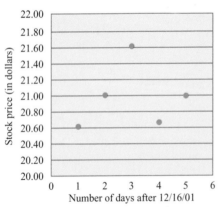 **Stock Price** Find P at $t = 2$, where P is a computer company's stock price at the end of the day, t days after December 16, 2001.

Source: Apple Computer Corporation.

In Exercises 10–11, determine whether the graphs represent functions by applying the Vertical Line Test.

10.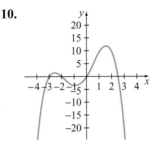

11.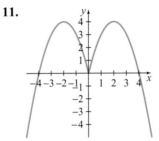

Section 1.2 *In Exercises 12–15, calculate the slope of the linear function passing through the points.*

12. $(2, 9)$ and $(4, 3)$

13. $(-3, -4)$ and $(0, -2)$

14. $(1.3, 3.5)$ and $(3.5, 1.3)$

15. $(7, 11)$ and $(8, 0)$

In Exercises 16–18, find the x-intercept and the y-intercept of the linear function.

16. $y = -5x + 10$ **17.** $y = -3x + 18$

18. $y = 2x - 12$

In Exercises 19–20, write the equation of the linear function passing through the points in slope-intercept form and in standard form.

19. $(2, 5)$ and $(4, 3)$ **20.** $(-3, 4)$ and $(0, -2)$

Section 1.3 *In Exercises 21–23, find a model for the data and answer the given questions.*

21. **Nutrition** The Recommended Daily Allowance (RDA) for fat for a person on a 2,000-calorie-per-day diet is less than 65 grams.

A McDonalds Big N' Tasty® sandwich contains 32 grams of fat. A super size order of French fries contains 29 grams of fat. (Source: www.mcdonalds.com.)

(a) Write the equation for fat grams consumed as a function of large orders of French fries eaten.

(b) Write the equation for fat grams consumed as a function of Big N' Tasty sandwiches eaten.

(c) How many combination meals (Big N' Tasty sandwich and super size order of French fries) can you eat without exceeding the RDA for fat?

22. **Solid Waste Disposal** The Enumclaw Transfer/Recycling Station accepts clean wood (stumps, branches, etc.) in addition to household garbage. The station charges $75 per ton with a minimum fee of $12.75 per entry for vehicles disposing of clean wood. (Source: www.dnr.metrokc.gov.)

(a) Write a piecewise linear function to model the cost of disposing of x pounds of clean wood. (Assume that the entire amount of wood is delivered in a single load.)

(b) How much does it cost to dispose of 300 pounds of clean wood? 700 pounds?

23. **Used Car Value** In 2001, the average retail price of a 1998 Mercedes-Benz Roadster two-door SL500 was $51,400. The average retail price of a 2000 Mercedes-Benz Roadster two-door SL500 was $66,025. (Source: www.nadaguides.com.)

(a) Find a linear model for the value of a Mercedes-Benz Roadster two-door SL500 in 2001 as a function of its production year.

(b) Use your linear model to predict the 2001 value of a 1999 Mercedes-Benz Roadster two-door SL500.

(c) A 1999 Mercedes-Benz Roadster two-door SL500 had an average retail price of $58,500 in 2001. How good was your linear model at predicting the value of the vehicle?

Make It Real

What to do

1. Find a set of at least six data points from an area of personal interest.
2. Draw a scatter plot of the data and explain why you do or do not believe that a linear model would fit the data well.
3. Find the equation of the line of best fit for the data.
4. Interpret the physical meaning of the slope and *y*-intercept of the model.
5. Use the model to predict the value of the function at an unknown point and explain why you do or do not think the prediction is accurate.
6. Explain how a consumer and/or a businessperson could benefit from the model.
7. Present your findings to the class and defend your conclusions.

Where to look for data

Box Office Guru
www.boxofficeguru.com
Look at historical data on movie revenues.

Nutri-Facts
www.nutri-facts.com
Compare the nutritional content of common foods based on serving size.

Quantitative Environmental Learning Project
www.seattlecentral.org/qelp
Look at environmental information in easy-to-access charts and tables.

U.S. Census Bureau
www.census.gov
Look at data on U.S. residents ranging from Internet usage to family size.

Local Gas Station or Supermarket
Track an item's price daily for a week.

School Registrar
Ask for historical tuition data.

Utility Bills
Look at electricity, water, or gas usage.

Employee Pay Statements
Look at take-home pay or taxes.

Body Weight Scale
Track your body weight for a week.

Nonlinear Models

Mathematical functions are commonly used to model real-world data. Although every model has its limitations, models are often used to forecast expected results. Selecting which mathematical model to use is relatively easy once you become familiar with the basic types of mathematical functions. Remarkably, these basic functions may be used to effectively model many real-world data sets. For example, based on data from 1993 to 2002, the sales revenue from Starbucks stores may be effectively modeled by a quadratic function.

2.1 Quadratic Function Models
- Use nonlinear functions to model real-life phenomena
- Interpret the meaning of mathematical models in their real-world context
- Model real-life data with quadratic functions

2.2 Higher-Order Polynomial Function Models
- Use nonlinear functions to model real-life phenomena
- Interpret the meaning of mathematical models in their real-world context
- Model real-life data with higher-order polynomial functions

2.3 Exponential Function Models
- Use nonlinear functions to model real-life phenomena
- Interpret the meaning of mathematical models in their real-world context
- Graph exponential functions
- Find the equation of an exponential function from a table
- Model real-life data with exponential functions

2.4 Logarithmic Function Models
- Use nonlinear functions to model real-life phenomena
- Interpret the meaning of mathematical models in their real-world context
- Graph logarithmic functions
- Model real-life data with logarithmic functions
- Apply the rules of logarithms to simplify logarithmic expressions
- Solve logarithmic equations

2.5 Choosing a Mathematical Model
- Use critical thinking skills in selecting a mathematical model

2.1 Quadratic Function Models

- Use nonlinear functions to model real-life phenomena
- Interpret the meaning of mathematical models in their real-world context
- Model real-life data with quadratic functions

GETTING STARTED   Many college-bound high school students enroll in Advanced Placement Program (AP®) courses to earn college credit while still in high school. The College Board® offers 35 courses in 19 disciplines ranging from Studio Art: 3D Design to Calculus AB. (**Source:** The College Board®.) The number of students taking the Calculus AB test has increased dramatically since 1969, as shown in Figure 2.1.

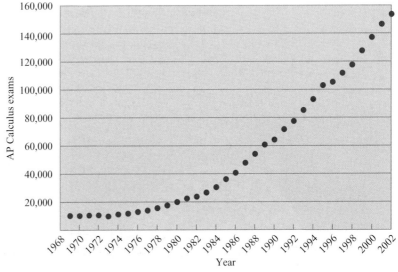

FIGURE 2.1

How many students will take the test in 2005? Nobody knows; however, using a mathematical model, we can predict what may happen.

In this section, we will discuss the identifying features of quadratic functions. We will then demonstrate how to use algebraic methods and quadratic regression to model data sets whose graphs open upward or downward over their domain.

A polynomial is a function that is the sum of terms of the form ax^n, where a is a real number and n is a nonnegative integer. For example, each of the following functions is a polynomial.

$$f(x) = 6x^4 - 3x^3 + 5x^2 - 2x + 9$$
$$g(x) = 4x^2 + 2x$$
$$h(x) = -1.3x + 9.7$$
$$j(x) = x^2 - 2x + 6$$

May a constant term like 9 be written in the form ax^m? Yes! Since $x^0 = 1$ for nonzero x, $9x^0 = 9$. Consequently, a constant function $s(x) = 9$ is also a polynomial.

The **degree of a polynomial** is the value of its largest exponent. For example, the degree of the polynomial $f(x) = 6x^4 - 3x^3 + 5x^2 - 2x + 9$ is 4, since 4 is the largest exponent. Since the equation of any line may be written as $y = ax^1 + b$, lines are polynomials of degree 1. We've worked extensively with first-degree polynomials (lines) in the preceding chapter. In this section, we are interested in polynomials of degree 2. Polynomials of degree 2 are called **quadratic functions.**

QUADRATIC FUNCTION

A polynomial function of the form $f(x) = ax^2 + bx + c$ with $a \neq 0$ is called a **quadratic function.** The graph of a quadratic function is a **parabola.**

When a parabola opens upward, we say that it is "concave up." When the parabola opens downward, we say that it is "concave down." The steepness of the sides of the parabola and its concavity are controlled by the value of a, the coefficient on the x^2 term in its equation. If $a > 0$, the graph is concave up. If $a < 0$, the graph is concave down. As the magnitude of a increases, the steepness of the graph increases. (For $a > 0$, the magnitude of a is a. For $a < 0$, the magnitude of a is $-a$.) Consider the graphs in Figures 2.2 and 2.3. These graphs have the same values for b and c ($b = -2$ and $c = 3$) but differing values of a.

In Figure 2.2, since a is the coefficient on the x^2 term, $a = 1$. The magnitude of a is 1. Since $a > 0$, the graph is concave up.

In Figure 2.3, since a is the coefficient on the x^2 term, $a = 2$. The magnitude of a is 2. Increasing the magnitude of a increased the graph's steepness.

Consider the graphs in Figures 2.4 and 2.5. These graphs have the same values for b and c ($b = 4$ and $c = 0$) but differing values of a.

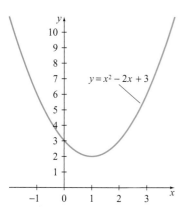

FIGURE 2.2

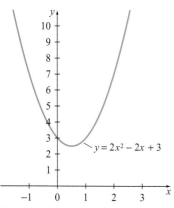

FIGURE 2.3

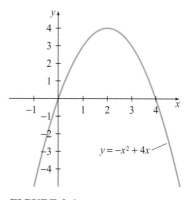

FIGURE 2.4

In Figure 2.4, since a is the coefficient on the x^2 term, $a = -1$. The magnitude of a is 1. Since $a < 0$, the graph is concave down.

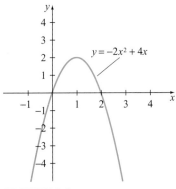

FIGURE 2.5

In Figure 2.5, since a is the coefficient on the x^2 term, $a = -2$. The magnitude of a is 2. Increasing the magnitude of a from 1 to 2 increased the graph's steepness.

The coefficient b in the equation affects the horizontal and vertical placement of the parabola. The constant term c in the equation indicates that the point $(0, c)$ is the y-intercept of the graph.

Recall that the graph of a function is said to be increasing if the value of y gets bigger as the value of x increases. Similarly, the graph of a function is said to be decreasing if the value of y gets smaller as the value of x increases. The vertex of a parabola is the point on the graph of a quadratic function where the curve changes from decreasing to increasing (or vice versa). The minimum y value of a concave up parabola occurs at the vertex (see Figure 2.6a). Similarly, the maximum y value of a concave down parabola occurs at the vertex (see Figure 2.6b).

For parabolas with x-intercepts, the x coordinate of the vertex always lies halfway between the x-intercepts. Recall that as a result of the quadratic formula, we know that the x coordinate of the vertex is $x = \dfrac{-b}{2a}$. The y coordinate of the vertex is obtained by evaluating the quadratic function at this x value. For example, for the quadratic function $y = x^2 - 2x + 3$, we know that $a = 1$ and $b = -2$. The x coordinate of the vertex is

$$x = \frac{-b}{2a}$$
$$= \frac{-(-2)}{2(1)}$$
$$= \frac{2}{2}$$
$$= 1$$

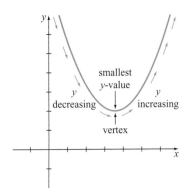

FIGURE 2.6a Concave Up Parabola

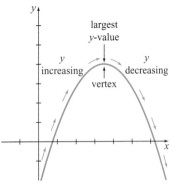

FIGURE 2.6b Concave Down Parabola

Evaluating the function at $x = 1$ yields

$$y = x^2 - 2x + 3$$
$$= (1)^2 - 2(1) + 3$$
$$= 1 - 2 + 3$$
$$= 2$$

Therefore, the vertex of the function is $(1, 2)$.

Parabolas are symmetrical. That is, if we draw a vertical line through the vertex of the parabola, the portion of the graph on the left of the line is the mirror image of the portion of the graph on the right of the line. The line is referred to as the **axis of symmetry** (see Figure 2.7).

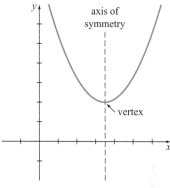

FIGURE 2.7

Since the axis of symmetry is a vertical line passing through the vertex, the equation of the axis of symmetry is $x = \dfrac{-b}{2a}$.

EXAMPLE 1

Describing the Graph of a Quadratic Function from Its Equation

Determine the concavity, the y-intercept, and the vertex of the quadratic function $y = 3x^2 + 6x - 1$.

SOLUTION We have $a = 3$, $b = 6$, and $c = -1$. Since $a > 0$, the parabola is concave up. Since $c = -1$, the y-intercept is $(0, -1)$. The x coordinate of the vertex is given by

$$x = \frac{-b}{2a}$$
$$= \frac{-6}{2(3)}$$
$$= \frac{-6}{6}$$
$$= -1$$

The y coordinate of the vertex is obtained by evaluating the function at $x = -1$.

$$y = 3x^2 + 6x - 1$$
$$= 3(-1)^2 + 6(-1) - 1$$
$$= 3 - 6 - 1$$
$$= -4$$

The vertex of the parabola is $(-1, -4)$.

EXAMPLE   2   **Determining the Equation of a Parabola from Its Graph**

Determine the equation of the parabola shown in Figure 2.8.

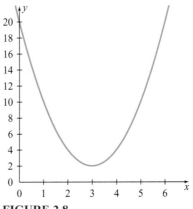

FIGURE 2.8

SOLUTION   The y-intercept is $(0, 20)$, so $c = 20$. The vertex is $(3, 2)$. Since the x coordinate of the vertex is $\frac{-b}{2a}$, we know that

$$\frac{-b}{2a} = 3$$
$$-b = 6a$$
$$b = -6a$$

Since $f(x) = ax^2 + bx + c$, we have

$$f(x) = ax^2 + (-6a)x + (20) \qquad \text{Since } b = -6a$$
$$= ax^2 - 6ax + 20$$

The vertex is $(3, 2)$, so $f(3) = 2$. Therefore,

$$f(x) = ax^2 - 6ax + 20$$
$$2 = a(3)^2 - 6a(3) + 20 \qquad \text{Substitute in } (3, 2)$$
$$2 = 9a - 18a + 20$$
$$-18 = -9a$$
$$a = 2$$

Since $b = -6a$, $b = -12$. The equation of the parabola is $f(x) = 2x^2 - 12x + 20$.

We can check our work by substituting in a different point from the graph. The parabola passes through the point $(4, 4)$. Consequently, this point should satisfy the equation $f(x) = 2x^2 - 12x + 20$.

$$f(x) = 2x^2 - 12x + 20$$
$$4 \overset{?}{=} 2(4)^2 - 12(4) + 20 \qquad \text{Substitute in } (4, 4)$$
$$4 \overset{?}{=} 2(16) - 48 + 20$$
$$4 \overset{?}{=} 32 - 48 + 20$$
$$4 = 4$$

Recognizing the relationship between the graph of a parabola and its corresponding quadratic function provides a quick way to evaluate the accuracy of a quadratic model. Quadratic models may be determined algebraically or by using quadratic regression. We will demonstrate both methods in the next several examples.

Let's return to the AP Calculus exam data introduced at the beginning of the section. At first glance, the Calculus AB exam data don't look at all like a parabola; however, we do observe that the data appear to be increasing at an ever-increasing rate. (It took six years [1980 to 1986] for the number of exams to increase from 20,000 to 40,000, but it took only three years [1986 to 1989] for the number of exams to increase from 40,000 to 60,000.) Plotting the quadratic equation $f(t) = 157.8486t^2 - 622,378t + 613,500,479$ together with the data set, we observe that, in fact, a quadratic equation fits the data very well (Figure 2.9). (Source: The College Board®.) This equation is found by performing quadratic regression on the data, a process that we will detail in the forthcoming Technology Tip.

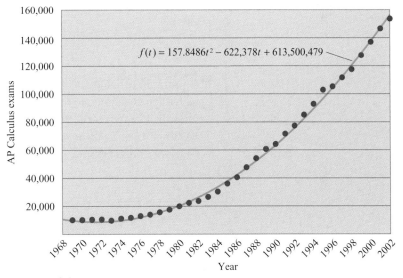

FIGURE 2.9

Observe that the coefficients of the variables in the quadratic equation are very large. We can come up with an equation with smaller coefficients by **aligning the data.** We'll let $t = 0$ in 1969, $t = 1$ in 1970, and so on. Doing quadratic regression on the aligned data yields the equation $f(t) = 157.85t^2 - 770.64t + 10,268$ with the coefficient of determination $r^2 = 0.9979$. Recall that the coefficient of determination is a measure of how well the model fits the data. The closer r^2 is to 1, the better the model fits the data. Although both of the models fit the data, the second model will make computations easier because of the smaller coefficients.

Using the model, we predict how many tests will be administered in 2005. In 2005, $t = 36$.

$$f(36) = 157.85(36)^2 - 770.64(36) + 10,268$$
$$= 187,099$$

We estimate that 187,099 AP Calculus AB exams will be administered in 2005.

A quadratic function model for a data set may be generated by using the quadratic regression feature on a graphing calculator, as demonstrated in the following Technology Tip. However, the fact that the calculator can create a quadratic model does not guarantee that the model will be a good fit for the data.

TECHNOLOGY TIP

Quadratic Regression

1. Enter the data using the Statistics Menu List Editor. (Refer to Section 1.3 if you've forgotten how to do this.)

```
L1      L2      L3      3
0       10280   ------
1       10273
2       10592
3       10611
4       9871
5       11213
6       11804
L3(1) =
```

2. Bring up the Statistics Menu Calculate feature by pressing [STAT] and using the blue arrows to move to the CALC menu. Then select item 5:QuadReg and press [ENTER].

```
EDIT CALC TESTS
1:1-Var Stats
2:2-Var Stats
3:Med-Med
4:LinReg(ax+b)
5QuadReg
6:CubicReg
7↓QuartReg
```

3. If you want to automatically paste the regression equation into the [Y=] editor, press the key sequence [VARS] Y-Vars; 1:Function; 1:Y1 and press [ENTER]. Otherwise press [ENTER].

```
QuadReg
 y=ax²+bx+c
 a=157.8485883
 b=-770.6397775
 c=10268.35154
 R²=.9979069591
```

Although the quadratic model fits the AP Calculus AB exam data very well from 1969 to 2002, we must be cautious in using it to predict future behavior. (Predicting the output value for an input value outside of the interval of the input data is called **extrapolation**.) For this data set, we may feel reasonably comfortable with an estimate two or three years beyond the last data point; however, we would doubt the accuracy of the model 100 years beyond 2002. For example, in 2102 ($t = 133$), the estimated number of exams is 2,702,587. This figure is more than 17 times greater than the maximum number of exams that have ever been administered!

We also need to look at the population of students who could possibly take the exam. Between 1970 and 1999, the number of twelfth graders (those who typically take the AP Calculus exam) fluctuated from a high of 3,026,000 in 1977 to a low of 2,381,000 in 1990. (**Source:** *Statistical Abstract of the United States, 2001, Table 232, p. 149.*) If the number of twelfth graders continues to fluctuate between these two values, at some point our model estimate for the exams would exceed the number of people who could conceivably take the exam. This illustrates the necessity to verify that a model makes sense in its real-world context.

It is often possible to model a data set with more than one mathematical model. When selecting a model, we should consider the following:

1. The graphical fit of the model to the data
2. The correlation coefficient (r) or the coefficient of determination (r^2)
3. The known behavior of the thing being modeled

Recall that the closer the correlation coefficient is to 1 or −1, the better the model fits the data. Similarly, the closer the coefficient of determination is to 1, the better the model fits the data.

EXAMPLE 3

Using Quadratic Regression to Forecast Prescription Drug Sales

Retail prescription drug sales in the United States increased from 1995 to 2000 as shown in Table 2.1.

TABLE 2.1

Years Since 1995 (t)	Retail Sales (billions of dollars) [$S(t)$]
0	68.6
2	89.1
3	103.0
4	121.7
5	140.7

Source: *Statistical Abstract of the United States, 2001*, Table 127, p. 94.

Model the data using a quadratic function. Then use the model to predict retail prescription drug sales in 1996 and 2001.

SOLUTION We observe from the scatter plot that the data appear concave up everywhere (Figure 2.10).

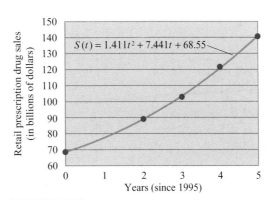

FIGURE 2.10

A quadratic function may fit the data well. We use quadratic regression to find the quadratic model that best fits the data.

Based on data from 1995 to 2000, retail prescription drug sales in the United States may be modeled by

$$S(t) = 1.411t^2 + 7.441t + 68.55 \text{ billion dollars}$$

where t is the number of years since 1995.

The coefficient of determination ($r^2 = 0.9997$) is extremely close to 1. The graph also appears to "touch" each data point. The model appears to fit the data well.

In 1996, $t = 1$, and in 2001, $t = 6$. Evaluating the function at each t value, we get

$$S(1) = 1.411(1)^2 + 7.441(1) + 68.55$$
$$= 77.402$$

$$S(6) = 1.411(6)^2 + 7.441(6) + 68.55$$
$$= 163.992$$

Since the original data were accurate to only one decimal place, we will round our solutions to one decimal place as well. We estimate that prescription drug sales were \$77.4 billion in 1996 and \$164.0 billion in 2001.

You may ask, "Is there a way to find a quadratic model without using quadratic regression?" There is. The model may not be the model of best fit, but it may still model the data effectively. In Example 4, we repeat Example 3 using an algebraic method to find a quadratic model.

EXAMPLE 4   **Using Algebraic Methods to Model Prescription Drug Sales**

Retail prescription drug sales in the United States increased from 1995 to 2000 as shown in Table 2.2.

TABLE 2.2

Years Since 1995 (t)	Retail Sales (billions of dollars) [$S(t)$]
0	68.6
2	89.1
3	103.0
4	121.7
5	140.7

Source: *Statistical Abstract of the United States, 2001*, Table 127, p. 94.

Model the data using a quadratic function. Then use the model to predict retail prescription drug sales in 1996 and 2001.

SOLUTION Given any three data points, we can find a quadratic function that passes through the points, provided that the points define a nonlinear function. We will pick the points $(0, 68.6), (3, 103.0)$, and $(5, 140.7)$ from the table. A quadratic function is of the form $S(t) = at^2 + bt + c$. Each of the points must satisfy this equation.

$$68.6 = a(0)^2 + b(0) + c \qquad \text{Substitute } t = 0, S(t) = 68.6$$
$$c = 68.6$$

$$103.0 = a(3)^2 + b(3) + c \qquad \text{Substitute } t = 3, S(t) = 103.0$$
$$103.0 = 9a + 3b + c$$

$$140.7 = a(5)^2 + b(5) + c \qquad \text{Substitute } t = 5, S(t) = 140.7$$
$$140.7 = 25a + 5b + c$$

Since we know $c = 68.6$, we can simplify the last two equations.

$$103.0 = 9a + 3b + 68.6$$
$$34.4 = 9a + 3b$$

$$140.7 = 25a + 5b + 68.6$$
$$72.1 = 25a + 5b$$

We can now find the values of a and b by solving the system of equations.

$$9a + 3b = 34.4$$
$$25a + 5b = 72.1$$

$$\begin{bmatrix} 9 & 3 & | & 34.4 \\ 25 & 5 & | & 72.1 \end{bmatrix}$$

$$\begin{bmatrix} 9 & 3 & | & 34.4 \\ -30 & 0 & | & -44.3 \end{bmatrix} \qquad 5R_1 - 3R_2$$

$$\begin{bmatrix} 0 & 30 & | & 211.1 \\ -30 & 0 & | & -44.3 \end{bmatrix} \qquad 10R_1 + 3R_2$$

$$-30a = -44.3 \qquad\qquad 30b = 211.1$$
$$a = 1.477 \qquad\qquad b = 7.037$$

A quadratic model for the data is

$$S(t) = 1.477t^2 + 7.037t + 68.6$$

Graphing this model with the data shows that it fits the data relatively well (Figure 2.11).

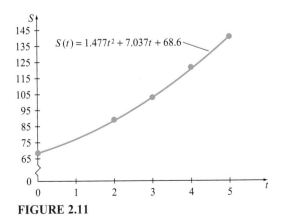

FIGURE 2.11

In 1996, $t = 1$, and in 2001, $t = 6$. Evaluating the function at each t value, we get

$$S(1) = 1.477(1)^2 + 7.037(1) + 68.6 \qquad S(6) = 1.477(6)^2 + 7.037(6) + 68.6$$
$$\approx 77.1 \text{ billion} \qquad\qquad\qquad \approx 164.0 \text{ billion}$$

These estimates are close to the estimates from Example 3. In Example 3, we estimated that prescription drug sales were $77.1 billion in 1996 and $164.0 billion in 2001.

EXAMPLE 5

Using Quadratic Regression to Forecast Nursing Home Care Costs

Because of their need for constant medical care, many elderly Americans are placed in nursing homes by their families. The amount of money spent on nursing home care has risen substantially since 1960, as shown in Table 2.3.

TABLE 2.3

Years Since 1960 (t)	Nursing Home Care (billions of dollars) [$N(t)$]
0	1
10	4
20	18
30	53
40	96

Source: *Statistical Abstract of the United States, 2001,* Table 119, p. 91.

The U.S. Centers for Medicare and Medicaid Services predict that by 2010 the cost of nursing home care will reach $183 billion.

Model the data with a quadratic function and calculate the cost of nursing home care in 2010. Compare your result to the U.S. Centers for Medicare & Medicaid Services estimate.

SOLUTION Using quadratic regression, we determine that the quadratic model of best fit is

$$N(t) = 0.07214t^2 - 0.4957t + 1.029 \text{ billion dollars}$$

where t is the number of years since 1960.

Since the coefficient of determination ($r^2 = 0.9987$) is extremely close to 1, we anticipate that the model fits the data well. Graphing the data and the model together yields Figure 2.12.

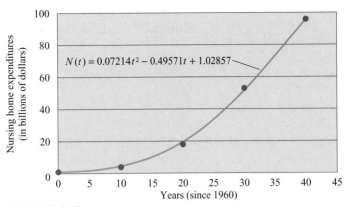

FIGURE 2.12

The graph passes near each data point. The model appears to fit the data fairly well. Zooming in, however, we see that the graph decreases from 1 billion to 0 billion between 1960 and 1963 (Figure 2.13). (We rounded to whole numbers, since the original data were in whole numbers.)

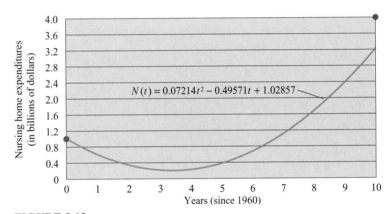

FIGURE 2.13

In reality, nursing home expenditures have increased every year since 1960. Nevertheless, despite this limitation in our model, we will use it to predict nursing home expenditures in 2010. In 2010, $t = 50$.

$$N(50) = 0.721(50)^2 - 0.4957(50) + 1.0286$$
$$= 156.60$$

We estimate that in 2010, $157 billion will be spent on nursing home care. Our model estimate is substantially less than the $183 billion that the U.S. Centers for Medicare & Medicaid Services estimate. Their estimate probably anticipated the health care needs of the aging population of baby boomers, while ours did not.

Models projecting health care costs are useful for legislators, insurance companies, and consumers. By considering future costs, people can prepare for the future and avert financial crises.

EXAMPLE 6

Using Algebraic Methods to Find a Quadratic Model for Net Sales

Based on the data in Table 2.4, find a quadratic model for the net sales of the Kellogg Company algebraically.

TABLE 2.4

Years Since 1999 (t)	Kellogg Company Net Sales (millions of dollars) $R(t)$
0	6,984.2
1	6,954.7
2	8,853.3

Source: Kellogg Company 2001 Annual Report, pp. 7, 27.

SOLUTION Since we are given the y-intercept, $(0, 6984.2)$, we know that $c = 6984.2$. We have

$$R(t) = at^2 + bt + c$$
$$= at^2 + bt + 6984.2$$

$$R(1) = a(1)^2 + b(1) + 6984.2$$
$$6954.7 = a + b + 6984.2$$
$$-29.5 = a + b$$

$$R(2) = a(2)^2 + b(2) + 6984.2$$
$$8853.3 = 4a + 2b + 6984.2$$
$$1869.1 = 4a + 2b$$

We must solve the system of equations

$$a + b = -29.5$$
$$4a + 2b = 1869.1$$

We will solve the system using the substitution method. Solving the first equation for a yields $a = -b - 29.5$. Substituting this result into the second equation $4a + 2b = 1869.1$ yields

$$4(-b - 29.5) + 2b = 1869.1 \qquad \text{Since } a = -b - 29.5$$
$$-4b - 118 + 2b = 1869.1$$
$$-2b = 1987.1$$
$$b = -993.55$$

Since $a = -b - 29.5$,

$$a = -(-993.55) - 29.5$$
$$= 964.05$$

The quadratic function that models the revenue of the Kellogg Company is $R(t) = 964.05t^2 - 993.55t + 6984.2$, where t is the number of years since the end of 1999 and $R(t)$ is the revenue from sales in millions of dollars.

In each of the preceding examples, the quadratic model fit the data well. This is not always the case, as demonstrated in Example 7.

EXAMPLE 7   Determining When a Quadratic Model Should Not Be Used

The per capita consumption of ready-to-eat and ready-to-cook breakfast cereal is shown in Table 2.5.

TABLE 2.5

Years Since 1980 (t)	Cereal Consumption (pounds) [$C(t)$]	Years Since 1980 (t)	Cereal Consumption (pounds) [$C(t)$]
0	12	10	15.4
1	12	11	16.1
2	11.9	12	16.6
3	12.2	13	17.3
4	12.5	14	17.4
5	12.8	15	17.1
6	13.1	16	16.6
7	13.3	17	16.3
8	14.2	18	15.6
9	14.9	19	15.5

Source: *Statistical Abstract of the United States, 2001*, Table 202, p. 129.

Explain why you do or do not believe that a quadratic function will model the data set well.

SOLUTION We first draw the scatter plot of the data set (Figure 2.14).

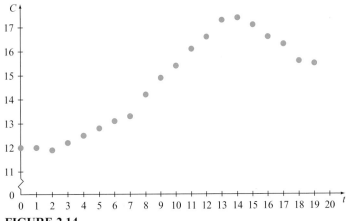

FIGURE 2.14

(*Note:* We have adjusted the viewing window so that we may better analyze the data.

Recall that a parabola is either concave up everywhere or concave down everywhere. This scatter plot appears to be concave up between $t = 1$ and $t = 10$, concave down between $t = 10$ and $t = 17$, and concave up between $t = 17$ and $t = 19$. The fact that the scatter plot changes concavity causes us to doubt that a quadratic model will fit the data well. In short, the scatter plot doesn't look like a parabola or a portion of a parabola.

The quadratic model that best fits the data is $C(t) = -0.0179t^2 + 0.637t + 10.8$ and is shown in Figure 2.15.

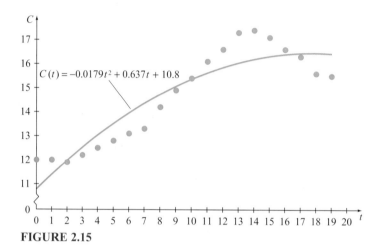

FIGURE 2.15

The coefficient of determination ($r^2 = 0.853$) is not as close to 1 as in our previous examples. Additionally, we can see visually that the model does not fit the data well.

Once a model has been created, we can often use it to make educated guesses about what may happen in the future. Forecasting the future is a key function for many businesses.

EXAMPLE 8

Using Quadratic Regression to Forecast Fuel Consumption

As vans, pickups, and SUVs have increased in popularity, the total fuel consumption of these types of vehicles has also increased (Table 2.6).

TABLE 2.6   Motor Fuel Consumption of Vans, Pickups, and SUVs

Years Since 1980 (t)	Fuel Consumption (billions of gallons) (F)
0	23.8
5	27.4
10	35.6
15	45.6
19	52.8

Source: *Statistical Abstract of the United States, 2001*, Table 1105, p. 691.

Model the fuel consumption with a quadratic function and forecast the year in which fuel consumption will reach 81.0 billion gallons.

SOLUTION   Using quadratic regression, we determine the model to be

$$F(t) = 0.0407t^2 + 0.809t + 23.3 \text{ billion gallons}$$

where t is the number of years since the end of 1980. We want to know at what value of t does $F(t) = 81.0$. This problem may be solved algebraically or graphically using technology. We will solve the problem twice (once with each method) and allow you to use the method of your choice in the exercises.

ALGEBRAIC SOLUTION

$$81.0 = 0.0407t^2 + 0.809t + 23.3$$
$$0 = 0.0407t^2 + 0.809t - 57.7$$

This is a quadratic function with $a = 0.0407$, $b = 0.809$, and $c = -57.7$.

Recall that the solution to a quadratic equation of the form $at^2 + bt + c = 0$ is given by the Quadratic Formula,

$$t = \frac{-b \pm \sqrt{b^2 - 4ac}}{2a}$$

Substituting our values of a, b, and c into the Quadratic Formula yields

$$t = \frac{-0.809 \pm \sqrt{(-0.809)^2 - 4(0.0407)(-57.7)}}{2(0.0407)}$$
$$= \frac{-0.809 \pm \sqrt{10.0}}{0.0814}$$
$$= \frac{-0.809 \pm 3.17}{0.0814}$$

In the context of the problem, we know that t must be nonnegative. Consequently, we will calculate only the nonnegative solution.

$$t = \frac{-0.809 + 3.17}{0.0814}$$

$$= \frac{2.36}{0.0814}$$

$$= 29.0$$

We anticipate that at the end of 2009 (29 years after the end of 1980), the fuel consumption will have reached 81.0 billion gallons.

GRAPHICAL SOLUTION

We graph the function $F(t) = 0.0407t^2 + 0.809t + 23.3$ and the horizontal line $y = 81.0$ simultaneously (Figure 2.16). $F(t) = 81.0$ at the point at which these two functions intersect.

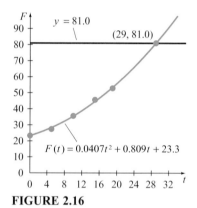

FIGURE 2.16

Using the intersect feature on the graphing calculator, we determine that the point of intersection is $(29, 81.0)$. The interpretation of the solution is the same as that given in the discussion of the algebraic method.

2.1 Summary

In this section, you learned how to use quadratic regression to model a data set. You also discovered the importance of analyzing a mathematical model before using it to calculate unknown values.

2.1 Exercises

In Exercises 1–5, determine the concavity, y-intercept, and vertex of the quadratic equation.

1. $y = x^2 - 2x + 1$ **2.** $f(x) = -2x^2 + 4$

3. $g(x) = 3x^2 + 3x$

4. $h(t) = -1.2t^2 + 2.4t + 4.5$

5. $f(t) = 2.8t^2 - 1.4t + 2.1$

In Exercises 6–10, determine the equation of the parabola from the graph.

6.
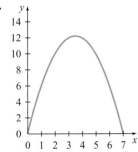

7.
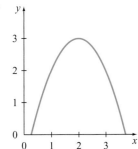

8.
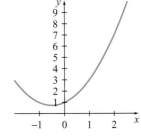

9.
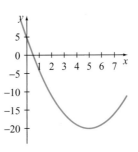

10.
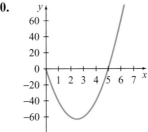

In Exercises 11–15, do the following:

(a) Find a quadratic model for the data algebraically.
(b) Graph the model together with the scatter plot.
(c) Explain why you do or do not believe that the function is a good model.

11.   **Basketball Game Attendance**

NCAA Basketball Game Attendance

Years Since 1985 (t)	Women's Games (thousands) (W)
0	2,072
5	2,777
10	4,962
14	8,698

Source: *Statistical Abstract of the United States, 2001*, Table 1241, p. 759.

12.   **Ice Cream Cost**

Prepackaged Ice Cream Cost

Years Since 1990 (t)	Ice Cream Cost (dollars per 1/2 gallon) (C)
0	2.54
3	2.59
4	2.62
5	2.68
6	2.94
7	3.02
8	3.30
9	3.40
10	3.66

Source: *Statistical Abstract of the United States, 2001*, Table 706, p. 468.

13.   **DVD Player Shipments**

DVD Player Shipments in North America

Years Since 1997 (t)	Shipments (thousands) (D)
1	1,080
2	4,100
3	9,700
4	16,700

Source: DVD Entertainment Group.

14.   **Health Care Spending**

Health Care Spending by Insurance Companies

Years Since 1960 (t)	Spending (billions of dollars) (I)
0	6
10	16
20	68
30	233
40	438

Source: *Statistical Abstract of the United States, 2001*, Table 119, p. 91.

15.   **VHS Tape and CD-R Sales**

VHS Tape Versus CD-R Disk Sales, 2000-2002

Millions of Blank VHS Tapes Sold Worldwide (v)	Millions of CD-R Disks Sold Worldwide (C)
1,409	2,730
1,322	3,730
1,195	4,225

Source: International Recording Media Association.

In Exercises 16–50, do the following:

(a) Draw a scatter plot for the data.

(b) Explain why a quadratic function might or might not fit the data.

If it appears that a quadratic function may fit the data:

(c) Use quadratic regression to find the quadratic function that best fits the data.

(d) Graph the model together with the scatter plot.

(e) Explain why you do or do not believe that the function is a good model.

(f) Answer any additional questions that may be given.

16. College Freshmen

Male College Freshmen

Years Since 1970 (*t*)	Percentage of Freshmen That Are Male (*F*)
0	52.1
10	48.8
15	48.9
20	46.9
25	45.6
27	45.5
28	45.5
29	45.3
30	45.2

According to the model, in what year were there an equal number of male and female freshmen?

  17. **Personal Income**   Because of inflation, the buying power of a dollar decreases over time. As a result, your personal income may increase while your buying power decreases. In the following table, personal income per capita is listed using constant (1996) dollars. That is, the actual income figures are adjusted for inflation.

Personal Income per Capita, Connecticut, Ranked #1 in 2000

Years Since 1980 (*t*)	Income (dollars) (*I*)
0	22,530
10	31,223
20	37,854

Source: *Statistical Abstract of the United States, 2001*, Table 652, p. 426.

 18. **Personal Income**   In the following table, personal income per capita is listed using constant (1996) dollars. That is, the actual income figures are adjusted for inflation.

Personal Income per Capita, Mississippi, ranked #50 in 2000

Years Since 1980 (*t*)	Income (dollars) (*I*)
0	12,817
10	15,373
20	19,554

Source: *Statistical Abstract of the United States, 2001*, Table 652, p. 426.

What is the vertex of the quadratic model, and what does it mean in the context of the problem?

  19. Game Sales

U.S. Video Game Industry Sales

Years Since 1997 (*t*)	Sales (billions of dollars) (*S*)
0	5.1
1	6.2
2	6.9
3	6.6
4	9.4

Source: www.npdfunworld.com.

20.   **Milk Consumption**

Per Capita Milk Beverage Consumption

Years Since 1980 (t)	Consumption (gallons) (M)
0	27.6
5	26.7
10	25.7
15	24.3
17	24.0
18	23.7
19	23.6

Source: *Statistical Abstract of the United States, 2001*, Table 202, p. 129.

 21.   **Conventional Mortgages**

Percentage of New Privately Owned One-Family Houses Financed with a Conventional Mortgage

Years Since 1970 (t)	Conventional Mortgages (percent) (M)
0	47
10	55
20	62
25	74
29	79

Source: *Statistical Abstract of the United States, 2001*, Table 938, p. 597.

According to the model, in what year were 50 percent of the new privately owned homes financed with a conventional mortgage?

 22.   **Home Square Footage**

Percentage of New One-Family Houses Under 1,200 Square Feet

Years Since 1970 (t)	Percentage of New Houses (percent) (H)
0	36
10	21
20	11
25	10
29	7

Source: *Statistical Abstract of the United States, 2001*, Table 938, p. 597.

What is the concavity of the graph of the model, and what does it mean in the context of the problem?

 23.   **Home Square Footage**

New One-Family Homes: Average Number of Square Feet

Years Since 1970 (t)	Houses (square feet) (H)
0	1,500
10	1,740
20	2,080
25	2,095
29	2,225

Source: *Statistical Abstract of the United States, 2001*, Table 938, p. 597.

24. **Air Conditioning in Homes**

**New One-Family Homes
with Central Air Conditioning**

Years Since 1970 (t)	Percentage of New Houses (percent) (H)
0	34
10	63
20	76
25	80
29	84

Source: *Statistical Abstract of the United States, 2001*, Table 938, p. 597.

 25. **Homes with Garages**

New One-Family Homes with a Garage

Years Since 1970 (t)	Percentage of New Houses (percent) (H)
0	58
10	69
20	82
25	84
29	87

Source: *Statistical Abstract of the United States, 2001*, Table 938, p. 597.

In what year were 90 percent of new one-family homes expected to have a garage?

 26. **Home Sales Price**

Median Sales Price of a New One-Family House in the Southern United States

Years Since 1980 (t)	Price (thousands of dollars) (P)
0	59.6
5	75.0
10	99.0
15	124.5
20	148.0

Source: *Statistical Abstract of the United States, 2001*, Table 940, p. 598.

 27. **Home Sales Price**

Median Sales Price of a New One-Family House in the Northeastern United States

Years Since 1980 (t)	Price (thousands of dollars) (P)
0	69.5
5	103.3
10	159.0
15	180.0
20	227.4

Source: *Statistical Abstract of the United States, 2001*, Table 940, p. 598.

28.   **Home Sales Price**

Median Sales Price of a New One-Family House in the Western United States

Years Since 1980 (*t*)	Price (thousands of dollars) (*P*)
0	72.3
5	92.6
10	147.5
15	141.4
20	196.4

Source: *Statistical Abstract of the United States, 2001*, Table 940, p. 598.

 29.   **Vehicle Leasing**

Percentage of Households Leasing Vehicles

Years Since 1989 (*t*)	Percentage of Households (percent) (*P*)
0	2.5
3	2.9
6	4.5
9	6.4

Source: *Statistical Abstract of the United States, 2001*, Table 1086, p. 680.

In what year did the percentage of households leasing vehicles reach 5 percent?

 30.   **Fuel Consumption**

Motor Fuel Consumption of Vans, Trucks, and SUVs

Years Since 1980 (*t*)	Fuel Consumption (billions of gallons) (*F*)
0	23.8
5	27.4
10	35.6
15	45.6
19	52.8

Source: *Statistical Abstract of the United States, 2001*, Table 1105, p. 691.

31.   **Coffee Sales**

Starbucks Corporation Sales

Years Since 09/93 (*t*)	Income from Sales (millions of dollars) (*S*)
0	163.5
1	284.9
2	465.2
3	696.5
4	966.9
5	1,308.7
6	1,680.1
7	2,169.2
8	2,649.0
9	3,288.9

Source: moneycentral.msn.com.

According to the model, when did Starbucks Corporation sales reach $5000 million?

32. **Organic Cropland**

Certified Organic Cropland

Years Since 1992 (t)	Acres (thousands) (A)
0	403
1	465
2	557
3	639
4	850

Source: *Statistical Abstract of the United States, 2001*, Table 805, p. 526.

 33. **Poultry Pricing**

Average Retail Price of Fresh Whole Chicken

Years Since 1985 (t)	Price (dollars per pound) (P)
0	0.78
5	0.86
6	0.86
7	0.88
8	0.91
9	0.90
10	0.94
11	1.00
12	1.00
13	1.06
14	1.05
15	1.08

Source: *Statistical Abstract of the United States, 2001*, Table 706, p. 468.

 34. **Prison Rate**

Federal and State Prison Rate

Years Since 1980 (t)	Rate (prisoners per 100,000 people) (R)
0	139
2	171
4	188
6	217
8	247
10	297
12	332
14	389
16	427
18	461

Source: *Statistical Abstract of the United States, 2001*, Table 332, p. 200.

 35. **College Enrollment**

Private College Enrollment

Years Since 1980 (t)	Students (in thousands) (S)
0	2,640
2	2,730
4	2,765
6	2,790
8	2,894
10	2,974
12	3,103
14	3,145
16	3,247
18	3,373

Source: *Statistical Abstract of the United States, 2001*, Table 205, p. 133.

36. **Military Personnel**

Active-Duty Military Personnel

Years Since 1990 (t)	Personnel (thousands) (P)
1	1,263
2	1,214
3	1,171
4	1,131
5	1,085
6	1,056
7	1,045
8	1,004
9	1,003

Source: *Statistical Abstract of the United States, 2001,* Table 499, p. 328.

   37. **Personal Income**

Per Capita Personal Income: California

Years Since 1993 (t)	Personal Income (dollars) (P)
0	22,833
1	23,348
2	24,339
3	25,373
4	26,521
5	28,240
6	29,772
7	32,149

Source: Bureau of Economic Analysis (www.bea.gov).

  38. **Personal Income**

Per Capita Personal Income: Colorado

Years Since 1993 (t)	Personal Income (dollars) (P)
0	22,196
1	23,055
2	24,289
3	25,514
4	27,067
5	28,764
6	30,206
7	32,434

Source: Bureau of Economic Analysis (www.bea.gov).

  39. **Personal Income**

Per Capita Personal Income: Louisiana

Years Since 1993 (t)	Personal Income (dollars) (P)
0	17,587
1	18,602
2	19,314
3	19,978
4	20,874
5	21,948
6	22,274
7	23,090

Source: Bureau of Economic Analysis (www.bea.gov).

40. **Personal Income**

Per Capita Personal Income: Minnesota

Years Since 1993 (t)	Personal Income (dollars) (P)
0	21,903
1	23,241
2	24,295
3	25,904
4	27,086
5	29,092
6	30,105
7	31,935

Source: Bureau of Economic Analysis (www.bea.gov).

 41. **Pharmaceutical Company Profit**

Johnson & Johnson Pharmaceutical Operating Profit

Years Since 1997 (t)	Operating Profit (millions of dollars) (P)
0	2,332
1	3,114
2	3,735
3	4,394
4	4,928

Source: Johnson & Johnson Annual Report, 2001, p. 8.

In what year were Johnson & Johnson profits expected to reach $6000 million?

42. **Pharmaceutical Income**

Johnson & Johnson Net Income

Years Since 1990 (t)	Net Income (millions of dollars) (I)
0	1,195
1	1,441
2	1,572
3	1,786
4	1,998
5	2,418
6	2,958
7	3,385
8	3,798
9	4,348
10	4,998
11	5,899

Source: Johnson & Johnson Annual Report, 2001, p. 18.

In what year was Johnson & Johnson net income projected to reach $7000 million?

43.   **Company Payroll**

Payroll of Private-Employer Firms*

Years Since 1990 (t)	Payroll (billions of dollars) (F)
0	2,104
1	2,145
2	2,272
3	2,363
4	2,488
5	2,666
6	2,849
7	3,048
8	3,309

*Firms are an aggregation of all establishments owned by a parent company.

Source: *Statistical Abstract of the United States, 2001*, Table 726, p. 486.

   44.   **Public-School Teachers**

Projected Number of Public-School Teachers

Years Since 2000 (t)	Public-School Teachers (thousands) (G)
0	2,850
1	2,865
2	2,877
3	2,891
4	2,905
5	2,914
6	2,919
7	2,927
8	2,932
9	2,937
10	2,940

Source: *Statistical Abstract of the United States, 2001*, Table 207, p. 134.

According to the model, in what year is the number of public-school teachers expected to reach 3 million?

  45.   **Private-School Teachers**

Projected Number of Private-School Teachers

Years Since 2000 (t)	Private-School Teachers (thousands) (G)
0	402
1	403
2	404
3	405
4	407
5	408
6	409
7	410
8	411
9	411
10	412

Source: *Statistical Abstract of the United States, 2001*, Table 207, p. 134.

46. **Apparel Production Wages**

Apparel Production Workers' Wages

Years Since 1980 (t)	Wage of Men's and Boys' Furnishings Production Workers (dollars per hour) (M)
0	4.23
10	6.06
15	7.19
16	7.40
17	7.72
18	7.97
19	8.27
20	8.54

Source: *Statistical Abstract of the United States, 2001*, Table 609, p. 394.

   **47.** **Apparel Production Wages**

Apparel Production Workers' Wages

Years Since 1980 (t)	Wage of Women's and Misses' Outerwear Production Workers (dollars per hour) (W)
0	4.61
10	6.26
15	7.27
16	7.49
17	7.84
18	8.15
19	8.41
20	8.40

Source: *Statistical Abstract of the United States, 2001*, Table 609, p. 394.

48. **McDonalds Restaurant Sales**

McDonalds Franchised Sales

Years Since 1990 (t)	Franchised Sales (millions of dollars) (S)
1	12,959
2	14,474
3	15,756
4	17,146
5	19,123
6	19,969
7	20,863
8	22,330
9	23,830
10	24,463
11	24,838

Source: www.mcdonalds.com.

49. **Advertising Expenditures**

Advertising Expenditures: Billboards

Years Since 1990 (t)	Advertising Expenditures (millions of dollars) (A)
0	1,084
1	1,077
2	1,030
3	1,090
4	1,167
5	1,263
6	1,339
7	1,455
8	1,576
9	1,725
10	1,870

Source: *Statistical Abstract of the United States, 2001*, Table 1272, p. 777.

In what year were billboard advertising expenditures projected to reach $2 billion?

In what year were direct mail advertising expenditures expected to reach $50 billion?

 50.   **Advertising Expenditures**

Advertising Expenditures: Direct Mail

Years Since 1990 (t)	Advertising Expenditures (millions of dollars) (A)
0	23,370
1	24,460
2	25,392
3	27,266
4	29,638
5	32,866
6	34,509
7	36,890
8	39,620
9	41,403
10	44,715

Source: *Statistical Abstract of the United States, 2001*, Table 1272, p. 777.

Exercises 51–55 are intended to challenge your understanding of quadratic functions.

51. Find two different quadratic functions that pass through the points $(0, 0)$ and $(4, 0)$.

52. A classmate claims that any group of three points defines a unique quadratic function. Is your classmate correct? Explain.

53. Is there a single quadratic function that passes through each of the following points: $(1, 3), (2, 5), (4, 9)$? Explain.

54. What is the graphical significance of the point with x coordinate $x = \dfrac{-b + \sqrt{b^2 - 4ac}}{2a}$ on the graph of $y = ax^2 + bx + c$?

55. If $ax^2 + bx + c > 3$ for all values of x, what do we know about a and the graph of $y = ax^2 + bx + c$?

2.2 Higher-Order Polynomial Function Models

- Use nonlinear functions to model real-life phenomena
- Interpret the meaning of mathematical models in their real-world context
- Model real-life data with higher-order polynomial functions

GETTING STARTED   Since 1975, the average wage of sailors working on the East Coast of the United States has trailed that of sailors working on the West Coast (Table 2.7). At one point, West Coast sailors made nearly $1\frac{1}{2}$ times the wage of East Coast sailors.

TABLE 2.7 Typical Basic Monthly Wage for Able-Bodied Seamen in Addition to Room and Board

Years Since 1980 (t)	East Coast Wage (dollars) (E)	West Coast Wage (dollars) (W)
0	967	1,414
5	1,419	2,029
10	1,505	2,218
13	1,721	2,438
14	1,790	2,536
15	1,918	2,637
16	2,014	2,769
17	2,094	2,879
18	2,178	2,994
19	2,265	3,114
20	2,453	3,114

Source: *Statistical Abstract of the United States, 2001*, Table 1072, p. 674.

Salaries of both groups of sailors have increased substantially since 1980, although pay increases have been somewhat irregular. Will East Coast sailors ever make as much as West Coast sailors?

In this section, we will use *cubic* and *quartic* polynomials to model data. To keep computations simple, we use these models only if we're dissatisfied with the fit of simpler models.

Cubic Functions

A constant function is a polynomial of degree 0. A nonconstant linear function is a polynomial of degree 1. A quadratic function is a polynomial of degree 2. A **cubic function** is a polynomial of degree 3 and has the form $f(x) = ax^3 + bx^2 + cx + d$. The graph of a cubic function has exactly one **inflection point** (a point where the graph of the function changes concavity). The term **concavity** refers to the curvature of the graph. Recall that when the graph curves upward, we say that it is "concave up," and when the graph curves downward, we say that it is "concave down." The following rhyme is helpful in remembering the meaning of the terms.

Concave up is like a cup.
Concave down is like a frown.

The graphs in Figure 2.17 are concave up.

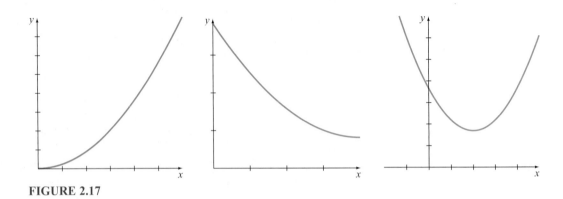

FIGURE 2.17

The graphs in Figure 2.18 are concave down.

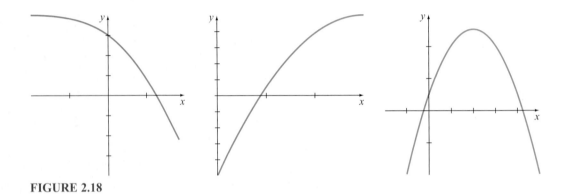

FIGURE 2.18

Cubic function graphs consist of two pieces: a concave up piece and a concave down piece. If $a > 0$, the graph of the cubic function is first concave down and then concave up. The graph of the cubic function will take on one of the basic shapes shown in Figure 2.19.

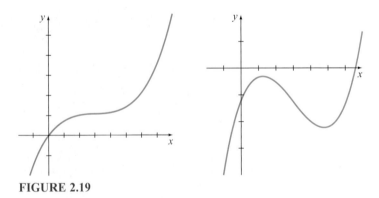

FIGURE 2.19

The size of the "bends" in the graph will vary depending upon the values of b and c. When a is positive, the graph of a cubic function will be increasing (rising from left to right) for sufficiently large values of x.

If $a < 0$, the graph of the cubic function is first concave up and then concave down. The graph of the cubic function will take on one of the shapes shown in Figure 2.20.

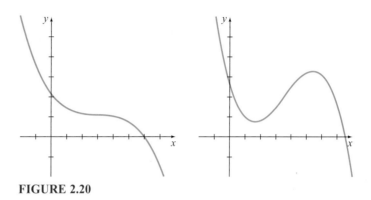

FIGURE 2.20

Again, the size of the "bends" in the graph will vary depending upon the values of b and c. When a is negative, the graph of a cubic function will be decreasing (falling from left to right) for sufficiently large values of x.

Returning to the sailors' wages, we can draw some conclusions by looking at a scatter plot of the data. Since a portion of the scatter plot appears linear, we first attempt to model the data with a linear function (Figure 2.21).

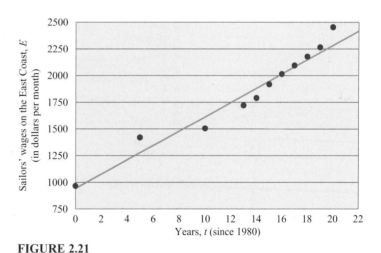

FIGURE 2.21

The linear model does not fit the data as well as we would like. We also note that from 1980 to 1990, the graph appears to be concave down, and from 1990 to 2000, the graph appears to be slightly concave up. We expect that a cubic model with $a > 0$ may fit the data well. Using the cubic regression feature on our calculator, we determine that the cubic function that best fits the data is

$$E(t) = 0.2792t^3 - 7.037t^2 + 102.6t + 983.6 \text{ dollars per month}$$

where t is the number of years since 1980. The model graph is shown in Figure 2.22.

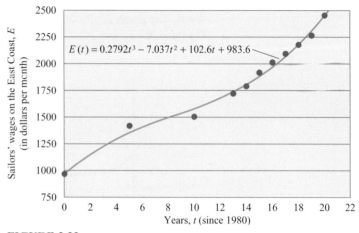

$$E(t) = 0.2792t^3 - 7.037t^2 + 102.6t + 983.6$$

FIGURE 2.22

The model fits the data fairly well, especially after 1993. The coefficient of determination of $r^2 = 0.9905$ further increases our confidence in the model. Let's now look at the scatter plot of the data from the West Coast (Figure 2.23).

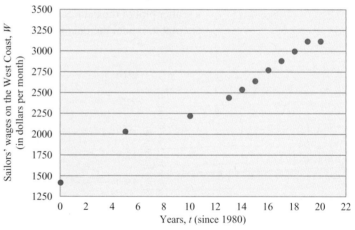

FIGURE 2.23

Again, the graph appears to be concave down from 1980 to 1990. From 1990 to 1999, the graph appears to be concave up. From 1999 to 2000, the wage did not change. Using the cubic regression feature on our calculator, we determine that the cubic equation that best fits the data is

$$W(t) = 0.2546t^3 - 7.277t^2 + 132.2t + 1437 \text{ dollars per month}$$

where t is the number of years since 1980. The model graph is shown in Figure 2.24.

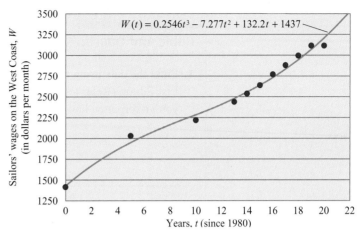

FIGURE 2.24

Will the two models ever be equal? In 2017, the graphs intersect with a monthly wage of nearly $9100 (Figure 2.25)! Since 2017 is 17 years beyond the last point in our data set, we are somewhat skeptical of the result. Nevertheless, we can conclude that the wage gap isn't likely to be eliminated in the next couple of years.

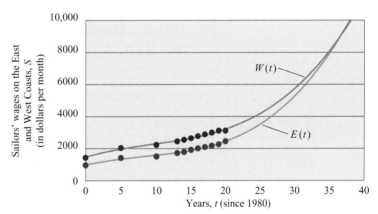

FIGURE 2.25

The following Technology Tip details how to use your TI-83 Plus calculator to do cubic and quartic regression.

TECHNOLOGY TIP

Cubic and Quartic Regression

1. Enter the data using the Statistics Menu List Editor. (Refer to Section 1.3 if you've forgotten how to do this.)

L1	L2	L3
0	612	------
5	967	
10	1419	
15	1505	
18	1721	
19	1790	
20	1918	

L1(1) = 0

2. Bring up the Statistics Menu Calculate feature by pressing STAT and using the blue arrows to move to the CALC menu. Then select item 6:CubicReg or item 7:QuadReg and press ENTER.

```
EDIT CALC TESTS
1:1-Var Stats
2:2-Var Stats
3:Med-Med
4:LinReg(ax+b)
5:QuadReg
6:CubicReg
7↓QuartReg
```

3. If you want to automatically paste the regression equation into Y1 in the Y= editor, press the key sequence VARS Y-VARS; 1:Function; 1:Y1 and press ENTER. Otherwise, press ENTER.

```
CubicReg
 y=ax³+bx²+cx+d
 a=.1643275034
 b=-5.657940597
 c=112.3153143
 d=594.9549798
 R²=.9927303408
```

EXAMPLE 1

Choosing a Polynomial Function to Model a Company's Revenue

Electronic Arts, Inc., is one of the premier producers of interactive electronic games playable on game platforms such as the Sony PlayStation 2®, Microsoft Xbox®, Nintendo GameCube™, and Sony PlayStation. Despite the sluggish economy in the early 2000s, Electronic Arts showed an overall strong growth in revenue (Table 2.8).

TABLE 2.8

Years Since End of Fiscal Year 1998 (t)	Net Revenues (millions of dollars) [$R(t)$]
0	908.852
1	1221.863
2	1420.011
3	1322.273
4	1724.675
5	2482.244

Source: Electronic Arts, Inc., Annual Reports, March 31, 2002 and 2004.

Find a polynomial model for the annual revenue of Electronic Arts, Inc., and forecast in what year the revenue will reach $3 billion.

SOLUTION From the scatter plot, it appears that a cubic function will best fit the data (Figure 2.26). Using cubic regression, we determine that the cubic model that best fits the data is

$$R(t) = 43.87t^3 - 274.8t^2 + 594.7t + 900.0 \text{ million dollars}$$

where t is the number of years since the end of fiscal year 1998.

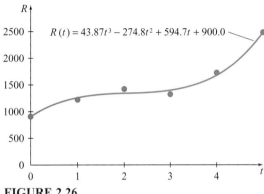

FIGURE 2.26

Since $3 billion is the same as $3000 million, we must determine when $R(t) = 3000$. To do this, we simultaneously graph $R(t) = 43.87t^3 - 274.8t^2 + 594.7t + 900.0$ and $y = 3000$. Then, using the intersection feature on our graphing calculator, we determine when the two functions will be equal (Figure 2.27).

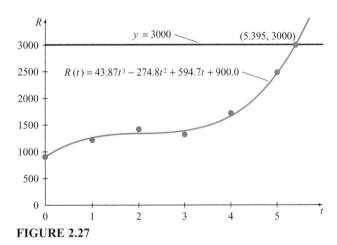

FIGURE 2.27

According to the model, 5.395 years after the end of fiscal year 1998, the annual revenue is expected to reach $3 billion. To add meaning to the result, we convert 0.395 years into months.

UNITS

$$0.395 \text{ years} \cdot \frac{12 \text{ months}}{1 \text{ year}} \approx 5 \text{ months}$$

During the one-year period prior to the end of the fifth month of fiscal year 2004, we anticipate that $3 billion in revenues will be earned. (*Note:* Since $t = 5$ is the end of fiscal year 2003, $t = 5.395$ is in fiscal year 2004.)

Quartic Functions

A fourth-degree polynomial is a function of the form $f(x) = ax^4 + bx^3 + cx^2 + dx + e$ and is called a **quartic** function. A quartic function may have zero or two inflection points. A quartic graph with $a > 0$ will have one of the basic shapes shown in Figure 2.28.

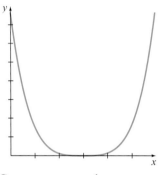

Concave up everywhere
No inflection points
FIGURE 2.28

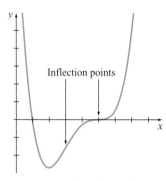

Concave up, then down, then up
Two inflection points

Concave up, then down, then up
Two inflection points

Notice that each graph opens upward. The size of the "bends" in the graph will vary depending upon the values of b, c, and d. When a is positive, the graph of a quartic function will be increasing (rising from left to right) for sufficiently large values of x.

A quartic graph with $a < 0$ will have one of the basic shapes shown in Figure 2.29.

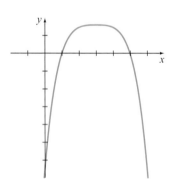

Concave down everywhere
No inflection points
FIGURE 2.29

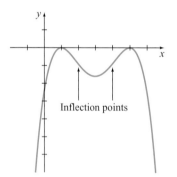

Concave down, then up, then down
Two inflection points

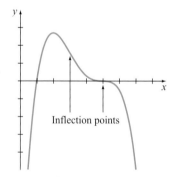

Concave down, then up, then down
Two inflection points

Notice that each graph opens downward. Again, the size of the "bends" in the graph will vary depending upon the values of b, c, and d. When a is negative, the graph of a quartic function will be decreasing (falling from left to right) for sufficiently large values of x.

EXAMPLE 2

Choosing a Nonlinear Function to Model Interest Rates

The average fixed interest rate for a conventional mortgage on a new home fluctuated between 1985 and 1995 as shown in Table 2.9.

TABLE 2.9   Average Fixed Interest Rates on New Home Conventional Mortgages

Years Since 1985 (t)	Percentage Rate (percent) (R)
0	11.90
2	9.50
4	10.20
6	9.32
8	7.27
10	7.95

Source: *Statistical Abstract of the United States, 2001*, Table 1185, p. 733.

Model the data with a nonlinear function. Do you think the model will be a good indicator of future interest rates? Explain.

SOLUTION

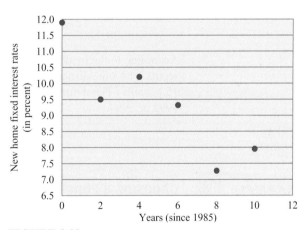

FIGURE 2.30

As shown in Figure 2.30, the graph appears to be concave up between 1985 and 1988, concave down from 1988 to 1992, and concave up from 1992 to 1995. A quartic model may work well (Figure 2.31).

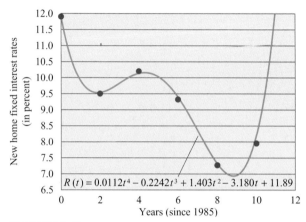

FIGURE 2.31

Using the quartic regression feature on our calculator, we determine that fixed interest rates for mortgages on new homes between 1985 and 1995 may be modeled by the equation

$$R(t) = 0.0112t^4 - 0.2242t^3 + 1.403t^2 - 3.1796t + 11.89 \text{ percent}$$

where t is the number of years since the end of 1985.

The model shows that interest rates will rise rapidly every year after 1995. Let's calculate the predicted rate at the end of 1997 ($t = 12$).

$$R(12) = 0.0112(12)^4 - 0.2242(12)^3 + 1.403(12)^2 - 3.1796(12) + 11.89$$
$$= 20.59$$

According to the model, interest rates reached 20.59 percent in 1997. This figure seems a bit unrealistic. (Between 1970 and 2000, interest rates remained below 15 percent.) The model may be a better indicator for intermediate years, such as 1992 or 1994.

EXAMPLE 3   **Choosing a Function to Model the Number of Mutual Funds**

The number of mutual funds available to investors has increased dramatically since 1970, as shown in Table 2.10.

TABLE 2.10   Number of Mutual Funds

Years Since 1970 (t)	Mutual Funds (M)
0	361
2	410
4	431
6	452
8	505
10	564
12	857
14	1,241
16	1,835
18	2,708
20	3,081
22	3,826
24	5,330
26	6,254
28	7,314
30	8,171

Source: *Statistical Abstract of the United States, 2001*, Table 1214, p. 744.

Find a function that models the data and estimate the number of mutual funds in 2005.

SOLUTION   We could model the data with a quadratic function (Figure 2.32); however, even though the correlation coefficient is close to 1, the quadratic model doesn't fit the data very well between 1970 and 1980.

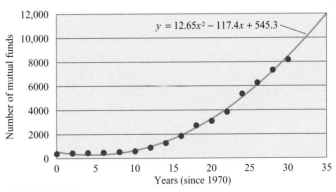

$$y = 12.65x^2 - 117.4x + 545.3$$

FIGURE 2.32

A quartic model will give a better fit (Figure 2.33).

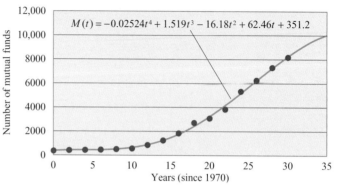

FIGURE 2.33

The number of mutual funds is modeled by the quartic equation

$$M(t) = -0.02524t^4 + 1.519t^3 - 16.18t^2 + 62.46t + 351.2 \text{ mutual funds}$$

where t is the number of years since 1970. For the year 2005, $t = 35$. To estimate the number of mutual funds in 2005, we evaluate $M(t)$ at $t = 35$.

$$M(35) = -0.02524(35)^4 + 1.519(35)^3 - 16.18(35)^2 + 62.46(35) + 351.2$$
$$= 9968$$

According to the model, in 2005 ($t = 35$), there will be 9968 mutual funds. Given the historically rapid growth in the number of mutual funds, this estimate seems reasonable.

2.2 Summary

In this section, you learned how to use cubic and quartic polynomials to model data. You also discovered the importance of being cautious in using model predictions.

5.   **School Internet Access**

**Percentage of Public-School Classrooms
with Internet Access**

Years Since 1994 (t)	Percentage of Classrooms (percent) (C)
0	3
1	8
2	14
3	27
4	51
5	64
6	77

Source: *Statistical Abstract of the United States, 2001,* Table 243, p. 155.

According to the model, what percentage of classrooms had Internet access in 2001? In 2002? Do the model estimates seem reasonable? Explain.

 6.   **Banks**

**Number of Different Banks
(Not Bank Branches)**

Years Since 1984 (t)	Banks (B)
0	17,900
1	18,033
2	17,876
3	17,325
4	16,562
5	15,829
6	15,192
7	14,517
8	13,891
9	13,261
10	12,641
11	12,002
12	11,478
13	10,923
14	10,463
15	10,221
16	9,908

Source: *Statistical Abstract of the United States, 2001,* Table 1173, p. 728.

According to the model, when will the number of banks drop below 9000? What do you think is causing the number of different banking companies to decrease?

7.   **Debit Cards**

Number of Debit Cards

Years Since 1990 (t)	Cards (millions) (C)
0	164
5	201
6	205
7	211
8	217
9	228

Source: *Statistical Abstract of the United States, 2001,* Table 1189, p. 734.

A consulting firm has projected that the number of debit cards will reach 270 million by 2005. Does this projection agree with your model?

 8.   **NASDAQ Volume**

NASDAQ Average Daily Volume

Years Since 1980 (t)	Shares (millions) (S)
0	27
2	33
4	60
6	114
8	123
10	132
12	191
14	295
16	544
18	802
20	1,757

Source: *Statistical Abstract of the United States, 2001,* Table 1205, p. 741.

According to your model, when will the NASDAQ market have an average daily volume of 2 billion shares? Visit www.marketdata.nasdaq.com to see if your model estimate is accurate.

 9.   **Stock Trading**

New York Stock Exchange Shares Traded

Years Since 1980 (t)	Shares (millions) (S)
0	11,562
2	16,669
4	23,309
6	36,009
8	41,118
10	39,946
12	51,826
14	74,003
16	105,477
18	171,188
20	265,499

Source: *Statistical Abstract of the United States, 2001,* Table 1207, p. 742.

According to the model, how many shares will be traded in 2005? Does this seem reasonable? Explain.

10. 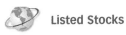   **Listed Stocks**

Companies with Stocks Listed on the NYSE

Years Since 1980 (t)	Companies (C)
0	1,570
2	1,526
4	1,543
6	1,575
8	1,681
10	1,774
12	2,088
14	2,570
16	2,907
18	3,114
20	2,862

Source: *Statistical Abstract of the United States, 2001,* Table 1207, p. 742.

According to the model, in which years is the number of companies listed on the New York Stock Exchange expected to exceed 2700 companies?

 11.   **Computer Retailers' Wages**

Average Annual Wage per Worker for Computer and Equipment Retailers

Years Since 1992 (t)	Wages (dollars) (W)
1	32,200
2	30,500
3	32,100
4	33,800
5	35,000
6	37,300

Source: *Statistical Abstract of the United States, 2001,* Table 1123, p. 703.

In what year do you expect the average wage to exceed $40,000?

 12.   **Computer Wholesalers' Wages**

Average Annual Wage per Worker for Computer and Equipment Wholesalers

Years Since 1992 (t)	Wages (dollars) (W)
1	52,500
2	52,900
3	52,900
4	54,300
5	56,700
6	62,200

Source: *Statistical Abstract of the United States, 2001,* Table 1123, p. 703.

Calculate the average wage in 2000. Does this seem reasonable? Explain.

 13.   **Software Developers' Wages**

Average Wage per Worker at Prepackaged Software Development Firms

Years Since 1992 (t)	Wages (dollars) (W)
0	57,000
1	54,500
2	57,000
3	63,700
4	70,100
5	79,200
6	94,100

Source: *Statistical Abstract of the United States, 2001,* Table 1123, p. 703.

In what year will the wage reach $120,000?

14. **Software Retailers' Wages**

Average Wage per Worker at Prepackaged Software Retailers

Years Since 1992 (t)	Wages (dollars) (W)
0	32,200
1	30,500
2	32,100
3	33,800
4	35,000
5	37,300
6	40,400

Source: *Statistical Abstract of the United States, 2001,* Table 1123, p. 703.

What do you expect the average worker's wage was in 1999?

   15. **Software Wholesalers' Wages**

Average Wage per Worker at Prepackaged Software Wholesalers

Years Since 1992 (t)	Wages (dollars) (W)
0	52,500
1	52,900
2	52,900
3	54,300
4	56,700
5	62,200
6	69,700

Source: *Statistical Abstract of the United States, 2001,* Table 1123, p. 703.

Compare the wholesalers' average wage model with the retailers' average wage model from Exercise 14. According to the models, will wages at retailers ever catch up with wages at wholesalers?

  16. **Radio Broadcasters' Wages**

Average Annual Wage per Worker in Radio Broadcasting

Years Since 1992 (t)	Annual Wage (dollars) (W)
0	23,500
1	24,300
2	26,000
3	27,200
4	29,300
5	31,300
6	34,200

Source: *Statistical Abstract of the United States, 2001,* Table 1123, p. 703.

According to the model, what was the average wage of a radio broadcasting employee in 2006? Does this estimate seem reasonable? Explain.

  17. **Television Broadcasters' Wages**

Average Annual Wage per Worker in Television Broadcasting

Years Since 1992 (t)	Annual Wage (dollars) (W)
0	41,400
1	42,200
2	43,700
3	47,200
4	51,100
5	51,000
6	54,600

Source: *Statistical Abstract of the United States, 2001,* Table 1123, p. 703.

According to the model, when will the average annual wage for television broadcasting employees exceed $60,000? Does this estimate seem reasonable? Explain.

18.  **Internet Usage**

Projected Internet Usage, Hours per Person per Year (Based on 1995–1999 Data)

Years Since 1995 (t)	Usage per Person (hours per year) (H)
0	5
1	10
2	34
3	61
4	99
5	135
6	162
7	187
8	208
9	228

Source: *Statistical Abstract of the United States, 2001*, Table 1125, p. 704.

What do you project the Internet usage will be in 2006? Do you believe your projection is a good estimate? Explain.

 19.   **Homes with Cable**

Percentage of TV Homes with Cable

Years Since 1970 (t)	TV Homes with Cable (percent) (C)
0	6.7
5	12.6
10	19.9
15	42.8
16	45.6
17	47.7
18	49.4
19	52.8
20	56.4
21	58.9

(Continued)

(Continued)

22	60.2
23	61.4
24	62.4
25	63.4
26	65.3
27	66.5
28	67.2
29	67.5

Source: *Statistical Abstract of the United States, 2001*, Table 1126, p. 705.

According to the model, what percentage of TV homes had cable in 2002? Does this seem reasonable? Explain.

20. **Homes with VCRs**

Percentage of TV Homes with a VCR

Years Since 1980 (t)	Homes with a VCR (percent) (C)
0	1.1
5	20.8
6	36.0
7	48.7
8	58.0
9	64.6
10	68.6
11	71.9
12	75.0
13	77.1
14	79.0
15	81.0
16	82.2
17	84.2
18	84.6
19	84.6

Source: *Statistical Abstract of the United States, 2001*, Table 1126, p. 705.

Using the model from Exercise 19, determine in what year the percentage of homes with VCRs and the percentage of homes with cable was the same. (*Hint:* The meaning of t in the two exercises is different.)

Exercises 21–25 are intended to challenge your understanding of higher-order polynomial functions.

21. A "bend" in a graph occurs when the graph changes from increasing to decreasing or from decreasing to increasing. A quadratic function has exactly one bend. A cubic function has zero or two bends. A quartic function has one or three bends. In general, how many bends may the graph of a polynomial of degree n have?

22. Can the number of x-intercepts of a graph of a polynomial function ever exceed the degree of the polynomial? Explain.

23. A function of the form $f(x) = x^n$ is called a **power function.** If n is a positive even integer, what is the relationship between $f(x)$ and $f(-x)$?

24. Find the equation of the cubic function that passes through the points $(-1, 0)$, $(0, 0)$, $(1, 0)$, and $(3, 24)$.

25. Find the equation of the quartic function that passes through the points $(-1, 0)$, $(0, 0)$, $(1, 0)$, $(2, 0)$, and $(3, 24)$.

2.3 Exponential Function Models

- Use nonlinear functions to model real-life phenomena
- Interpret the meaning of mathematical models in their real-world context
- Graph exponential functions
- Find the equation of an exponential function from a table
- Model real-life data with exponential functions

GETTING STARTED   Michael Jordan is arguably the most renowned athlete of all time. With millions of dollars in endorsements in addition to his multimillion-dollar salary, he is one of the wealthiest athletes in history. Since his entry into the NBA in 1984, the average salary for NBA players has increased *exponentially*.

In this section, we will demonstrate how exponential functions can be used to model rapidly increasing data sets such as the average salary for NBA players. We will develop exponential models from tables of data and verbal descriptions. In addition, we will show what an exponential function graph looks like.

The average annual salary of an NBA player increased from $170 thousand in 1980 to $2.6 million in 1998, as shown in Table 2.11.

If we plot the NBA salary data together with the graph of the function $S(t) = 161.4(1.169)^t$, we see that the graph of S fits the data fairly well (Figure 2.34).

TABLE 2.11   Average NBA Athlete's Salary

Years Since 1980 (t)	Annual Salary (thousands of dollars) (S)
0	170
5	325
10	750
15	1,900
16	2,000
17	2,200
18	2,600

Source: *Statistical Abstract of the United States, 2001*, Table 1324, p. 829.

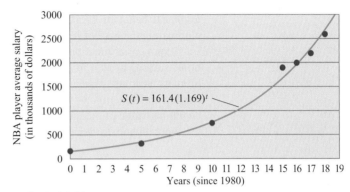

FIGURE 2.34

Notice that the independent variable, t, appears as an exponent. The function S is called an **exponential function.**

EXPONENTIAL FUNCTION

If a and b are real numbers with $a \neq 0$, $b > 0$, and $b \neq 1$, then the function

$$y = ab^x$$

is called an **exponential function.** The value b is called the **base** of the exponential function.

Why must b be positive? Consider the function $y = (-1)^x$. For integer values, y oscillates between -1 and 1 (Table 2.12). However, the function is undefined for numerous noninteger values of x (Table 2.13).

TABLE 2.12

x	y
-2	1
-1	-1
0	1
1	-1
2	1

TABLE 2.13

x	y
-0.5	Undefined
-0.3	Undefined
0.1	Undefined
1.4	-1
1.5	Undefined

On the other hand, if $b > 0$, then the value of b^x is defined for *all* integer and noninteger values of x.

Why don't we allow b to be 1? If $b = 1$, then

$$y = ab^x$$
$$y = a(1)^x$$
$$y = a \qquad \text{Since } (1)^x = 1 \text{ for all } x$$

The graph of the function $y = a$ is a horizontal line and does not exhibit the same graphical behavior as all other functions of the form $y = ab^x$ with $b > 0$. By eliminating the case of $b = 1$, we are able to talk about a family of like functions.

Exponential functions are used frequently to model growth and decay situations, such as growth in population, depreciation of a vehicle, or growth in a retirement account. When the growth or decay of a quantity is modeled by an exponential function $y = ab^x$, the independent variable is frequently time. The beginning value of the quantity at time $x = 0$ is referred to as the **initial value** of the function and, as demonstrated below, is equal to a.

$$y = ab^0$$
$$= a(1) \qquad \text{Since } b^0 = 1 \text{ for all nonzero real numbers } b$$
$$= a$$

Graphically speaking, the constant a in $y = ab^x$ corresponds with the y-intercept of the exponential graph. In most real-life applications, a will be positive. The initial value of the NBA player salary function $S(t) = 161.4(1.169)^t$ is 161.4. That is, the model estimates that the average NBA player's salary was about $161 thousand in the year $t = 0$ (1980).

The base b of $y = ab^x$ is often referred to as the **growth factor** of the function. Increasing the value of x by one unit increases y by a factor of b. The growth factor of the NBA player salary function $S(t) = 161.4(1.169)^t$ is 1.169, since $b = 1.169$. That is, the average NBA player's salary is increasing by a factor of 1.169 annually. Next year's salary is forecast to be 1.169 times this year's salary.

Recall that the x-intercept of a function occurs when $y = 0$. For what values of x does $ab^x = 0$? Let's consider the exponential function $y = 2^x$. We'll generate a table of values and plot a few points to get an idea about what is happening graphically (Figure 2.35).

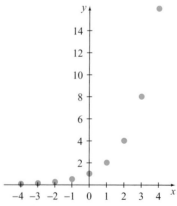

FIGURE 2.35

x	$y = 2^x$
-4	$\frac{1}{16}$
-3	$\frac{1}{8}$
-2	$\frac{1}{4}$
-1	$\frac{1}{2}$
0	1
1	2
2	4
3	8
4	16

Observe that as x increases, y also increases. Also, observe that y is positive for all values of x. That is, *there are no x-intercepts.* This remains true even if we pick a negative number with a larger magnitude, say $x = -33$.

$$2^{-33} = \frac{1}{2^{33}}$$

$$2^{-33} = \frac{1}{8,589,934,592}$$

We call the line $y = 0$ a **horizontal asymptote** of the function $y = 2^x$. For sufficiently small values of x, the graph of $y = 2^x$ approaches the graph of the line $y = 0$. All exponential functions of the form $y = ab^x$ have a horizontal asymptote at $y = 0$.

Exponential Function Graphs

Exponential function graphs will take on one of the four basic shapes specified in Table 2.14. Each graph has a horizontal asymptote at $y = 0$ and a y-intercept at $(0, a)$.

TABLE 2.14 Exponential Function Graphs: $y = ab^x$

Value of a	Value of b	Concavity of Graph	Increasing/ Decreasing	Sample Graph
$a > 0$	$b > 1$	Concave up	Increasing	
$a > 0$	$0 < b < 1$	Concave up	Decreasing	
$a < 0$	$b > 1$	Concave down	Decreasing	
$a < 0$	$0 < b < 1$	Concave down	Increasing	

EXAMPLE 1

Comparing Exponential Graphs

Compare and contrast the graphs of $f(x) = 3(2)^x$ and $g(x) = 4(0.5)^x$. (You may sketch the graphs by hand by plotting several points, or you may use technology to graph the functions.)

SOLUTION Both f and g are exponential functions. In the function equation of f, $a = 3$ and $b = 2$. The graph of f is concave up, since $a > 0$, and is increasing, since $b > 1$. The graph of f has a y-intercept at $(0, 3)$.

In the function equation of g, $a = 4$ and $b = 0.5$. The graph of g is concave up, since $a > 0$, and is decreasing, since $b < 1$. The graph of g has a y-intercept at $(0, 4)$. The graphs of both functions have a horizontal asymptote at $y = 0$ (Figure 2.36).

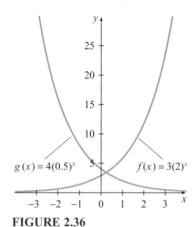

FIGURE 2.36

Properties of Exponents

As seen earlier in this section, it is often necessary to apply the properties of exponents when working with exponential functions. Since these properties are typically covered in depth in an algebra course, we will only summarize them here.

PROPERTIES OF EXPONENTS

If b, m, and n are real numbers with $b > 0$, the following properties hold.

Property	Example
1. $b^{-n} = \dfrac{1}{b^n}$	1. $2^{-3} = \dfrac{1}{2^3}$
2. $b^m \cdot b^n = b^{m+n}$	2. $3^2 \cdot 3^4 = 3^6$
3. $\dfrac{b^m}{b^n} = b^{m-n}$	3. $\dfrac{5^6}{5^4} = 5^2$
4. $b^{mn} = (b^m)^n = (b^n)^m$	4. $6^{2 \cdot 3} = (6^2)^3 = (6^3)^2$

Finding an Exponential Function from a Table

We can easily determine if a table of values models an exponential function by calculating the ratio of consecutive outputs for evenly spaced inputs. The ratio will be constant if the table of values models an exponential function. Consider Table 2.15.

TABLE 2.15

x	y
0	6
2	24
4	96
6	384

The domain values are equally spaced (two units apart). Calculating the ratios of the consecutive range values, we get

$$\frac{24}{6} = 4 \qquad \frac{96}{24} = 4 \qquad \frac{384}{96} = 4$$

In each case, the ratio was 4. Therefore, the table of values models an exponential function. But what is the equation of the function?

We can find the equation of the function algebraically. We know that an exponential function must be of the form $y = ab^x$. Substituting the point $(0, 6)$ into the equation and solving, we get

$$y = ab^x$$
$$6 = ab^0 \qquad \text{Substitute } x = 0 \text{ and } y = 6$$
$$6 = a$$

Since $a = 6$, $y = 6b^x$. Substituting the point $(2, 24)$ into the equation $y = 6b^x$, we get

$$y = 6b^x$$
$$24 = 6b^2 \qquad \text{Substitute } x = 2 \text{ and } y = 24$$
$$4 = b^2$$
$$b = 2 \qquad b \neq -2 \text{ since the base of an exponential function}$$
$$\qquad\qquad \text{must be positive}$$

Therefore, the exponential function that models the table data is $y = 6(2)^x$.

EXAMPLE 2 | ## Finding an Exponential Equation from a Table

Find the equation of the exponential function modeled by Table 2.16.

TABLE 2.16

x	y
1	21
2	63
3	189
4	567

SOLUTION We know that an exponential function must be of the form $y = ab^x$. Substituting the point $(2, 63)$ into the equation, we get

$$y = ab^x$$
$$63 = ab^2 \qquad \text{Substitute } x = 2 \text{ and } y = 63$$

Substituting the point $(1, 21)$ into the equation $y = ab^x$, we get

$$y = ab^x$$
$$21 = ab^1 \qquad \text{Substitute } x = 1 \text{ and } y = 21$$

We can eliminate the a variable by dividing the first equation by the second equation. (*Note:* What we are really doing is dividing both sides of the first equation by the same nonzero quantity, expressed in two different forms.)

$$\frac{63}{21} = \frac{ab^2}{ab^1}$$

$$3 = \frac{a}{a} \cdot b^{2-1} \qquad \text{Since } \frac{b^m}{b^n} = b^{m-n}$$

$$3 = 1 \cdot b$$

$$b = 3$$

We may then substitute the value of b into either equation to find a.

$$21 = ab^1$$

$$21 = a \cdot 3^1 \qquad \text{Substitute } b = 3$$

$$\frac{21}{3} = a$$

$$a = 7$$

The exponential function that models the table data is $y = 7(3)^x$.

When using the method of dividing one equation by the other, computations will tend to be easier if we divide the equation with the largest exponent by the equation with the smallest exponent.

The technique shown in Example 2 may be generalized for all exponential functions. Using the result from the generalized solution will allow us to determine the exponential equation more quickly.

Consider an exponential function $y = ab^x$ whose graph goes through the points (x_1, y_1) and (x_2, y_2). We have

$$y = ab^x$$

$$y_1 = ab^{x_1} \qquad \text{Substitute } x = x_1 \text{ and } y = y_1$$

and

$$y = ab^x$$

$$y_2 = ab^{x_2} \qquad \text{Substitute } x = x_2 \text{ and } y = y_2$$

Dividing the second equation by the first equation yields a quick way to calculate b.

$$\frac{y_2}{y_1} = \frac{ab^{x_2}}{ab^{x_1}}$$

$$\frac{y_2}{y_1} = \frac{a}{a} b^{x_2 - x_1}$$

$$\frac{y_2}{y_1} = b^{x_2 - x_1}$$

$$\left(\frac{y_2}{y_1}\right)^{\frac{1}{x_2 - x_1}} = \left(b^{x_2 - x_1}\right)^{\frac{1}{x_2 - x_1}}$$

$$\left(\frac{y_2}{y_1}\right)^{\frac{1}{x_2 - x_1}} = b^{\frac{x_2 - x_1}{x_2 - x_1}}$$

$$b = \left(\frac{y_2}{y_1}\right)^{\frac{1}{x_2 - x_1}}$$

Using the points (1, 21) and (3, 189) from Example 2, we get

$$b = \left(\frac{y_2}{y_1}\right)^{\frac{1}{x_2-x_1}}$$

$$b = \left(\frac{189}{21}\right)^{\frac{1}{3-1}} \qquad \text{Substitute } x_1 = 0, x_2 = 2, y_1 = 21, \text{ and } y_2 = 189$$

$$= 9^{1/2}$$

$$= \sqrt{9} \qquad \text{Recall that for } b \geq 0,\ b^{1/2} = \sqrt{b}$$

$$= 3$$

This method is especially useful when checking to see if a data table with unequally spaced inputs is an exponential function. If it is an exponential function, the value of b will be constant regardless of which two points we substitute into the formula.

Using Exponential Regression to Model Data

Data sets with near-constant ratios of change may be modeled using the exponential regression feature on our graphing calculator. It is often helpful to do a scatter plot of the data first to see if the graph looks like an exponential function.

EXAMPLE 3

Using Exponential Regression to Model the Population of Akron, Ohio

TABLE 2.17

Years Since 1970 (x)	Population (thousands) (y)
0	275
10	237
20	223
30	217

Source: *Statistical Abstract of the United States, 2001,* Table 34, p. 34.

The population of Akron, Ohio, is shown in Table 2.17. Model the population with an exponential function and forecast the population of Akron in 2010.

SOLUTION We first draw a scatter plot so that we can visually predict whether an exponential function will fit the data well (Figure 2.37). (Note that because we have zoomed in on the data, the "x axis" is given by $y = 210$ instead of $y = 0$.)

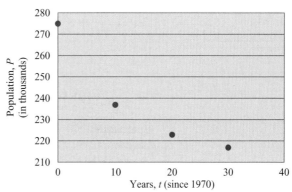

FIGURE 2.37

Since the scatter plot is decreasing and concave up, an exponential function may fit the data well. Using techniques given in the Technology Tip following this example, we can calculate an exponential model for the data set (Figure 2.38).

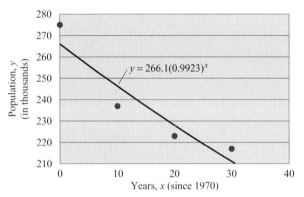

FIGURE 2.38

TABLE 2.18

Year (x)	Population -210 (thousands) (y)
0	65
10	27
20	13
30	7

The model $y = 266.1(0.9923)^x$ is a terrible fit for the data. What happened? Recall that an exponential function of the form $y = ab^x$ has a horizontal asymptote at $y = 0$. From the scatter plot of the data, it appears that the population is approaching a constant value of $y = 210$ instead of $y = 0$. (This value is not precise. We are *guessing* what the horizontal asymptote would be based on the scatter plot of the data set.) If we subtract 210 from each of the y values, we get a new data set with a horizontal asymptote of $y = 0$. This is referred to as an **aligned data set** (Table 2.18).

Using exponential regression to find a model for the aligned data set yields $y = 60.81(0.9285)^x$, a function that fits the aligned data well (Figure 2.39).

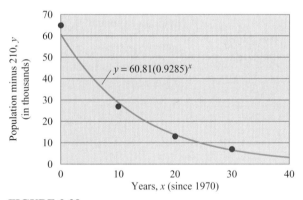

FIGURE 2.39

Subtracting 210 from each of the y values of the original data set had the graphical implication of moving the data down 210 units. To return the aligned data model to the position of the original data, we need to move the model up 210 units. To do this, we add back the 210.

$$P(x) = 60.81(0.9285)^x + 210 \qquad \text{Add 210 to shift the data graph upward}$$

The model created from the aligned data fits those data much better than the unaligned data model (Figure 2.40).

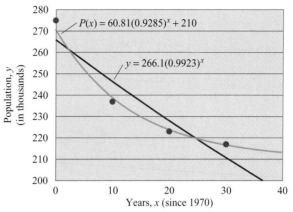

FIGURE 2.40

To estimate the population of Akron, Ohio, in 2010, we evaluate $P(x) = 60.81(0.9285)^x + 210$ at $x = 40$.

$$P(40) = 60.81(0.9285)^{40} + 210$$

$$\approx 213$$

We forecast that the population of Akron, Ohio, in 2010 will be 213,000.

The methods of aligning data and of performing exponential regression are detailed in the following Technology Tips.

TECHNOLOGY TIP

Aligning a Data Set

1. Enter the data using the Statistics Menu List Editor. (Refer to Section 1.3 if you've forgotten how to do this.)

2. Move the cursor to the top of L3. We want the entries in L3 to equal the entries in L2 minus the amount of the vertical shift (in this case 210). To do this, we must enter the equation L3=L2-210.

(Continued)

3. Press [2nd] then [2] to place L2 on the equation line at the bottom of the viewing window. Then press [−] and [2][1][0] to subtract 210.

L1	L2	**L3**	3
0	275	------	
10	237		
20	223		
30	217		
------	------	------	

L3 =L2−210

4. Press [ENTER] to display the list of aligned values in L3.

L1	L2	L3	3
0	275	65	
10	237	27	
20	223	13	
30	217	7	
------	------	------	

L3(5) =

TECHNOLOGY **TIP**

Exponential Regression

1. Enter the data using the Statistics Menu List Editor. (Refer to Section 1.3 if you've forgotten how to do this.)

L1	L2	L3	3
0	275	65	
10	237	27	
20	223	13	
30	217	7	
------	------	------	

L3(5) =

2. Bring up the Statistics Menu Calculate feature by pressing [STAT] and using the blue arrows to move to the CALC menu. Then select item Ø:ExpReg and press [ENTER].

```
EDIT CALC TESTS
7↑QuartReg
8:LinReg(a+bx)
9:LnReg
Ø:ExpReg
A:PwrReg
B:Logistic
C:SinReg
```

3. If the data to be evaluated are in L1 and L2, press [ENTER]. Otherwise, go to step 4. (If you want to automatically paste the regression equation into the [Y=] editor, press the key sequence [VARS] Y-VARS; 1:Function; 1:Y1 before pressing [ENTER].)

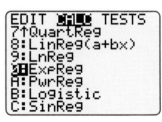

```
ExpReg
 y=a*b^x
 a=266.0545948
 b=.9923145889
 r²=.8858866457
 r=-.941215515
```

(Continued)

4. If the data to be evaluated are in L1 and L3, press the key sequence [2nd] [1] [,] [2nd] [3] to place the entries L1 and L3 on the home screen.

```
ExpReg  L1, L3
```

5. Press [ENTER] to display the exponential model. (If you want to automatically paste the regression equation into the [Y=] editor, press the key sequence [VARS] Y-VARS; 1:Function; 1:Y1 before pressing [ENTER].)

```
ExpReg
 y=a*b^x
 a=60.8078408
 b=.9285201575
 r²=.9938920452
 r=-.9969413449
```

Finding an Exponential Function from a Verbal Description

Much of the information we encounter in the media is given to us in terms of percentages. Consider these typical news headlines. "Gasoline Prices Flare Up 9 Percent" and "Home Values Increase by 6 Percent." Anything that increases or decreases at a constant percentage rate may be modeled with an exponential function. The value of the growth factor b is $1 + r$, where r is the decimal form of the percentage (i.e., 5 percent = 0.05). Note that if $r < 0$, then $b < 1$. A negative rate of growth is referred to as **depreciation,** while a positive rate of growth is referred to as **appreciation.**

ANNUAL GROWTH RATE

Let $y = ab^t$ model the amount of a quantity at time t years. The annual growth rate r of the quantity y is given by $r = b - 1$. Similarly, the annual growth factor is $b = r + 1$.

For example, the author ordered an appraisal of his home in August 2004. In her report, the appraiser asserted that home values in the area were increasing at a rate of 5 percent per year. In other words, $r = 0.05$. The corresponding growth factor is $b = 1.05$. To forecast the value of his home in August 2005, the author multiplied the appraised value by the growth factor.

EXAMPLE 4

Using an Exponential Model to Forecast a Car's Value

New cars typically lose 50 percent of their value in the first three years. Many cars lose more than 25 percent of their value in the first year alone. (**Source:** Runzheimer International.) Depreciating approximately 13.4 percent annually, the Saturn

Sedan 2 holds its value better than most vehicles. Estimate the value of a $12,810 Saturn Sedan 2 three years after it is purchased.

SOLUTION Since the car is depreciating at a constant percentage rate, an exponential function may be used to model the data. The initial value of the car is $12,810, so $a = 12,810$. The car is losing value at 13.4 percent per year, so $r = -0.134$. Since $b = 1 + r$, the exponential function is

$$V(t) = 12,810(b)^t$$
$$= 12,810[1 + (-0.134)]^t \qquad \text{Since } b = 1 + (-0.134)$$
$$= 12,810(0.866)^t \text{ dollars}$$

where t is the age of the car in years.

We want to know the value of the car after three years, so $t = 3$.

$$V(3) = 12,810(0.866)^3$$
$$= 12,810(0.649)$$
$$= 8320$$

We estimate that the car will be valued at $8320 when it is three years old.

EXAMPLE   5

Using an Exponential Function to Model Inflation

Inflation (rising prices) causes money to lose its buying power. In the United States, inflation hovers around 3 percent annually.

(a) If a candy bar costs $0.59 today, what will it cost 10 years from now?
(b) When will a candy bar cost $1.00?

SOLUTION Since we're assuming a constant percentage increase in the price of the candy bar, an exponential model should be used. The initial value is $0.59, so $a = 0.59$. The price of the candy bar is increasing by 3 percent annually, so $r = 0.03$. The cost of the candy bar is given by

$$C(t) = 0.59(1 + 0.03)^t \text{ dollars}$$
$$= 0.59(1.03)^t \text{ dollars}$$

where t is the number of years from today.

(a) Evaluating the function at $t = 10$, we get

$$C(10) = 0.59(1.03)^{10}$$
$$= 0.59(1.344)$$
$$= 0.79$$

We estimate that the candy bar will cost $0.79 ten years from now.
(b) We want to know when $C(t) = 1$.

$$1 = 0.59(1.03)^t$$
$$\frac{1}{0.59} = 1.03^t$$
$$1.695 = 1.03^t$$

At this point, we are stuck. We haven't yet developed the mathematical machinery (logarithms) to get the variable out of the exponent. We can, however, attack the problem graphically by breaking the equation $1.695 = 1.03^t$ into two separate functions: $y_1 = 1.695$ and $y_2 = 1.03^t$. By graphing these functions simultaneously, we can determine the point of intersection of the two functions using our graphing calculator.

Using the graphing calculator, we generate the graph in Figure 2.41 and determine the point of intersection.

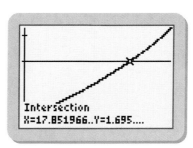

Intersection
X=17.851966.. Y=1.695....

FIGURE 2.41

The functions intersect when $t = 17.85$. In about 18 years, we estimate that the cost of a candy bar will be $1.00.

We also could have used the guess-and-check method to approximate the result by evaluating $C(t) = 0.59(1.03)^t$ at various values of t. Since

$$C(17) = 0.59(1.03)^{17} \qquad \text{and} \qquad C(18) = 0.59(1.03)^{18}$$
$$= 0.9752 \qquad\qquad\qquad\qquad = 1.004$$

we determine that the price reaches $1.00 between the seventeenth and eighteenth years.

2.3 Summary

In this section, you learned how exponential functions can be used to model rapidly increasing (or decreasing) data sets. You graphed exponential functions and developed exponential models from tables of data and verbal descriptions.

2.3 Exercises

In Exercises 1–10, do the following:

(a) Determine if the graph of the function is increasing or decreasing.

(b) Determine if the graph of the function is concave up or concave down.

(c) Identify the coordinate of the *y*-intercept.

(d) Graph the function to verify your conclusions.

1. $y = 4(0.25)^x$ **2.** $y = -2(0.5)^x$

3. $y = 0.5(2)^x$ **4.** $y = 6(0.1)^x$

5. $y = 0.4(5)^x$ **6.** $y = -0.1(0.2)^x$

7. $y = -1.2(2.3)^x$ **8.** $y = 5(0.4)^x$

9. $y = 3(0.9)^x$ **10.** $y = -5(3)^x$

In Exercises 11–20, use algebraic methods to find the equation of the exponential function that fits the data in the table.

11.

x	y
0	2
1	6
2	18
3	54

12.

x	y
0	3
1	12
2	48
3	192

13.

x	y
1	10
2	20
3	40
4	80

14.

x	y
1	10
2	50
3	250
4	1,250

15.

x	y
2	16
4	4
6	1
8	0.25

16.

x	y
2	100
4	1
6	0.01
8	0.0001

17.

x	y
1	1
2	4
4	64
5	256

18.

x	y
2	4
5	32
9	512
11	2,048

19.

x	y
0	256
5	8
7	2
10	0.25

20.

x	y
0	10,000
2	900
5	24.3
7	2.187

In Exercises 21–25, use exponential regression to model the data in the table. Use the model to predict the value of the function when t = 25, and interpret the real-world meaning of the result.

The Consumer Price Index is used to measure the increase in prices over time. In each of the following tables, the index is assumed to have the value 100 in the year 1984.

 21. **Dental Prices**

Price of Dental Services

Years Since 1980 (t)	Price Index (I)
0	78.9
5	114.2
10	155.8
15	206.8
20	258.5

Source: *Statistical Abstract of the United States, 2001,* Table 694, p. 455.

22. **Wine Prices**

Price of Wine Consumed at Home

Years Since 1980 (t)	Price Index (I)
0	89.5
5	100.2
10	114.4
15	133.6
20	151.6

Source: *Statistical Abstract of the United States, 2001*, Table 694, p. 455.

23. **Television Prices**

Price of a Television Set

Years Since 1980 (t)	Price Index (I)
0	104.6
5	88.7
10	74.6
15	68.1
20	49.9

Source: *Statistical Abstract of the United States, 2001*, Table 694, p. 455.

24. **Alcohol Prices**

Price of Distilled Spirits Consumed at Home

Years Since 1980 (t)	Price Index (I)
0	89.8
5	105.3
10	125.7
15	145.7
20	162.3

Source: *Statistical Abstract of the United States, 2001*, Table 694, p. 455.

25. **Entertainment Prices**

Price of Admission to Entertainment Venues

Years Since 1980 (t)	Price Index (I)
0	83.8
5	112.8
10	151.2
15	181.5
20	230.5

Source: *Statistical Abstract of the United States, 2001*, Table 694, p. 455.

In Exercises 26–30, find the exponential function that fits the verbal description.

26. Salaries Instructors' salaries are $45,000 per year and are expected to increase by 3.5 percent annually. What will instructors' salaries be five years from now?

27. Savings Account A savings account balance is currently $235 and is earning 2.32 percent per year. When will the balance reach $250?

28. **Television Price** The cost of a 27-inch flat-screen television in 2002 was $599.99. (Source: www.bestbuy.com.) Television prices are expected to decrease by 16 percent per year. How much is the flat-screen television expected to cost in 2007?

29. **Concert Admission** Reserved seating at a Dave Matthews Band concert at the Gorge Amphitheater in George, Washington, cost $59.90 in September 2004. (Source: www.ticketmaster.com.) If admission fees for concerts are expected to increase by 28 percent per year, what is the expected price of a Dave Matthews Band ticket in September 2006?

30. Tuition Tuition is currently $2024 per year and is increasing by 12 percent annually. In how many years from now will tuition reach $3567?

In Exercises 31–35, find the solution by solving the equation graphically.

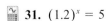 **31.** $(1.2)^x = 5$ **32.** $(0.9)^x = 0.5$

33. $(1.05)^x = 2$ **34.** $(1.12)^x = 2$

35. $(0.3)^x = 0.09$

Exercises 36–40 are intended to challenge your understanding of exponential functions.

36. **Retirement Investments** An investor has $3000 to invest in two different investment accounts: CREF Social Choice and TIAA Retirement Annuity. Based on data from January 1, 1993, through December 31, 2002, the average annual return on the CREF Social Choice account was 8.67 percent, and the average annual return on the TIAA Retirement Annuity was 6.93 percent. (Source: TIAA-CREF.) The investor plans to invest twice as much in the Retirement Annuity as in the Social Choice account. Assuming that the future rates of return will be the same as the past rates of return, do the following:

(a) Find the equation of the function for the predicted value of the $3000 investment *t* years from now.

(b) What is the predicted value of the investment 20 years from now?

37. **Retirement Investments** Repeat Exercise 36, except this time invest twice as much in the Social Choice account as in the Retirement Annuity. What factors should the investor consider before changing the amount invested in each of the accounts?

 38. **Retirement Investments** Based on data from January 1, 2002, through December 31, 2002, the annual rate of return on the CREF Social Choice account was −2.98 percent, substantially lower than the 10-year average return of 8.67 percent. Over the same period, the CREF Stock account earned −20.73 percent, substantially lower than its 10-year average return of 7.69 percent. (Source: TIAA-CREF.)

An investor invested $1000 in each of the two accounts at the start of 2002.

(a) What was the value of the investment in each account at the end of 2002?

(b) Assuming that in each year following 2002, the investments earned an annual return equal to the 10-year average, predict how long it will take for the combined value of the accounts to reach $2000.

 39. **Retirement Investments** Based on data from January 1, 2002, through December 31, 2002, the annual rate of return of the CREF Growth account was −30.06 percent, substantially lower than the five-year average return of −5.48 percent. Over the same period, the CREF Inflation-linked Bond account earned 16.32 percent, substantially higher than its five-year average return of 8.33 percent. (Source: TIAA-CREF.)

An investor invested $1000 in each of the two accounts at the start of 2002.

(a) What was the value of the investment in each account at the end of 2002?

(b) Assuming that in each year following 2002 the investments earned an annual return equal to the five-year average, predict how long it will take for the combined $2000 investment in the accounts to double.

 40. **Investments** How long will it take for a $1000 investment earning 10 percent annually to equal the value of a $2000 investment earning 6 percent annually?

2.4 Logarithmic Function Models

- Use nonlinear functions to model real-life phenomena
- Interpret the meaning of mathematical models in their real-world context
- Graph logarithmic functions
- Model real-life data with logarithmic functions
- Apply the rules of logarithms to simplify logarithmic expressions
- Solve logarithmic equations

GETTING STARTED   In Section 2.3, we showed that the NBA players' average salary could be modeled by

$$S(t) = 161.4(1.169)^t \text{ thousand dollars}$$

where t is the number of years since 1980. We could use the model to estimate the average salary in any year t. But what if we wanted to algebraically determine the year in which the salary would be \$1,000,000?

In this section, we will demonstrate how logarithmic functions can be used to model data sets that are increasing at a decreasing rate. We will show how to graph logarithmic functions and will illustrate graphically the inverse relationship between logarithmic and exponential functions. We will explain how the rules of logarithms can be used to simplify logarithmic expressions and solve logarithmic equations. We will also show how to develop logarithmic models from tables of data, such as NBA players' salaries, by using logarithmic regression.

We want to determine the year algebraically in which the NBA players' average salary reached \$1 million. We can estimate the year by looking at the data in Table 2.19.

TABLE 2.19   NBA Players' Average Salary

Salary (thousands of dollars) (s)	Years Since 1980 (t)
170	0
325	5
750	10
1,900	15
2,000	16
2,200	17
2,600	18

Source: *Statistical Abstract of the United States, 2001,* Table 1324, p. 829.

Notice that *salary* is the independent variable and *years* is the dependent variable. The average salary was \$750,000 in 1990 and \$1,900,000 in 1995, so we know that \$1,000,000 was reached sometime between 1990 and 1995.

Returning to the exponential model for the NBA players' average salary, we have

$$s(t) = 161.4(1.169)^t \text{ thousand dollars}$$

where t is the number of years since 1980. Recall that \$1,000,000 is equal to \$1000 thousand. We must determine when $s(t) = 1000$. In other words, we must solve the equation

$$1000 = 161.4(1.169)^t \qquad \text{Since the units of the salary are thousands of dollars}$$

$$\frac{1000}{161.4} = \frac{161.4(1.169)^t}{161.4}$$

$$6.196 = 1.169^t$$

If we want to solve this equation algebraically, at this point we are stuck. We must come up with some way to get the t out of the exponent. Observe that t is the exponent that we place on 1.169 in order to get 6.196. We need to create a symbol to represent the phrase "is the exponent we place on." The chosen symbol, as bizarre as it may seem, is **log**.

$$\underbrace{t \text{ is}}_{t\,=} \quad \underbrace{\text{the exponent we place on}}_{\log} \quad \underbrace{1.169}_{1.169} \quad \underbrace{\text{in order to get 6.196}}_{6.196}$$

Compressing these terms into a single equation yields

$$t = \log_{1.169} 6.196$$

Using the calculator $\boxed{\text{LOG}}$ button and some log rules that will be introduced later, we determine that $t \approx 11.7$ years. That is, between 1991 ($t = 11$) and 1992 ($t = 12$), the NBA players' average salary reached \$1 million.

Logarithmic Functions

The notation **log** is short for **logarithm.** It is helpful to think of **log** as representing the phrase "the exponent we place on." The equation $y = \log_b(x)$, which is read "y equals log base b of x," means "y is the exponent we place on b in order to get x." For example, $y = \log_2(8)$ means "y is the exponent we place on 2 in order to get 8." That is, $2^y = 8$. Since $2^3 = 8$, we conclude that $y = 3$.

EXAMPLE 1 **Calculating a Logarithm**

Find the value of y given $y = \log_3(9)$.

SOLUTION $y = \log_3(9)$ means "y is the exponent we place on 3 in order to get 9." That is, $3^y = 9$. Since $3^2 = 9$, we conclude that $y = 2$.

LOGARITHMIC FUNCTION

Let b and x be real numbers with $b > 0$ and $b \neq 1$ and $x > 0$. The function

$$y = \log_b(x)$$

is called a **logarithmic function.** The value b is called the **base** of the logarithmic function. We read the expression $\log_b(x)$ as "log base b of x."

The use of parentheses around the x in $y = \log_b(x)$ is optional. That is, $y = \log_b(x)$ is equivalent to $y = \log_b x$. However, we will consistently use parentheses throughout this text to remind ourselves that the expression "$\log_b$" is meaningless by itself.

Although any positive number not equal to 1 may be used as the base of a logarithmic function, the most commonly used bases are 10 and e. The irrational number e is approximately equal to 2.71828. The number e, like π, has many wonderful uses. Although we will see some of the power of e in this chapter, we will witness its full strength when we move into calculus. The number e is generated by calculating $\left(1 + \frac{1}{n}\right)^n$ for infinitely large n. You might think that $\left(1 + \frac{1}{n}\right)^n$ will get infinitely big as n goes to infinity. Surprisingly, $\left(1 + \frac{1}{n}\right)^n \approx 2.718281828$ for large values of n (see Table 2.20).

TABLE 2.20

n	$\left(1 + \frac{1}{n}\right)^n$
1	2.000
10	2.594
100	2.705
1,000	2.717
10,000	2.718
100,000	2.718

The value of e, 2.718281828. . . , may be accessed on the TI-83 Plus calculator by pressing the key sequence [2nd] and [÷].

The expression $\log_{10}(x)$ is often written $\log(x)$ and is called the **common log.** We will assume that the base is 10 when no base is written.

The expression $\log_e(x)$ is typically written $\ln(x)$ and is called the **natural log.**

Graphs of Logarithmic Functions

Like polynomial and exponential function graphs, logarithmic function graphs share many common characteristics. Consider the logarithmic functions $f(x) = \log_2(x)$, $g(x) = \log_3(x)$, and $h(x) = \log_4(x)$ (Figure 2.42).

Each graph is concave down and increasing. Each graph passes through the point $(1, 0)$ and appears to have a vertical asymptote at $x = 0$ (the y axis). In fact, for all positive values of b not equal to 1, the graph of $y = \log_b(x)$ has a vertical asymptote at $x = 0$ and an x-intercept at $(1, 0)$. Logarithmic functions are undefined for $x < 0$. The shape of the graph is determined by b. The base b controls how rapidly the graph increases (or decreases). For each of the graphs in Figure 2.42, $b > 1$. What if $b < 1$? Consider the logarithmic functions $f(x) = \log_{0.3}(x)$, $g(x) = \log_{0.5}(x)$, and $h(x) = \log_{0.7}(x)$ (Figure 2.43).

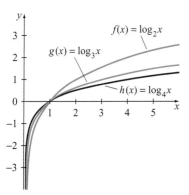

FIGURE 2.42

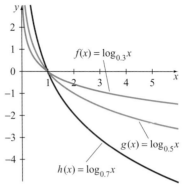

FIGURE 2.43

Each graph is concave up and decreasing. However, these graphs also have a vertical asymptote at $x = 0$ and pass through the point $(1, 0)$. Our graphical observations are summarized in Table 2.21.

TABLE 2.21   Logarithmic Function Graphs: $y = \log_b(x)$

Value of b	Concavity	Increasing/Decreasing	Graph
$b > 1$	Concave down	Increasing	
$0 < b < 1$	Concave up	Decreasing	

EXAMPLE **2**

Determining the Shape of a Logarithmic Function

Determine the concavity and increasing/decreasing behavior of $y = \log_2(x)$ and $y = \log_{0.5}(x)$. Then graph both functions to verify your results.

SOLUTION   Since $2 > 1$, $y = \log_2(x)$ will be concave down and increasing. In order to graph the function by hand, we must generate a table of values for the function. You may find it helpful to write $y = \log_2(x)$ as $x = 2^y$, then select the values of y and calculate the values of x (Table 2.22). Plotting the points and connecting the dots yields the graph shown in Figure 2.44.

TABLE 2.22

$x = 2^y$	y
1	0
2	1
4	2
8	3
16	4

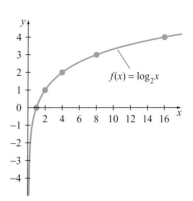

FIGURE 2.44

Since $0.5 < 1$, $y = \log_{0.5}(x)$ will be concave up and decreasing. You may find it helpful to write $y = \log_{0.5}(x)$ as $x = (0.5)^y$, then select the values of y and calculate the values of x (Table 2.23). Plotting the points and connecting the dots yields the graph shown in Figure 2.45.

TABLE 2.23

$x = (0.5)^y$	y
16	-4
8	-3
4	-2
2	-1
1	0

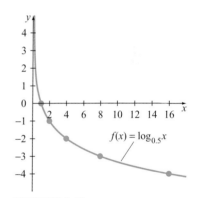

FIGURE 2.45

EXAMPLE 3

Comparing the Graphs of Logarithmic Functions

Graph $f(x) = \ln(x)$ and $g(x) = \log(x)$ simultaneously. Then determine where $f(x) < g(x)$ and where $g(x) > f(x)$.

SOLUTION

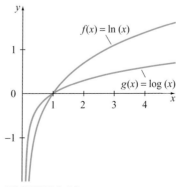

FIGURE 2.46

(To graph the function on the TI-83 Plus, use the $\boxed{\text{LOG}}$ and $\boxed{\text{LN}}$ keys as appropriate.) Based on the graphs in Figure 2.46, we conclude that for $x < 1$, $\ln(x) < \log(x)$. For $x > 1$, $\ln(x) > \log(x)$.

Relationship Between Logarithmic and Exponential Functions

Exponential and logarithmic functions are intimately related. Any logarithmic function may be rewritten as an exponential function (and vice versa).

LOGARITHMIC AND EXPONENTIAL FUNCTION RELATIONSHIP

For $b > 0$ with $b \neq 1$ and $x > 0$, the following statements are equivalent:

1. $y = \log_b(x)$
2. $b^y = x$

Consider tables of values for the functions $f(x) = 10^x$ and $g(x) = \log(x)$ (Table 2.24).

TABLE 2.24

x	$f(x) = 10^x$	x	$g(x) = \log(x)$
-2	0.01	0.01	-2
-1	0.1	0.1	-1
0	1	1	0
1	10	10	1
2	100	100	2

Notice that the *input* of the exponential function is the *output* of the logarithmic function. Similarly, the *output* of the exponential function is the *input* of the logarithmic function. A point (x_1, y_1) is on the graph of f if and only if the point (y_1, x_1) is on the graph of g. Functions with this property are called **inverse** functions. Exponential functions and logarithmic functions are inverse functions of each other. The inverse relationship of logarithmic and exponential functions may also be seen by graphing $y = \ln(x)$ and $y = e^x$ together with the graph of $y = x$ (Figure 2.47). The graph of $y = \ln(x)$ is the reflection of $y = e^x$ about the line $y = x$. That is, if we folded our paper along the line $y = x$, the graphs of the two functions would lie directly on top of each other.

When we solve a logarithmic function $y = \log_b(x)$ for x, we get the exponential function, $x = b^y$. This relationship between logarithmic and exponential functions will be critical to solving the logarithmic and exponential equations.

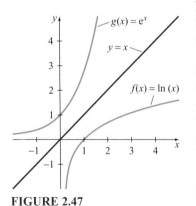

FIGURE 2.47

Rules of Logarithms

There are several rules that are used to manipulate logarithms. Each rule is important; however, we will use Rule 7 most extensively. It is so essential that we comically call it "the mother of all log rules."

RULES OF LOGARITHMS

If x, y, and n are real numbers with $x > 0$ and $y > 0$, then the following rules apply.

Common Logarithm	Natural Logarithm
1. $\log(1) = 0$	1. $\ln(1) = 0$
2. $\log(10) = 1$	2. $\ln(e) = 1$
3. $\log(10^x) = x$	3. $\ln(e^x) = x$
4. $10^{\log(x)} = x$	4. $e^{\ln(x)} = x$
5. $\log(xy) = \log(x) + \log(y)$	5. $\ln(xy) = \ln(x) + \ln(y)$
6. $\log\left(\dfrac{x}{y}\right) = \log(x) - \log(y)$	6. $\ln\left(\dfrac{x}{y}\right) = \ln(x) - \ln(y)$
7. $\log(x^p) = p\log(x)$	7. $\ln(x^p) = p\ln(x)$

In fact, Rules 1, 5, 6, and 7 apply for any base, not just 10 and e. Rules similar to Rules 2 through 4 may easily be formulated for any other base. In our applications, we will be using base 10 and base e predominantly. This is because the following change-of-base formula allows us to convert any logarithm into a logarithm with one of these two bases.

CHANGE-OF-BASE FORMULA

$$\log_b(a) = \frac{\log(a)}{\log(b)} = \frac{\ln(a)}{\ln(b)}$$

The change-of-base formula is easily derived.

$$\text{Let } y = \log_b(a)$$

$$b^y = a \qquad \text{Rewrite in exponential form}$$

$$\log(b^y) = \log(a) \qquad \text{Take the log of both sides}$$

$$y\log(b) = \log(a) \qquad \text{Apply log Rule 7}$$

$$y = \frac{\log(a)}{\log(b)} \qquad \text{Divide both sides by } \log(b)$$

$$\text{But } y = \log_b(a)$$

so

$$\log_b(a) = \frac{\log(a)}{\log(b)}$$

The ability to apply the log rules to manipulate expressions and equations is one of the critical skills required to be successful in solving exponential and logarithmic equations. To assist you in mastering those skills, we will do several "skill-and-drill" examples.

EXAMPLE 4

Using Log Rules to Simplify a Logarithmic Expression

Use the log rules to rewrite the logarithmic expression as a single term.

$$\log(3) - \log(27) + \log(12)$$

SOLUTION

$$
\begin{aligned}
\log(3) - \log(27) + \log(12) &= \log\left(\frac{3}{27}\right) + \log(12) &&\text{Rule 6}\\[6pt]
&= \log\left(\frac{3}{27} \cdot 12\right) &&\text{Rule 5}\\[6pt]
&= \log\left(\frac{36}{27}\right)\\[6pt]
&= \log\left(\frac{4}{3}\right)
\end{aligned}
$$

EXAMPLE 5

Writing a Logarithmic Expression as a Single Logarithm

Rewrite the logarithmic expression as a single logarithm.

$$\log(x) + \log(x^3) - 4\log(x^2)$$

SOLUTION

$$
\begin{aligned}
&\log(x) + \log(x^3) - 4\log(x^2)\\
&= \log(x) + 3\log(x) - 2 \cdot 4\log(x) &&\text{Rule 7}\\
&= 4\log(x) - 8\log(x) &&\text{Combine like terms}\\
&= -4\log(x) &&\text{Combine like terms}
\end{aligned}
$$

The solution may also be written as

$$-\log(x^4),\ \log(x^{-4}),\ \text{or}\ \log\left(\frac{1}{x^4}\right)$$

When checking a solution with the answers in the back of the book, it is important to recognize that the correct solution may be written in a variety of forms.

EXAMPLE 6

Writing a Logarithmic Expression as a Single Logarithm

Rewrite the expression $2\log\left(\frac{2}{3x}\right) - \log(x^3)$ as a single logarithm.

SOLUTION

$$
\begin{aligned}
&2\log\left(\frac{2}{3x}\right) - \log(x^3)\\[6pt]
&= \log\left(\frac{2}{3x}\right)^2 - \log(x^3) &&\text{Rule 7}\\[6pt]
&= \log\left[\frac{(2)^2}{(3x)^2}\right] + \log(x^3)^{-1} &&\text{Rule 7, rules of exponents}
\end{aligned}
$$

$$= \log\left(\frac{4}{9x^2}\right) + \log(x^{-3}) \qquad \text{Rules of exponents}$$

$$= \log\left(\frac{4}{9x^2} \cdot x^{-3}\right) \qquad \text{Rule 5}$$

$$= \log\left(\frac{4}{9x^2} \cdot \frac{1}{x^3}\right) \qquad \text{Rules of exponents}$$

$$= \log\left(\frac{4}{9x^5}\right) \qquad \text{Rules of exponents and fractions}$$

EXAMPLE 7   Solving a Logarithmic Equation

Solve the equation $\log_3(x) = 4$ for x.

SOLUTION

$$\log_3(x) = 4$$
$$3^4 = x \qquad \text{Rewrite in exponential form}$$
$$x = 81$$

EXAMPLE 8   Solving an Exponential Equation

Solve the equation $3^x = \frac{1}{12}$ for x.

SOLUTION

$$3^x = \frac{1}{12}$$

$$\ln(3^x) = \ln\left(\frac{1}{12}\right) \qquad \text{Take the natural log of both sides}$$

$$x \ln(3) = \ln(1) - \ln(12) \qquad \text{Rule 7 and Rule 6}$$

$$x \ln(3) = 0 - \ln(12) \qquad \text{Rule 1}$$

$$x \ln(3) = -\ln(12)$$

$$x = -\frac{\ln(12)}{\ln(3)} \qquad \text{Divided by } \ln(3)$$

$$x \approx -2.262$$

It is important to note that $-\frac{\ln(12)}{\ln(3)} \neq -\ln\left(\frac{12}{3}\right)$.

Logarithmic Models

Functions that are increasing at a decreasing rate may often be modeled by logarithmic functions. Using the logarithmic regression feature, **LnReg,** on our calculator returns a function of the form $y = a + b\ln(x)$. The a shifts the graph of $y = b\ln(x)$ vertically by $|a|$ units. The resultant graph passes through the point $(1, a)$ instead of $(1, 0)$.

EXAMPLE 9

Using Logarithmic Regression to Forecast the Population of Hawaii

TABLE 2.25 Projected Population of Hawaii

People (thousands) (p)	Years Since 1995 (T)
1,187	0
1,257	5
1,342	10
1,553	20
1,812	30

Source: www.census.gov.

Find the logarithmic model that best fits the data in Table 2.25. Then evaluate the function at $p = 1700$ and interpret your result.

SOLUTION Using the logarithmic regression feature on our calculator (as shown in the following Technology Tip), we get

$$y = -496.0 + 70.17 \ln(x)$$

or, in terms of our variables,

$$T(p) = -496.0 + 70.17 \ln(p)$$

Evaluating the function when $p = 1700$, we get

$$T(1700) = -496.0 + 70.17 \ln(1700)$$
$$= -496.0 + 70.17(7.438)$$
$$= -496.0 + 522.0$$
$$= 26.0$$

The population of Hawaii is projected to reach 1,700,000 in 2021 (26 years after 1995).

TECHNOLOGY TIP

Logarithmic Regression

1. Enter the data using the Statistics Menu List Editor. (Refer to Section 1.3 if you've forgotten how to do this.)

L1	L2	L3 3
1187	0	■■■■■
1257	5	
1342	10	
1553	20	
1812	30	

L3(1) =

---

2. Bring up the Statistics Menu Calculate feature by pressing [STAT] and using the blue arrows to move to the CALC menu. Then select item 9:LnReg and press [ENTER].

```
EDIT CALC TESTS
7↑QuartReg
8:LinReg(a+bx)
9■LnReg
0:ExpReg
A:PwrReg
B:Logistic
C:SinReg
```

---

3. If you want to automatically paste the regression equation into the [Y=] editor, press the key sequence [VARS] Y-VARS; 1:Function; 1:Y1 and press [ENTER]. Otherwise, press [ENTER].

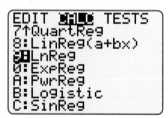

```
LnReg
y=a+blnx
a=-496.0111246
b=70.17413909
r²=.9976542757
r=-.9988264492
```

2.4 Summary

In this section, you learned how to graph logarithmic functions. You discovered the inverse relationship between logarithmic and exponential functions and practiced using rules of logarithms to simplify logarithmic expressions and solve logarithmic equations. You also developed logarithmic models from data tables by using logarithmic regression.

2.4 Exercises

In Exercises 1–5, solve the logarithmic equation for y without using a calculator.

1. $y = \log_5(25)$ **2.** $y = \log_3(81)$

3. $y = \log_2(64)$ **4.** $y = \log_2(4^{-1})$

5. $y = \log_3\left(\dfrac{1}{9}\right)$

In Exercises 6–10, determine the concavity and increasing/decreasing behavior of the graph of the function. Then graph the function. (Hint: You will have to use the change-of-base formula if you graph the function on your calculator. That is,

$$y = \log_b(x) \Rightarrow y = \frac{\ln(x)}{\ln(b)}.)$$

6. $y = \log_2(x)$ **7.** $y = \log_4(x)$

8. $y = \log_{0.2}(x)$ **9.** $y = \log_{0.7}x$

10. $y = \log_{0.8}x$

In Exercises 11–20, use the inverse relationship between logarithmic and exponential functions to solve the equations for x. Simplify your answers.

11. $2^x = 64$ **12.** $5^x = 125$

13. $2^x = \dfrac{1}{2}$ **14.** $3^x = \dfrac{1}{9}$

15. $4^x = 64$ **16.** $\log_5(x) = 3$

17. $\log_4(x) = -2$ **18.** $\log_2(x) = 6$

19. $\log_3(x) = 5$ **20.** $\log_4(x) = \dfrac{1}{2}$

In Exercises 21–40, use the rules of logarithms to rewrite each expression as a single logarithmic expression.

21. $\log(2) + \log(8)$ **22.** $\log(2) - \log(8)$

23. $\log(2x) + \log(x^3)$ **24.** $\log(5x) + \log(20)$

25. $\log(j) + \log(a) + \log(m)$

26. $3 \log(x^2) - 4 \log(x^2) + \log(2x^3)$

27. $7 \log(x^{-1}) - 4 \log(3x^2) - \log(4x)$

28. $2 \log(3x)^2 + \log(x)^{-12}$

29. $4 \log\left(\dfrac{1}{x}\right) + 3 \log(x)$

30. $3 \log\left(\dfrac{2}{5x}\right) - \log(x^3)$

31. $\ln(x) - \ln\left(\dfrac{1}{x}\right)$

32. $2 \ln(2x)^2 + \ln\left(\dfrac{1}{x^2}\right)$

33. $\ln(1) - \log(1) + \ln(x^2)$

34. $-2 \ln\left(\dfrac{x}{2}\right) + \ln(2)$

35. $3 \ln(3x^2) - \ln(x^6)$

36. $2 \ln(3x) - \ln(3x^2) + 2 \ln(3)$

37. $-\ln(4x^2)^3 + 2 \ln(x^6) - 6 \ln(x)$

38. $-\ln(3x)^2 - 3 \ln(x^{-2}) - 5 \ln(x)$

39. $4 \ln(3x) + \ln(81x^2) - \ln(9x)$

40. $\ln(3x) - \ln(9x^2) + \ln(3) - \ln(x)$

In Exercises 41–45, use logarithmic regression to model the data in the table. Use the model to predict the value of the function when i = 125, and interpret the real-world meaning of the result.

The Consumer Price Index is used to measure the increase in prices over time. In each of the following tables, the index is assumed to have the value 100 in the year 1984.

  41. **Dental Services**

Price of Dental Services

Price Index (i)	Years Since 1980 (t)
78.9	0
114.2	5
155.8	10
206.8	15
258.5	20

Source: *Statistical Abstract of the United States, 2001,* Table 694, p. 455.

  42. **Price of Wine**

Price of Wine Consumed at Home

Price Index (i)	Years Since 1980 (t)
89.5	0
100.2	5
114.4	10
133.6	15
151.6	20

Source: *Statistical Abstract of the United States, 2001,* Table 694, p. 455.

  43. **Price of a TV**

Price of a Television Set

Price Index (i)	Years Since 1980 (t)
104.6	0
88.7	5
74.6	10
68.1	15
49.9	20

Source: *Statistical Abstract of the United States, 2001,* Table 694, p. 455.

  44. **Price of Alcohol**

Price of Distilled Spirits Consumed at Home

Price Index (i)	Years Since 1980 (t)
89.8	0
105.3	5
125.7	10
145.7	15
162.3	20

Source: *Statistical Abstract of the United States, 2001,* Table 694, p. 455.

 45. **Entertainment Admission Price**

Price of Admission to Entertainment Venues

Price Index (i)	Years Since 1980 (t)
83.8	0
112.8	5
151.2	10
181.5	15
230.5	20

Source: *Statistical Abstract of the United States, 2001,* Table 694, p. 455.

Exercises 46–50 are intended to challenge your understanding of logarithmic and exponential functions.

 46. **Homes with VCRs** Given the following table, perform each of the tasks identified as **(a)** through **(c)**.

Percentage of TV Homes with a VCR

Years Since 1984 (t)	Homes with a VCR (percent) (V)
1	20.8
2	36.0
3	48.7
4	58.0
5	64.6
6	68.6
7	71.9
8	75.0
9	77.1
10	79.0
11	81.0
12	82.2
13	84.2
14	84.6

Source: *Statistical Abstract of the United States, 2001*, Table 1126, p. 705.

(a) Use logarithmic regression to find the logarithmic function that best fits the data.
(b) Solve the equation for the input variable, t.
(c) Determine the year in which 90 percent of TV homes were expected to have VCRs. Do you think this prediction will be accurate? Explain.

47. The domain of the function $y = \log_b(x)$ is the set of all positive real numbers. Explain why the negative real numbers and zero are not in the domain of the function.

48. If $b > c > 1$, for what values of x is $\log_b(x) > \log_c(x)$?

49. Does the equation $e^x = \ln(x)$ have a solution? Explain.

50. In order for the equation $\log_b(x) = 2\log_c(x)$ to be true for all positive values of x, what must be the relationship between b and c?

2.5 Choosing a Mathematical Model

- Use critical thinking skills in selecting a mathematical model

GETTING **STARTED** Many data sets may be effectively fitted with more than one mathematical model. Determining which model to use requires critical thinking. In this section, we will discuss strategies and techniques for choosing a mathematical model. However, before delving into model selection strategies, we will *briefly* introduce one other type of mathematical model: the logistic function. More detailed coverage of logistic functions is given in Chapter 8.

Logistic Functions

A logistic function graph is an *s*-shaped curve that is bounded above and below by horizontal asymptotes (Figures 2.48 and 2.49). It may be either an increasing function or a decreasing function.

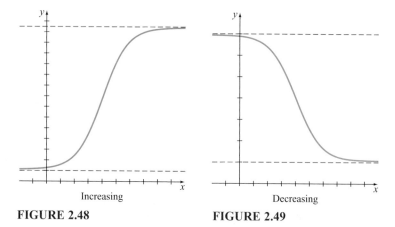

Increasing	Decreasing
FIGURE 2.48	**FIGURE 2.49**

The equation of a logistic function is of the form

$$f(x) = \frac{c}{1 + ae^{-bx}} + k$$

where *a*, *b*, *c*, and *k* are all constants. Logistic functions exhibit slow growth, then rapid growth, followed by slow growth or slow decay, then rapid decay, followed by slow decay.

Selecting a Mathematical Model

Graphing calculators and computer spreadsheets are extremely useful tools for mathematical modeling. When you are given a table of data to model, use the following model selection strategies.

> **HOW TO** **Model Selection Strategies to Use When Given a Table of Data**
>
> 1. Draw a scatter plot.
> 2. Determine whether the scatter plot exhibits the behavior of the graph of one or more of the standard mathematical functions: linear, quadratic, cubic, quartic, exponential, logarithmic, or logistic.
> 3. Find a mathematical model for each function type selected in Step 2.
> 4. Use all available information to anticipate the expected behavior of the thing being modeled outside of the data set. Eliminate models that don't exhibit the expected behavior. (Sometimes it is convenient to switch the order of Steps 3 and 4.)
> 5. Choose the simplest model from among the models that meet your criteria.

In accomplishing Step 2, it is helpful to recognize key graphical features exhibited by the data set. Often we can eliminate one or more model types by noting the concavity of the scatter plot. Table 2.26 identifies key graphical features to look for.

TABLE 2.26

Function Type	Model Equation	Key Graphical Features
Linear	$y = mx + b$	Straight line
Quadratic	$y = ax^2 + bx + c$	Concave up everywhere or concave down everywhere
Cubic	$y = ax^3 + bx^2 + cx + d$	Changes concavity exactly once; no horizontal asymptotes
Quartic	$y = ax^4 + bx^3 + cx^2 + dx + f$	Changes concavity zero or two times; no horizontal asymptotes
Exponential	$y = ab^x$ with $a > 0$	Concave up; horizontal asymptote at $y = 0$
Logarithmic	$y = a + b \ln(x)$ with $b > 0$	Concave down; vertical asymptote at $x = 0$
Logistic	$y = \dfrac{c}{1 + ae^{-bx}} + k$	Changes concavity exactly once; horizontal asymptotes at $y = k$ and $y = c + k$

EXAMPLE 1

Choosing a Mathematical Model to Forecast a Room Rate

As shown in Table 2.27, the average room rate for a hotel/motel increased between 1990 and 1999. Find a mathematical model for the data and forecast the average hotel/motel room rate for 2004.

TABLE 2.27 Average Hotel/Motel Room Rate

Years Since 1990 (t)	Room Rate (dollars) (R)
0	57.96
1	58.08
2	58.91
3	60.53
4	62.86
5	66.65
6	70.93
7	75.31
8	78.62
9	81.33

Source: *Statistical Abstract of the United States, 2001,* Table 1266, p. 774.

SOLUTION We first draw a scatter plot of the data on our graphing calculator (see Figure 2.50).

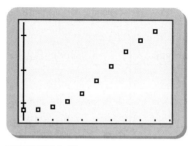

FIGURE 2.50

The graph is initially concave up, but it changes concavity near $t = 6$. Since the graph changes concavity exactly once, a cubic or a logistic model (Figure 2.51) may fit the data.

R as a function of t

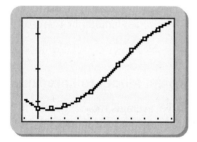

$R(t) = -0.04815t^3 + 0.8868t^2 - 1.495t + 58.32$

R as a function of t

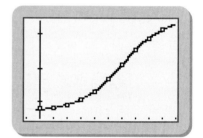

$$R(t) = \frac{27.80}{(1 + 47.23e^{-0.6425t})} + 57$$

FIGURE 2.51

Both the cubic model and the logistic model appear to fit the data extremely well. Which model will be best for forecasting R in 2004 ($t = 14$)? To answer this question, we extend the viewing rectangle for each graph to include the interval $[-1, 15]$ (see Figure 2.52).

R as a function of t

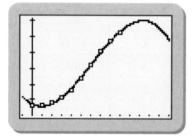

$R(t) = -0.04815t^3 + 0.8868t^2 - 1.495t + 58.32$

R as a function of t

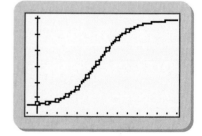

$$R(t) = \frac{27.80}{(1 + 47.23e^{-0.6425t})} + 57$$

FIGURE 2.52

The cubic model forecasts a decrease in the room rate by 2004 ($t = 14$). According to the cubic model, the rate will drop to $79.07. Since room rates have increased every year since 1990, it is unlikely that the room rate will drop at any point in the future.

The logistic model forecasts a leveling off of the room rate. According to the logistic model, the 2004 room rate will be $84.64. This forecast is likely to be more accurate than the cubic model projection.

It is important to note that there is a certain degree of uncertainty when picking a model to use from a group of models that seem to fit the data. It is possible that two different people may select different models as the "best" model. For this reason, it is important to always explain the reasoning behind the selection of a particular model.

EXAMPLE 2

Choosing a Mathematical Model to Forecast a Movie Ticket Price

As shown in Table 2.28, the average price of a movie ticket increased between 1975 and 1999. Find a mathematical model for the data and forecast the average price of a movie ticket in 2004.

TABLE 2.28

Years Since 1975 (t)	Movie Ticket Price (dollars) (P)
0	2.05
5	2.69
10	3.55
15	4.23
20	4.35
24	5.08

Source: *Statistical Abstract of the United States, 2001*, Table 1244, p. 761.

SOLUTION We first draw the scatter plot (Figure 2.53).

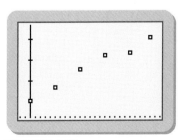

FIGURE 2.53

The first four data points appear to be somewhat linear, so our initial impression is that a linear model may fit the data well. However, the fifth data point is not aligned with the first four, so we know that the data set isn't perfectly linear. However, a linear model may still fit the data fairly well.

Alternatively, we may look at the scatter plot and conclude that the function is concave down on $[0, 15]$ and concave up on $[15, 24]$. Since the scatter plot appears to change concavity once and does not have any horizontal asymptotes, a cubic model may work well. Figure 2.54 shows the two models.

P as a function of t P as a function of t

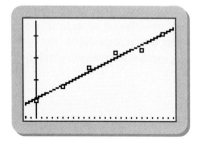 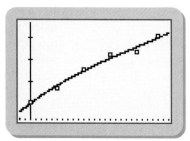

$P(t) = 0.1224t + 2.148$ $P(t) = 0.00007426t^3 - 0.004278t^2 + 0.1846t + 2.001$

FIGURE 2.54

In 2004, $t = 29$. Evaluating each function at $t = 29$ yields the following:

$$P(29) = 0.1224(29) + 2.148$$
$$\approx \$5.70$$

$$P(29) = 0.00007426(29)^3 - 0.004278(29)^2 + 0.1846(29) + 2.001$$
$$\approx \$5.57$$

It is difficult to know which of these estimates of the average price of a movie ticket is more accurate. The models seem to fit the data equally well. Furthermore, we have no other information that leads us to believe that one model would be better than the other. Therefore, we pick the simpler model: $P(t) = 0.1224t + 2.148$.

Certain verbal descriptions often hint at the particular mathematical model to use. By watching for key phrases, we can narrow the model selection process. Table 2.29 presents some examples of typical phrases, their interpretation, and possible models.

TABLE 2.29   Finding a Mathematical Model from a Verbal Description

Phrase	Interpretation	Possible Model
Salaries are projected to increase by $800 per year for the next several years.	The graph of the salary function will have a constant slope, $m = 800$.	Linear $S(t) = 800t + b$
The company's revenue has decreased by 5 percent annually for the past six years.	Since the annual decrease is a *percentage* of the previous year's revenue, the dollar amount by which revenue decreases annually is decreasing. The revenue graph is decreasing and concave up.	Exponential $R(t) = a(1 - 0.05)^t$ $= a(0.95)^t$
Product sales were initially slow when the product was introduced, but sales increased rapidly as the popularity of the product increased. Sales are continuing to increase, but not as quickly as before.	The graph of the sales function may have a horizontal asymptote at or near $y = 0$ and a horizontal asymptote slightly above the maximum projected sales amount.	Logistic
Enrollments have been dropping for years. Each year we lose more students than we did the year before.	The graph of the enrollment function is decreasing, since enrollments are dropping. It will also be concave down, since the rate at which students are dropping continues to increase in magnitude.	Quadratic
Company profits increased rapidly in the early 1990s but leveled off in the late 1990s. In the early 2000s, profits again increased rapidly.	The graph of the profit function increases rapidly, then levels off, then increases rapidly again.	Cubic

EXAMPLE 3

Choosing a Mathematical Model to Forecast Coca-Cola Production

In its 2001 Annual Report, the Coca-Cola Company reported the following:

> Our worldwide unit case volume increased 4 percent in 2001, on top of a 4 percent increase in 2000. The increase in unit case volume reflects consistent performance across certain key operations despite difficult global economic conditions. Our business system sold 17.8 billion unit cases in 2001.
>
> (**Source:** Coca-Cola Company 2001 Annual Report, p. 46.)

Find a mathematical model for the unit case volume of the Coca-Cola Company.

SOLUTION Since the unit case volume is increasing at a constant percentage rate (4 percent), an exponential model will fit the data well. Furthermore, the initial number of unit cases sold was 17.8 billion.

$$V(t) = ab^t \text{ where } t \text{ is the number of years since the end of 2001}$$
$$= 17.8(1 + 0.04)^t \text{ billion unit cases}$$
$$= 17.8(1.04)^t \text{ billion unit cases}$$

The mathematical model for the unit case volume is $V(t) = 17.8(1.04)^t$.

Sometimes a data set may not be effectively modeled by any of the aforementioned functions. In these cases, we look to see if we can model the data with a piecewise function. That is, we use one function to model a portion of the data and a different function to model another portion of the data. Even with this approach, there are some data sets (e.g., daily stock prices) that can rarely be modeled by one of the standard functions.

EXAMPLE **4**

Finding a Piecewise Model for AIDS Deaths in the United States

From 1981 to 1995, the number of adult and adolescent AIDS deaths in the United States increased dramatically. However, from 1995 to 2001, the annual death rate plummeted, as shown in Table 2.30.

TABLE 2.30   Adult and Adolescent AIDS Deaths in the United States

Years Since 1981 (t)	Number of Deaths During Year (D)
0	122
1	453
2	1,481
3	3,474
4	6,877
5	12,016
6	16,194
7	20,922
8	27,680
9	31,436
10	36,708

(Continued)

(Continued)

Years Since 1981 (*t*)	Number of Deaths During Year (*D*)
11	41,424
12	45,187
13	50,071
14	50,876
15	37,646
16	21,630
17	18,028
18	16,648
19	14,433
20	8,963

Source: Centers for Disease Control and Prevention, "HIV/AIDS Surveillance Report," December 2001, p. 30.

Find the mathematical model that best models the data and forecast the number of adult and adolescent AIDS deaths in 2004.

SOLUTION We first draw a scatter plot of the data (Figure 2.55).

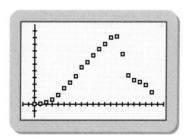

FIGURE 2.55

The data set appears to exhibit logistic behavior up until 1995. After 1995, the graph appears to exhibit cubic behavior. We will use a piecewise function to model the data set. We determine each of the model pieces by using logistic and cubic regression. (Although the data point associated with $t = 14$ was used in finding both pieces of the model, we must assign the domain value $t = 14$ to one piece or the other. We choose to assign $t = 14$ to the logistic piece of the model.)

$$P(t) = \begin{cases} \dfrac{53,955}{1 + 38.834e^{-0.45127t}} & 0 \leq t \leq 14 \\ -381.06t^3 + 20,770t^2 - 379,469t + 2,339,211 & t > 14 \end{cases}$$

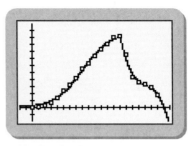

FIGURE 2.56

Our piecewise model (Figure 2.56) appears to fit the data set very well. We are asked to forecast the number of AIDS deaths in 2004 ($t = 23$). Since $23 > 14$, we will use the second function in the piecewise model.

$$P(23) = -381.06(23)^3 + 20{,}770(23)^2 - 379{,}469(23) + 2{,}339{,}211$$
$$= -37{,}603 \text{ adult and adolescent AIDS deaths}$$

It is impossible to have a negative number of deaths! Despite the fact that the model fit the data well, using the model to forecast the 2004 mortality rate yielded an unreasonable result. Returning to the data set, we estimate that the number of AIDS deaths in the years beyond 2001 will be somewhere between 0 and 8963 (the 2001 figure).

If a data set cannot be effectively modeled by one of the standard mathematical functions or a piecewise function, we may conclude that we don't know how to effectively model the data. In such cases, we may estimate a future result by identifying a range of seemingly reasonable values.

EXAMPLE 5

Analyzing Data Not Easily Modeled with a Common Function

The number of firearms detected during airport passenger screening is shown in Table 2.31.

TABLE 2.31

Years Since 1980 (t)	Firearms Detected (F)	Years Since 1980 (t)	Firearms Detected (F)
0	1,914	14	2,994
5	2,913	15	2,390
10	2,549	16	2,155
11	1,644	17	2,067
12	2,608	18	1,515
13	2,798	19	1,552

Source: *Statistical Abstract of the United States, 2001*, Table 1062, p. 669.

Estimate the number of firearms detected by airport screeners in 2002.

SOLUTION

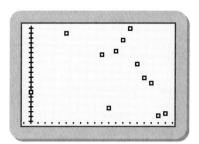

FIGURE 2.57

The scatter plot (Figure 2.57) does not resemble any of the standard mathematical functions. It also does not appear that a piecewise model will fit the data well. The number of firearms detected in a given year appears to be somewhat random, ranging from about 1500 firearms to roughly 3000 firearms.

  After September 11, 2001, airline screening became much more thorough. The increased security may be having a deterrent effect. According to the Bureau of Transportation Statistics (www.bts.gov), there were 1071 firearms detected in 2001. As of March 2004, airline screening data for 2002 had not yet been published on the bureau's web site. We estimate that the number of firearms detected in 2002 will be in the 1000–2000 range. Our estimate is based in part on the additional data we discovered through research.

2.5 Summary

In this section, you learned strategies and techniques for selecting mathematical models. You also discovered that oftentimes more than one function may be used to model the same data set.

2.5 Exercises

In Exercises 1–10, find the equation of the mathematical model (if possible) that you believe will most accurately forecast the indicated result. Justify your conclusions.

1.   **Gaming Hardware Sales**

Electronic Gaming Hardware Factory Sales

Years Since 1990 (t)	Sales (millions of dollars) (S)
0	975
1	1,275
2	1,575
3	1,650
4	1,575
5	1,500
6	1,600
7	1,650
8	1,980
9	2,250

Source: *Statistical Abstract of the United States, 2001*, Table 1005, p. 634.

Forecast gaming hardware sales for 2002.

2.   **Number of Farms**

Years Since 1978 (t)	Farms (thousands) (f)
0	1,015
4	987
9	965
14	946
19	932

Source: *Statistical Abstract of the United States, 2001*, Table 803, p. 523.

Forecast the number of farms in 2000.

3.   **Community College Education Costs**

Maricopa Community College District Tuition and Fees

Years Since 1997–98 (t)	Cost per Credit (dollars) (C)
0	37
1	38
2	40
3	41
4	43
5	46
6	51

Source: www.dist.maricopa.edu.

Forecast the cost per credit for students attending college in the district in 2005–06.

4.   **Breakfast Cereal Consumption**

Per Capita Consumption of Breakfast Cereals

Years Since 1980 (t)	Consumption (pounds) (C)
0	12
1	12
2	11.9
3	12.2
4	12.5
5	12.8
6	13.1
7	13.3
8	14.2
9	14.9
10	15.4
11	16.1

(Continued)

(Continued)

12	16.6
13	17.3
14	17.4
15	17.1
16	16.6
17	16.3
18	15.6
19	15.5

Source: *Statistical Abstract of the United States, 2001*, Table 202, p. 129.

Forecast the per capita consumption of breakfast cereal in 2001.

5. **State University Enrollment**

Washington State Public University Enrollment

Years Since 1990 (t)	Students (S)
0	81,401
1	81,882
2	83,052
3	84,713
4	85,523
5	86,080
6	87,309
7	89,365
8	90,189
9	91,543
10	92,821

Source: Washington State Higher Education Coordinating Board, Higher Education Statistics, September 2001.

Forecast the Washington state public university enrollment in 2010.

6. **Community College Education Costs**

Average Annual Undergraduate Tuition and Fees at Washington State Community Colleges

Years Since 1984-85 (t)	Tuition and Fees (dollars) (F)
0	581
1	699
2	699
3	759
4	780
5	822
6	867
7	945
8	999
9	1,125
10	1,296
11	1,350
12	1,401
13	1,458
14	1,515
15	1,584
16	1,641
17	1,743

Source: Washington State Higher Education Coordinating Board, Higher Education Statistics, September 2001.

Forecast the annual tuition and fees at a Washington state community college in 2005.

7. **AIDS Incidence in Children**

**Estimated Pediatric AIDS Incidence
(United States only)**

Years Since 1992 (t)	Number of Cases (C)
0	954
1	927
2	821
3	687
4	515
5	329
6	235
7	179
8	120
9	101

Source: Centers for Disease Control and Prevention, "HIV/AIDS Surveillance Report," December 2001, p. 36.

Forecast the number of pediatric AIDS cases in the United States in 2005.

8. **Per Capita Personal Income**

Per Capita Personal Income-Florida

Years Since 1993 (t)	Personal Income (dollars) (P)
0	21,320
1	21,905
2	22,942
3	23,909
4	24,869
5	26,161
6	26,593
7	27,764

Source: Bureau of Economic Analysis (www.bea.gov).

Forecast the per capita personal income in Florida in 2004.

9.  **Aviation**

Air Carrier Accidents

Year (t)	Accidents (A)
1992	18
1993	23
1994	23
1995	36
1996	37
1997	49
1998	50
1999	52
2000	54

Source: *Statistical Abstract of the United States, 2001,* Table 1063, p. 669.

Forecast the number of air carrier accidents in 2002.

10. **Federal Funds for Elections**

Federal Funds for Presidential Election Campaigns

Years Since 1980 (t)	Federal Funds (millions of dollars) (F)
0	62.7
4	80.3
8	92.2
12	110.4
16	152.6

Source: *Statistical Abstract of the United States, 2001,* Table 409, p. 255.

Forecast the amount of federal funds spent in the 2004 presidential election.

In Exercises 11–20, find mathematical models for each of the verbal descriptions.

11. **Candy Bar Prices** Candy bars currently cost $0.60 each. The price of a candy bar is expected to increase by 3 percent per year in the future.

12. Housing Prices   On March 12, 2004, a builder priced a new home in Queen Creek, Arizona, at $198.9K. The builder's sales representative told the author that the price for that home style would increase on March 16, March 30, and April 13 to $205.9K, $209.9K, and $212.9K, respectively. (**Source:** Fulton Homes.)

Find a mathematical model for the price of the new home style.

13. Calculator Prices   A calculator currently costs $87. The price of the calculator is expected to increase by $3 per year.

Find a mathematical model for the price of the calculator.

14. Mortality Rates   There are presently 95 members of a high school graduating class who are still living. The number of surviving class members is decreasing at a rate of 4 percent per year.

Find a mathematical model for the number of surviving class members.

15. Product Sales Growth   We are introducing a new product next year. We anticipate that sales will initially be slow but will increase rapidly once people become aware of our product. We anticipate that our monthly sales will start to level off in 18 months at about $200,000. We predict that sales for the first two months will be $12,000 and $19,000, respectively.

Develop a mathematical model to forecast monthly product sales.

16. Club Membership   We are concerned about the decreasing number of members of our business club. Two years ago, we had 200 members. Last year we had 165 members, and this year we have 110 members. If something doesn't change, we expect to lose even more members next year than we lost this year.

Develop a mathematical model for the club membership.

17. Population Growth   The town of Queen Creek, Arizona, was founded in 1989. In 1990, there were 2667 people living in the town. The town grew rapidly in the 1990s, in large part because of new home construction in the area. There were 4316 people living in the town in 2000 and 4940 people in 2001. The

Arizona Department of Commerce estimated the 2002 population of Queen Creek at 5555 people.

If a logistic model is used to model the population of Queen Creek, what is the maximum projected population of the city? (*Hint:* First align the data.)

The Arizona Department of Commerce estimated the 2003 population of Queen Creek at 7480. In light of this additional information, is there a different type of model that would have better predicted the 2003 population of Queen Creek? Explain.

18. Housing Prices   On March 12, 2004, a builder priced a new home in Queen Creek, Arizona, at $143.9K. The builder's sales representative told the author that the price for that home style would increase on March 16, March 30, and April 13 to $148.9K, $151.9K, and $153.9K, respectively. (**Source:** Fulton Homes.)

Find a mathematical model for the price of the new home style.

19. Federal Tax Rates   Federal income tax rates are dependent upon the amount of taxable income received. In 2003, federal income taxes were calculated as follows. For single filers, the first $7000 earned was taxed at 10 percent. The next $21,400 earned was taxed at 15 percent. The next $40,400 earned was taxed at 25 percent.

For example, the tax of a single woman who earned $25,000 would be calculated as follows:

10% tax on the first $7,000

$7,000 \times 0.10 = 700

Amount to be taxed at a higher rate

$25,000 - $7000 = $18,000$

15% tax on the next $18,000

$18,000 \times 0.15 = $2,700$

The person's total tax is

$700 + $2,700 = $3,400$

Find a mathematical model for income tax as a function of taxable income for single filers.

20. Federal Tax Rates   In 2003, federal income taxes for married individuals filing jointly were calculated as follows. The first $14,000 was taxed at 10 percent. The next $42,800 was taxed at 15 percent. The next $57,850 was taxed at 25 percent.

The tax of a married couple who earned $65,000 would be calculated as follows:

10% tax on the first $14,000

$14,000 × 0.10 = $1,400

Amount to be taxed at a higher rate

$65,000 − $14,000 = $51,000

15% tax on the next $42,800

$42,800 × 0.15 = $6420

Amount to be taxed at a higher rate

$51,000 − $42,800 = $8,200

25% tax on next $8,200

$8,200 × 0.25 = $2,050

The couple's total tax is

$1,400 + $6,420 + $2,050 = $9,870

Find a mathematical model for income tax as a function of taxable income for couples filing jointly.

Exercises 21–25 are intended to challenge your understanding of mathematical modeling.

21. The graph of a mathematical model passes through all of the points of a data set. A student claims that the model is a perfect forecaster of future results. How would you respond?

22. A scatter plot is concave up and increasing on [0, 5], concave down and increasing on [5, 8], and decreasing at a constant rate on [8, 15]. Describe two different mathematical models that may fit the data set.

23. Describe how a business owner can benefit from mathematical modeling, despite the imprecision of a model's results.

24. Daily fluctuations in the stock market make the share price of a stock very difficult to model. What approach would you take if you wanted to model the long-term performance of a particular stock?

25. You are asked by your boss to model the data shown in the following scatter plot. How would you respond?

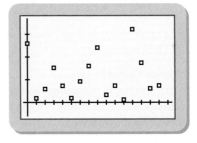

Chapter 2 Review Exercises

Section 2.1   *In Exercises 1–13, find the model that best fits the data. Use the model to answer the given questions.*

1.   **Advertising Expenditures: Magazines**

Years Since 1990 (t)	Advertising Expenditures (millions of dollars) (A)
0	6,803
1	6,524
2	7,000
3	7,357
4	7,916
5	8,580
6	9,010
7	9,821
8	10,518
9	11,433
10	12,348

Source: *Statistical Abstract of the United States, 2001,* Table 1272, p. 777.

According to the model, how much money was spent on magazine advertising in 2002?

2.

Advertising Expenditures: Cable TV

Years Since 1990 (t)	Advertising Expenditures (millions of dollars) (A)
0	2,457
1	2,728
2	3,201
3	3,678
4	4,302
5	5,108
6	6,438
7	7,237
8	8,301
9	10,429
10	12,364

Source: *Statistical Abstract of the United States, 2001,* Table 1272, p. 777.

According to the model, how much money was spent on cable television advertising in 2002?

3.   Using the models from Exercises 1 and 2, determine in what year cable television advertising expenditures are expected to exceed magazine advertising expenditures.

4. **Advertising Expenditures: Radio**

Years Since 1990 (*t*)	Advertising Expenditures (millions of dollars) (*A*)
0	8,726
1	8,476
2	8,654
3	9,457
4	10,529
5	11,338
6	12,269
7	13,491
8	15,073
9	17,215
10	19,585

Source: *Statistical Abstract of the United States, 2001,* Table 1272, p. 777.

According to the model, when will radio advertising exceed $25 billion?

5. **Advertising Expenditures: Yellow Pages**

Years Since 1990 (*t*)	Advertising Expenditures (millions of dollars) (*A*)
0	8,926
1	9,182
2	9,320
3	9,517
4	9,825
5	10,236
6	10,849
7	11,423
8	11,990
9	12,652
10	13,367

Source: *Statistical Abstract of the United States, 2001,* Table 1272, p. 777.

According to the model, when will Yellow Pages advertising exceed $15 billion?

Section 2.2

6. **Federal Credit Unions**

Years Since 1975 (*t*)	Credit Unions (*C*)
0	12,737
5	12,440
10	10,125
15	8,511
20	7,329
25	6,336

Source: *Statistical Abstract of the United States, 2001,* Table 1184, p. 732.

According to the model, how many federal credit unions were there in 1999? Does this seem reasonable? Explain.

7. **Advertising Expenditures: Newspapers**

Years Since 1990 (*t*)	Advertising Expenditures (millions of dollars) (*A*)
0	32,281
1	30,409
2	30,737
3	32,025
4	34,356
5	36,317
6	38,402
7	41,670
8	44,292
9	46,648
10	49,246

Source: *Statistical Abstract of the United States, 2001,* Table 1272, p. 777.

According to the model, how much money was spent on newspaper advertising in 2001?

8. **Full-Service Restaurant Sales**

Years Since 1980 (t)	Sales (millions of dollars) (S)
0	39,307
2	46,443
4	54,815
6	61,474
8	69,356
10	77,811
12	83,561
14	91,457
16	100,830
18	117,774
20	134,461

Source: *Statistical Abstract of the United States, 2001,* Table 1268, p. 775.

According to the model, when will full-service restaurant sales exceed $150 billion?

9. **Advertising Expenditures: Broadcast TV**

Years Since 1990 (t)	Advertising Expenditures (millions of dollars) (A)
0	26,616
1	25,461
2	27,249
3	28,020
4	31,133
5	32,720
6	36,046
7	36,893
8	39,173
9	40,011
10	44,438

Source: *Statistical Abstract of the United States, 2001,* Table 1272, p. 777.

According to the model, when will broadcast television advertising expenditures exceed $50 billion?

10. **Manufacturing Full-Time Employees: Leather and Leather Products**

Years Since 1995 (t)	Employees (thousands) (N)
0	106
1	95
2	89
3	84
4	76

Source: *Statistical Abstract of the United States, 2001,* Table 979, p. 622.

According to the model, in what year will the number of full-time employees in the leather and leather products manufacturing industry drop below 65,000?

11.    **Wages**

Average Hourly Earnings in Manufacturing Industries: Michigan

Years Since 1980 (t)	Average Earnings (dollars per hour) (E)
0	9.52
1	10.53
2	11.18
3	11.62
4	12.18
5	12.64
6	12.80
7	12.97
8	13.31
9	13.51
10	13.86
11	14.52
12	14.81
13	15.36
14	16.13
15	16.31
16	16.67
17	17.18
18	17.61
19	18.38
20	19.20

Source: *Statistical Abstract of the United States, 2001*, Table 978, p. 622.

According to the model, what will the average hourly wage in Michigan manufacturing industries be in 2003?

12.    **Wages**

Average Hourly Earnings in Manufacturing Industries: Florida

Years Since 1980 (t)	Average Earnings (dollars per hour) (E)
0	5.98
1	6.53
2	7.02
3	7.33
4	7.62
5	7.86
6	8.02
7	8.16
8	8.39
9	8.67
10	8.98
11	9.30
12	9.59
13	9.76
14	9.97
15	10.18
16	10.55
17	10.95
18	11.43
19	11.83
20	12.28

Source: *Statistical Abstract of the United States, 2001*, Table 978, p. 622.

According to the model, what will the average hourly wage in Florida manufacturing industries be in 2003?

13. Based on the wage data in Exercises 11–12, do you think it would be better to start up a manufacturing business in Florida or in Michigan? Justify your answer and explain what other issues might affect your decision.

Section 2.3

In Exercises 14–15, determine if the graph of the function is increasing or decreasing and if it is concave up or concave down. Identify the coordinates of the y-intercept. Then graph the function to verify your conclusions.

14. $y = 2(0.75)^x$ **15.** $y = -0.3(2.8)^x$

In Exercises 16–17, find the equation of the exponential function that fits the data in the table algebraically.

16.

x	y
0	3
1	15
2	75
3	375

17.

x	y
2	36
4	144
6	576
8	2,304

In Exercises 18–19, use exponential regression to model the data in the table. Use the model to predict the value of the function when $t = 10$, and interpret the real-world meaning of the result.

  18. **Number of Subway Restaurants**

Years Since 1996 (t)	Restaurants (N)
0	12,516
1	13,066
2	13,600
3	14,162
4	14,662

Source: www.subway.com.

19.   **Number of McDonalds Restaurants**

Years Since 1997 (t)	Restaurants (N)
0	22,928
1	24,513
2	26,309
3	28,707
4	30,093

Source: www.mcdonalds.com.

In Exercises 20–21, find the exponential function that fits the verbal description. Calculate the value of the function five years from now and interpret its real-world meaning.

20. My car is depreciating at a rate of 17 percent per year. It is currently valued at $6000.

21. My monthly household expenses are increasing by 2.3 percent annually. It currently costs $4500 per month to maintain my household.

In Exercises 22–23, find the solution by graphically solving the equation.

22. $1.9(2.6)^x = 10$ **23.** $9.7(0.4)^x = 2$

Section 2.4

In Exercises 24–25, solve the logarithmic equation for y without using a calculator.

24. $y = \log_6(36)$ **25.** $y = \log_5(0.2)$

In Exercises 26–27, determine the concavity and increasing/decreasing behavior of the graph of the function. Then graph the function to verify your results. (Hint: You will have to use the change-of-base formula if you graph the function on your calculator. That is, $y = \log_b(x) \Rightarrow y = \frac{\ln(x)}{\ln(b)}$)

26. $y = \log_5(x)$ **27.** $y = \log_{0.4}(x)$

In Exercises 28–31, use the inverse relationship between logarithmic and exponential functions to solve the equations for x. Simplify your answers.

28. $\left(\frac{1}{2}\right)^x = 64$ **29.** $5^x = 625$

30. $\log_3(x) = 4$ **31.** $\log_5(x) = -2$

In Exercises 32–35, use the rules of logarithms to rewrite each expression as a single logarithm.

32. $\log(5) + \log(3)$ **33.** $2\log(4x) - \log(8)$

34. $3\log(2x)^2 - \log(2x^3)$

35. $-\log(3x)^2 + \log(3x^2)$

In Exercises 36–37, use logarithmic regression to model the data in the table. Use the model to predict the value of the function when t = 10, and interpret the real-world meaning of the result.

  36.   **McDonalds' Systemwide Sales**

Years Since 1997 (t)	Sales (millions of dollars) (S)
1	33,638
2	35,979
3	38,491
4	40,181
5	40,630

Source: www.mcdonalds.com.

 37.   **Non-Alcohol-Related Auto Accident Fatalities**

Years Since 1989 (t)	Fatalities (percentage) (F)
1	50.5
3	54.5
4	56.5
5	59.3
6	58.8
7	59.1
8	61.5
9	61.4
10	61.7

Source: *Statistical Abstract of the United States, 2001,* Table 1099, p. 688.

Does the model estimate at *t* = 10 agree with the raw data value?

Section 2.5   *In Exercises 38–40, find the equation of the mathematical model (if possible) that you believe will most accurately forecast the future behavior of the thing being modeled. Justify your conclusions.*

38.   **Projected Teacher Salaries**

Years Since 1990 (t)	Public School Teacher Average Annual Salary (thousands of dollars) (S)
0	31.4
1	33.1
2	34.1
3	35.0
4	35.7
5	36.7
6	37.7
7	38.5
8	39.5
9	40.6
10	41.7

Source: *Statistical Abstract of the United States, 2001,* Table 237, p. 151.

39. **Cassette Tape Market Share**

Cassette Tape Sales

Years Since 1993 (t)	Percent of Music Market (percentage points) (P)
0	38.0
1	32.1
2	25.1
3	19.3
4	18.2
5	14.8
6	8.0
7	4.9
8	3.4
9	2.4

Source: Recording Industry Association of America.

40. **Music Market Size**

Music Market Size

Years Since 1997 (t)	Dollar Volume (millions) (P)
0	$12,236.80
1	$13,723.50
2	$14,584.50
3	$14,323.00
4	$13,740.89
5	$12,614.21

Source: Recording Industry Association of America.

What to do

1. Visit the Bureau of Labor Statistics web site (www.bls.gov/cpi) and access the most recent Consumer Price Index news report.

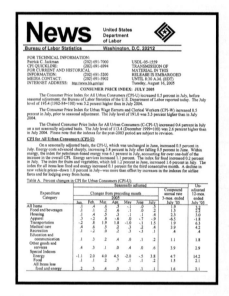

2. Select an expenditure category.

3. Record the annual percentage rate of change in that category. (This is called the compound annual rate in the report.)

4. Find the price of a product that belongs to the category you selected at a local retailer.

5. Using the Consumer Price Index information, estimate the price of the item five years from now.

Consumer Price Index explained

"The Consumer Price Index (CPI) is a measure of the average change in prices over time of goods and services purchased by households. The CPI for All Urban Consumers (CPI-U) and the Chained CPI for All Urban Consumers (C-CPI-U), which cover approximately 87 percent of the total population and include in addition to wage earners and clerical worker households, groups such as professional, managerial, and technical workers, the self-employed, short-term workers, the unemployed, and retirees and others not in the labor force." (**Source:** Bureau of Labor Statistics News, July 2002, USDL-020-480.)

The Derivative

It is impossible to determine how quickly a person is running from a single photograph, since speed is calculated as a change in distance over a change in time. Nevertheless, we may estimate a person's speed at a particular instant in time by determining the distance traveled over a small interval of time (e.g., one second). A runner's speed may be classified as a rate of change in distance. A key component of calculus is the study of rates of change.

3.1 Average Rate of Change

- Calculate the average rate of change of a function over an interval

GETTING **STARTED**   Colleges and universities periodically raise their tuition rates in order to cover rising staffing and facilities costs. As a result, it is often difficult for students to know how much money they should save to cover future tuition costs. By calculating the *average rate of change* in the tuition price over a period of years, we can estimate projected increases in tuition costs. In this section, we will demonstrate how to calculate the average rate of change in the value of a function over a specified interval [a, b]. (The interval notation [a, b] is equivalent to $a \le x \le b$.)

THE DIFFERENCE QUOTIENT: AN AVERAGE RATE OF CHANGE

The average rate of change of a function $y = f(x)$ over an interval $[a, b]$ is

$$\frac{f(b) - f(a)}{b - a}$$

This expression is referred to as the **difference quotient.** For a linear function, the difference quotient gives the slope of the line.

In calculating the difference quotient, we answer the question, "Over the interval $[a, b]$, on average, how much does a one-unit increase in the x value change the y value of the function?"

EXAMPLE   **1**   **Calculating an Average Rate of Change**

The quarterly cost of tuition for full-time resident students at Green River Community College is shown in Table 3.1.

TABLE 3.1

Years (since 1994–1995) (t)	Quarterly Tuition Cost (dollars) [$f(t)$]	Change in Tuition Cost from Prior Year (dollars)
0	432	
1	450	18
2	467	17
3	486	19
4	505	19
5	528	23
6	547	19
7	581	34

Source: Green River Community College.

What is the average rate of change in the quarterly tuition cost from the 1994–1995 academic year to the 2001–2002 academic year? Rounded to the nearest dollar, what do you estimate the 2002–2003 quarterly tuition cost will be?

SOLUTION One way to calculate the average rate of change is to average the annual rates of change over the entire period. To calculate the average, we sum each of the seven rate of change data values and divide by the number of data values.

UNITS

$$\text{Average rate of change } = \frac{(18 + 17 + 19 + 19 + 23 + 19 + 34) \text{ dollars}}{7 \text{ years}}$$

$$= \frac{149}{7} \text{ dollars per year}$$

$$= 21.29 \text{ per year}$$

It is significant to note that the same result may be obtained more rapidly by applying the difference quotient formula. For the period 1994–1995 to 2001–2002, the interval $[a, b] = [0, 7]$. The average rate of change of the tuition is

UNITS

$$\frac{f(b) - f(a)}{b - a} = \frac{f(7) - f(0)}{7 - 0}$$

$$= \frac{581 - 432 \text{ dollars}}{7 \text{ years}}$$

$$= \frac{149 \text{ dollars}}{7 \text{ years}}$$

$$= 21.29 \text{ dollars per year}$$

From 1994–1995 to 2001–2002, the tuition increased by an average of approximately $21 per year. We estimate that the 2002–2003 quarterly tuition will be $21 more than the 2001–2002 tuition, or $602.

When determining the meaning of an average rate of change in a real-life problem, it is essential to find the units of measurement of the result. Fortunately, the units are easily determined. **The units of the rate of change are the units of the output divided by the units of the input.** In Example 1, the units of the output were *dollars* and the units of the input were *years*. Consequently, the units of the average rate of change were *dollars* divided by *years*, or *dollars per year*.

EXAMPLE 2

Calculating an Average Rate of Change

A company's stock price often fluctuates dramatically from day to day. If an investor is preoccupied with daily price changes, she may sell too early or buy too late to get a good return on her investment. One way an investor can gauge the change in a company's stock price is to calculate the average rate of change in the price of the stock over a period of time.

Table 3.2 shows the closing stock price of Apple Computer Corporation over a nine-month period.

TABLE 3.2

Months (since August 2001) (t)	Stock Price at the End of the First Trading Day of the Month (Dollars) [$f(t)$]
August 2001 ($t = 0$)	19.06
September ($t = 1$)	18.25
October ($t = 2$)	15.54
November ($t = 3$)	18.59
December ($t = 4$)	21.05
January 2002 ($t = 5$)	23.30
February ($t = 6$)	24.41
March ($t = 7$)	24.29
April ($t = 8$)	24.46

Source: www.quicken.com.

What was the average rate of change in the stock price between

(a) August 2001 and October 2001?
(b) October 2001 and April 2002?
(c) August 2001 and April 2002?

SOLUTION

(a) In the table, August is month 0 and October is month 2. The average rate of change of the stock over the interval $[0, 2]$ is

 UNITS

$$\frac{f(b) - f(a)}{b - a} = \frac{f(2) - f(0)}{2 - 0}$$

$$= \frac{15.54 - 19.06 \text{ dollars}}{2 \text{ months}}$$

$$= -1.76 \text{ dollars per month}$$

Between August and October 2001, the share price of Apple Computer stock decreased by an average of $1.76 per month.

(b) In the table, October is month 2 and April is month 8. The average rate of change of the stock over the interval $[2, 8]$ is

UNITS

$$\frac{f(b) - f(a)}{b - a} = \frac{f(8) - f(2)}{8 - 2}$$

$$= \frac{24.46 - 15.54 \text{ dollars}}{6 \text{ months}}$$

$$= 1.49 \text{ dollars per month}$$

Between October 2001 and April 2002, the share price of Apple Computer stock increased by an average of $1.49 per month.

c. In the table, August is month 0 and April is month 8. The average rate of change of the stock over the interval $[0, 8]$ is

UNITS

$$\frac{f(8) - f(0)}{8 - 0} = \frac{24.46 - 19.06 \text{ dollars}}{8 \text{ months}}$$

$$= 0.68 \text{ dollar per month}$$

Between October 2001 and April 2002, the share price of Apple Computer stock increased by an average of $0.68 per month.

Some investors choose to buy or sell a stock once it hits a certain limit price. In fact, many investment companies allow investors to activate a limit order for a period of 30 days. If the stock price hits the limit price within the 30-day period, the order is executed. If the stock price fails to hit the limit price during the 30-day period, the order expires with no trading action. Observing the overall trend in a stock's price helps the investor to pick a limit price and predict whether the order will be executed.

EXAMPLE 3   **Calculating an Average Rate of Change**

The population of Washington state may be modeled by the function $P(t) = 0.8981(1.019)^t$, where P is the population in millions of people and t is the number of years since 1900. (**Source:** Modeled from data at www.ofm.wa.gov.) What is the average rate of change in the population from 1990 to 2000?

SOLUTION   We observe that in the model, t is the number of years since 1900. So for the model, $t = 90$ represents 1990 and $t = 100$ represents 2000. The average rate of change in the population over the interval $[90, 100]$ is

UNITS

$$\frac{P(100) - P(90)}{100 - 90} = \frac{5.90 - 4.89 \text{ million people}}{10 \text{ years}}$$

$$= 0.101 \text{ million people per year}$$

Between 1990 and 2000, the population of Washington state increased by an average of 101,000 people per year.

Graphical Interpretation of the Difference Quotient

A line connecting any two points on a graph is referred to as a **secant line.** Graphically speaking, the difference quotient for a function $y = f(x)$ is the slope of the secant line connecting $(a, f(a))$ and $(b, f(b))$ (Figure 3.1).

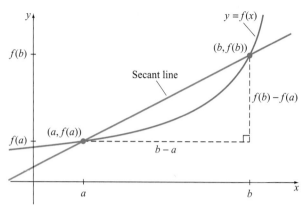

FIGURE 3.1 Secant line slope $= \dfrac{f(b) - f(a)}{b - a}$

EXAMPLE 4

Finding the Slope of a Secant Line

The graph of the function $f(x) = 3x^2 - 6x + 5$ is shown in Figure 3.2. Calculate the slope of the secant line of f that passes through $(1, 2)$ and $(2, 5)$.

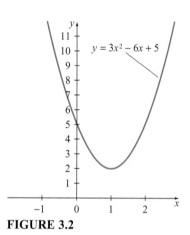

FIGURE 3.2

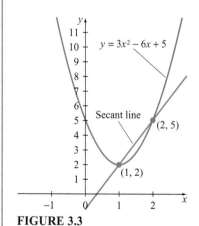

FIGURE 3.3

SOLUTION We first plot the points $(1, 2)$ and $(2, 5)$ and draw the line connecting them (Figure 3.3).

This line is the secant line of the graph between $x = 1$ and $x = 2$. The slope of the secant line is given by

$$m = \frac{f(b) - f(a)}{b - a}$$

$$= \frac{f(2) - f(1)}{2 - 1}$$

$$= \frac{5 - 2}{1}$$

$$= 3$$

The slope of the secant line is 3. That is, on the interval $[1, 2]$, a one-unit increase in x results in a three-unit increase in y, on average.

EXAMPLE 5

Calculating the Slope of a Secant Line

Calculate the slope of the secant line of the function f shown in Figure 3.4.

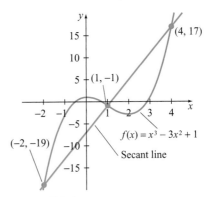

FIGURE 3.4

SOLUTION From the graph, we see that the line intersects the graph of f at $(-2, -19)$, $(1, -1)$, and $(4, 17)$. The slope of the given secant line may be determined by using any pair of these points.

$$m = \frac{f(4) - f(-2)}{4 - (-2)}$$

$$= \frac{17 - (-19)}{6}$$

$$= \frac{36}{6}$$

$$= 6$$

For this function f, the average rate of change in f over the interval $[-2, 1]$ is the same as the average rate of change in f over the intervals $[1, 4]$ and $[-2, 4]$.

3.1 Summary

In this section, you learned how to calculate the average rate of change of a function over a specified interval. You also learned that the units of the average rate of change are the units of the *output* divided by the units of the *input*. You saw that, graphically speaking, the difference quotient is the slope of a secant line. In the next section, we will demonstrate how to use the difference quotient to estimate an *instantaneous* rate of change.

3.1 Exercises

In Exercises 1–10, calculate the average rate of change of the function over the given interval.

1. $f(x) = 2x - 5$ over the interval $[3, 5]$

2. $g(x) = x^2 - 5x$ over the interval $[-2, 4]$

3. $v(x) = -x^3$ over the interval $[-1, 1]$

4. $g(t) = 3^t - \dfrac{4}{t}$ over the interval $[2, 4]$

5. $v(m) = m^2 - m$ over the interval $[-3, 4]$

6. $h(s) = \ln(s)$ over the interval $[4, 7]$

7. $z = \dfrac{\ln(x)}{x}$ over the interval $[1, 5]$

8. $f(t) = \dfrac{2t - 5}{t + 5}$ over the interval $[-2, 2]$

9. $q(x) = \sqrt{x + 2}$ over the interval $[0, 6]$

10. $w(t) = 2t^{-2} + 5$ over the interval $[1, 4]$

In Exercises 11–20, calculate the average rate of change in the designated quantity over the given interval(s).

11. **Air Temperature** Temperature between 11:00 a.m. and 3:00 p.m.

Time of Day	Temperature (°F)
11:00 a.m.	68
1:00 p.m.	73
3:00 p.m.	75

12. **Internet Access** People with Internet access between 1997 and 2000.

Year	Number of People with Internet Access at Home or Work (in thousands)
1997	46,305
1998	62,273
1999	83,677
2000	112,949

Source: www.census.gov.

13. **Dow Jones Industrial Average** Dow Jones Industrial Average between 1996 and 1998 and between 1997 and 1999.

Year	Dow Jones Industrial Average Closing Value at the End of the Year (points)
1996	6,448.30
1997	7,908.30
1998	9,181.40
1999	11,497.10

Source: www.census.gov.

14. **Nasdaq Composite Index** Nasdaq Composite Index closing value between 1996 and 1998 and between 1997 and 1999.

Year	Nasdaq Composite Index Closing Value at the End of the Year (points)
1996	1,291.0
1997	1,570.4
1998	2,192.7
1999	4,069.3

Source: www.census.gov.

15. **Reading Scores** A third-grade student's reading score between the first and third quarter.

Quarter	Reading Score (words per minute)
1	69
2	107
3	129

Source: Author's data.

16. 🌐 **Owner-Occupied Homes** The percentage of U.S. housing units that are owner-occupied when the number of housing units increased from 94.2 million (in 1990) to 107.0 million (in 2001).

Number of Housing Units (millions)	Percent of Housing Units That Are Owner-Occupied
94.2	63.9
100.0	64.7
104.9	66.8
105.7	67.4
107.0	67.8

Source: www.census.gov.

17. 🌐 **Newspaper Subscriptions** Daily newspaper subscriptions as the number of cable TV subscribers increased from 50.5 million (in 1990) to 67.7 million (in 2000).

Cable TV Subscribers (millions)	Daily Newspaper Circulation (millions)
50.5	62.3
60.9	58.2
66.7	56.0
67.7	55.8

Source: www.census.gov.

18. 🌐 **Personal Income** Disposable personal income between 1990 and 2000.

Year	Disposable Personal Income (dollars per person)
1990	17,176
1995	20,358
1999	23,708
2000	24,889

Source: www.census.gov.

19. 🌐 **Poverty Level** Percent of people below poverty level when the unemployment rate decreased from 5.6 percent (in 1990) to 4.0 percent (in 2000).

Unemployment Rate (percent)	People Below Poverty Level (percent)
5.6	13.5
4.2	11.8
4.0	11.3

Source: www.census.gov.

20. 🌐 **Net Farm Income** Net farm income when land in farms decreased from 987 million acres (in 1990) to 943 million acres (in 2000).

Land in Farms (millions of acres)	Farm Income (billions of dollars)
987	44.6
943	46.4

Source: www.census.gov.

In Exercises 21–25, graph each function. Then use the difference quotient, $\dfrac{f(b) - f(a)}{b - a}$, to calculate the slope of the secant line through the points $(1, f(1))$ and $(3, f(3))$.

21. $f(x) = 2^x$ **22.** $f(x) = 5 - x^2$

23. $f(x) = 5$ **24.** $f(x) = 3x + 1$

25. $f(x) = (x - 2)^2$

In Exercises 26–30, use the difference quotient,
$\dfrac{f(b) - f(a)}{b - a}$, *to calculate the slope of the secant line through the points* $(1, f(1))$ *and* $(3, f(3))$ *for the given graph of f.*

26.
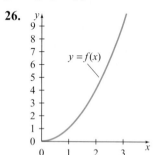

27.
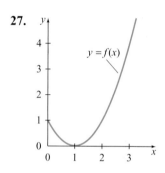

28.
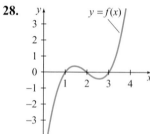

29.
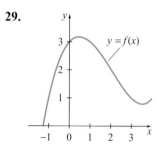

30.
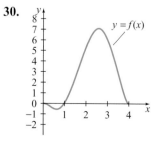

In Exercises 31–40, calculate the average rate of change in the state populations (in people per month) between April 1, 2000, and July 1, 2001.

31. **Population**

Montana	
Date	Population (in thousands)
April 2000	902.2
July 2001	904.4

Source: www.census.gov.

32. **Population**

North Dakota	
Date	Population (in thousands)
April 2000	642.2
July 2001	634.4

Source: www.census.gov.

33. **Population**

Massachusetts	
Date	Population (in thousands)
April 2000	6,349.1
July 2001	6,379.3

Source: www.census.gov.

34. **Population**

Utah	
Date	Population (in thousands)
April 2000	2,233.2
July 2001	2,269.8

Source: www.census.gov.

35. **Population**

Missouri	
Date	Population (in thousands)
April 2000	5,595.2
July 2001	5,629.7

Source: www.census.gov.

36. **Population**

Louisiana	
Date	Population (in thousands)
April 2000	4,469.0
July 2001	4,465.4

Source: www.census.gov.

37. **Population**

West Virginia	
Date	Population (in thousands)
April 2000	1,808.3
July 2001	1,801.9

Source: www.census.gov.

38. **Population**

Wyoming	
Date	Population (in thousands)
April 2000	493.8
July 2001	494.4

Source: www.census.gov.

39. **Population**

Pennsylvania	
Date	Population (in thousands)
April 2000	12,281.1
July 2001	12,287.2

Source: www.census.gov.

40. **Population**

Arizona	
Date	Population (in thousands)
April 2000	5,130.6
July 2001	5,307.3

Source: www.census.gov.

Exercises 41–45 are intended to challenge your understanding of average rates of change.

41. Let $f(x) = x^2 - 4x$. Compare and contrast the average rate of change of $f(x)$ and of $2(f(x))$ on the interval $[1, 4]$.

42. Given $f(x) = x^2 - 4$, determine the average rate of change of f over the interval $[a, a + h]$.

43. The slope of the secant line passing through $(2, f(2))$ and $(2.01, f(2.01))$ is 4. Determine the vertical distance between $(2, f(2))$ and $(2.01, f(2.01))$.

44. The slope of a secant line of the function $f(x) = x^2$ passes through the points $(0, 0)$ and (h, h^2). Determine the slope of the secant line when $h = 1$, $h = 0.1$, $h = 0.01$, and $h = 0.001$. As h gets smaller, what value does the slope of the secant line appear to be approaching?

45. A secant line of the function $f(x) = x^{1/3}$ passes through the points $(0, 0)$ and $(h, h^{1/3})$. Determine the slope of the secant line when $h = 1$, $h = 0.1$, $h = 0.01$, and $h = 0.001$. As h gets smaller, what value does the slope of the secant line appear to be approaching?

3.2 Limits and Instantaneous Rates of Change

- Estimate the instantaneous rate of change of a function at a point
- Use derivative notation and terminology to describe instantaneous rates of change

In the July 1988 Olympic trials, Florence Griffith Joyner shattered the world record for the 100-meter dash with a time of 10.49 seconds. How fast was she running (in meters per second) when the picture at the left was taken?

From a single photo, it is impossible for us to determine her speed. However, if we knew how long it took her to reach various checkpoints during the race, we could approximate her speed at the finish line. In this section, we will demonstrate how to use the difference quotient to estimate and the derivative to calculate an *instantaneous rate of change*.

While an average rate of change is calculated over an interval $[a, b]$, an instantaneous rate of change is calculated at a single value a. For example, the average highway speed of a person over a 200-mile trip may have been 59 miles per hour. However, when he passed a state patrol car exactly 124 miles into the trip, he was speeding at 84 miles per hour. His *average* speed on the interval $[0, 200]$ was 59 miles per hour; however, his *instantaneous* speed at $d = 124$ was 84 miles per hour.

EXAMPLE **1**

Estimating an Instantaneous Rate of Change

Suppose that a runner in the 100-meter dash recorded the times shown in Table 3.3.

TABLE 3.3

Time (seconds) (t)	Total Distance Traveled (meters) $[D(t)]$
0	0
5.24	50
9.46	90
9.97	95
10.39	99
10.49	100

Estimate her speed when she crossed the finish line.

SOLUTION Her average speed over various distances may be calculated using the difference quotient, $\dfrac{D(b) - D(a)}{b - a}$.

Her average speed over the 100-meter distance was

UNITS

$$\text{Average speed} = \frac{100 - 0 \text{ meters}}{10.49 - 0 \text{ seconds}}$$

$$= 9.53 \text{ meters per second}$$

Her average speed over the last 50 meters was

UNITS

$$\text{Average speed} = \frac{100 - 50 \text{ meters}}{10.49 - 5.24 \text{ seconds}}$$

$$= \frac{50 \text{ meters}}{5.25 \text{ seconds}}$$

$$= 9.52 \text{ meters per second}$$

Her average speed over the last 10 meters was

UNITS

$$\text{Average speed} = \frac{100 - 90 \text{ meters}}{10.49 - 9.46 \text{ seconds}}$$

$$= \frac{10 \text{ meters}}{1.03 \text{ seconds}}$$

$$= 9.71 \text{ meters per second}$$

Her average speed over the last 5 meters was

UNITS

$$\text{Average speed} = \frac{100 - 95 \text{ meters}}{10.49 - 9.97 \text{ seconds}}$$

$$= \frac{5 \text{ meters}}{0.52 \text{ second}}$$

$$= 9.62 \text{ meters per second}$$

And her average speed over the last meter was

UNITS

$$\text{Average speed} = \frac{100 - 99 \text{ meters}}{10.49 - 10.39 \text{ seconds}}$$

$$= \frac{1 \text{ meter}}{0.10 \text{ second}}$$

$$= 10.0 \text{ meters per second}$$

Although each calculation yielded a different result, all of these difference quotients estimate the runner's finish-line speed. Which of the estimates do you think is most accurate?

The last calculation best estimates her finish-line speed because it measures the change in distance over the smallest interval of time: 0.1 second. (Reducing the time interval to an even smaller amount of time, say 0.01 second, would further improve the estimate.) We estimate that the runner's speed when she crossed the finish line was 10.0 meters per second.

To estimate the *instantaneous rate of change* of a function $y = f(x)$ at a point $(a, f(a))$, we calculate the average rate of change of the function over a very small interval $[a, b]$. If we let the variable h represent the distance between

$x = a$ and $x = b$, then $b = a + h$. Consequently, the difference quotient may be rewritten as

$$\frac{f(b) - f(a)}{b - a} = \frac{f(a + h) - f(a)}{(a + h) - a}$$

$$= \frac{f(a + h) - f(a)}{h}$$

THE DIFFERENCE QUOTIENT AS AN ESTIMATE OF AN INSTANTANEOUS RATE OF CHANGE

The *instantaneous rate of change* of a function $y = f(x)$ at a point $(a, f(a))$ may be *estimated* by calculating the **difference quotient of f at a**,

$$\frac{f(a + h) - f(a)}{h}$$

using an h arbitrarily close to 0. (If $h = 0$, the difference quotient is undefined.)

EXAMPLE 2

Estimating an Instantaneous Rate of Change

The average income of an elementary or secondary teacher may be modeled by

$$S(t) = 0.00713t^3 - 0.111t^2 + 1.42t + 31.6, \quad 0 \le t \le 10$$

where $S(t)$ is the annual salary in thousands of dollars and t is the number of years since 1990. (**Source:** Modeled from *Statistical Abstract of the United States, 2001*, Table 237, p. 151.) According to the model, how quickly was the average income of a teacher increasing in 1999?

SOLUTION In 1999, $t = 9$. Using the difference quotient

$$\frac{S(9 + h) - S(9)}{h}$$

and selecting increasingly small values of h, we generate Table 3.4.

TABLE 3.4

h	$\dfrac{S(9 + h) - S(9)}{h}$
1.000	1.2432
0.100	1.1628
0.010	1.1554
0.001	1.1547

We conclude that in 1999 the average annual salary of a teacher was increasing by about $1.155 thousand ($1155) per year.

The following Technology Tip details how to generate a table of values for the difference quotient using the TI-83 Plus.

TECHNOLOGY **TIP**

Calculating the Difference Quotient for Different Values of *h*

1. Enter the function $S(t)$ as Y1 by pressing the Y= button and typing the equation.

2. Press the MATH button, scroll to 0:Solver..., and press ENTER. (*Hint:* It is quicker to scroll up than to scroll down.)

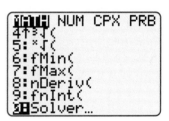

3. If the Solver already contains an equation, you may see something similar to the graphic shown. Press the blue up arrow and then press CLEAR to delete the equation.

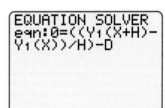

4. Notice that the equation is set equal to zero. We may rewrite the difference quotient $D = \dfrac{Y_1(x + h) - Y_1(x)}{h}$ as

   $0 = \dfrac{Y_1(x + h) - Y_1(x)}{h} - D$. Type in the second equation using the VARS; Y-VARS; 1:Function menu sequence to enter the function Y1. To enter the variable *H*, press the ALPHA ^ key sequence. To enter the variable *D*, press the ALPHA X^{-1} key sequence. Press ENTER.

   ```
   EQUATION SOLVER
   eqn:0=((Y1(X+H)-
   Y1(X))/H)-D
   ```

5. Enter the value of *x* where you want to evaluate the difference quotient and enter a small positive value of *H*. Move your cursor to the *D* variable and press the ALPHA ENTER key sequence to solve for the difference quotient *D*. Repeat for various values of *H*.

   ```
   ((Y1(X+H)-Y1(...=0
    X=9
    H=.001
   ■D=1.154671517
    bound={-1E99,1...
   ■left-rt=0
   ```

EXAMPLE 3 **Using Technology to Estimate an Instantaneous Rate of Change**

Based on data from 1990–2000, the cost of 1/2 gallon of prepackaged ice cream may be modeled by

$$C(t) = 0.0142t^2 - 0.0272t + 2.53 \text{ dollars}$$

where t is the number of years since 1990. (**Source:** Modeled from *Statistical Abstract of the United States, 2001*, Table 706, p. 468.) Use the Solver on your calculator to estimate the instantaneous rate of change of the cost in 1999.

SOLUTION We begin by entering $C(t)$ into the $\boxed{Y=}$ editor.

```
Plot1 Plot2 Plot3
\Y1◼0.0142X^2-0.
0272X+2.53
\Y2=
\Y3=
\Y4=
\Y5=
\Y6=
```

In 1999, $t = 9$. Entering $x = 9$ and $H = 0.001$ and solving for D yields

```
(Y1(X+H)-Y1(X...=0
 X=9
 H=.001
◼D=.2284142
 bound={-1E99,1...
◼left-rt=0
```

Repeating this process for various values of H, we create Table 3.5.

TABLE 3.5

h	$\dfrac{C(9+h) - C(9)}{h}$
0.001	0.22841420
0.0001	0.22840142
0.00001	0.22840014
0.000001	0.22840000

In 1999, the cost of 1/2 gallon of prepackaged ice cream was increasing by about $0.23 per year. That is, from 1999 to 2000, the price was expected to increase by about 23 cents.

Limits

In Examples 1 through 3, we estimated the instantaneous rate of change at a point by calculating the average rate of change over a "short" interval by picking "small" values of h. The terms *short* and *small* are vague. Numerically, what does "small" mean? Mathematicians struggled with this dilemma for years before developing the concept of the **limit.** We will explore the limit concept graphically before giving a formal definition.

Consider the graph of the function $f(x) = -x^2 + 4$ over the interval $[-3, 3]$ (Figure 3.5).

We ask the question, "As x gets close to 2, to what value does $f(x)$ get close?"

Observe from the graph of f that as the value of x moves from 0 to 2, the value of $f(x)$ moves from 4 to 0. We represent this behavior symbolically with the notation

$$\lim_{x \to 2^-} f(x) = 0$$

which is read, "the limit of $f(x)$ as x approaches 2 *from the left* is 0." This is commonly referred to as a **left-hand limit** because we approach $x = 2$ through values to the *left* of 2.

Observe from the graph of f that as the value of x moves from 3 to 2, the value of $f(x)$ moves from -5 to 0. We write

$$\lim_{x \to 2^+} f(x) = 0$$

which is read, "the limit of $f(x)$ as x approaches 2 *from the right* is 0." This is commonly referred to as a **right-hand limit** because we approach $x = 2$ through values to the *right* of 2.

On a number line, $-\infty$ lies to the left of $+\infty$. For the left-hand limit, $\lim\limits_{x \to a^-} f(x)$, the minus sign is used to indicate that we are approaching $x = a$ from the direction of $-\infty$. For the right-hand limit, $\lim\limits_{x \to a^+} f(x)$, the plus sign is used to indicate that we are approaching $x = a$ from the direction of $+\infty$.

The left- and right-hand limit behavior can also be seen from a table of values for $f(x)$ (Table 3.6).

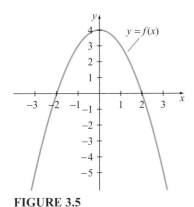

FIGURE 3.5

TABLE 3.6

	x	$f(x)$	
	0.00	4.000	
$\downarrow$	1.00	3.000	$f(x)$ gets close to 0
Left of $x = 2$	1.90	0.390	as x nears 2
$\downarrow$	1.99	0.040	
	2.00	**0.000**	
$\uparrow$	2.01	-0.040	
Right of $x = 2$	2.10	-0.410	$f(x)$ gets close to 0
$\uparrow$	3.00	-5.000	as x nears 2

When the left- and right-hand limits of $f(x)$ approach the same finite value, we say that "the limit exists." In this case, the left- and right-hand limits of $f(x)$ neared the same value ($y = 0$) as x approached 2. We say that "the limit of $f(x)$ as x approaches 2 is 0" and write

$$\lim_{x \to 2} f(x) = 0$$

Sometimes the left- and right-hand limits of a function at a point are not equal. Consider the graph of the piecewise function

$$f(x) = \begin{cases} -x^2 + 2 & x \le 1 \\ x + 1 & x > 1 \end{cases}$$

on the interval $[-3, 3]$ (see Figure 3.6).

We ask the question, "As x gets close to 1, to what value does $f(x)$ get close?"

Observe from the graph that as the value of x moves from 0 to 1, the value of $f(x)$ moves from 2 to 1. We write

$$\lim_{x \to 1^-} f(x) = 1$$

That is, the left-hand limit of $f(x)$ as x approaches 1 is 1. Similarly, as the value of x moves from 2 to 1, the value of $f(x)$ moves from 3 to 2. We write

$$\lim_{x \to 1^+} f(x) = 2$$

That is, the right-hand limit of $f(x)$ as x approaches 1 is 2. Since for this function, the left- and right-hand limits are not equal, we say that "the limit of $f(x)$ as x approaches 1 does not exist" or, simply, "the limit does not exist." This can also be seen from a table of values for $f(x)$ (Table 3.7).

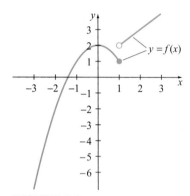

FIGURE 3.6

TABLE 3.7

		x	$f(x)$	
	↓	0.00	2.000	
Left of $x = 1$		0.90	1.190	$f(x)$ gets close to 1 as x nears 1
	↓	0.99	1.020	
		1.00	1.000	
	↑	1.01	2.010	
Right of $x = 1$		1.10	2.100	$f(x)$ gets close to 2 as x nears 1
	↑	2.00	3.000	
		3.00	4.000	

EXAMPLE 4   **Finding Left-hand and Right-hand Limits**

Given Table 3.8, determine $\lim_{x \to 0^-} f(x)$ and $\lim_{x \to 0^+} f(x)$.

TABLE 3.8

x	$f(x)$
-1.00	0.84147
-0.10	0.99833
-0.01	0.99998
0.00	Undefined
0.01	0.99998
0.10	0.99833
1.00	0.84147

SOLUTION We observe that as x moves from -1 to -0.01, the value of $f(x)$ nears 1. That is,

$$\lim_{x \to 0^-} f(x) = 1$$

Similarly, as x moves from 1 to 0.01, the value of $f(x)$ nears 1. That is,

$$\lim_{x \to 0^+} f(x) = 1$$

The limit of $f(x)$ as x approaches 0 is 1, since the left- and right-hand limits of $f(x)$ are both equal to 1. That is,

$$\lim_{x \to 0} f(x) = 1$$

It is important to note that even though $f(x)$ is undefined at $x = 0$, as x gets close to 0, $f(x)$ gets close to 1.

As illustrated in Example 4, one of the most powerful features of limits is that the limit of a function may exist at a point where the function itself is undefined. This feature will be used often when we introduce the limit definition of the derivative later in this section.

THE LIMIT OF A FUNCTION

If $f(x)$ is defined for all values of x near c, then

$$\lim_{x \to c} f(x) = L$$

means that as x approaches c, $f(x)$ approaches L.

We say that the limit exists if

1. L is a finite number and
2. Approaching c from the left or right yields the same value of L.

The theory surrounding limits is rich and worthy of study; however, in this text, we are most interested in the limit of the difference quotient as h approaches zero. That is,

$$\lim_{h \to 0} \frac{f(a + h) - f(a)}{h}$$

Recall that the difference quotient represents the average rate of change of $f(x)$ over the interval $[a, a + h]$. When we place the limit on the difference quotient, we are symbolically asking, "As the distance between the two x values (a and $a + h$) gets smaller, what happens to the average rate of change of $f(x)$ on the interval $[a, a + h]$?" The limit of the difference quotient as h approaches zero (if the limit exists) is the **instantaneous rate of change in $f(x)$ at $x = a$**. Note that even though the difference quotient is undefined when $h = 0$, the limit may still exist.

EXAMPLE 5

Calculating an Instantaneous Rate of Change

Let $f(x) = x^2$. What is the instantaneous rate of change of $f(x)$ when $x = 3$?

SOLUTION We can calculate the instantaneous rate of change by taking the limit of the difference quotient as h approaches zero. The instantaneous rate of change of $f(x)$ at $x = 3$ is given by

$$\lim_{h \to 0} \frac{f(3 + h) - f(3)}{h} = \lim_{h \to 0} \frac{(3 + h)^2 - (3)^2}{h} \quad \text{Since } f(x) = x^2$$

$$= \lim_{h \to 0} \frac{(9 + 6h + h^2) - 9}{h}$$

$$= \lim_{h \to 0} \frac{6h + h^2}{h}$$

$$= \lim_{h \to 0} \frac{h(6 + h)}{h}$$

$$= \lim_{h \to 0} (6 + h) \text{ for } h \neq 0 \quad \text{Since } \frac{h}{h} = 1 \text{ for } h \neq 0$$

As h nears 0, what happens to the value of $6 + h$? Let's pick values of h near 0 (Table 3.9).

TABLE 3.9

h	$6 + h, h \neq 0$
-0.100	5.900
-0.010	5.990
-0.001	5.999
0.000	Undefined
0.001	6.001
0.010	6.010
0.100	6.100

As seen from the table, even though the difference quotient is undefined when $h = 0$, the value of the simplified difference quotient, $6 + h$, gets close to 6 as h approaches 0. In fact, by picking sufficiently small values of h, we can get as close to 6 as we would like. So the instantaneous rate of change of $f(x)$ when $x = 3$ is 6.

Observe that we can attain the same result by plugging in $h = 0$ after canceling out the h in the denominator of the difference quotient. That is,

$$\lim_{h \to 0}(6 + h) = 6 + 0$$
$$= 6$$

Throughout the rest of this chapter, we will substitute in $h = 0$ after eliminating the h in the denominator of the difference quotient. This process will simplify our computations while still giving the correct result.

The limit of the difference quotient as h approaches zero is used widely throughout calculus and is called **the derivative.**

THE DERIVATIVE OF A FUNCTION AT A POINT

The **derivative** of a function $y = f(x)$ at a point $(a, f(a))$ is

$$f'(a) = \lim_{h \to 0} \frac{f(a + h) - f(a)}{h}$$

$f'(a)$ is read "f prime of a" and is the *instantaneous rate of change* of the function f at the point $(a, f(a))$.

EXAMPLE 6

Calculating the Derivative of a Function at a Point

Given $f(x) = 3x + 1$, find $f'(2)$.

SOLUTION

$$f'(2) = \lim_{h \to 0} \frac{f(2 + h) - f(2)}{h}$$
$$= \lim_{h \to 0} \frac{[3(2 + h) + 1] - [3(2) + 1]}{h} \quad \text{Since } f(2 + h) = [3(2 + h) + 1] \text{ and } f(2) = 3(2) + 1$$
$$= \lim_{h \to 0} \frac{(6 + 3h + 1) - (7)}{h}$$
$$= \lim_{h \to 0} \frac{(3h + 7) - (7)}{h}$$
$$= \lim_{h \to 0} \frac{3h}{h}$$
$$= \lim_{h \to 0} 3 \quad \text{Since } \frac{h}{h} = 1$$
$$= 3$$

In this case, the difference quotient turned out to be a constant value of 3, so taking the limit of the difference quotient as h approached 0 did not alter the value of the difference quotient.

For linear functions, the slope of the line is the instantaneous rate of change of the function at any value of x. Consequently, the derivative of a linear function will always be a constant value that is equal to the slope of the line.

Recall that in Example 1, we estimated the runner's finish-line speed by finding the average rate of change in distance over a small interval of time. In other words, the distance traveled divided by the amount of time used to travel the distance yielded the average speed. One of the drawbacks of using the term *speed* is that it does not indicate direction. For example, consider two race cars traveling at 75 mph on the same road (Figure 3.7).

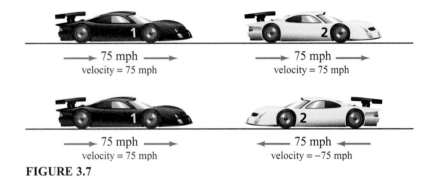

FIGURE 3.7

If the cars are traveling in the same direction, we don't foresee any problems. However, if the cars are traveling in opposite directions, a head-on collision may be imminent. The direction of the cars is a key piece of information that is not captured by the term *speed*. Fortunately, the term *velocity* combines the concepts of speed and direction. A negative sign placed on the speed is used to distinguish a difference in direction. In Figure 3.7, a positive velocity indicates that the car is traveling from left to right. A negative velocity indicates that the car is traveling from right to left.

When talking about the velocity of objects that are rising or falling, it is customary to use a positive velocity to indicate that the object is rising and a negative velocity to indicate that the object is falling.

It is common to model the vertical position of a free-falling object on earth with the equation $s(t) = -16t^2 + v_0t + s_0$ feet, where v_0 is the velocity of the object (in feet per second) and s_0 is the vertical position of the object (in feet) at time $t = 0$ seconds. (Although this model neglects air resistance, it does give us an idea of what is happening physically.) To determine the velocity of the object at time t seconds, we simply need to find $s'(t)$.

EXAMPLE 7

Calculating the Velocity of a Falling Object

A water balloon is dropped from the top of a 20-foot roof. How fast is it falling 1 second into its flight? How fast is it falling when it hits the ground?

SOLUTION The *initial velocity* of the balloon, v_0, is 0 feet per second because the balloon was dropped, not thrown. The *initial position* of the balloon, s_0, is 20 feet, since the balloon is being dropped from 20 feet above the ground. Therefore, the balloon's position may be modeled by

$$s(t) = -16t^2 + v_0 t + s_0$$
$$= -16t^2 + 0t + 20$$
$$= -16t^2 + 20$$

To calculate the velocity of the balloon at time $t = 1$ second, we must find $s'(1)$.

$$s'(1) = \lim_{h \to 0} \frac{s(1 + h) - s(1)}{h}$$

$$= \lim_{h \to 0} \frac{[-16(1 + h)^2 + 20] - [-16(1)^2 + 20]}{h} \qquad \text{Since } s(1 + h) = -16(1 + h)^2 + 20$$

$$= \lim_{h \to 0} \frac{[-16(1 + 2h + h^2) + 20] - [-16 + 20]}{h}$$

$$= \lim_{h \to 0} \frac{(-16 - 32h - 16h^2 + 20) - (4)}{h}$$

$$= \lim_{h \to 0} \frac{(-32h - 16h^2 + 4) - (4)}{h}$$

$$= \lim_{h \to 0} \frac{-32h - 16h^2}{h}$$

$$= \lim_{h \to 0} \frac{h(-32 - 16h)}{h} \qquad \text{Factor } h \text{ out of the numerator}$$

$$= \lim_{h \to 0} (-32 - 16h) \qquad \text{Since } \frac{h}{h} = 1 \text{ for } h \neq 0$$

$$= -32 - 16(0) \qquad \text{Plug in } h = 0$$

UNITS

$$= -32 \text{ feet per second}$$

So 1 second after the balloon is dropped, it is traveling toward the ground at a speed of 32 feet per second. (As previously noted, a positive velocity typically indicates that an object is moving upward, while a negative velocity signifies that an object is moving downward.)

To determine how fast the balloon is falling when it hits the ground, we must first determine when it will hit the ground. Since $s(t)$ is the height of the balloon above the ground, we must determine when $s(t) = 0$.

$$s(t) = -16t^2 + 20$$
$$0 = -16t^2 + 20$$
$$16t^2 = 20$$
$$t^2 = \frac{20}{16}$$
$$t^2 = 1.25$$
$$t \approx 1.1 \text{ seconds}$$

We estimate that the balloon will hit the ground in 1.1 seconds. To determine the velocity of the balloon at that time, we calculate $s'(1.1)$.

$$s'(1.1) = \lim_{h \to 0} \frac{s(1.1 + h) - s(1.1)}{h}$$

$$= \lim_{h \to 0} \frac{[-16(1.1 + h)^2 + 20] - [-16(1.1)^2 + 20]}{h}$$

$$= \lim_{h \to 0} \frac{[-16(1.21 + 2.2h + h^2) + 20] - [-16(1.21) + 20]}{h}$$

$$= \lim_{h \to 0} \frac{(-19.36 - 35.2h - 16h^2 + 20) - (-19.36 + 20)}{h}$$

$$= \lim_{h \to 0} \frac{(-35.2h - 16h^2 + 0.64) - (0.64)}{h}$$

$$= \lim_{h \to 0} \frac{-35.2h - 16h^2}{h}$$

$$= \lim_{h \to 0} \frac{h(-35.2 - 16h)}{h} \qquad \text{Factor } h \text{ out of the numerator}$$

$$= \lim_{h \to 0} (-35.2 - 16h) \qquad \text{Since } \frac{h}{h} = 1 \text{ for } h \neq 0$$

$$= -35.2 - 16(0) \qquad \text{Plug } in \ h = 0$$

$$= -35.2 \text{ feet per second}$$

UNITS

We estimate that the balloon will be falling at a rate of a little over 35 feet per second when it strikes the ground.

EXAMPLE 8   Finding and Interpreting the Meaning of the Derivative of a Function at a Point

The population of Washington state may be modeled by the function

$$P(t) = 0.8981(1.019)^t \text{ million people}$$

where t is the number of years since 1900. (**Source:** Modeled from www.ofm.wa.gov data.) Find and interpret the meaning of $P'(104)$.

SOLUTION   Since t is the number of years since 1900, $t = 104$ is the year in 2004. $P'(104)$ is the instantaneous rate of change in the population in 2004, given in millions of people per year.

$$P'(104) = \lim_{h \to 0} \frac{P(104 + h) - P(104)}{h}$$

$$= \lim_{h \to 0} \frac{[0.8981(1.019)^{104 + h}] - [0.8981(1.019)^{104}]}{h}$$

$$= \lim_{h \to 0} \frac{0.8981[(1.019)^{104 + h} - (1.019)^{104}]}{h} \qquad \text{Factor out } 0.8981$$

$$= \lim_{h \to 0} \frac{0.8981[(1.019)^{104}(1.019)^h - (1.019)^{104}]}{h} \qquad \text{Since } (1.019)^{104 + h} = (1.019)^{104}(1.019)^h$$

$$= \lim_{h \to 0} \frac{(0.8981)(1.019)^{104}\left[(1.019)^h - 1\right]}{h} \qquad \text{Factor out } (1.019)^{104}$$

$$= \lim_{h \to 0} \frac{6.360\left[(1.019)^h - 1\right]}{h}$$

Unlike in Examples 5, 6, and 7, we are unable to eliminate the h in the denominator algebraically and calculate the exact value of $P'(104)$. Nevertheless, by picking a small value for h (say $h = 0.001$), we can estimate the instantaneous rate of change.

$$\frac{6.360\left[(1.019)^{0.001} - 1\right]}{0.001} = 0.1197 \text{ million people per year}$$

$$= 119.7 \text{ thousand people per year}$$

$$\approx 120 \text{ thousand people per year}$$

According to the model, the population of Washington was increasing by approximately 120 thousand people per year in 2004.

Although we were unable to obtain the exact value of the derivative of the exponential function in Example 8, we will develop the theory in later sections that will allow us to calculate the exact value of the derivative of an exponential function.

3.2 Summary

In this section, you learned how to use the difference quotient to estimate an instantaneous rate of change. You also discovered that the derivative may be used to calculate the exact value of an instantaneous rate of change of a function at a point.

3.2 Exercises

In Exercises 1–10, use the difference quotient
$\frac{f(a + h) - f(a)}{h}$ *(with $h = 0.1$, $h = 0.01$, and $h = 0.001$) to estimate the instantaneous rate of change of the function at the given input value.*

1. $f(x) = x^2; x = 2$

2. $h(x) = -x^2 + 1; x = 0$

3. $s(t) = -16t^2 + 64; t = 2$

4. $s(t) = -16t^2 + 64t + 32; t = 0$

5. $w(t) = 4t + 2; t = 5$

6. $P(t) = t^2 - 2t + 1; t = 1$

7. $P(t) = 5; t = 25$

8. $g(x) = 5x - x^2; x = -1$

9. $P(r) = 500(1 + r)^2; r = 0.07$

10. $P(r) = 100(1 + r)^2; r = 0.12$

In Exercises 11–20, use the derivative to calculate the instantaneous rate of change of the function at the given input value. (In each exercise, you can eliminate the h algebraically.) Compare your answers to the solutions of Exercises 1–10.

11. $f(x) = x^2; x = 2$ **12.** $g(x) = -x^2 + 1; x = 2$

13. $s(t) = -16t^2 + 64$; $t = 2$

14. $s(t) = -16t^2 + 64t + 32$; $t = 0$

15. $w(t) = 4t + 2$; $t = 5$

16. $P(t) = t^2 - 2t + 1$; $t = 1$

17. $P(t) = 5$; $t = 25$

18. $g(x) = 5x - x^2$; $x = -1$

19. $P(r) = 500(1 + r)^2$; $r = 0.07$

20. $P(r) = 100(1 + r)^2$; $r = 0.12$

In Exercises 21–30, use the difference quotient (with $h = 0.1$, $h = 0.01$, and $h = 0.001$) to estimate the instantaneous rate of change of the function at the given input value. You may find it helpful to apply the Technology Tip demonstrated in the section.

21. $f(x) = 2x^{-3}$; $x = 3$

22. $P(t) = 50(1.02)^t$; $t = 10$

23. $P(t) = 230(0.9)^t$; $t = 25$

24. $g(x) = -2(3)^x$; $x = -1$

25. $P(r) = 500(1 + r)^{10}$; $r = 0.07$

26. $P(r) = 100(1 + r)^5$; $r = 0.12$

27. $y = \ln(x)$; $x = 2$

28. $y = x \ln(x)$; $x = 1$

29. $g(x) = e^{3x}$; $x = 1$

30. $g(x) = 8 - 2x$; $x = 5$

In Exercises 31–40, determine the instantaneous rate of change of the function at the indicated input value. (You may find it helpful to apply the Technology Tip demonstrated in the section.) Then explain the real-life meaning of the result.

31. 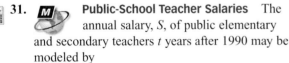 **Public-School Teacher Salaries** The annual salary, S, of public elementary and secondary teachers t years after 1990 may be modeled by

$$S(t) = 0.00713t^3 - 0.111t^2 + 1.42t + 31.6$$

thousand dollars. (**Source:** Modeled from *Statistical Abstract of the United States, 2001*, Table 237, p. 151.) Find and interpret the meaning of $S'(10)$.

32. **Apparel and Textile Industry Pay** The annual payroll, P, of the apparel and textile manufacturing industry between 1992 and 1997 may be modeled by

$$P = -1560x + 24{,}850$$

billion dollars, where x is the number of paid employees (in thousands). (**Source:** Modeled from *Statistical Abstract of the United States, 2001*, Table 722, p. 482.) Find and interpret the meaning of $P'(15)$.

33. **Prescription Drug Spending** The amount of money spent in the United States on prescription drugs may be modeled by

$$D(t) = 0.003667t^3 - 0.0986t^2 + 0.8976t + 2.829$$

billion dollars, where t is the number of years after 1960. (**Source:** Modeled from *Statistical Abstract of the United States, 2001*, Table 119, p. 91.) Find and interpret the meaning of $D'(45)$.

34. **Video-Game Sales** The annual sales, S, of the U.S. video-game industry from 1997 to 2001 may be modeled by

$$S(t) = 0.1958t^4 - 1.275t^3 + 2.254t^2 - 0.0750t + 5.100$$

billion dollars, where t is the number of years since 1997. (**Source:** Modeled from data at www.npdfunworld.com.) Find and interpret the meaning of $S'(5)$.

35. **Video-Game Sales** In the first quarter of 2001, U.S. video-game industry sales totaled \$1.6 billion. Sales increased by 20 percent in the first quarter of 2002. The number one video game in the first quarter of 2002 was Grand Theft Auto 3. (**Source:** www.npdfunworld.com.) If this pattern continues, video game industry sales in the first quarter, S, may be modeled by

$$S = 1.6(1.2)^t$$

billion dollars, where t is the number of years since 2001. Find and interpret the meaning of $S'(6)$.

36. **Nursing Home Care** The amount of money spent annually in the United States on nursing home care from 1960 to 2000 may be modeled by

$$P(t) = 0.07214t^2 - 0.4957t + 1.029$$

billion dollars, where t is the number of years after 1960. (**Source:** Modeled from *Statistical Abstract of the United States, 2001*, Table 119, p. 91). Find and interpret the meaning of $D'(45)$.

37. **College Tuition** The cost of full-time resident tuition at Green River Community College from 1994 to 2001 may be modeled by

$$E(t) = 0.1869t^3 - 1.081t^2 + 19.57t + 431.6$$

dollars, where t is the number of years since 1994. (**Source:** Modeled from Green River Community College data.) Find and interpret the meaning of $E'(7)$.

38. **Demand for VHS Tapes** From 2000 to 2002, the worldwide demand for blank VHS tapes fell. The demand for the tapes may be modeled by

$$V(t) = -20t^2 - 67t + 1409$$

million tapes, where t is the number of years since 2000. (**Source:** Modeled from International Recording Media Association data.) Find and interpret the meaning of $V'(2)$.

39. **Demand for CD-R Disks** From 2000 to 2002, the worldwide demand for CD-R disks increased substantially. The demand for the disks may be modeled by

$$C(t) = -252.5t^2 + 1252.5t + 2730$$

million disks, where t is the number of years since 2000. (**Source:** Modeled from International Recording Media Association data.) Find and interpret the meaning of $C'(2)$.

40. **VHS Tapes and CD-R Disks** From 2000 to 2002, the worldwide demand for blank VHS tapes fell, while the demand for CD-R disks increased. The demand for CD-R disks may be modeled by

$$C(v) = -0.03550v^2 + 85.45v - 47,200$$

million disks, where v is the number of blank VHS tapes sold (in millions). (**Source:** Modeled from International Recording Media Association data.) Find and interpret the meaning of $C'(500)$. Do you believe that the sales of blank VHS tapes and CD-R disks are related? Justify your answer.

Exercises 41–45 deal with the velocity of a free-falling object on earth. As explained in the section, the vertical position of a free-falling object may be modeled by $s(t) = -16t^2 + v_0 t + s_0$ feet, where v_0 is the velocity of the object and s_0 is the vertical position of the object at time $t = 0$ seconds.

41. **Velocity of a Dropped Object** A can of soda is dropped from a diving board 40 feet above the bottom of an empty pool. How fast is the can traveling when it reaches the bottom of the pool?

42. **Velocity of a Dropped Object** The Space Needle in Seattle, Washington, is a popular tourist attraction. The observation deck is approximately 520 feet above the ground. Dropping anything from that height is extremely dangerous to people on the ground. How fast would a penny dropped from the observation deck be traveling when it hit the ground?

43. **Velocity of a Ball** A small rubber ball is thrown into the air by a child at a velocity of 20 feet per second. The child releases the ball 4 feet above the ground. What is the velocity of the ball after 1 second?

44. **Velocity of a Diving Ring** A diving ring is thrown from a diving board at a velocity of -10 feet per second toward the surface of a diving pool. If the surface of the pool is 40 feet below the point of release, what will be the velocity of the ring when it hits the surface of the water?

45. **Velocity of a Juggling Pin** A juggler tosses a bowling pin into the air at a velocity of 15 feet per second. If she catches the pin at the same point she released it, what will be the velocity of the pin when she catches it?

Exercises 46–50 are intended to challenge your understanding of instantaneous rates of change and derivatives.

46. The derivative of a continuous function f is undefined at the point $[a, f(a)]$. Draw a possible graph of f, including the point $[a, f(a)]$.

47. A function f has the property that $f'(a) = f'(b)$. Does $a = b$? Explain.

48. The instantaneous rate of change of a function f is equal to 3 for all values of x. How many x-intercepts does the graph of f have?

49. The instantaneous rate of change of a function f is equal to 0 for all values of x. How many x-intercepts does the graph of f have?

50. The Mean Value Theorem states that if f is continuous and smooth (no breaks, no sharp points) on an interval $[a, b]$, then there exists a value c with $a < c < b$ such that

$$f'(c) = \frac{f(b) - f(a)}{b - a}$$

Given $f(x) = x^2$ on the interval $[2, 4]$, find a value c with $2 < c < 4$ such that

$$f'(c) = \frac{f(4) - f(2)}{4 - 2}$$

3.3 The Derivative as a Slope: Graphical Methods

- Find the equation of the tangent line of a curve at a given point
- Numerically approximate derivatives from a table of data
- Graphically interpret average and instantaneous rates of change
- Use derivative notation and terminology to describe instantaneous rates of change

GETTING **STARTED**   Based on data from 1990 to 2000, the average price for a half gallon of prepackaged ice cream may be modeled by

$$C(t) = 0.0142t^2 - 0.0272t + 2.53 \text{ dollars}$$

where t is the number of years since 1990. (**Source:** Modeled from *Statistical Abstract of the United States, 2001*, Table 706, p. 468.)

To help project future revenue, a creamery wants to know what the average annual increase in the price of ice cream was from 1990 to 2000 and at what rate the price will be increasing at the end of 2000. We will answer these questions by calculating the average and instantaneous rates of change in the price of ice cream.

In Sections 3.1 and 3.2, you learned how to calculate the average rate of change of a function over an interval and the instantaneous rate of change of a function at a point. In this section, we will revisit these concepts from a graphical standpoint. We will also demonstrate how to use tangent-line approximations to estimate the value of a function.

As shown in Section 3.1, the difference quotient formula, $\frac{f(a + h) - f(a)}{h}$, gives the average rate of change in the value of the function between the points $(a, f(a))$ and $(a + h, f(a + h))$. If we let $a = c - h$, then the difference quotient formula becomes $\frac{f(c) - f(c - h)}{h}$. We will use this modified form of the difference quotient in our exploration of the price of ice cream, since we will be approaching $t = c$ through values to the left of $t = c$. Recall from Section 3.1 that, graphically speaking, the difference quotient is the **slope of the secant line** connecting the two points on the graph of a function. According to the ice cream price model, $C(0) = 2.530$ and $C(10) = 3.678$. We're interested in the slope of the secant line between $(0, 2.530)$ and $(10, 3.678)$. The slope of this line represents the average rate of change in the price of a half gallon of ice cream between 1990 and 2000 (Figure 3.8).

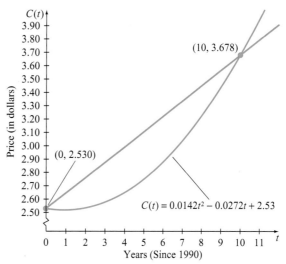

FIGURE 3.8

The ice cream graph secant line between $(0, 2.530)$ and $(10, 3.678)$ has the slope

$$m = \frac{C(10) - C(0)}{10} \frac{\text{dollars}}{\text{years}}$$

$$= \frac{3.678 - 2.530}{10} \frac{\text{dollars}}{\text{years}}$$

$$= 0.115 \text{ dollar per year}$$

Between 1990 and 2000, the average price of a half gallon of ice cream increased by $0.115 (about 12 cents) per year. Does this mean that from 2000 to 2001, the price will increase by about 12 cents? No. Looking at the graph of the model, we notice that the price of ice cream is rising at an increasing rate as time progresses (the steeper the graph, the greater the magnitude of the rate of change). We can approximate the instantaneous rate of change at the end of 2000 ($t = 10$) by calculating the slope of a secant line through $(10, 3.678)$ and a "nearby" point, as shown in Figure 3.9.

The ice cream graph secant line between $(5, 2.749)$ and $(10, 3.678)$ has the slope

$$m = \frac{C(10) - C(5)}{5} \frac{\text{dollars}}{\text{years}}$$

$$= \frac{3.678 - 2.749}{5} \frac{\text{dollars}}{\text{years}}$$

$$= 0.186 \text{ dollar per year}$$

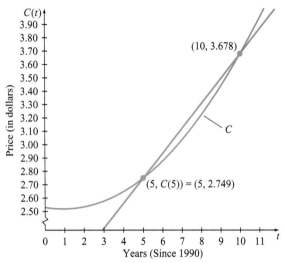

FIGURE 3.9

As shown in Figure 3.10, the ice cream graph secant line between $(9, 3.435)$ and $(10, 3.678)$ has the slope

$$m = \frac{C(10) - C(9)}{10 - 9} \frac{\text{dollars}}{\text{years}}$$

$$= \frac{3.678 - 3.435}{1} \frac{\text{dollars}}{\text{years}}$$

$$= 0.243 \text{ dollar per year}$$

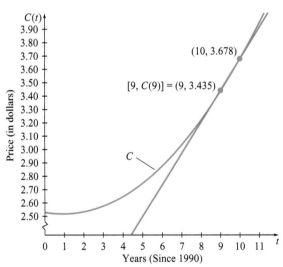

FIGURE 3.10

By zooming in (Figure 3.11), we can see that between $t = 9$ and $t = 10$, the secant line and the graph of the function are extremely close together.

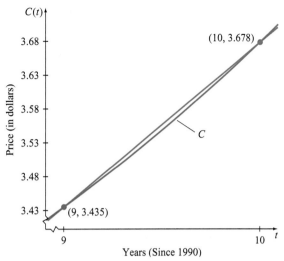

FIGURE 3.11

The ice cream graph secant line between (9.9, 3.653) and (10, 3.678) has the slope

$$m = \frac{C(10) - C(9.9)}{10 - 9.9} \frac{\text{dollars}}{\text{years}}$$

$$= \frac{3.678 - 3.652}{0.1} \frac{\text{dollars}}{\text{years}}$$

$$= 0.26 \text{ dollar per year}$$

Through (9.9, 3.652) and (10, 3.678), the slope of the secant line is \$0.26 per year.

Again zooming in, we see that the secant line and the graph of the function are nearly identical between $t = 9.9$ and $t = 10$ (Figure 3.12). In fact, we are unable to visually distinguish between the two.

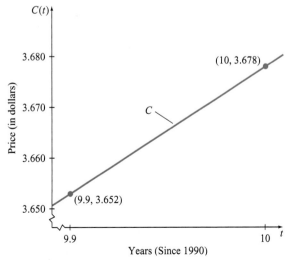

FIGURE 3.12

As t gets closer and closer to 10, the slope of the secant line will approach $C'(10)$. This occurs because $\frac{C(10) - C(10 - h)}{h} \approx C'(10)$ for small values of h. Since h is the horizontal distance between the points, h becomes increasingly small as t nears 10. Using the methods covered in Section 3.2, we determine algebraically that $C'(10) = 0.2568$. At the end of 2000, ice cream prices were increasing at a rate of roughly $0.26 per year. In other words, we anticipate that ice cream prices increased by *approximately* $0.26 between 2000 and 2001.

As the preceding example demonstrates, to find the instantaneous rate of change of a function $y = f(x)$ at a point $P = (a, f(a))$, we can find the limit of the slope of the secant line through P and a nearby point Q as the point Q gets closer and closer to P. Graphically speaking, we select values $Q_1, Q_2, Q_3, \ldots$, with each consecutive value of Q_i being a point on the curve that is closer to the point P than the one before it. The limit of the slope of the secant lines (imagine the points P and Q finally coinciding) is the line *tangent* to the curve at the point $P = (a, f(a))$ (Figure 3.13).

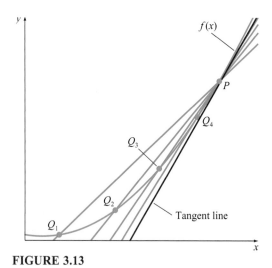

FIGURE 3.13

The **tangent line** to a graph f at a point $(a, f(a))$ is the line that passes through $(a, f(a))$ and has slope $f'(a)$.

THE GRAPHICAL MEANING OF THE DERIVATIVE

The derivative of a function f at a point $(a, f(a))$ is the **slope of the tangent line** to the graph of f at that point. The slope of the tangent line at a point is also referred to as the **slope of the curve** at that point.

EXAMPLE 1

Finding the Equation of a Tangent Line

Find the equation of the tangent line to the graph of $f(x) = x^2$ that passes through $(2, 4)$. Then graph the tangent line and the graph of f.

SOLUTION The slope of the tangent line is $f'(2)$.

$$f'(2) = \lim_{h \to 0} \frac{f(2 + h) - f(2)}{h}$$

$$= \lim_{h \to 0} \frac{(2 + h)^2 - 2^2}{h} \qquad \text{Since } f(x) = x^2$$

$$= \lim_{h \to 0} \frac{4 + 4h + h^2 - 4}{h}$$

$$= \lim_{h \to 0} \frac{4h + h^2}{h}$$

$$= \lim_{h \to 0} \frac{h(4 + h)}{h}$$

$$= \lim_{h \to 0}(4 + h)$$

$$= 4 + 0$$

$$= 4$$

The slope of the tangent line is 4. Using the slope-intercept form of a line, we have $y = 4x + b$. Substituting in the point $(2, 4)$, we get

$$4 = 4(2) + b$$

$$4 = 8 + b$$

$$b = -4$$

The equation of the tangent line is $y = 4x - 4$.

We generate a table of values for $y = 4x - 4$ and $f(x) = x^2$. Then we graph the results (Figure 3.14).

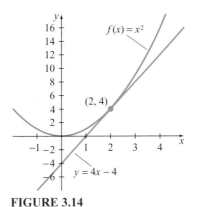

x	y	$f(x)$
-1	-8	1
0	-4	0
1	0	1
2	4	4
3	8	9
4	12	16

FIGURE 3.14

EXAMPLE 2

Finding the Equation of a Tangent Line

Find the equation of the tangent line of the graph of $f(x) = -x^2 + 2x + 2$ that passes through $(1, 3)$. Then plot the tangent line and the graph of f.

SOLUTION The slope of the tangent line is $f'(1)$. Because of the complex nature of f, we will calculate $f(1 + h)$ and $f(1)$ individually, then substitute the simplified values into the derivative formula.

$$\begin{aligned} f(1+h) &= -(1+h)^2 + 2(1+h) + 2 \\ &= -(1 + 2h + h^2) + 2 + 2h + 2 \\ &= -1 - 2h - h^2 + 2h + 4 \\ &= -h^2 + 3 \end{aligned} \qquad \begin{aligned} f(1) &= -(1)^2 + 2(1) + 2 \\ &= -1 + 2 + 2 \\ &= 3 \end{aligned}$$

$$\begin{aligned} f'(1) &= \lim_{h\to 0} \frac{f(1+h) - f(1)}{h} \\ &= \lim_{h\to 0} \frac{(-h^2 + 3) - 3}{h} \\ &= \lim_{h\to 0} \frac{-h^2}{h} \\ &= \lim_{h\to 0}(-h) \\ &= 0 \end{aligned}$$

The slope of the tangent line is 0. Using the slope-intercept form of a line, we have $y = 0x + b$. Substituting in the point $(1, 3)$, we get

$$\begin{aligned} 3 &= 0(1) + b \\ b &= 3 \end{aligned}$$

The equation of the tangent line is $y = 3$. We generate a table of values for $y = 3$ and $f(x) = -x^2 + 2x + 2$. Then we graph both functions (Figure 3.15).

FIGURE 3.15

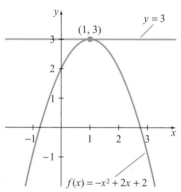

x	y	$f(x)$
−1	3	−1
0	3	2
1	3	3
2	3	2
3	3	−1

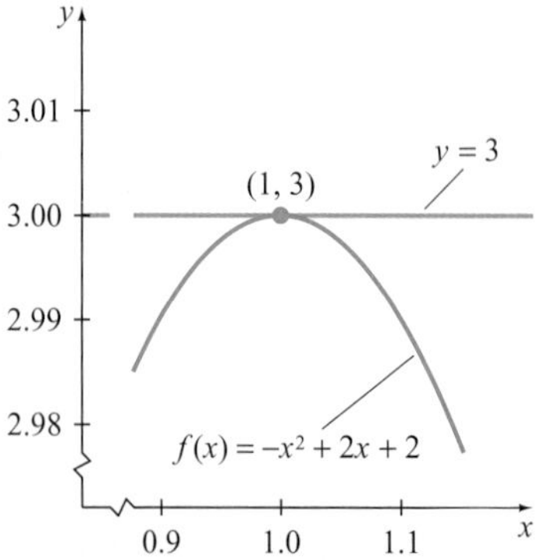

FIGURE 3.16

As shown in Example 2, a function $y = f(x)$ will have a horizontal tangent line at $(a, f(a))$ if $f'(a) = 0$.

Tangent-Line Approximations

In Examples 1 and 2, you may have noticed that the tangent line lies very near to the graph of f for values of x near a. In fact, if we zoom in to the region immediately surrounding $(a, f(a))$, the graph of f and the tangent line to the graph of f at $(a, f(a))$ appear nearly identical. Zooming in on the graph in Example 2, we can see that for values of x near 1, the tangent line y value is a good approximation of the actual function value (see Figure 3.16).

Since it is frequently easier to calculate the values of the tangent line than the values of the function, sometimes the tangent line is used to estimate the value of the function. For example, suppose we wanted to estimate $f(0.9)$ given $f(x) = -x^2 + 2x + 2$. Since $x = 0.9$ is near $x = 1$, we may use the equation of the tangent line, $y = 3$, to estimate $f(0.9)$. That is, $f(0.9) \approx 3$. The actual value is

$$f(0.9) = -(0.9)^2 + 2(0.9) + 2$$
$$= -0.81 + 1.8 + 2$$
$$= 2.99$$

Our tangent-line estimate ($y = 3$) was remarkably close to the actual value of the function ($f(0.9) = 2.99$) and required no additional computations.

EXAMPLE 3

Using a Tangent Line Approximation to Forecast DVD Player Demand

Based on data from 1998 to 2001, the number of DVD players shipped in North America may be modeled by

$$P(t) = 995t^2 + 271t - 245 \text{ thousand players}$$

where t is the number of years since the end of 1997. (Model based on data from DVD Entertainment Group.) According to the model, 16,759 thousand DVD players were shipped in 2001.

Determine how quickly DVD player shipments were increasing at the end of 2001. Then use a tangent line to estimate how many DVD players were shipped in 2002 and compare the estimate to the actual value predicted by the model.

SOLUTION Since $t = 4$ corresponds with the year 2001, the instantaneous rate of change in DVD player shipments at the end of 2001 is given by

UNITS

$$P'(4) = \lim_{h \to 0} \frac{P(4 + h) - P(4)}{h} \frac{\text{thousand players}}{\text{years}}$$

We already know that $P(4) = 16,759$. We'll calculate $P(4 + h)$ and then substitute the simplified value into the derivative formula.

$$P(4 + h) = 995(4 + h)^2 + 271(4 + h) - 245$$
$$= 995(16 + 8h + h^2) + 1084 + 271h - 245$$
$$= 15,920 + 7960h + 995h^2 + 271h + 839$$
$$= 995h^2 + 8231h + 16,759$$

$$P'(4) = \lim_{h \to 0} \frac{P(4 + h) - P(4)}{h}$$
$$= \lim_{h \to 0} \frac{(995h^2 + 8231h + 16,759) - 16,759}{h}$$
$$= \lim_{h \to 0} \frac{995h^2 + 8231h}{h}$$
$$= \lim_{h \to 0} \frac{h(995h + 8231)}{h}$$

$$= \lim_{h \to 0} (995h + 8231)$$

$$= 995(0) + 8231$$

$$= 8231 \frac{\text{thousand players}}{\text{year}}$$

At the end of 2001, DVD player shipments were increasing at a rate of 8231 thousand (8.231 million) players per year. That is, between 2001 and 2002, the number of DVD players shipped annually was expected to increase by *approximately* 8.231 million.

The slope-intercept form of the tangent line is $y = 8231t + b$. Using the point $(4, 16{,}759)$, we determine

$$16{,}759 = 8231(4) + b$$

$$16{,}759 = 32{,}924 + b$$

$$b = -16{,}165$$

The tangent-line equation is $y = 8231t - 16{,}165$. At $t = 5$, we have

$$y = 8231(5) - 16{,}165$$

$$= 41{,}155 - 16{,}165$$

$$= 24{,}990$$

Using the tangent-line equation, we estimate that 24.990 million DVD players were shipped in 2002. According to the model, the actual number of DVD players shipped was slightly greater.

$$P(5) = 995(5)^2 + 271(5) - 245$$

$$= 25{,}985$$

The model indicates that 25.985 million DVD players were shipped in 2002.

Why was there a discrepancy between the two estimates in Example 3? Look at the graph of the model and the tangent line (Figure 3.17).

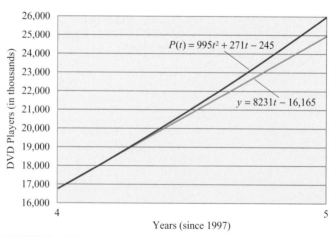

FIGURE 3.17

Although both functions were equal at $t = 4$, by the time t reached 5, the value of the model exceeded the tangent-line estimate by about 1,000 thousand DVD players. Although tangent-line estimates of a function's value are not exact, they are often good enough for their intended purpose.

Numerical Derivatives

TABLE 3.10

x	$f(x)$
0	0
1	1
2	4
3	9
4	16

Often we encounter real-life data in tables or charts. Is it possible to calculate a derivative from a table of data? We'll investigate this question by looking at a table of data for $f(x) = x^2$ (see Table 3.10). For this function, $f'(2) = 4$.

We can estimate $f'(2)$ by calculating the slope of the secant line through points whose x values are equidistant from $x = 2$. That is,

$$f'(2) \approx \frac{f(3) - f(1)}{3 - 1}$$

$$= \frac{9 - 1}{2}$$

$$= \frac{8}{2}$$

$$= 4$$

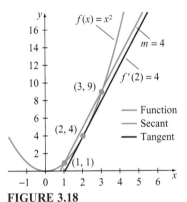

FIGURE 3.18

In this case, our estimate was equal to the tangent-line slope. Let's look at the situation graphically (see Figure 3.18).

When estimating a derivative numerically, we typically select the two closest data points that are horizontally equidistant from our point of interest. Doing so often yields a line that is parallel to the tangent line. Picking two points that are equidistant from the point of interest will tend to give the best estimate of the derivative and can be used as long as the point of interest is not an endpoint. (If the point of interest is an endpoint, we find the slope of the secant line between the endpoint and the next closest point.) If we assume that each output in a table of data represents $f(a)$ for a corresponding input a, then we can symbolically represent the process of numerically estimating a derivative as follows.

NUMERICAL ESTIMATE OF THE DERIVATIVE

The derivative of a function f at a point $(a, f(a))$ may be approximated from a table by

$$f'(a) \approx \frac{f(a + h) - f(a - h)}{(a + h) - (a - h)}$$

$$\approx \frac{f(a + h) - f(a - h)}{2h}$$

where h is the horizontal distance between a and $a + h$.

If a is the largest domain value in the data set, then

$$f'(a) \approx \frac{f(a) - f(a - h)}{(a) - (a - h)}$$

$$\approx \frac{f(a) - f(a - h)}{h}$$

If a is the smallest domain value in the data set, then

$$f'(a) \approx \frac{f(a + h) - f(a)}{(a + h) - (a)}$$

$$\approx \frac{f(a + h) - f(a)}{h}$$

In both cases, h is the distance between a and the next closest domain value.

EXAMPLE   4

Estimating the Derivative from a Table of Data

Use Table 3.11 to estimate how quickly home sales prices in the southern United States were increasing at the end of 1990. [That is, estimate the slope of the tangent line at $(10, 99.0)$.]

TABLE 3.11   Median Sales Price of a New One-Family House in the Southern United States

Years Since 1980 (t)	Price (thousands of dollars) (P)
0	59.6
5	75.0
10	99.0
15	124.5
20	148.0

Source: *Statistical Abstract of the United States, 2001,* Table 940, p. 598.

SOLUTION   We'll estimate the slope of the tangent line at $(10, 99.0)$ by calculating the slope of the secant line between $(5, 75.0)$ and $(15, 124.5)$.

$$f'(10) \approx \frac{f(15) - f(5)}{15 - 5} \frac{\text{thousand dollars}}{\text{years}}$$

$$\approx \frac{124.5 - 75.0}{10}$$

$$\approx 4.95 \text{ thousand dollars per year}$$

We estimate that home prices in the southern United States were increasing at a rate of 5.0 thousand dollars per year at the end of 1990. (Since the original data were accurate to one decimal place, we rounded our result to one decimal place.)

EXAMPLE 5   **Estimating the Derivative from a Table of Data**

Given the data in Table 3.12, how quickly was the percentage of households leasing vehicles increasing in 1995 and 1998?

TABLE 3.12   Percentage of Households Leasing Vehicles

Years Since 1989 (t)	Percent (P)
0	2.5
3	2.9
6	4.5
9	6.4

Source: *Statistical Abstract of the United States, 2001,* Table 1086, p. 680.

SOLUTION   We want to find $f'(6)$ and $f'(9)$.

$$f'(6) \approx \frac{f(9) - f(3)}{9 - 3} \; \frac{\text{percentage points}}{\text{years}}$$

$$\approx \frac{6.4 - 2.9}{6}$$

$$\approx 0.583 \text{ percentage point per year}$$

In 1995, the percentage of households with leased vehicles was increasing by about 0.6 percentage point per year.

Since an endpoint occurs at $t = 9$, we will estimate $f'(9)$ by calculating the slope of the secant line between the endpoint and the next closest point.

$$f'(9) \approx \frac{f(9) - f(6)}{9 - 6} \; \frac{\text{percentage points}}{\text{years}}$$

$$\approx \frac{6.4 - 4.5}{3}$$

$$\approx 0.633 \text{ percentage point per year}$$

In 1998, the percentage of households with leased vehicles was also increasing by about 0.6 percentage point per year.

3.3 Summary

In this section, you learned that the derivative at a point is the slope of the tangent line to the graph at that point. You discovered that the tangent line may be used to approximate the value of a function. You also learned how to estimate a derivative from a table of data.

3.3 Exercises

In Exercises 1–10, determine the equation of the tangent line of the function at the given point. Then graph the tangent line and the function together.

1. $f(x) = x^2 - 4x; (1, -3)$

2. $f(x) = -x^2 + 6; (2, 2)$

3. $g(x) = x^2 + 2x + 1; (0, 1)$

4. $g(x) = x^2 - 4; (3, 5)$

5. $g(x) = x^2 - 4x - 5; (4, -5)$

6. $h(x) = x^3; (2, 8)$

7. $h(x) = x^3; (0, 0)$

8. $h(x) = x^3; (1, 1)$

9. $f(x) = (x - 3)^2; (3, 0)$

10. $f(x) = (x + 2)^2; (-1, 1)$

In Exercises 11–20, answer the questions by calculating the slope of the tangent line and the tangent-line equation, as appropriate.

11. **Motor Fuel Consumption** Based on data from 1980 to 1999, the amount of motor fuel consumed annually by pickups, SUVs, and vans may be modeled by

$$F(t) = 0.0407t^2 + 0.8091t + 23.33$$

billion gallons, where t is the number of years since 1980. (**Source:** Modeled from *Statistical Abstract of the United States, 2001*, Table 1105, p. 691.)

How quickly was fuel consumption increasing in 1999? What was the estimated fuel consumption in 2000? (Use a tangent-line approximation.)

12. **High School Principal Salaries** Based on data from 1980 to 2000, the average salary of a high school principal may be modeled by

$$S(t) = 2499t + 30,039$$

dollars, where t is the number of years since 1980. (**Source:** Modeled from *Statistical Abstract of the United States, 2001*, Table 238, p. 152.)

How quickly were high school principals' salaries increasing in 2000? What was the average high school principal's salary in 2001? (Calculate the tangent-line approximation for the 2001 salary

and compare it to the model calculation for the 2001 salary.)

13. **New One-Family Home Size** Based on data from 1970 to 1999, the average number of square feet in new one-family homes may be modeled by

$$H(t) = -0.1669t^2 + 29.95t + 1490$$

square feet, where t is the number of years after 1970. (**Source:** Modeled from *Statistical Abstract of the United States, 2001*, Table 938, p. 597.)

At what rate was home size increasing in 1999? What was the estimated size of a new home in 2000? (Use a tangent-line approximation.)

14. **New One-Family Home Size** How could a real estate developer or home builder use the results of Exercise 13?

15. **Student-to-Teacher Ratio** Based on data from 1960 to 1998, the student-to-teacher ratio at private elementary and secondary schools may be modeled by

$$R(t) = 0.01484t^2 - 1.000t + 31.60$$

students per teacher, where t is the number of years since 1960. (**Source:** Modeled from *Statistical Abstract of the United States, 2001*, Table 235, p. 150.)

According to the model, how quickly was the student-to-teacher ratio changing in 1998? What was the estimated student-to-teacher ratio for 1999? (Use a tangent-line approximation.)

16. **Student-to-Teacher Ratio** Private schools succeed financially by convincing families that children taught at their school will obtain a better educational experience than that which is available at other public and private schools in the surrounding area. How could a private elementary school use the model and results of Exercise 15 in its marketing plan?

17. **Cassette Tape Shipment Value** Based on data from 1990–2004, the value of music cassette tapes shipped may be modeled by

$$V(s) = -0.00393s^2 + 9.74s - 63.0 \text{ million dollars}$$

where s represents the number of music cassette tapes shipped (in millions). (**Source:** Modeled from *Statistical Abstract of the United States, 2006*, Table 1131.)

According to the model, at what rate is the value of the cassette tapes shipped changing when the number of cassette tapes shipped is 250 million? What was the estimated shipment value when 251 million cassette tapes were shipped? (Use a tangent-line approximation.)

18. **Decreasing Popularity of the Cassette Tape** Since the advent of the compact disk and mp3 player, cassette tape sales have dropped off dramatically. In fact, in 2004 there were 5.2 million cassettes shipped compared to 442.2 million in 1990. (**Source:** *Statistical Abstract of the United States, 2006*, Table 1131.) Use the model in Exercise 17 to calculate the rate of change in cassette tape shipment value in 1990 and in 2004.

19. **College Attendance** Based on data from 2000–2002 and Census Bureau projections for 2003–2013, *private* college enrollment may be modeled by

$$P(x) = 0.340x - 457 \text{ thousand students}$$

where x is the number of students (in thousands) enrolled in *public* colleges. (**Source:** Modeled from *Statistical Abstract of the United States, 2006*, Table 204.) Calculate the rate of change in *private* college enrollment when there are 12,752 thousand *public* college students enrolled. Use a tangent line approximation to estimate the number of students enrolled in *private* colleges when there are 12,753 thousand *public* college students.

20. **Radio Broadcasting Wages** Based on data from 1992 to 1998, the average wage of a worker in the radio broadcasting industry may be modeled by

$$W(t) = 164.3t^2 + 778.6t + 23,500$$

dollars, where t is the number of years since 1992. (**Source:** Modeled from *Statistical Abstract of the United States, 2001*, Table 1123, p. 703.)

How quickly were wages of radio broadcasting employees increasing at the end of 1998? What was the average salary in 1999? (Use a tangent-line approximation.)

In Exercises 21–30, estimate the specified derivative by using the data in the table. Then interpret the result.

21. **Television Broadcasting Wages**

Average Annual Wage per Worker: Television Broadcasting

Year Since 1992 (t)	Annual Wage (dollars) $[W(t)]$
0	41,400
1	42,200
2	43,700
3	47,200
4	51,100
5	51,000
6	54,600

Source: *Statistical Abstract of the United States, 2001*, Table 1123, p.703.

Estimate $W'(5)$.

22. Using the table in Exercise 21, estimate $W'(6)$.

23. **Daily Newspapers**

Different Daily Newspapers

Years Since 1970 (t)	Newspapers $[N(t)]$
0	1,748
5	1,756
10	1,745
15	1,676
20	1,611
25	1,533
30	1,480

Source: *Statistical Abstract of the United States, 2001*, Table 1130, p. 706.

Estimate $N'(25)$.

24. Use the table in Exercise 23 to estimate $N'(30)$.

25. **Tuition at Green River Community College**

Full-Time Resident Tuition

Years Since 1994 (t)	Quarterly Tuition (dollars) [E(t)]
0	432
1	450
2	467
3	486
4	505
5	528
6	547
7	581

Source: Green River Community College.

Estimate $E'(6)$ and $E'(7)$. What can you conclude about the rate of increase in tuition at Green River Community College?

26. **Bread Prices**

Whole Wheat Bread Cost

Years Since 1990 (t)	Cost (dollars per loaf) [C(t)]
3	1.12
4	1.12
5	1.15
6	1.30
7	1.30
8	1.32
9	1.36
10	1.36

Source: Statistical Abstract of the United States, 2001, Table 706, p. 468.

Estimate $C'(4)$ and $C'(5)$.

27. **Baseball Game Attendance**

Major League Baseball Attendance

Years Since 1994 (t)	American League Games (millions of people) [A(t)]
0	24.2
1	25.4
2	29.7
3	31.3
4	31.9
5	31.8

Source: Statistical Abstract of the United States, 2001, Table 1241, p. 759.

Estimate $A'(3)$.

28. **Baseball Game Attendance**

Major League Baseball Attendance

Years Since 1994 (t)	National League Games (millions of people) [N(t)]
0	25.8
1	25.1
2	30.4
3	31.9
4	38.4
5	38.3

Source: Statistical Abstract of the United States, 2001, Table 1241, p. 759.

Estimate $N'(3)$.

29. **Major League Baseball**   Using the tables from Exercises 27 and 28, estimate the overall rate of change in major league baseball game attendance in 1997. (*Hint:* First make a table with the combined attendance of both leagues.)

30.   **Movie Prices**

Movie Ticket Prices

Years Since 1975 (t)	Price (dollars) $[P(t)]$
0	2.05
5	2.69
10	3.55
15	4.23
20	4.35
24	5.08

Source: *Statistical Abstract of the United States, 2001*, Table 1244, p. 761.

Estimate $P'(24)$.

Exercises 31–35 are intended to challenge your understanding of derivatives.

31. Given $f(x) = x^2$, find two pairs of points $(a, f(a))$ and $(b, f(b))$ such that $f'(2) = \frac{f(b) - f(a)}{b - a}$. What do both soltuions have in common?

32. Given $f(x) = x^2 - 2x$, find two pairs of points $(a, f(a))$ and $(b, f(b))$ such that $f'(2) = \frac{f(b) - f(a)}{b - a}$. What do both solutions have in common?

33. Given $f(x) = x^3$, calculate $f'(-1)$ and $f'(1)$. What can you conclude about the graph of f at $(-1, -1)$ and $(1, 1)$?

34. Given $f'(3) = 4$ and $g(x) = x^2 - f(x)$, find $g'(3)$.

35. Given $p(x) = f(x) + g(x)$, what is the relationship between $p'(0), f'(0)$, and $g'(0)$?

3.4 The Derivative as a Function: Algebraic Method

- Use the limit definition of the derivative to find the derivative of a function
- Use derivative notation and terminology to describe instantaneous rates of change

GETTING STARTED   Based on data from 1990 to 2000, the cumulative number of homicides resulting from a romantic triangle may be modeled by $R(t) = -20.45t^2 + 471.8t\ t - 152.5$ homicides between the start of 1991 and the end of year t, where t is the number of years since 1990. (**Source:** Modeled from *Crime in the United States 2000*, Uniform Crime Report, FBI.)

How quickly was the cumulative number of homicides increasing at the end of 1997, 1998, and 1999? Although we could calculate the derivative at $t = 7, t = 8$, and $t = 9$, we can save time by finding the *derivative function* and then substituting in the various values of t.

In this section, we will introduce the derivative function and show how to find it algebraically. Your skill in finding the derivative at a point will prove to be especially useful in this section. We'll begin with a couple of simple examples before returning to the romantic triangle problem.

EXAMPLE 1

Calculating the Derivative of a Function at Multiple Points

Determine the instantaneous rate of change of $f(x) = 3x^2$ at $(1, 3), (3, 27)$, and $(10, 300)$.

SOLUTION   Although we could calculate $f'(1)$, $f'(3)$, and $f'(10)$ individually, it will be more efficient to find the derivative function itself and then substitute in the different values of x.

We begin with the derivative formula; however, instead of substituting a specific value for a, we replace a with the variable x.

$$f'(x) = \lim_{h \to 0} \frac{f(x+h) - f(x)}{h}$$

$$= \lim_{h \to 0} \frac{[3(x+h)^2] - (3x^2)}{h} \qquad \text{Since } f(x) = 3x^2$$

$$= \lim_{h \to 0} \frac{[3(x^2 + 2hx + h^2)] - (3x^2)}{h}$$

$$= \lim_{h \to 0} \frac{(3x^2 + 6hx + 3h^2) - (3x^2)}{h}$$

$$= \lim_{h \to 0} \frac{6hx + 3h^2}{h}$$

$$= \lim_{h \to 0} \frac{h(6x + 3h)}{h}$$

$$= \lim_{h \to 0} (6x + 3h)$$

$$= 6x + 3(0)$$

$$= 6x$$

The result $f'(x) = 6x$ is the derivative function for $f(x) = 3x^2$. It can be used to calculate the instantaneous rate of change of f at any point $(a, f(a))$.

$$f'(1) = 6(1)$$

$$= 6$$

The instantaneous rate of change of f at $(1, 3)$ is 6.

$$f'(3) = 6(3)$$

$$= 18$$

The instantaneous rate of change of f at $(3, 27)$ is 18.

$$f'(10) = 6(10)$$

$$= 60$$

The instantaneous rate of change of f at $(10, 300)$ is 60.

As demonstrated in Example 1, the techniques used to find the derivative function are virtually identical to the procedures used to find the derivative at a point. However, knowing the derivative function allows us to calculate the derivative at a number of different points more quickly than calculating the derivative at each point separately. The derivative function for a function f is typically called the **derivative of f.**

THE DERIVATIVE FUNCTION

The **derivative** of a function f is given by

$$f'(x) = \lim_{h \to 0} \frac{f(x+h) - f(x)}{h}$$

if the limit exists.

EXAMPLE 2

Finding the Derivative of a Function

Find the derivative of $f(x) = 4x^3 - 2x + 1$. Then calculate the slope of the tangent line at $x = 0$, $x = 1$, and $x = 2$.

SOLUTION

$$f'(x) = \lim_{h \to 0} \frac{f(x+h) - f(x)}{h}$$

$$= \lim_{h \to 0} \frac{[4(x+h)^3 - 2(x+h) + 1] - (4x^3 - 2x + 1)}{h} \qquad \text{Since } f(x) = 4x^3 - 2x + 1$$

$$= \lim_{h \to 0} \frac{[4(x^3 + 3hx^2 + 3h^2x + h^3) - 2x - 2h + 1] - (4x^3 - 2x + 1)}{h}$$

$$= \lim_{h \to 0} \frac{(4x^3 + 12hx^2 + 12h^2x + 4h^3 - 2x - 2h + 1) - (4x^3 - 2x + 1)}{h}$$

$$= \lim_{h \to 0} \frac{(12hx^2 + 12h^2x + 4h^3 - 2h)}{h}$$

$$= \lim_{h \to 0} \frac{h(12x^2 + 12hx + 4h^2 - 2)}{h}$$

$$= \lim_{h \to 0} (12x^2 + 12hx + 4h^2 - 2)$$

$$= 12x^2 + 12(0)x + 4(0)^2 - 2 \qquad \text{Replace } h \text{ with } 0$$

$$= 12x^2 - 2$$

The derivative of $f(x) = 4x^3 - 2x + 1$ is $f'(x) = 12x^2 - 2$.

Graphically speaking, the derivative is the slope of the tangent line. Therefore,

$$f'(0) = 12(0)^2 - 2$$
$$= -2$$

means that at $x = 0$ the slope of the tangent line is -2. Similarly,

$$f'(1) = 12(1)^2 - 2$$
$$= 10$$

means that at $x = 1$ the slope of the tangent line is 10. Likewise,

$$f'(2) = 12(2)^2 - 2$$
$$= 46$$

means that at $x = 2$ the slope of the tangent line is 46.

EXAMPLE 3

Calculating the Instantaneous Rate of Change of a Function at Multiple Points

Based on data from 1990 to 2000, the cumulative number of homicides resulting from a romantic triangle may be modeled by

$$R(t) = -20.45t^2 + 471.8t - 152.5 \text{ homicides}$$

between the start of 1991 and the end of year t, where t is the number of years since 1990. (**Source:** Modeled from *Crime in the United States 2000*, Uniform Crime Report, FBI.)

How quickly was the cumulative number of homicides increasing at the end of 1997, 1998, and 1999?

SOLUTION We will begin by finding the derivative $R'(t)$.

$$R'(t) = \lim_{h \to 0} \frac{R(t + h) - R(t)}{h}$$

Because of the complex nature of $R(t)$, we will first calculate $R(t + h)$ and then substitute the result into the derivative formula.

$$R(t + h) = -20.45(t + h)^2 + 471.8(t + h) - 152.5$$
$$= -20.45(t^2 + 2ht + h^2) + 471.8t + 471.8h - 152.5$$
$$= -20.45t^2 - 40.90ht - 20.45h^2 + 471.8t + 471.8h - 152.5$$

We already know that $R(t) = -20.45t^2 + 471.8t - 152.5$. Substituting both of these quantities into the derivative formula yields

$$R'(t) = \lim_{h \to 0} \frac{R(t + h) - R(t)}{h}$$

$$= \lim_{h \to 0} \frac{(-20.45t^2 - 40.90ht - 20.45h^2 + 471.8t + 471.8h - 152.5) - (-20.45t^2 + 471.8t - 152.5)}{h}$$

$$= \lim_{h \to 0} \frac{-40.90ht - 20.45h^2 + 471.8h}{h}$$

$$= \lim_{h \to 0} \frac{h(-40.90t - 20.45h + 471.8)}{h} \quad \text{Every term in the numerator without an } h \text{ cancelled out}$$

$$= \lim_{h \to 0} (-40.90t - 20.45h + 471.8) \quad \text{Since } \frac{h}{h} = 1 \text{ for } h \neq 0$$

$$= -40.90t - 20.45(0) + 471.8$$

$$= -40.90t + 471.8$$

So $R'(t) = -40.90t + 471.8$ homicides per year. We can now compute the instantaneous rate of change in the cumulative number of homicides in 1997, 1998, and 1999.

$$R'(7) = -40.90(7) + 471.8$$
$$= 185.5$$
$$\approx 186 \text{ homicides per year}$$

$$R'(8) = -40.90(8) + 471.8$$
$$= 144.6$$
$$\approx 145 \text{ homicides per year}$$
$$R'(9) = -40.90(9) + 471.8$$
$$= 103.7$$
$$\approx 104 \text{ homicides per year}$$

The cumulative number of homicides resulting from a romantic triangle was increasing at a rate of 186 homicides per year in 1997, 145 homicides per year in 1998, and 104 homicides per year in 1999. According to the model, although the cumulative number of homicides continued to increase, the rate at which these homicides were increasing slowed between the end of 1997 and the end of 1999.

EXAMPLE 4

Finding the Derivative of a Function

Find the derivative of $g(t) = 2t^3 - 4t + 3$.

SOLUTION

We must find $g'(t) = \lim\limits_{h \to 0} \dfrac{g(t+h) - g(t)}{h}$. We'll first find $g(t+h)$ and then substitute the result into the derivative formula.

$$g(t+h) = 2(t+h)^3 - 4(t+h) + 3$$
$$= 2(t^3 + 3t^2h + 3th^2 + h^3) - 4t - 4h + 3$$
$$= 2t^3 + 6t^2h + 6th^2 + 2h^3 - 4t - 4h + 3$$

$$g'(t) = \lim_{h \to 0} \frac{g(t+h) - g(t)}{h}$$
$$= \lim_{h \to 0} \frac{(2t^3 + 6t^2h + 6th^2 + 2h^3 - 4t - 4h + 3) - (2t^3 - 4t + 3)}{h}$$
$$= \lim_{h \to 0} \frac{(6t^2h + 6th^2 + 2h^3 - 4h)}{h} \quad \text{Every term in the numerator without an } h \text{ cancelled out}$$
$$= \lim_{h \to 0} \frac{h(6t^2 + 6th + 2h^2 - 4)}{h}$$
$$= \lim_{h \to 0}(6t^2 + 6th + 2h^2 - 4)$$
$$= 6t^2 + 6t(0) + 2(0)^2 - 4$$
$$= 6t^2 - 4$$

The derivative of $g(t) = 2t^3 - 4t + 3$ is $g'(t) = 6t^2 - 4$.

Estimating Derivatives

For polynomial functions, all terms in the numerator of the derivative formula without an h will cancel out. This allows us always to eliminate the h in the denominator. However, with some other types of functions, the h in the denominator cannot be eliminated algebraically. In this case, we can estimate the derivative function by substituting in a small positive value (i.e. 0.001) for h. The closer the value of h is to zero, the more accurate the estimate of the derivative will be.

EXAMPLE 5   **Calculating the Instantaneous Rate of Change of a Function at Multiple Points**

The per capita consumption of bottled water in the United States may be modeled by

$$W = 2.593(1.106)^t \text{ gallons}$$

where t is the number of years since the end of 1980. (**Source:** Modeled from *Statistical Abstract of the United States, 2001*, Table 204, p. 130.)

Determine how quickly bottled water consumption was increasing at the end of 2002, 2004, and 2006.

SOLUTION We must evaluate $W'(t)$ at $t = 22, 24,$ and 26.

$$W'(t) = \lim_{h \to 0} \frac{W(t + h) - W(t)}{h}$$

$$= \lim_{h \to 0} \frac{[2.593(1.106)^{t+h}] - [2.593(1.106)^t]}{h}$$

$$= \lim_{h \to 0} \frac{[2.593(1.106)^t (1.106)^h] - [2.593(1.106)^t]}{h} \qquad \text{Since } (1.106)^{t+h} = (1.106)^t (1.106)^h$$

$$= \lim_{h \to 0} \frac{2.593(1.106)^t [(1.106)^h - 1]}{h} \qquad \text{Factor out } 2.593(1.106)^t$$

$$= 2.593(1.106)^t \cdot \lim_{h \to 0} \frac{[(1.106)^h - 1]}{h}$$

We can move the expression $2.593(1.106)^t$ to the other side of the limit because it does not contain an h. Since

$$\lim_{h \to 0} \frac{[(1.106)^h - 1]}{h} \approx \frac{[(1.106)^{0.001} - 1]}{0.001}$$

$$\approx 0.1008$$

we have

$$W'(t) = 2.593(1.106)^t \cdot \lim_{h \to 0} \frac{[(1.106)^h - 1]}{h}$$

$$\approx 2.593(1.106)^t \cdot (0.1008)$$

$$\approx 0.2614(1.106)^t$$

We will now evaluate the derivative function at $t = 22, 24,$ and 26.

$$W'(22) \approx 0.2614(1.106)^{22}$$

$$\approx 2.398$$

According to the model, bottled water consumption was increasing by 2.398 gallons per year at the end of 2002.

$$W'(24) \approx 0.2614(1.106)^{24}$$

$$\approx 2.934$$

Bottled water consumption was increasing by 2.934 gallons per year at the end of 2004.

$$W'(26) \approx 0.2614(1.106)^{26}$$
$$\approx 3.589$$

Bottled water consumption is expected to be increasing at a rate of 3.589 gallons per year at the end of 2006.

3.4 Summary

In this section, you learned how to find the derivative function algebraically. You discovered that it is often easier to expand and simplify $f(x + h)$ before substituting it into the derivative formula. Additionally, you found that when the h in the denominator cannot be eliminated algebraically, you can estimate the derivative numerically.

3.4 Exercises

In Exercises 1–10, find the derivative of the function.

1. $f(x) = x^2 - 4x$

2. $f(x) = -x^2 + 6$

3. $g(x) = x^2 + 2x + 1$

4. $g(x) = x^2 - 4$

5. $g(x) = x^2 - 4x - 5$

6. $w(x) = x^3$

7. $j(x) = x^3 + 2$

8. $j(x) = x^3 + x$

9. $f(t) = (t - 3)^2$

10. $f(t) = (t + 2)^2$

In Exercises 11–20, find the slope of the tangent line of the function at $x = 1$, $x = 3$, and $x = 5$.

11. $g(x) = 2x^2 + x - 1$

12. $g(x) = 3x^2 - 4x - 5$

13. $f(x) = x^2 - 2x$

14. $f(x) = 4x - 5$

15. $j(x) = -5$

16. $j(x) = -2x + 4$

17. $W(x) = -4x + 9$

18. $W(x) = x^3 - 1$

19. $S(x) = 3x^2 - 2x + 1$

20. $S(x) = -4x^2 + 9$

In Exercises 21–25, estimate the derivative of the function. When you are unable to eliminate the h in the denominator of the derivative formula algebraically, use $h = 0.001$.

21. $P(x) = 3^x$

22. $R(x) = 2 \cdot 3^x$

23. $C(x) = -3 \cdot 4^x$

24. $P(x) = 2.03^x$

25. $R(x) = 5.042 \cdot (0.98)^x$

In Exercises 26–35, use the derivative function to answer the questions.

26. **New Home Size** Based on data from 1970 to 1999, the average size of a new one-family home may be modeled by

$$H(t) = -0.1669t^2 + 29.95t + 1490$$

square feet, where t is the number of years since 1970. (**Source:** Model based on *Statistical Abstract of the United States, 2001*, Table 938, p. 597.)

Was the average size of a new home growing faster at the end of 1989 or the end of 1999?

27. **New Home Prices** Based on data from 1980 to 2000, the median sales price of a new one-family home in the southern United States may be modeled by

$$P(t) = 0.0506t^2 + 3.515t + 58.49$$

thousand dollars, where t is the number of years since 1980. (**Source:** Modeled from *Statistical Abstract of the United States, 2001*, Table 940, p. 598.)

According to the model, was the median sales price increasing more quickly at the end of 1990 or the end of 2000?

28. **[M]** **New Home Prices** Based on data from 1980 to 2000, the median sales price of a new one-family home in the northeastern United States may be modeled by

$$P(t) = -0.01242t^2 + 8.279t + 68.27$$

thousand dollars, where t is the number of years since 1980. (**Source:** Modeled from *Statistical Abstract of the United States, 2001*, Table 940, p. 598.)

According to the model, was the median sales price increasing more quickly at the end of 1990 or the end of 2000?

29. New Home Prices Using the models from Exercises 27 and 28, determine whether median new home prices were increasing faster in the South or the Northeast in 1999.

30. **[M]** **Restaurant Sales** Based on data from 1980 to 2000, full-service restaurant sales may be modeled by

$$S(t) = 12.77t^2 - 276.9t + 5222t + 38,203$$

million dollars, where t is the number of years since 1980. (**Source:** Modeled from *Statistical Abstract of the United States, 2001*, Table 1268, p. 775.)

Were restaurant sales growing faster at the end of 1984 or the end of 1994? Explain.

31. **[M]** **Gaming Software Sales** Based on data from 1990 to 1999, electronic gaming software factory sales may be modeled by

$$S(t) = 39.79t^2 - 48.41t + 2378$$

million dollars, where t is the number of years since the end of 1990. (**Source:** Modeled from *Statistical Abstract of the United States, 2001*, Table 1005, p. 634.)

Were software sales growing faster at the end of 1997 or the end of 1999?

32. **[M]** **Certified Organic Cropland** Based on data from 1992 to 1996, the amount of certified organic cropland may be modeled by

$$A(t) = 20.57t^2 - 24.51t + 410.3$$

thousand acres, where t is the number of years since the end of 1992. (**Source:** Modeled from *Statistical Abstract of the United States, 2001*, Table 805, p. 526.)

In addition to showing that the amount of organic farmland increased between 1992 and 1996, how could you use the model to convince organic farming investors that there is still room for growth in the organic farming market?

33. **[M]** **Prescription Drug Spending** Based on data from 1990–2003, per capita prescription drug spending may be modeled by

$$P(t) = 2.889t^2 - 2.613t + 158.7 \text{ dollars}$$

where t is the number of years since 1990. (**Source:** *Statistical Abstract of the United States, 2006*, Table 121.)

In what year was prescription drug spending increasing twice as fast as it was increasing in 2000?

34. **[M]** **Federal and State Prison Rate** Based on data from 1980 to 1998, the federal and state prison rate may be modeled by

$$R(t) = 0.3750t^2 + 11.70t + 138.7$$

prisoners per 100,000 people, where t is the number of years since the end of 1980. (**Source:** Modeled from *Statistical Abstract of the United States, 2001*, Table 332, p. 200.)

State lawmakers are debating whether or not the state should invest more money in a crime prevention program. An opponent of the measure argues that between 1980 and 1998, the prison rate increased by an average of 18 prisoners per hundred thousand people per year. She contends that additional money should not be spent on crime prevention unless the crime rate is increasing by at least 25 prisoners per 100,000 people per year. How could a proponent of the crime prevention measure use the model to convince her that crime prevention legislation should be passed now? (Assume that the debate takes place in 2000.)

35. **[M]** **Private College Enrollment** Based on data from 1980 to 1998, the number of students enrolled at private colleges may be modeled by

$$S(t) = 0.9195t^2 + 23.25t + 2652$$

thousand students, where t is the number of years since the end of 1980. (**Source:** Modeled from *Statistical Abstract of the United States, 2001*, Table 205, p. 133.)

How much more rapidly were enrollments increasing in 1998 than they were increasing in 1995?

Exercises 36–40 are intended to challenge your understanding.

36. Given $f(x) = |x|$, find $f'(x)$.

37. Given $f'(x) = 3x$ and $g(x) = x^2 + 3 + f(x)$, find $g'(x)$.

38. Given $f(x) = x^2 - 2x$, determine where $f'(x) = 0$.

39. Given $f(x) = x^3 - 3x$, determine where $f'(x) = 0$.

40. Graph $f(x) = x^3 - 3x$. What does the result of Exercise 39 tell you about the graph of f?

3.5 Interpreting the Derivative

- Interpret the meaning of the derivative in the context of a word problem

GETTING STARTED   Many of us feel inundated by the advertisements we are sent through the mail. Don't expect this to let up anytime soon: Spending on direct-mail advertising has risen every year since 1990. The amount of money spent on direct-mail advertising may be modeled by

$$A(t) = 70.54t^2 + 1488t + 22{,}828 \text{ million dollars}$$

where t is the number of years since 1990. (**Source**: Modeled from *Statistical Abstract of the United States, 2001*, Table 1272, p. 777.)

According to the model, $A(9) = 41{,}934$ and $A'(9) = 2758$. But what does this mean? In this section, we will discuss how to interpret the meaning of a derivative in the context of a real-life problem.

Recall that the units of the derivative are the units of the output divided by the units of the input. In this case, the units of A' are $\frac{\text{millions of dollars}}{\text{year}}$ or millions of dollars per year. Note that $t = 9$ corresponds to the year 1999. We conclude that in 1999, 41,934 million dollars were spent on direct mail advertising, and spending was increasing by 2758 million dollars per year. In other words, according to the model, 41,934 million dollars were spent on direct mail advertising in 1999 and spending increased by *about* 2758 million dollars between 1999 and 2000. The term *about*, *approximately*, or *roughly* must be used when using the second interpretation of the derivative, since we are using a tangent-line approximation to estimate the increase over the next year. (For a graphical discussion of tangent-line approximations, refer to Section 3.3.)

INTERPRETING THE DERIVATIVE

Let $f(x)$ be a function. The meaning of $f'(a) = c$ may be written in either of the following two ways:

- When $x = a$, the value of the function f is increasing (decreasing) by c units of output per unit of input.
- The value of the function f will increase (decrease) by *about* c units of output between a units of input and $a + 1$ units of input.

EXAMPLE 1

Interpreting the Meaning of the Derivative

Based on data from 1969 to 2002, the number of students taking the AP Calculus AB exam may be modeled by

$$E(t) = 157.8t^2 - 770.6t + 10{,}268 \text{ students}$$

where t is the number of years since 1969. (**Source:** Modeled from College Board data.)

Interpret the meaning of $E(33) = 156{,}682$ and $E'(33) = 9644$. Then estimate $E(34)$.

SOLUTION Since t is the number of years since 1969, $t = 33$ corresponds to 2002. $E(33) = 156{,}682$ means that 156,682 students took the exam in 2002.

The units of the derivative are $\dfrac{\text{students}}{\text{year}}$. $E'(33) = 9644$ means that in 2002, the number of students taking the exam was increasing by 9644 students per year. In other words, the number of students taking the exam will increase by *about* 9644 students between 2002 and 2003.

We use a tangent-line approximation to estimate $E(34)$. Since

$$E'(33) \approx \frac{E(34) - E(33)}{34 - 33}$$

$$E'(33) \approx \frac{E(34) - E(33)}{1}$$

$$E'(33) \approx E(34) - E(33)$$

$$E(34) \approx E(33) + E'(33)$$

$$\approx 156{,}682 + 9644$$

$$\approx 166{,}326$$

We estimate that 166,326 students took the exam in 2003.

Each year the College Board hires high school and college educators from around the globe to score the free-response portion of the Calculus AB exam. If an average grader can score 270 exams during the week-long grading, $\frac{9644}{270} = 35.7 \approx 36$ additional graders were needed to score the exams in 2003.

EXAMPLE 2

Interpreting the Meaning of the Derivative

Based on data from 1940 to 2000, the average monthly Social Security benefit for men may be modeled by

$$P(t) = 0.3486t^2 - 5.505t + 35.13 \text{ dollars}$$

where t is the number of years since 1940. (**Source:** Modeled from Social Security Administration data.)

Interpret the meaning of $P(60) = 959.79$ and $P'(60) = 36.33$. Then use a tangent-line approximation to estimate $P(61)$.

SOLUTION Since t is the number of years since 1940, $t = 60$ corresponds to 2000. $P(60) = 959.79$ means that (according to the model) the average Social Security benefit for men in 2000 was $959.79.

The units of the derivative are $\frac{\text{dollars}}{\text{year}}$. $P'(60) = 36.33$ means that in 2000, the average Social Security benefit for men was increasing by \$36.33 per year. In other words, the average Social Security benefit for men was expected to increase by *about* \$36.33 between 2000 and 2001.

We estimate the benefit in 2001 by approximating $P(61)$.

$$P(61) \approx P(60) + P'(60)$$
$$\approx 959.79 + 36.33$$
$$\approx 996.12$$

In 2001, we estimate that the average Social Security benefit for men was \$996.12.

EXAMPLE 3

Using a Tangent Line Approximation to Estimate a Function Value

The average weight of a boy between the ages of 2 and 13 years old may be modeled by

$$W(a) = 0.215a^2 + 2.993a + 23.78 \text{ pounds}$$

where a is the age of the boy in years. (**Source:** Modeled from www.babybag.com data.)

Interpret the meaning of $W(10) = 75.21$ and $W'(10) = 7.29$. Then use a tangent-line approximation to estimate $W(11)$.

SOLUTION Since a is the age of the boy, $a = 10$ corresponds to a 10-year-old boy. $W(10) = 75.21$ means that the average weight of a 10-year-old boy is 75.21 pounds.

The units of the derivative are $\frac{\text{pounds}}{\text{year of his age}}$. $W'(10) = 7.29$ means that the average weight of a 10-year-old boy is increasing by 7.29 pounds per year of his age. In other words, the average weight of a boy will increase by *about* 7.29 pounds between his tenth and eleventh years.

We use a tangent-line approximation to estimate $W(11)$.

$$W(11) \approx W(10) + W'(10)$$
$$\approx 75.21 + 7.29$$
$$\approx 82.5$$

We estimate that the average weight of an 11-year-old boy is 82.5 pounds.

EXAMPLE 4

Interpreting the Meaning of the Derivative

Based on data from 1970 to 1999, the percentage of new one-family homes that are 1200 square feet or less may be modeled by

$$H(t) = 0.02676t^2 - 1.753t + 35.93 \text{ percentage points}$$

where t is the number of years since 1970. (**Source:** Modeled from *Statistical Abstract of the United States, 2001*, Table 938, p. 597.)

Interpret the meaning of $H(14) = 16.6$ and $H'(14) = -1.00$.

SOLUTION Since t is the number of years since 1970, $t = 14$ corresponds to 1984. $H(14) = 16.6$ means that in 1984, 16.6 percent of new homes were 1200 square feet or less in size.

The units of the derivative are $\frac{\text{percentage points}}{\text{year}}$. $H'(14) = -1.00$ means that in 1984, the percentage of new homes that are 1200 square feet or less in size was decreasing by 1 percentage point per year. In other words, the percentage of new homes that are 1200 square feet or less in size was expected to decrease by *about* 1 percentage point (from 16.6 percent to 15.6 percent) from 1984 to 1985.

3.5 Summary

In this section, you learned how to interpret the meaning of the derivative verbally in its real-world context. You also discovered two different ways of representing the concept.

3.5 Exercises

In Exercises 1–30, interpret the real-life meaning of the indicated values. Answer additional questions as appropriate.

1. **Body Weight**   The weight of a 2-year-old to 13-year-old girl may be modeled by

$$W(a) = 0.289a^2 + 2.464a + 23.10$$

pounds, where a is the age of the girl. (**Source:** Modeled from www.babybag.com data.)

Interpret the meaning of $W(10) = 76.64$ and $W'(10) = 8.24$. Then estimate $W(11)$.

2. **Body Weight**   Compare the results of Example 3 and Exercise 1. Were boys or girls expected to gain more weight between their tenth and eleventh years? Explain.

3. **Per Capita Income**   Based on data from 1959 to 1989, the annual per capita income of residents of Washington state may be modeled by

$$P(t) = -2.98t^2 + 327.4t + 7881$$

dollars, where t is the number of years since 1959. (**Source:** Modeled from www.census.gov.)

Interpret the meaning of $P(25) = 14{,}204$ and $P'(25) = 178.4$. Then estimate $P(26)$.

4. **Per Capita Income**   Using the model from Exercise 3, $P'(20) = 208.2$, $P'(25) = 178.4$, $P'(30) = 148.6$, and $P'(35) = 118.8$. Based on these results, what conclusions can you draw about per capita income in the state of Washington?

5. **NASDAQ Market Volume**   Based on data from 1980 to 2000, the average daily volume of the NASDAQ market may be modeled by

$$V(t) = 0.06754t^4 - 1.936t^3 + 17.97t^2 \\ - 41.27t + 39.03$$

million shares, where t is the number of years since 1980. (**Source:** Model based on *Statistical Abstract of the United States, 2001*, Table 1205, p. 741.)

Interpret the meaning of $V(18) = 918$ and $V'(18) = 299$. Then estimate $V(19)$.

6. **NASDAQ Market**   Using the model from Exercise 5, $V'(18) = 299$, $V'(19) = 398$, and $V'(20) = 516$. Based on these results, what conclusions can you draw about the average daily volume of the NASDAQ market?

7. **New York Stock Exchange Volume**
Based on data from 1980 to 2000, the annual volume of the New York Stock Exchange may be modeled by

$$V(t) = 4.807t^4 - 98.56t^3 + 541.3t^2 + 2936t + 10,612$$

million shares, where t is the number of years since 1980. (**Source:** Modeled from *Statistical Abstract of the United States, 2001*, Table 1207, p. 742.)

Interpret the meaning of $V(18) = 168,659$ and $V'(18) = 38,760$. Then estimate $V(19)$.

8. **New York Stock Exchange** Using the model from Exercise 7, $V'(18) = 38,760$, $V'(19) = 48,650$, and $V'(20) = 60,140$. Based on these results, what conclusions can you draw about the annual volume of the New York Stock Exchange?

9. **Radio Broadcasting Wages** Based on data from 1992 to 1998, the average annual wage of an employee in the radio broadcasting industry may be modeled by

$$W(t) = 164.3t^2 + 778.6t + 23,500$$

dollars, where t is the number of years since 1992. (**Source:** Modeled from *Statistical Abstract of the United States, 2001*, Table 1123, p. 703.)

Interpret the meaning of $W(6) = 34,086$ and $W'(6) = 2750$. Estimate $W(7)$.

10. **Restaurant Sales** Based on data from 1989 to 1998, restaurant sales may be modeled by

$$S(r) = 0.9291r - 160.0$$

billion dollars, where r is the number of restaurants (and other eating places) in thousands. (**Source:** Modeled from *Statistical Abstract of the United States, 2001*, Table 1268, p. 775.)

Interpret the meaning of $S(400) = 212$ and $S'(400) = 0.9291$. (*Hint:* Convert the units of the derivative into thousands of dollars per restaurant.)

11. **Billboard Advertising** Based on data from 1990 to 2000, the amount of money spent on billboard advertising may be modeled by

$$A(t) = 9.396t^2 - 11.58t + 1063$$

million dollars, where t is the number of years since 1990. (**Source:** Modeled from *Statistical Abstract of the United States, 2001*, Table 1272, p. 777.)

Interpret the meaning of $A(10) = 1887$ and $A'(10) = 176$.

12. **Cassette Tapes** Based on data from 1992 to 2000, the net number of cassette tapes shipped by recording industry manufacturers may be modeled by

$$C(t) = 0.6527t^3 - 8.348t^2 - 10.30t + 367.5$$

million cassettes, where t is the number of years since 1992. (**Source:** Modeled from Recording Industry Association of America data.)

Interpret the meaning of $C(8) = 85.0$ and $C'(8) = -18.5$.

13. **Compact Disks** Based on data from 1992 to 2000, the value of CDs shipped by recording industry manufacturers may be modeled by

$$V(x) = 0.008086x^2 + 3.308x + 2735$$

million dollars, where x is the net number of CDs shipped in millions. (*Net* means "after returns.") (**Source:** Modeled from Recording Industry Association of America data.)

Interpret the meaning of $V(900) = 12,261.9$ and $V'(900) = 17.9$.

14. **DVDs** Based on data from 1998 to 2001, the value of DVD videos shipped by recording industry manufacturers may be modeled by

$$V(x) = -0.2173x^2 + 25.84x - 0.02345$$

million dollars, where x is the net number of DVDs shipped in millions. (*Net* means "after returns.") (**Source:** Modeled from Recording Industry Association of America data.)

Interpret the meaning of $V(7.5) = 181.6$ and $V'(7.5) = 22.6$.

15. **Summer Olympics** Based on data from 1904 to 1992, the number of women in the Summer Olympics may be modeled by

$$W(t) = 0.0002344t^4 - 0.03440t^3 + 1.701t^2 - 19.69t + 70.80$$

women, where t is the number of years since 1900. (**Source:** Modeled from www.olympicwomen.co.uk data.)

Interpret the meaning of $W(92) = 2662$ and $W'(92) = 150$.

16. **Summer Olympics** Refer to the results of Exercise 15. Do you think $W(96)$ will be greater than 2662 and $W'(96)$ will exceed 150? Defend your conclusions.

17. **U.S. Population** Based on data from 1790 to 2000, the population of the United States may be modeled by

$$P(t) = 0.006702t^2 - 24.11t + 21{,}696$$

million people, where t is the calendar year.
(**Source:** Modeled from U.S. Bureau of the Census data.)
Interpret the meaning of $P(2000) = 284$ and $P'(2000) = 2.70$.

18. **Homicide Rate** Based on data from 1990 to 2000, the homicide rate (deaths per 100,000 people) in the United States may be modeled by

$$H(t) = 0.01129t^3 - 0.2002t^2 + 0.4826t + 9.390$$

people, where t is the number of years since 1990.
(**Source:** Modeled from *Crime in the United States 2000*, Uniform Crime Report, FBI).
Interpret the meaning of $H(10) = 5.5$ and $H'(10) = -0.1$.

19. **Deadly Alcohol-Related Brawls** Based on data from 1991 to 2000, the number of homicides resulting from an alcohol-related brawl may be modeled by

$$H(t) = -0.3551t^3 + 9.867t^2 - 104.7t + 600.4$$

deaths, where t is the number of years since 1990.
(**Source:** Modeled from *Crime in the United States 2000*, Uniform Crime Report, FBI.)
Interpret the meaning of $H(10) = 185$ and $H'(10) = -14$.

20. **Body Height** The average height of a girl between the ages of 2 and 13 years may be modeled by

$$H(a) = -0.0392a^2 + 2.987a + 29.69$$

inches, where a is the age of the girl. (**Source:** Modeled from www.babybag.com data.)
Interpret the meaning of $H(5) = 43.6$ and $H'(5) = 2.6$.

21. **Company Revenue** Based on data from 1999 to 2001, the gross revenue from sales of Johnson & Johnson and its subsidiaries may be modeled by

$$R(t) = 665t^2 + 1150t + 27{,}357$$

million dollars, where t is the number of years since the end of 1999. (**Source:** Modeled from Johnson & Johnson 2001 Annual Report.)
Interpret the meaning of $R(2) = 32{,}317$ and $R'(2) = 3810$.

22. **Company Costs** Based on data from 1999 to 2001, the cost of goods sold for Johnson & Johnson and its subsidiaries may be modeled by

$$C(t) = 103t^2 + 315t + 8539$$

million dollars, where t is the number of years since the end of 1999. (**Source:** Modeled from Johnson & Johnson 2001 Annual Report.)
Interpret the meaning of $C(2) = 9581$ and $C'(2) = 727$.

23. **Company Profit** Based on data from 1999 to 2001, the operating profit of Frito-Lay North America may be modeled by

$$P(t) = -47.5t^2 + 283.5t + 1679$$

million dollars, where t is the number of years since the end of 1999. (**Source:** Modeled from PepsiCo 2001 Annual Report.)
Interpret the meaning of $P(2) = 2056$ and $P'(2) = 93.5$.

24. **Employee Earnings** Based on data from 1991 to 2001, the average hourly earnings for a Ford Motor Company employee may be modeled by

$$W(t) = -0.003931t^4 + 0.1005t^3 - 0.8295t^2 + 3.188t + 16.48$$

dollars, where t is the number of years since the end of 1990. (**Source:** Ford Motor Company 2001 Annual Report).
Interpret the meaning of $W(11) = 27.39$ and $W'(11) = 0.49$.

25. **River Flow** Based on data from November 22 to November 24, 2002, the flow of the Snoqualmie River near Snoqualmie Falls, Washington, may be modeled by

$$f(t) = 25t^2 - 145t + 490$$

cubic feet per second, where t is the number of 24-hour periods since noon on November 22, 2002.
(**Source:** Modeled from www.dreamflows.com data.)
Interpret the meaning of $f(2) = 300$ and $f'(2) = -45$. How could a river rafting company use this type of information?

26. **Personal Income: California** Based on data from 1993 to 2000, the per capita personal income of California may be modeled by

$$P(t) = 123.73t^2 - 445.58t + 22{,}847$$

dollars, where t is the number of years since the end of 1993. (**Source:** Modeled from Bureau of Economic Analysis data.)

Interpret the meaning of $P(5) = 23{,}712$ and $P'(5) = 791.7$.

27. 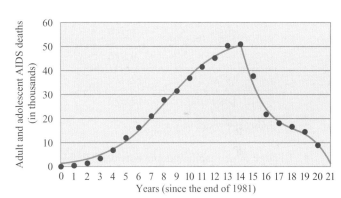 **Personal Income: Colorado** Based on data from 1993 to 2000, the per capita personal income of Colorado may be modeled by

$$P(t) = 80.994t^2 + 890.16t + 22{,}158$$

dollars, where t is the number of years since the end of 1993. (**Source:** Modeled from Bureau of Economic Analysis data.)

Interpret the meaning of $P(5) = 28{,}634$ and $P'(5) = 1700.1$.

28. **Personal Income: Connecticut** Based on data from 1993 to 2000, the per capita personal income of Connecticut may be modeled by

$$P(t) = -19.652t^3 + 296.96t^2 \\ + 502.89t + 29{,}285$$

dollars, where t is the number of years since the end of 1993. (**Source:** Modeled from Bureau of Economic Analysis data.)

Interpret the meaning of $P(5) = 36{,}767$ and $P'(5) = 1998.6$.

29. **Income Analysis** If a high per capita personal income is considered to be one indicator of a strong economy, compare and contrast the income and projected income growth in California, Colorado, and Connecticut between 1998 and 1999, using the results of Exercises 26–28.

30. **Price of a VCR** Based on data from 1997 to 2003, the average price of a VCR may be modeled by

$$P(t) = \frac{123.0}{1 + 0.2649e^{0.9733t}} + 60$$

dollars, where t is the number of years since the end of 1997. (**Source:** Modeled from Consumer Electronics Association data.)

Interpret the meaning of $P'(1) = -29.00$, $P'(3) = -16.82$, and $P'(5) = -3.285$.

Exercises 31–35 are intended to challenge your understanding of the meaning of derivatives.

31. **Market Analysis** The DVD player was introduced in 1997. Based on the results

of Exercise 30, what effect did the introduction of the DVD player have on the price of a VCR? As t gets larger, what value do you think $P'(t)$ will approach?

32. **Company Sales Income** Based on data from 1993 to 2002, the annual sales income of Starbucks Corporation may be modeled by

$$S(t) = 29.23t^2 + 79.33t + 177.4$$

million dollars, where t is the number of years since the end of 1993. According to the model, in what year (between 1993 and 2002) was annual sales income increasing most rapidly? Justify your answer.

33. **AIDS Deaths in the United States** Based on data from 1981 to 2001, the number of adult and adolescent AIDS deaths in the United States may be modeled by a piecewise function f, whose graph is shown in the following figure along with a scatter plot of the number of deaths as a function of years since the end of 1981.

The highest number of AIDS deaths (50,876) occurred in 1995. (**Source:** Centers for Disease Control and Prevention.)

Estimate from the graph the year in which f' was the greatest and the year in which f' was the most negative.

34. Let f be any continuous, smooth function (no breaks or sharp points) that is defined on the interval $(-\infty, \infty)$. If $f(a) \leq f(x)$ for all x, what is the value of $f'(a)$?

35. Let f be any continuous, smooth function (no breaks or sharp points) that is defined on the interval $(-\infty, \infty)$. If $f(a) \geq f(x)$ for all x, what is the value of $f'(a)$?

Chapter 3 Review Exercises

Section 3.1   *In Exercises 1–2, calculate the average rate of change of the function over the given interval.*

1. $f(x) = 6x - 12; [1, 8]$

2. $g(x) = x^2 - 2x + 1; [0, 2]$

In Exercises 3–4, calculate the average rate of change in the designated quantity over the given interval.

3.   Change in per capita income between 1969 and 1989.

United States Per Capita Income, 1959–1989

Year (t)	Income (dollars) (P)
1959	7,259
1969	9,816
1979	12,229
1989	14,420

Source: www.census.gov.

4. Change in per capita income between 1959 and 1989.

Washington Per Capita Income, 1959–1989

Year Since 1959 (t)	Income (dollars) (P)
0	7,978
10	10,565
20	13,528
30	14,923

Source: www.census.gov.

In Exercises 5–6, use the difference quotient to calculate the slope of the secant line through the points $(1, f(1))$ and $(3, f(3))$.

5. $f(x) = 3^x$

6. The graph of f is shown.

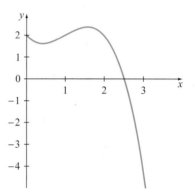

Section 3.2   *In Exercises 7–8, use the difference quotient (with $h = 0.1$, $h = 0.01$, and $h = 0.001$) to estimate the instantaneous rate of change of the function at the given input value.*

7. $f(x) = 2x^2 + 3; x = 1$

8. $h(x) = -x^2 + 5x - 2; x = 4$

In Exercises 9–10, use the derivative to calculate the instantaneous rate of change of the function at the given input value. (In each exercise, you can eliminate h algebraically.) Compare your answers to the solutions of Exercises 7–8.

9. $f(x) = 2x^2 + 3; x = 1$

10. $h(x) = -x^2 + 5x - 2; x = 4$

In Exercises 11–12, use the difference quotient (with $h = 0.1$, $h = 0.01$, and $h = 0.001$) to estimate the instantaneous rate of change of the function at the given input value. You may find it helpful to apply the Technology Tip demonstrated in Section 3.2.

11. $f(x) = 2x^{-3} + x; x = 4$

12. $P(t) = 16(1.04)^t; t = 20$

In Exercises 13–14, determine the instantaneous rate of change of the function at the indicated input value. (You may find it helpful to apply the Technology Tip demonstrated in Section 3.2.) Explain the real-life meaning of the result.

13. **Olympic Runners** Based on the run times of the top seven runners of the 100-meter men's race in the 2000 Olympics, a runner's time may be modeled by

$$T(p) = 0.003056p^3 - 0.04274p^2 + 0.2178p + 9.694$$

seconds, where p is the place (rank) of the runner. (**Source:** Modeled from www.sporting-heroes.net data.)

Find and interpret the meaning of $T'(7)$.

14. **Presidential Election Funding** Based on data from 1980 to 1996 the amount of federal funds used for the presidential election may be modeled by

$$F(t) = 0.03867t^3 - 0.6804t^2 + 6.604t + 62.62$$

million dollars, where t is the number of years since 1980. (**Source:** Modeled from *Statistical Abstract of the United States, 2001;* Table 409, p. 255.)

Find and interpret the meaning of $F'(16)$.

Section 3.3 *In Exercises 15–16, determine the equation of the line tangent to the function at the given point. Then graph the tangent line and the function together.*

15. $f(x) = 2x^2 - 2x; (1, 0)$

16. $f(x) = -x^2 + 6x; (2, 8)$

In Exercises 17–18, answer the questions by calculating the slope of the tangent line and the tangent-line equation, as appropriate.

17. **Social Security Benefit** Based on data from 1940 to 2000, the average monthly Social Security benefit for a retired man may be modeled by

$$P(t) = 0.3486t^2 - 5.505t + 35.13$$

dollars, where t is the number of years since 1940.

How quickly was the average monthly Social Security benefit for a retired man increasing in 2000? What was the average monthly Social Security benefit for a retired man in 2001? (Use a tangent-line approximation.)

18. **Weight of an Average Girl** The weight of a 2-year-old to 13-year-old girl may be modeled by

$$W(t) = 0.289a^2 + 2.464a + 23.10$$

pounds, where a is the age of the girl. (**Source:** Modeled from www.babybag.com data.)

How quickly is the average girl gaining weight when she is 12 years old? What is the estimated weight of a 13-year-old girl? (Use a tangent-line approximation.)

In Exercises 19–20, estimate the specified derivative by using the data in the table. Then interpret the result.

19. **Highway Fatalities**

Highway Fatalities

Years Since 1970 (t)	Fatalities (F)
0	52,627
10	51,091
20	44,599
25	41,817
30	41,821

Source: U.S. Department of Transportation, Bureau of Transportation Statistics.

Estimate $F'(15)$.

20. **Aviation Fatalities**

General Aviation Fatalities

Years Since 1970 (t)	Fatalities (F)
0	1,310
10	1,239
20	767
25	734
30	592

Source: U.S. Department of Transportation, Bureau of Transportation Statistics.

Estimate $F'(25)$.

Section 3.4   *In Exercises 21–22, find the derivative of the function.*

21. $f(x) = 2x^2 - 4$ **22.** $f(x) = -x^2 + 3x$

In Exercises 23–24, find the slope of the line tangent to the function at $x = 1$, $x = 3$, and $x = 5$.

23. $g(x) = 3x^2 + 2x - 5$

24. $g(x) = x^2 - 6x + 9$

In Exercises 25–26, estimate the derivative of the function, using $h = 0.001$ if you are unable to eliminate the h in the denominator of the derivative formula algebraically.

25. $P(x) = 2^x$ **26.** $R(x) = 3 \cdot 4^x$

In Exercises 27–28, use the derivative function to answer the questions.

27. **Major League Baseball Salaries**   Based on data from 1980 to 2000, the average annual salary of a major league baseball player may be modeled by

$$B(t) = 2.699t^2 + 20.76t + 158.4$$

thousand dollars, where t is the number of years since 1980. (**Source:** Modeled from *Statistical Abstract of the United States, 2001*, Table 1241, p. 759.)

Was the average player salary growing faster at the end of 1989 or the end of 1999?

28. **Women's NCAA Basketball**   Based on data from 1985 to 1999, the number of people attending women's NCAA basketball games may be modeled by

$$W(t) = 17.1t^2 + 359t + 135$$

thousand people, where t is the number of years since 1985. (**Source:** Modeled from *Statistical Abstract of the United States, 2001*, Table 1241, p. 759).

Was attendance increasing faster at the end of 1995 or the end of 1999?

Section 3.5   *In Exercises 29–30, interpret the real-life meaning of the function and of the derivative evaluated at the given point.*

29. **McDonald's Sales**   Based on data from 1990 to 2001, franchised sales at McDonald's fast-food restaurants may be modeled by

$$S(t) = 1233.5t + 12{,}213$$

million dollars, where t is the number of years since 1990. (**Source:** Modeled from McDonald's July 26, 2002, Financial Report.)

Interpret the meaning of $S(12) = 27{,}015$ and $S'(12) = 1233.5$.

30. **Church Membership**   Based on data from 1978 to 2000, the number of congregations (wards and branches) in the United States of the Church of Jesus Christ of Latter-Day Saints may be modeled by

$$C(t) = 8679(1.023)^t$$

congregations, where t is the number of years since 1978. (**Source:** Modeled from data compiled by Mark Davies, Brigham Young University.)

Interpret the meaning of $C(14) = 11{,}932$ and $C'(14) = 271$.

Make It Real

What to do

1. Collect a minimum of six data points related to a public or privately owned business (for example, the profit or revenue of a company over a six-year period).
2. Model the data with a polynomial function.
3. Find the derivative of the model function.
4. Evaluate the derivative at three different points.
5. Interpret the real-life meaning of the results of Step 4.
6. Discuss the benefits and the drawbacks of using the model.
7. Explain how company management could use the results of your analysis.

Where to look for data

Company Web Sites

Many company web site addresses are of the form *www.companyname.com*, where *"companyname"* is the name of the company. Look for the area of the web site with investor information. Some web sites to try are
www.mcdonalds.com
www.levistrauss.com
www.albertsons.com
www.gm.com
www.kelloggs.com

Company Financial Reports

Financial reports are typically mailed annually to investors who own shares in a company.

Financial Newspapers and Magazines

The *Wall Street Journal, BusinessWeek, the New York Times, Forbes* magazine, and others contain a wealth of business information.

Differentiation Techniques

A mathematical model may be used to forecast a company's revenue at a given point in time. However, business executives and investors aren't interested only in the dollar amount of a company's revenue; they are also interested in the direction in which revenue is heading. As a rate of change, the derivative of the revenue function shows whether revenues are increasing or decreasing.

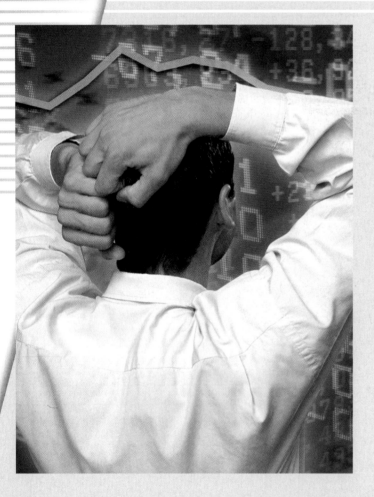

4.1 Basic Derivative Rules

- Use various forms of derivative notation
- Use the Constant Rule, Power Rule, Constant Multiple Rule, and Sum and Difference Rule to find the derivative of a function

GETTING STARTED Home video game systems have become increasingly popular over the last two decades. Based on data from 1990 to 1999, electronic game system factory sales may be modeled by

$$S(t) = 8.451t^3 - 111.9t^2 + 470.0t + 965.0 \text{ million dollars}$$

where t is the number of years since 1990. How quickly were game system sales increasing in 1999? How quickly were sales expected to increase in 2000?

In the previous chapter, you learned how to find the derivative function by using the limit definition of the derivative. Unfortunately, calculating the difference quotient for a complex function (such as the electronic gaming function) is cumbersome and prone to error. In this section, we will introduce some alternative forms of derivative notation and then demonstrate several shortcuts that will greatly enhance our efficiency in calculating derivatives.

Derivative Notation

Up to this point, we have used $f'(x)$ to represent the derivative of a function $f(x)$. Due in part to the fact that calculus was first developed by two people working independently (Newton and Leibniz), there are multiple ways to represent the same concept.

DERIVATIVE NOTATION

The derivative of a function $y = f(x)$ may be represented by any of the following:

- $f'(x)$, read "f prime of x"
- y', read "y prime"
- $\dfrac{dy}{dx}$, read "dee why dee ex" or "the derivative of y with respect to x"

Note that $\dfrac{d}{dx}$ means "find the derivative with respect to x," whereas $\dfrac{dy}{dx}$ is "the derivative of y with respect to x." The form $\dfrac{dy}{dx}$ is referred to as **Leibniz notation** for the derivative. Throughout this section, we will use various forms of notation for the derivative. Additionally, we will use the term **differentiate** to mean "find the derivative of" and the term **differentiation** to refer to the process of finding the derivative. A function whose derivative exists for all values of x in its domain is said to be **differentiable.**

The Constant Rule

Recall that the derivative represents the slope of the tangent line of the function. Consider the horizontal line, $f(x) = c$. The slope of the tangent line of f at (a, c) is

$$f'(a) = \lim_{h \to 0} \frac{f(a + h) - f(a)}{h}$$

$$= \lim_{h \to 0} \frac{c - c}{h}$$

$$= \lim_{h \to 0} \frac{0}{h}$$

$$= 0 \qquad\qquad \text{Since dividing 0 by any}$$
$$\text{nonzero number yields 0}$$

CONSTANT RULE

The derivative of a constant function $y = c$ is

$$y' = 0$$

EXAMPLE 1   **Finding the Derivative of a Constant Function**

Find the derivative of $y = 5$.

SOLUTION

$$y' = 0$$

The Power Rule

Recall that a power function is a function of the form $f(x) = x^n$, where x is a variable and n is a constant. In Exercises 3.4, you found that the power function $y = x^3$ had derivative $y' = 3x^2$. Similarly, the derivative of $y = x^4$ is $y' = 4x^3$, and the derivative of $y = x^5$ is $y' = 5x^4$. Although the derivative methods previously introduced allow us to find the derivative of any power function, the Power Rule provides a remarkably quick and easy way to calculate the derivative.

POWER RULE

The derivative of a function $y = x^n$, where x is a variable and n is a nonzero constant, is

$$y' = nx^{n-1}$$

EXAMPLE 2

Finding the Derivative of a Power Function

Find $\dfrac{dy}{dx}$ given $y = x^3$.

SOLUTION

$$\frac{dy}{dx} = 3x^{3-1}$$
$$= 3x^2$$

EXAMPLE 3

Finding the Derivative of a Power Function

Differentiate $g(x) = x^{-5}$.

SOLUTION

$$g'(x) = -5x^{-5-1}$$
$$= -5x^{-6}$$

EXAMPLE 4

Finding the Derivative of a Power Function

Find y' given $y = \dfrac{1}{x}$.

SOLUTION Initially, this function doesn't look like a power function. However, recall that $y = \dfrac{1}{x}$ is equivalent to $y = x^{-1}$, which is of the form $y = x^n$.

$$y' = -1x^{-1-1}$$
$$= -x^{-2}$$
$$= -\frac{1}{x^2}$$

EXAMPLE 5

Finding the Derivative of a Power Function with a Rational Exponent

Find $f'(x)$ given $f(x) = \sqrt[3]{x}$.

SOLUTION Again, $f(x) = \sqrt[3]{x}$ doesn't appear to be of the form $y = x^n$. However, any function of the form $f(x) = \sqrt[n]{x}$ may be rewritten as $f(x) = x^{1/n}$. Therefore, $f(x) = \sqrt[3]{x}$ is equivalent to $f(x) = x^{1/3}$, which is of the form $y = x^n$.

$$f'(x) = \frac{1}{3}x^{1/3-1}$$
$$= \frac{1}{3}x^{-2/3}$$

This result may be written in a variety of alternative forms, such as the following:

$$f'(x) = \frac{1}{3x^{2/3}}$$

$$f'(x) = \frac{1}{3\sqrt[3]{x^2}}$$

When working with functions with rational exponents, it is helpful to recall that

$$y = \sqrt[n]{x^m}$$
$$= x^{m/n}$$
$$= (x^m)^{1/n}$$
$$= (x^{1/n})^m$$

Each of these forms is a correct way to represent the function $y = \sqrt[n]{x^m}$.

Constant Multiple Rule

The Constant Multiple Rule is used when a constant is multiplied by a function. This rule allows us to factor out the constant, take the derivative of the function, and then multiply the result by the factored-out constant.

CONSTANT MULTIPLE RULE

The derivative of a differentiable function $g(x) = kf(x)$ with constant k is given by

$$g'(x) = kf'(x)$$

EXAMPLE 6   **Finding a Derivative of a Power Function**

Differentiate $g(x) = 3x^2$.

SOLUTION

$$g'(x) = \frac{d}{dx}(3x^2)$$

$$= 3 \cdot \frac{d}{dx}(x^2) \qquad \text{Constant Multiple Rule}$$

$$= 3(2x) \qquad \text{Power Rule}$$

$$= 6x$$

EXAMPLE 7 **Finding the Derivative of a Power Function**

Find the derivative of $y = 4x^3$.

SOLUTION

$$y' = \frac{d}{dx}(4x^3)$$

$$= 4 \cdot \frac{d}{dx}(x^3) \qquad \text{Constant Multiple Rule}$$

$$= 4(3x^2) \qquad \text{Power Rule}$$

$$= 12x^2$$

Sum and Difference Rule

The Sum and Difference Rule allows us to calculate the derivative of a function with multiple terms by adding together the derivatives of each of the terms.

SUM AND DIFFERENCE RULE

The derivative of a differentiable function $h(x) = f(x) \pm g(x)$ is given by

$$h'(x) = f'(x) \pm g'(x)$$

EXAMPLE 8 **Finding the Derivative of a Sum of Functions**

Differentiate $h(x) = 3x^{-2} + 5x$.

SOLUTION

$$h'(x) = \frac{d}{dx}(3x^{-2} + 5x)$$

$$= \frac{d}{dx}(3x^{-2}) + \frac{d}{dx}(5x) \qquad \text{Sum and Difference Rule}$$

$$= 3 \cdot \frac{d}{dx}(x^{-2}) + 5 \cdot \frac{d}{dx}(x) \qquad \text{Constant Multiple Rule}$$

$$= 3(-2x^{-3}) + 5(1) \qquad \text{Power Rule}$$

$$= -6x^{-3} + 5$$

Although we showed multiple steps in this example, as you become skilled in using the rules, you will frequently be able to calculate the derivative in one or two steps.

EXAMPLE 9 **Finding the Derivative of a Sum of Functions**

Find $\dfrac{dy}{dx}$ given $y = 5x^3 - 6x^2 + 1$.

SOLUTION

$$\frac{dy}{dx} = \frac{d}{dx}(5x^3 - 6x^2 + 1)$$

$$= \frac{d}{dx}(5x^3) - \frac{d}{dx}(6x^2) + \frac{d}{dx}(1) \qquad \text{By the Sum and Difference Rule}$$

$$= 15x^2 - 12x + 0 \qquad\qquad \text{By the Constant Multiple, Constant, and Power Rules}$$

$$= 15x^2 - 12x$$

We will now return to the electronic games sales function introduced at the beginning of this section.

EXAMPLE 10 **Using Derivative Rules to Quickly Calculate an Instantaneous Rate of Change**

Based on data from 1990 to 1999, electronic game system factory sales may be modeled by

$$S(t) = 8.451t^3 - 111.9t^2 + 470.0t + 965.0 \text{ million dollars}$$

where t is the number of years since 1990. (**Source:** Modeled from *Statistical Abstract of the United States, 2001*, Table 1005, p. 634.)

How quickly were game system sales increasing in 1999? How quickly were sales expected to be increasing in 2000?

SOLUTION We are asked to evaluate the derivative of the function at $t = 9$ and $t = 10$.

$$S'(t) = \frac{d}{dt}(8.451t^3 - 111.9t^2 + 470.0t + 965.0)$$

$$= \frac{d}{dt}(8.451t^3) - \frac{d}{dt}(111.9t^2) + \frac{d}{dt}(470.0t) + \frac{d}{dt}(965.0) \qquad \text{Sum and Difference Rule}$$

$$= 3(8.451)t^2 - 2(111.9)t + 470.0 + 0 \qquad\qquad \text{Power, Constant Multiple, and Constant Rules}$$

$$= 25.353t^2 - 223.8t + 470.0$$

$$S'(9) = 25.353(9)^2 - 223.8(9) + 470.0$$

$$= 2053.593 - 2014.2 + 470.0$$

$$\approx 509.4$$

$$S'(10) = 25.353(10)^2 - 223.8(10) + 470.0$$

$$= 2535.3 - 2238.0 + 470.0$$

$$= 767.3$$

In 1999, game system sales were increasing by $509.4 million per year. In 2000, game system sales were expected to be increasing at a rate of $767.3 million per year.

EXAMPLE 11

Using Derivative Rules to Quickly Calculate an Instantaneous Rate of Change

In Example 10, we saw that game system sales were increasing rapidly between 1990 and 1999. Game software sales were also increasing rapidly over the same time period. Based on data from 1990 to 1999, electronic game software factory sales may be modeled by

$$S(t) = 39.79t^2 - 48.41t + 2378 \text{ million dollars}$$

where t is the number of years since 1990. (**Source:** Modeled from *Statistical Abstract of the United States, 2001*, Table 1005, p. 634.)

Were electronic game software sales increasing faster in 1995 or 1999?

SOLUTION   We need to evaluate $S'(t)$ at $t = 5$ and $t = 9$.

$$S'(t) = \frac{d}{dt}(39.79t^2 - 48.41t + 2378)$$

$$= 2(39.79)t - 48.41 + 0 \qquad \text{Constant Multiple, Power, Constant, and Sum and Difference Rules}$$

$$= 79.58t - 48.41$$

$$S'(5) = 79.58(5) - 48.41$$
$$= 397.9 - 48.41$$
$$= 349.49$$

$$S'(9) = 79.58(9) - 48.41$$
$$= 716.22 - 48.41$$
$$= 667.81$$

In 1995, game software sales were increasing by $349.5 million per year. In 1999, sales were increasing by $667.8 million per year. Sales were increasing much faster in 1999 than in 1995.

4.1 Summary

In this section you learned how to use the Constant Rule, the Power Rule, the Constant Multiple Rule, and the Sum and Difference Rule. You also learned alternative ways of representing the derivative.

4.1 Exercises

In Exercises 1–10, use the Constant Rule, the Power Rule, the Constant Multiple Rule, and the Sum and Difference Rule (as appropriate) to find the derivative of the function.

1. $f(x) = 5$

2. $g(x) = 2x + 4$

3. $v(t) = 5t^3 - 10t^{-2}$

4. $s(t) = 7t^4 - 12t^2 + 2t + 43$

5. $g(x) = -2x + 4x^3$

6. $s(n) = 44n^2 - n$

7. $f(t) = 4t^{-1}$

8. $h(t) = 15t^5 - 75t^4 + 300t^3$

9. $g(t) = t^3 + 3t^2 + 6t + 6$

10. $f(x) = 0$

In Exercises 11–20, find the equation of the tangent line at the indicated domain value.

11. $y = \sqrt{x}; x = 1$

12. $y = 6.9x^4 - 3.2x^2 + 9.1; x = 0$

13. $y = -2.6x^5 + 3x^{-2}; x = -1$

14. $y = -7.62x^2 - 5.21x + 6.11; x = 1$

15. $y = 1235.3t^3 + 551.23t - 1203.9; t = 0$

16. $y = -t^{-1} + 2t^{-2} - 3t^{-3}; t = 4$

17. $f(n) = 9n^{1.3} + 5n^{2.1} + 92n^{-0.2}; n = 1$

18. $w(t) = 3.2t - t^3; t = 5$

19. $g(x) = x^{-3} + 2.22x^3 - 12.3; x = 1$

20. $r(x) = \frac{1}{2}x^{1.5}; x = 0$

In Exercises 21–30, answer the question by evaluating the derivative using the derivative rules introduced in this section.

21. **Organic Cropland** Based on data from 1992 to 1996, the amount of certified organic cropland in the United States may be modeled by

$$A(t) = 20.57t^2 - 24.51t + 410.3$$

thousand acres, where t is the number of years since 1992. (**Source:** Modeled from *Statistical Abstract of the United States, 2001*, Table 805, p. 526.)

How quickly was the amount of organic cropland increasing in 1996?

22. **Traffic Fatalities** Based on data from 1970 to 2000, the number of highway traffic fatalities may be modeled by

$$F(t) = 1.554t^3 - 72.90t^2 + 427.0t + 52,615$$

fatalities, where t is the number of years since 1970. (**Source:** Modeled from U.S. Department of Transportation, Bureau of Transportation Statistics data.)

At what rate was the number of highway fatalities changing in 1995?

23. **Average Height of a Girl** The average height of a 2- to 13-year-old girl may be modeled by

$$H(a) = -0.0392a^2 + 2.987a + 29.69$$

inches, where a is the age of the girl. (**Source:** Modeled from www.babybag.com data.)

Does the average height of a 4-year-old girl increase faster than the average height of an 11-year-old girl?

24. **Average Height of a Boy** The average height of a 2- to 13-year-old boy may be modeled by

$$H(a) = -0.0507a^2 + 2.997a + 30.44$$

inches, where a is the age of the boy. (**Source:** Modeled from www.babybag.com data.)

Does the average height of a 4-year-old boy increase faster than the average height of an 11-year-old boy?

25. Using the models in Exercises 23 and 24, does a 9-year-old boy or a 9-year-old girl grow at a quicker rate?

26. **Average Weight of a Girl** The average weight of a 2- to 13-year-old girl may be modeled by

$$W(a) = 0.289a^2 + 2.464a + 23.10$$

pounds, where a is the age of the girl. (**Source:** Modeled from www.babybag.com data.)

Does the average weight of a 4-year-old girl increase faster than that of an 11-year-old girl?

27. **Average Weight of a Boy** The average weight of a 2- to 13-year-old boy may be modeled by

$$W(a) = 0.215a^2 + 2.993a + 23.78$$

pounds, where a is the age of the boy. (**Source:** Modeled from www.babybag.com data.)

Does the average weight of a 4-year-old boy increase faster than the average weight of an 11-year-old boy?

28. Using the models in Exercises 26 and 27, does a 9-year-old boy or a 9-year-old girl gain weight at a quicker rate?

29. **Billboard Advertising** Based on data from 1990 to 2000, the amount of money spent on billboard advertising may be modeled by

$$A(t) = 34.99t^2 + 94.91t + 8963$$

million dollars, where t is the number of years since 1990. (**Source:** Modeled from *Statistical Abstract of the United States, 2001*, Table 1272, p. 777.)

Was billboard advertising expected to increase more rapidly in 1999 or in 2000?

30. **Television Advertising** Based on data from 1990 to 2000, the amount of money spent on broadcast TV advertising may be modeled by

$$A(t) = 10.22t^4 - 221.5t^3 + 1616t^2 - 2488t + 26{,}671$$

million dollars, where t is the number of years since 1990. (**Source:** Modeled from *Statistical Abstract of the United States, 2001*, Table 1272, p. 777.)

At what rate was television advertising expected to increase in 2000?

Exercises 31–35 are intended to challenge your understanding of the basic rules of differentiation.

31. Let f, g, and h be differentiable functions with $h(x) = f(x) \cdot g(x)$. May the Constant Multiple Rule be used to find $h'(x)$? Explain.

32. Let $f(x) = 2^x$. Is $f'(x) = x \cdot 2^{x-1}$? Defend your conclusions.

33. Use the limit definition of the derivative to prove that if s is differentiable and $s(x) = f(x) + g(x)$, then $s'(x) = f'(x) + g'(x)$.

34. Use the limit definition of the derivative to prove that if f is differentiable, k is a constant, and $f(x) = k \cdot g(x)$, then $f'(x) = k \cdot g'(x)$.

35. Given $f(x) = x^{1/3}$, for what values of x is $f'(x)$ undefined?

4.2 The Product and Quotient Rules

- Use the Product Rule and Quotient Rule to find the derivative of a function

GETTING STARTED Based on data from 1985 to 1999, the per capita consumption of chicken may be modeled by

$$C(t) = 0.002623t^4 - 0.07545t^3 + 0.6606t^2 - 0.4688t + 36.92$$

pounds of *boneless, trimmed* chicken, where t is the number of years since 1985. (**Source:** Modeled from *Statistical Abstract of the United States, 2001*, Table 202, p. 129.)

Based on data from 1985 to 2000, the average retail price of *fresh whole* chicken may be modeled by

$$P(t) = 0.0007346t^2 - 0.009701t + 0.7803$$

dollars per pound, where t is the number of years since 1985. (**Source:** Modeled from *Statistical Abstract of the United States, 2001*, Table 706, p. 468.) A whole chicken loses about 30 percent of its weight when it is boned and trimmed.

Was per capita spending on chicken increasing or decreasing in 2000? By how much? We will use the Product Rule to answer this question in Example 4 of this section.

In this section, we will demonstrate how to find the derivative of a product by using the Product Rule and the derivative of a quotient by using the Quotient Rule.

PRODUCT RULE

The derivative of a function $h(x) = f(x) \cdot g(x)$ is given by

$$h'(x) = f'(x) \cdot g(x) + g'(x) \cdot f(x)$$

A common error among beginning calculus students is to assume that $\dfrac{d}{dx}[f(x) \cdot g(x)] = f'(x) \cdot g'(x)$. We can easily convince ourselves that this is not true by calculating the derivative of $f(x) = x^2$ using the Power Rule and then by using the erroneous rule.

Using the Power Rule,

$$\frac{d}{dx}(x^2) = 2x$$

Using the erroneous rule,

$$\begin{aligned}
\frac{d}{dx}(x^2) &= \frac{d}{dx}(x \cdot x) \\
&= \frac{d}{dx}(x) \cdot \frac{d}{dx}(x) \\
&= 1 \cdot 1 \\
&= 1
\end{aligned}$$

Since $2x \ne 1$, we know that the erroneous rule is not valid. Let's calculate the derivative using the Product Rule.

Using the Product Rule,

$$\begin{aligned}
\frac{d}{dx}(x^2) &= \frac{d}{dx}(x \cdot x) \\
&= \frac{d}{dx}(x) \cdot x + \frac{d}{dx}(x) \cdot x \\
&= 1 \cdot x + 1 \cdot x \\
&= 2x
\end{aligned}$$

The Product Rule yielded the same result as the Power Rule.

EXAMPLE 1 **Finding the Derivative of a Product of Functions**

Find the derivative of $h(x) = (2x + 4)(5x^3 - 3x)$.

SOLUTION

$$\begin{aligned}
h'(x) &= \frac{d}{dx}[(2x + 4)(5x^3 - 3x)] \\
&= \left[\frac{d}{dx}(2x + 4)\right] \cdot (5x^3 - 3x) + \left[\frac{d}{dx}(5x^3 - 3x)\right] \cdot (2x + 4) \qquad \text{Product Rule}
\end{aligned}$$

$$= (2) \cdot (5x^3 - 3x) + (15x^2 - 3) \cdot (2x + 4) \qquad \text{Power, Constant Multiple, and Constant Rules}$$
$$= (10x^3 - 6x) + (30x^3 + 60x^2 - 6x - 12)$$
$$= 40x^3 + 60x^2 - 12x - 12$$

We can check our work by multiplying out the factors of $h(x)$ and differentiating the result using the Power Rule.

$$h(x) = (2x + 4)(5x^3 - 3x)$$
$$= 10x^4 + 20x^3 - 6x^2 - 12x$$
$$h'(x) = 40x^3 + 60x^2 - 12x - 12$$

The result is the same as that obtained by using the Product Rule. We are confident that we have done the problem correctly.

In the previous two examples, we saw that there are multiple ways to calculate the derivative of a function. In general, we will use the simplest techniques possible to find the derivative. However, in order to demonstrate the Product Rule, in this section we may use the Product Rule in places where other methods are typically preferred.

EXAMPLE 2 **Finding the Derivative of a Product of Functions**

Differentiate $y = (2x^3 - 6x + 5)(x^4 - 2x^2 + 1)$.

SOLUTION

$$\frac{dy}{dx} = \frac{d}{dx}(y)$$

$$= \frac{d}{dx}[(2x^3 - 6x + 5)(x^4 - 2x^2 + 1)]$$

$$= \left[\frac{d}{dx}(2x^3 - 6x + 5)\right](x^4 - 2x^2 + 1) + \left[\frac{d}{dx}(x^4 - 2x^2 + 1)\right](2x^3 - 6x + 5)$$

$$= (6x^2 - 6)(x^4 - 2x^2 + 1) + (4x^3 - 4x)(2x^3 - 6x + 5)$$

$$= (6x^6 - 12x^4 + 6x^2 - 6x^4 + 12x^2 - 6) + (8x^6 - 24x^4 + 20x^3 - 8x^4 + 24x^2 - 20x)$$

$$= 14x^6 - 50x^4 + 20x^3 + 42x^2 - 20x - 6$$

EXAMPLE 3 **Finding the Derivative of a Product of Three Functions**

Differentiate $R(t) = (5t - 1)(4t + 3)(2t + 6)$.

SOLUTION This function is different because it is the product of three factors instead of two. Nevertheless, the Product Rule may be generalized to work here. If we have a function of the form $w(x) = f(x) \cdot g(x) \cdot h(x)$, then $w'(x) = f'(x)g(x)h(x) + g'(x)f(x)h(x) + h'(x)f(x)g(x)$. To find each term of the derivative, we calculate the derivative of each factor and then multiply it by the remaining factors. The derivative of $w(x)$ is the sum of the individual terms.

$$
\begin{aligned}
R'(t) &= 5(4t + 3)(2t + 6) + 4(5t - 1)(2t + 6) + 2(5t - 1)(4t + 3) \\
&= 5(8t^2 + 24t + 6t + 18) + 4(10t^2 + 30t - 2t - 6) + 2(20t^2 + 15t - 4t - 3) \\
&= 5(8t^2 + 30t + 18) + 4(10t^2 + 28t - 6) + 2(20t^2 + 11t - 3) \\
&= (40t^2 + 150t + 90) + (40t^2 + 112t - 24) + (40t^2 + 22t - 6) \\
&= 120t^2 + 284t + 60
\end{aligned}
$$

EXAMPLE 4 **Applying the Product Rule in a Real-World Context**

Based on data from 1985 to 1999, the per capita consumption of chicken may be modeled by

$$C(t) = 0.0026t^4 - 0.075t^3 + 0.66t^2 - 0.47t + 37$$

pounds of *boneless, trimmed* chicken, where t is the number of years since 1985. (**Source:** Modeled from *Statistical Abstract of the United States, 2001*, Table 202, p. 129.)

Based on data from 1985 to 2000, the average retail price of *fresh whole* chicken may be modeled by

$$P(t) = 0.00073t^2 - 0.0097t + 0.78$$

dollars per pound, where t is the number of years since 1985. (**Source:** Modeled from *Statistical Abstract of the United States, 2001*, Table 706, p. 468.) A whole chicken loses about 30 percent of its weight when it is boned and trimmed.

Was per capita spending on chicken increasing or decreasing in 2000? At what rate?

SOLUTION We first need to find a function to model per capita spending on chicken. The units of $C(t)$ are pounds of *boneless, trimmed* chicken and the units of $P(t)$ are dollars per pound of *fresh whole chicken*. Since 30% of a whole chicken's weight is lost when it is boned and trimmed, 1 pound of *fresh whole* chicken yields 0.7 pound of *boneless, trimmed* chicken. We define a new function $S(t) = C(t)P(t)$. What are the units of $S(t)$? We have

$$(C \text{ pounds of boneless chicken})\left(\frac{P \text{ dollars}}{\text{pound of whole chicken}}\right)$$

Dividing the pounds of boneless chicken by 0.7 will yield the number of pounds of whole chicken that were used to get C pounds of boneless chicken and allow us to cancel units. Thus

UNITS

$$\left(\frac{C}{0.7} \text{ pounds of whole chicken}\right)\left(\frac{P \text{ dollars}}{\text{pound of whole chicken}}\right) = \frac{10}{7} CP \text{ dollars}$$

Therefore, the units of $S(t)$ are dollars.

$$
\begin{aligned}
S(t) &= \frac{10}{7} C(t) \cdot P(t) \\
&= \frac{10}{7}(0.0026t^4 - 0.075t^3 + 0.66t^2 - 0.47t + 37)(0.00073t^2 - 0.0097t + 0.78)
\end{aligned}
$$

We are asked to calculate the instantaneous rate of change of $S(t)$ in 2000. That is, we are to find $S'(15)$. We know from the Product Rule that

UNITS

$$S'(t) = \frac{10}{7}[C'(t) \cdot P(t) + P'(t) \cdot C(t)] \frac{\text{dollars}}{\text{year}}$$

Although we could calculate $S'(15)$ algebraically, the complexity of $S(t)$ would make the process extremely tedious. In cases such as this, it is appropriate to use the power of a graphing calculator, as demonstrated in the following Technology Tip. Using the Technology Tip, we determine that

$$S'(15) \approx 5.33 \frac{\text{dollars}}{\text{year}}$$

In 2000, per capita spending on chicken was increasing by $5.33 per year.

TECHNOLOGY TIP

Calculating a Derivative with *nDeriv(*

1. Enter the function equation using the $\boxed{\text{Y=}}$ editor. (In this case, we entered C in Y_1, P in Y_2, and S in Y_3. We input the variables Y_1 and Y_2 into equation Y_3 by using the $\boxed{\text{VARS}}$ Y-VARS; 1:Function key sequence and selecting the respective functions.)

```
Plot1 Plot2 Plot3
\Y1■0.0026X^4-.0
75X^3+.66X^2-.47
X+37
\Y2■.00073X^2-.0
097X+.78
\Y3■10/7*Y1*Y2
\Y4=
```

2. Press $\boxed{\text{2ND}}$ $\boxed{\text{MODE}}$ to quit and return to the home screen. Then press $\boxed{\text{MATH}}$ and select 8:nDeriv(. Then press $\boxed{\text{ENTER}}$.

```
MATH NUM CPX PRB
3↑³√
4: ³√(
5: ×√(
6:fMin(
7:fMax(
8∎nDeriv(
9↓fnInt(
```

3. The nDeriv(function requires three input values: the function name, the variable of differentiation, and the value of the variable. Enter the appropriate values.

```
nDeriv(Y3,X,15
```

4. Press $\boxed{\text{ENTER}}$ to display the result.

```
nDeriv(Y3,X,15
        5.334334044
```

Although any quotient of two functions may be rewritten as a product, using the **Quotient Rule** to find the derivative of a quotient is often easier than using the Product Rule.

QUOTIENT RULE

The derivative of a function $h(x) = \dfrac{f(x)}{g(x)}$ is given by

$$h'(x) = \frac{f'(x) \cdot g(x) - g'(x) \cdot f(x)}{[g(x)]^2}$$

EXAMPLE 5 **Finding the Derivative of a Quotient of Functions**

Find $\dfrac{dy}{dx}$ for $y = \dfrac{x^2 + 1}{x}$.

SOLUTION

$$\frac{dy}{dx} = \frac{(2x)(x) - (1)(x^2 + 1)}{(x)^2}$$

$$= \frac{2x^2 - x^2 - 1}{x^2}$$

$$= \frac{x^2 - 1}{x^2}$$

EXAMPLE 6 **Finding the Derivative of a Quotient of Functions**

Differentiate $y = \dfrac{6x + 1}{3x - 1}$.

SOLUTION

$$y' = \frac{(6)(3x - 1) - (3)(6x + 1)}{(3x - 1)^2}$$

$$= \frac{18x - 6 - 18x - 3}{(3x - 1)^2}$$

$$= \frac{-9}{(3x - 1)^2}$$

4.2 Summary

In this section, you learned how to use the Product and Quotient Rules to find the derivative of a function that is written as a product or quotient of factors. You also learned how technology may be used to differentiate functions with complex factors.

4.2 Exercises

In Exercises 1–10, use the Product or the Quotient Rule to find the derivative of the function. Don't simplify the result.

1. $f(x) = 2x(3x + 4)$ 2. $g(x) = 5x(9x - 2)$

3. $f(t) = (2t - 6)(10t + 5)$

4. $s(t) = 2t^3(3t^2 + 7t)$

5. $w(n) = (3n^2 + 8)(n^3 - n)$

6. $D(p) = (p^2 - 7p)(1 - p^3)$

7. $Q(t) = (t^2 + 2t + 1)(t^2 + 4t + 4)$

8. $P(t) = \dfrac{t^2 + 5t + 6}{t^2 + 5t + 4}$ 9. $f(x) = \dfrac{x^3 - 1}{x^4 - x^3}$

10. $g(x) = \dfrac{x^3 + 5x^2}{x^2 - 4x + 2}$

In Exercises 11–20, determine the slope of the graph at the indicated domain value.

11. $h(t) = \dfrac{t^2 + 5t + 6}{t}; t = -2$

12. $g(t) = \dfrac{t^3 - 12t^2}{t}; t = 1$

13. $s(t) = \dfrac{t^4 - 6t^2 + 8}{2t}; t = -1$

14. $p(t) = \dfrac{7t^2 - 2t + 9}{t}; t = 2$

15. $f(x) = 2x(3x + 4)(5x + 7); x = 0$

16. $R(p) = 5p(3p + 1)(4p - 1); p = 1$

17. $r(x) = (2x + 11)(x - 4)(7x + 3); x = 4$

18. $T(n) = (n^2 + 4)(2n - 8)(n + 9); n = 3$

19. $f(x) = (x + 1)(x - 1)(x^2 + 1); x = -1$

20. $g(x) = (x - 2)(x + 2)(x^2 + 4); x = 2$

In Exercises 21–25, use the Technology Tip to answer the questions from the real-life scenarios.

21.   **Employer Labor Costs**   Based on data from 1995 to 1999, the average annual earnings in the lumber and wood products manufacturing industry may be modeled by

$$S(t) = -63.57t^2 + 1253t + 25,066$$

dollars per employee, and the average number of employees in the lumber and wood products manufacturing industry may be modeled by

$$E(t) = 3.143t^2 + 5.029t + 772.5$$

thousand employees, where t is the number of years since 1995. (**Source:** Modeled from *Statistical Abstract of the United States, 2001*, Table 979, p. 622.)

In 1999, at what rate was employer spending on lumber and wood products manufacturing industry employee earnings increasing?

22. **Employer Labor Costs**   Based on data from 1995 to 1999, the average annual earnings in the rubber and plastics manufacturing industry may be modeled by

$$E(t) = 1196t + 29,825$$

dollars per employee, and the average number of employees in the rubber and plastics manufacturing industry may be modeled by

$$N(t) = -2.917t^3 + 15.29t^2 - 10.73t + 962.9$$

thousand employees, where t is the number of years since 1995. (**Source:** Modeled from *Statistical Abstract of the United States, 2001*, Table 979, p. 622.)

In 1999, at what rate was employer spending on rubber and plastics manufacturing industry employee earnings increasing?

23. **Employer Labor Costs**   Based on data from 1995 to 1999, the average annual earnings in the paper and allied products manufacturing industry may be modeled by

$$E(t) = 1335t + 39,408$$

dollars per employee, and the average number of employees in the same industry may be modeled by

$$N(t) = -t^3 + 5.571t^2 - 12.29t + 684.9$$

thousand employees, where t is the number of years since 1995. (**Source:** Modeled from *Statistical Abstract of the United States, 2001*, Table 979, p. 622.)

In 1999, at what rate was employer spending on paper and allied products manufacturing industry employee earnings increasing?

24. **Employer Labor Costs**   Based on data from 1995 to 1999, the average annual earnings in the apparel and other textile products manufacturing industry may be modeled by

$$E(t) = 1118t + 18{,}729$$

dollars per employee, and the average number of employees in the same industry may be modeled by

$$N(t) = -3.5t^3 + 20.21t^2 - 85.86t + 918.2$$

thousand employees, where t is the number of years since 1995. (**Source:** Modeled from *Statistical Abstract of the United States, 2001*, Table 979, p. 622.)

In 1999, at what rate was employer spending on apparel and other textile products manufacturing industry employee earnings changing?

25. **Employer Labor Costs**   Based on data from 1995 to 1999, the average annual earnings in the printing and publishing industry may be modeled by

$$E(t) = 88.36t^2 + 1291t + 34{,}528$$

dollars per employee, and the average number of employees in the printing and publishing industry may be modeled by

$$N(t) = -4.417t^3 + 24.93t^2 - 25.30t + 1450$$

thousand employees, where t is the number of years since 1995. (**Source:** Modeled from *Statistical Abstract of the United States, 2001*, Table 979, p. 622.)

In 1999, at what rate was employer spending on printing and publishing industry employee earnings increasing?

Exercises 26–30 are intended to challenge your understanding of the Product Rule. Solve the problems without using the Technology Tip.

26. **Apple Farming**   Historically, many apple farmers spaced trees 40 feet by 40 feet apart (27 trees per acre). Trees typically took 25 years to reach their maximum production of 500 bushels per acre. (A bushel is about 44 pounds.) In recent years, agriculturists have created dwarf and semidwarf varieties that allow trees to be spaced 10 feet by 10 feet apart. (**Source:** USDA.)

Suppose that a farmer has an apple orchard with 40 trees per acre. The orchard yields 10 bushels per tree. The farmer estimates that for each additional tree planted per acre, the average yield per tree is reduced by 0.1 bushel.

If $y = f(x)$ is the total number of bushels of apples produced per acre when an additional x trees per acre are planted, calculate and interpret the meaning of $f'(25)$, $f'(30)$, and $f'(35)$.

27. Apple Farming   Suppose that a farmer has an apple orchard with 30 trees per acre. The orchard yields 12 bushels per tree. The farmer estimates that for each additional tree planted per acre, the average yield per tree is reduced by 0.1 bushel.

If $y = f(x)$ is the total number of bushels of apples produced per acre when an additional x trees per acre are planted, calculate and interpret the meaning of $f'(40)$, $f'(45)$, and $f'(50)$.

28. Based on the results of Exercises 26 and 27, do you think that planting a particular number of trees will maximize the yield per acre? Justify your answer.

29. Apple Supplier Prices   A fruit farmer sells apples to a grocery store chain. The amount of apples the store buys depends linearly upon the price per pound that the farmer charges. The farmer estimates that for every $0.02 per pound increase in the price, the store will reduce its order by 44 pounds. The store currently orders 440 pounds per week and pays $0.18 per pound. What price should the farmer charge in order to maximize her revenue from apple sales?

30. Apple Retailer Prices   A grocery store has priced apples at $0.65 per pound and sells 1000 pounds per week. The amount of apples the store sells depends linearly upon the price per pound that the store charges. The store manager estimates that for every $0.05 per pound increase in the price, the store will reduce its sales by 100 pounds. What price should the store charge for its apples in order to maximize revenue from apple sales?

4.3 The Chain Rule

- Use the Chain Rule to find the derivative of a function

GETTING **STARTED**   Based on data from 1995 to 1999, the average annual earnings of an employee in the lumber and wood products industry may be modeled by

$$S(n) = -0.7685n^2 + 1295n - 516{,}596 \text{ dollars}$$

where n is the number of employees in thousands. (**Source:** Modeled from *Statistical Abstract of the United States, 2001*, Table 979, p. 622.)

Based on data from 1995 to 1999, the number of employees in the lumber and wood products industry may be modeled by

$$n(t) = 3.143t^2 + 5.029t + 772.5 \text{ thousand employees}$$

where t is the number of years since 1995. (**Source:** Modeled from *Statistical Abstract of the United States, 2001*, Table 979, p. 622.)

How quickly was the average annual earnings of an employee increasing in 1998? We will answer this question using the Chain Rule.

In this section, we will demonstrate how to find the composition of two functions. We will then demonstrate how to use the Chain Rule to find the derivative of a composition of functions.

Composition of Functions

A composition of two functions occurs when the outputs of one function are the inputs of a second function. Let's consider a common example. Let f represent the freezer function and b represent the blender function.

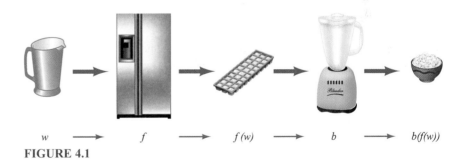

$$w \quad\longrightarrow\quad f \quad\longrightarrow\quad f(w) \quad\longrightarrow\quad b \quad\longrightarrow\quad b(f(w))$$

FIGURE 4.1

$b(f(w))$ means that first w is plugged into the function f, and then the result is substituted into b. In our example, water was placed in the freezer, and the result, ice, was put in the blender.

EXAMPLE 1

Finding a Composition of Functions

Let $f(t) = 3t + 2$ and $g(t) = t^2 + 1$. Find $f(g(t))$.

SOLUTION

$$
\begin{aligned}
f(g(t)) &= f(t^2 + 1) && \text{The output of the function } g \text{ is the input for } f\\
&= 3(t^2 + 1) + 2 && \text{The variable in } f \text{ is replaced with the output of } g\\
&= 3t^2 + 3 + 2\\
&= 3t^2 + 5
\end{aligned}
$$

Is $f(g(t))$ the same as $g(f(t))$? Let's return to the freezer and blender example and see.

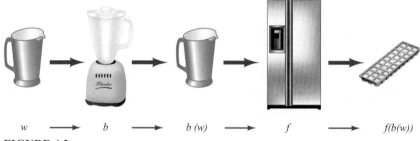

$$w \longrightarrow b \longrightarrow b\,(w) \longrightarrow f \longrightarrow f(b(w))$$

FIGURE 4.2

When we reversed the order of the functions, we got a different result. *Ice* is not the same as *chopped ice*. There are certain functions for which the two different compositions are equal, but, in general, $f(g(t)) \neq g(f(t))$.

EXAMPLE 2

Finding a Composition of Functions

Let $f(t) = 3t + 2$ and $g(t) = t^2 + 1$. Find $g(f(t))$.

SOLUTION

$$
\begin{aligned}
g(f(t)) &= g(3t + 2)\\
&= (3t + 2)^2 + 1\\
&= 9t^2 + 12t + 4 + 1\\
&= 9t^2 + 12t + 5
\end{aligned}
$$

Note that although Examples 1 and 2 used the same functions f and g, $f(g(t)) \neq g(f(t))$.

The Chain Rule

The Chain Rule is used whenever we have to find the derivative of a composition of functions.

CHAIN RULE

The derivative of a function $h(x) = f(g(x))$ is given by

$$h'(x) = f'(g(x)) \cdot g'(x)$$

The Chain Rule tells us to find $f'(x)$, substitute $g(x)$ in for the variable x, and then multiply the result by $g'(x)$.

EXAMPLE 3

Finding a Derivative by Using the Chain Rule

Given $h(x) = (2x + 1)^2 - 4$, find $h'(x)$.

SOLUTION Because there is a nonconstant function $(2x + 1)$ inside the parentheses, we know that we will need the Chain Rule. The function $h(x) = (2x + 1)^2 - 4$ is a function of the form $h(x) = f(g(x))$. We must determine the functions f and g. Often, selecting the function inside the parentheses as g gives the desired result. In this case, we pick $g(x) = 2x + 1$. Then

$$f(g(x)) = (2x + 1)^2 - 4$$
$$f(g(x)) = [g(x)]^2 - 4$$

To find the function f, we simply replace the $g(x)$ in the equation with the variable x. Thus we have

$$f(x) = x^2 - 4$$

We'll first find $f'(x)$ and then use it to calculate $f'(g(x))$.

$$f'(x) = \frac{d}{dx}(x^2 - 4)$$
$$= 2x$$
$$f'(g(x)) = 2 \cdot g(x) \qquad \text{By substitution}$$
$$= 2(2x + 1)$$

Next we'll find $g'(x)$.

$$g'(x) = \frac{d}{dx}(2x + 1)$$
$$= 2$$

The derivative of $f(g(x))$ is

$$f'(g(x)) \cdot g'(x) = [2(2x + 1)] \cdot (2)$$
$$= 4(2x + 1)$$
$$= 8x + 4$$

We'll rework the problem using a slightly different but somewhat easier approach.

$$h(x) = (2x + 1)^2 - 4$$

$$\frac{dh}{dx} = \frac{d}{dx}[(2x + 1)^2] - \frac{d}{dx}(4) \qquad \text{Sum and Difference Rule}$$

To use the Chain Rule to differentiate $(2x + 1)^2$, we treat the expression $2x + 1$ inside the parentheses as if it were a single variable and use the Power Rule. Then we multiply the result by the derivative of $2x + 1$, the expression inside the parentheses.

$$\frac{dh}{dx} = 2(2x + 1)^{2-1} \cdot \frac{d}{dx}(2x + 1) - 0 \qquad \text{Chain, Power, and Constant Rules}$$

$$= 2(2x + 1) \cdot (2)$$

$$= 4(2x + 1)$$

$$= 8x + 4$$

The second technique shown in Example 3 is a demonstration of the Generalized Power Rule.

GENERALIZED POWER RULE

Let $u = g(x)$. The derivative of a function $f(x) = u^n$ is given by

$$f'(x) = nu^{n-1}u'$$

The Generalized Power Rule is a special application of the Chain Rule. We'll use this rule in the next two examples.

EXAMPLE 4

Finding a Derivative by Using the Generalized Power Rule

Differentiate $f(x) = 4(2x^2 + x)^3$.

SOLUTION We pick $u = 2x^2 + x$. Then $f(x) = 4(2x^2 + x)^3$ may be rewritten as $f(x) = 4u^3$. From the Generalized Power Rule, we know that

$$\frac{d}{dx}(f(x)) = \frac{d}{dx}(4u^3)$$

$$= 4\frac{d}{dx}(u^3) \qquad \text{Constant Multiple Rule}$$

$$f'(x) = 12u^2u' \qquad \text{Generalized Power Rule}$$

We need to write the result in terms of x, not u. Since $u = 2x^2 + x$ and $u' = 4x + 1$, the derivative may be rewritten as

$$f'(x) = 12u^2u'$$
$$= 12(2x^2 + x)^2(4x + 1)$$
$$= 12[x(2x + 1)]^2(4x + 1)$$
$$= 12(x^2)(2x + 1)^2(4x + 1)$$
$$= 12x^2(2x + 1)^2(4x + 1)$$

We will often write derivatives in factored form. In real-life optimization problems, we are often interested in the values of x that make the derivative equal to zero. These values are easiest to find if the function is written in factored form.

EXAMPLE 5

Finding a Derivative by Using the Generalized Power Rule

Find $\dfrac{dy}{dx}$ given $y = \dfrac{1}{2x^3 + 4x + 1}$.

SOLUTION We first rewrite $y = \dfrac{1}{2x^3 + 4x + 1}$ as $y = (2x^3 + 4x + 1)^{-1}$. If we pick $u = 2x^3 + 4x + 1$, then $y = (2x^3 + 4x + 1)^{-1}$ may be rewritten as $y = u^{-1}$. From the Generalized Power Rule, we know that

$$\frac{d}{dx}(y) = \frac{d}{dx}(u^{-1})$$

$$\frac{dy}{dx} = -u^{-2}u' \qquad \text{Generalized Power Rule}$$

$$= -\frac{u'}{u^2}$$

Since $u = 2x^3 + 4x + 1$, $u' = 6x^2 + 4$. Thus

$$\frac{dy}{dx} = -\frac{u'}{u^2}$$

$$= -\frac{6x^2 + 4}{(2x^3 + 4x + 1)^2}$$

$$= -\frac{2(3x^2 + 2)}{(2x^3 + 4x + 1)^2}$$

As demonstrated in the next example, it is necessary to apply the Chain Rule repeatedly for some functions.

EXAMPLE 6

Finding a Derivative by Using the Chain Rule Repeatedly

Find the derivative of $S(t) = [3(2t + 1)^2 + 5]^3$.

SOLUTION We let $u = 3(2t + 1)^2 + 5$. Notice that this function also contains an expression inside parentheses. If we let $v = 2t + 1$, then $u = 3v^2 + 5$.

$$S(t) = u^3$$

$$\frac{dS}{dt} = 3u^2 u' \qquad\qquad\qquad\qquad\qquad \text{Generalized Power Rule}$$

$$= 3u^2 \frac{d}{dt}(u) \qquad\qquad\qquad\qquad \text{Since } u' = \frac{d}{dt}(u)$$

$$= 3(3v^2 + 5)^2 \cdot \frac{d}{dt}(3v^2 + 5) \qquad\quad \text{Since } u = 3v^2 + 5$$

$$= 3(3v^2 + 5)^2 \cdot 6vv' \qquad\qquad\qquad \text{Generalized Power Rule}$$

$$= 3(3v^2 + 5)^2 \cdot 6v \cdot \frac{d}{dt}(v) \qquad\qquad \text{Since } v' = \frac{d}{dt}(v)$$

$$= 3[3(2t + 1)^2 + 5]^2 \cdot 6(2t + 1) \cdot \frac{d}{dt}(2t + 1) \qquad \text{Since } v = 2t + 1$$

$$= 3[3(2t + 1)^2 + 5]^2 \cdot 6(2t + 1) \cdot (2)$$

At this point, we have calculated the derivative. If we want to rewrite the derivative in its simplest factored form, we continue as follows:

$$\frac{dS}{dt} = 3\,[3(2t + 1)^2 + 5]^2 \cdot 6(2t + 1) \cdot (2)$$

$$= 36\,[3(2t + 1)^2 + 5]^2\,(2t + 1)$$

$$= 36\,[3(4t^2 + 4t + 1) + 5]^2\,(2t + 1)$$

$$= 36\,(12t^2 + 12t + 3 + 5)^2\,(2t + 1)$$

$$= 36\,(12t^2 + 12t + 8)^2\,(2t + 1)$$

$$= 36\,[4(3t^2 + 3t + 2)]^2\,(2t + 1)$$

$$= 36\,(4)^2\,(3t^2 + 3t + 2)^2\,(2t + 1)$$

$$= 576\,(3t^2 + 3t + 2)^2\,(2t + 1)$$

Writing the derivative in factored form required a significant number of additional steps. Although there are some instances that require us to write derivatives in factored form, we will consider the additional steps optional at this point.

Chain Rule: Alternative Form

The Chain Rule may also be expressed using Leibniz notation.

CHAIN RULE: ALTERNATIVE FORM

If $y = f(x)$ and $x = g(t)$, then $y = f(g(t))$ and

$$\frac{dy}{dt} = \frac{dy}{dx} \cdot \frac{dx}{dt}$$

The units of $\dfrac{dy}{dt}$ are obtained by multiplying the units of $\dfrac{dy}{dx}$ by the units of $\dfrac{dx}{dt}$.

EXAMPLE 7

Finding a Derivative by Using the Alternative Form of the Chain Rule

Based on data from 1989 to 1998, restaurant sales in the United States may be modeled by

$$S(r) = 0.9291r - 160.0 \text{ billion dollars}$$

where r is the number of restaurants (in thousands).

The number of restaurants may be modeled by

$$r(t) = 0.3405t^2 + 6.910t + 330.9 \text{ thousand restaurants}$$

where t is the number of years since 1989. (**Source:** Modeled from *Statistical Abstract of the United States, 2001*, Table 1268, p. 775.)

How quickly were restaurant sales increasing in 1998?

SOLUTION We are asked to find $\dfrac{dS}{dt}$. Recall that the alternative form of the Chain Rule states that $\dfrac{dS}{dt} = \dfrac{dS}{dr} \cdot \dfrac{dr}{dt}$. We will first find $\dfrac{dS}{dr}$ and $\dfrac{dr}{dt}$.

$$S(r) = 0.9291r - 160.0$$

UNITS
$$\frac{dS}{dr} = 0.9291 \frac{\text{billion dollars}}{\text{thousand restaurants}}$$

$$r(t) = 0.3405t^2 + 6.910t + 330.9$$

UNITS
$$\frac{dr}{dt} = 0.681t + 6.910 \frac{\text{thousand restaurants}}{\text{year}}$$

UNITS
$$\frac{dS}{dt} = \frac{dS}{dr} \cdot \frac{dr}{dt}$$

$$= \left(0.9291 \frac{\text{billion dollars}}{\text{thousand restaurants}}\right) \cdot \left(0.681t + 6.910 \frac{\text{thousand restaurants}}{\text{year}}\right)$$

$$= (0.9291) \cdot (0.681t + 6.910) \frac{\text{billion dollars}}{\text{thousand restaurants}} \cdot \frac{\text{thousand restaurants}}{\text{year}}$$

$$= (0.9291) \cdot (0.681t + 6.910) \text{ billion dollars per year}$$

$$\approx 0.6327t + 6.420 \text{ billion dollars per year}$$

We need to determine the rate of change in sales for 1998 ($t = 9$).

$$\left.\frac{dS}{dt}\right|_{t=9} = 0.6327(9) + 6.420$$

$$\approx 12.11 \text{ billion dollars per year}$$

(Note: The notation $\left.\dfrac{dS}{dt}\right|_{t=9}$ means the value of $\dfrac{dS}{dt}$ when $t = 9$.)

In 1998, restaurant sales in the United States were increasing at an estimated rate of 12.11 billion dollars per year.

We'll now return to the problem introduced at the beginning of this section.

EXAMPLE 8

Finding a Derivative by Using the Alternative Form of the Chain Rule

Based on data from 1995 to 1999, the average annual earnings of an employee in the lumber and wood products industry may be modeled by

$$S(n) = -0.7685n^2 + 1295n - 516,596 \text{ dollars}$$

where n is the number of employees in thousands. (**Source:** Modeled from *Statistical Abstract of the United States, 2001*, Table 979, p. 622.)

Based on data from 1995 to 1999, the number of employees in the lumber and wood products industry may be modeled by

$$n(t) = 3.143t^2 + 5.029t + 772.5 \text{ thousand employees}$$

where t is the number of years since 1995. (**Source:** Modeled from *Statistical Abstract of the United States, 2001*, Table 979, p. 622.)

How quickly was the average annual earnings of an employee increasing in 1998?

SOLUTION We are asked to find $\dfrac{dS}{dt}$. By the Chain Rule, $\dfrac{dS}{dt} = \dfrac{dS}{dn} \cdot \dfrac{dn}{dt}$. We will first find $\dfrac{dS}{dn}$ and $\dfrac{dn}{dt}$.

$$S(n) = -0.7685n^2 + 1295n - 516,596$$

UNITS

$$\frac{dS}{dn} = -1.537n + 1295 \; \frac{\text{dollars}}{\text{thousand employees}}$$

$$n(t) = 3.143t^2 + 5.029t + 772.5$$

UNITS

$$\frac{dn}{dt} = 6.286t + 5.029 \; \frac{\text{thousand employees}}{\text{year}}$$

UNITS

$$\frac{dS}{dt} = \frac{dS}{dn} \cdot \frac{dn}{dt}$$

$$= \left(-1.537n + 1295 \; \frac{\text{dollars}}{\text{thousand employees}} \right) \cdot \left(6.286t + 5.029 \; \frac{\text{thousand employees}}{\text{year}} \right)$$

$$= (-1.537n + 1295) \cdot (6.286t + 5.029) \frac{\text{dollars}}{\text{thousand employees}} \; \frac{\text{\sout{thousand employees}}}{\text{year}}$$

$$= [-1.537(3.143t^2 + 5.029t + 772.5) + 1295] \cdot (6.286t + 5.029) \text{ dollars per year}$$

$$\approx (-4.8308t^2 - 7.7296t - 1187.3325 + 1295) \cdot (6.286t + 5.029)$$

$$= (-4.8308t^2 - 7.7296t + 107.6675) \cdot (6.286t + 5.029) \text{ dollars per year}$$

We must find $\dfrac{dS}{dt}\Big|_{t=3}$, since $t = 3$ in the year 1998.

$$\frac{dS}{dt}\Big|_{t=3} = [-4.8308(3)^2 - 7.7296(3) + 107.6675] \cdot [6.286(3) + 5.029]$$

$$= (-43.4772 - 23.1888 + 107.6675) \cdot (23.887)$$

$$= (41.0015) \cdot (23.887)$$

$$\approx 979.40 \text{ dollars per year}$$

According to the model, the average annual earnings of a lumber and wood products industry employee was increasing at a rate of roughly $979 per year in 1998.

4.3 Summary

In this section you learned how to compose two functions. You also discovered how to use the Chain Rule to find the derivative of a composition of functions.

4.3 Exercises

In Exercises 1–5, find the function $f(g(x))$. Then simplify the result.

1. $f(x) = 2x + 3, g(x) = x^2 + 4$

2. $f(x) = -2x^2 + 7x, g(x) = x - 10$

3. $f(x) = 7x - 4, g(x) = x^2 - 2x + 1$

4. $f(x) = x^2 - 1, g(x) = x^2 + 1$

5. $f(x) = 3x^2 - 11x, g(x) = x^3$

In Exercises 6–10, the function is of the form $h(x) = f(g(x))$. Determine $f(x)$ and $g(x)$.

6. $h(x) = 5(x^2 - 4)^3 + 1$

7. $h(x) = (x + 1)^2 - 6(x + 1) + 9$

8. $h(x) = e^{2x-4}$ **9.** $h(x) = 2\ln(x^2 + 4)$

10. $h(x) = 4(x^3 - 1)^2 - x^3 + 1$

In Exercises 11–20, use the Chain Rule in finding the derivative of the function.

11. $f(x) = 5(3x + 1)^2$

12. $g(t) = 4(6t - 7)^3 + 2$

13. $h(t) = 3(t^3 - t^2 + 1)^4$

14. $S(n) = (n^3 - 3n^2)^2$

15. $f(n) = 5(n^2 + 2n + 1)^2 + 6$

16. $g(x) = [4(3x + 5)^2 + x]^2$

17. $f(x) = [-5(4x - 1)^2 + 2x]^2$

18. $s(t) = (2t + 1)^2(6t)$

19. $f(t) = (5t^2 + t)(3t + 2)^2$

20. $h(x) = (4x + 1)^2(-x + 2)^2$

In Exercises 21–25, determine the slope of the graph at the indicated domain value.

21. $g(x) = \dfrac{x^2 + 3}{x^2 - x}; x = 2$

22. $r(x) = \dfrac{2x^2 - 8}{2x^2 + 8}; x = 0$

23. $R(x) = \dfrac{3x + 1}{2x + 5}; x = 2$

24. $P(t) = (3t - 1)^{-3}(5t - 2); t = 0.4$

25. $f(x) = [5(4x + 1)^2 + 2]^2; x = -1$

In Exercises 26–30, find the equation of the tangent line at the indicated domain value.

26. $R(n) = [5(-n + 5)^3 - 2n]^2; n = 5$

27. $s(t) = (3t + 1)^2[2(-4t - 1) + 6]^{-2}; t = 0$

28. $g(x) = [5x(x^2 + 1)^3 + (x + 1)^3]^2; x = -1$

29. $p(x) = \{[(x + 1)^2 + 2]^3 + 3\}^4; x = -1$

30. $r(t) = \{[(2t + 1)^2 - 1]^3 + 2\}^4; t = 0$

In Exercises 31–32, use the Chain Rule to answer the questions.

31. **Average Height of a Girl** The average height of a 2- to 13-year-old girl may be modeled by
$$F(m) = 1.071m - 3.554 \text{ inches}$$
where m is the average height of a boy of the same age.

The average height of a 2- to 13-year-old boy may be modeled by

$$m(a) = -0.0507a^2 + 2.997a + 30.44 \text{ inches}$$

where a is the age of the boy in years. (**Source:** Modeled from www.babybag.com data.)

According to the models, how quickly is the height of a girl increasing when she is 10 years old?

32.   **Average Weight of a Boy**   The average weight of a 2- to 13-year-old boy may be modeled by

$$M(f) = 0.9099f + 5.065 \text{ pounds}$$

where f is the average weight of a girl of the same age.

The average weight of a 2- to 13-year-old girl may be modeled by

$$f(a) = 0.289a^2 + 2.464a + 23.10 \text{ pounds}$$

where a is the age of the girl in years. (**Source:** Modeled from www.babybag.com data.)

According to the models, how quickly is the average weight of a boy increasing when he is 10 years old?

Exercises 33–35 are intended to challenge your understanding of the Chain Rule.

33. Find the derivative of
$$f(x) = [(x^2 + 1)^2 + 1]^2 + 1$$

34. Differentiate $g(t) = (t^2 + 1)^3(t^2 + 3)^4$

35. Given that $\dfrac{dy}{dt} = 5$ people per day,

$\dfrac{dt}{dx} = 2$ days per year, and

$\dfrac{dp}{dy} = 4$ sales per person, calculate

$\dfrac{dp}{dx}$, including units.

4.4 Exponential and Logarithmic Rules

- Use the Exponential Rule and Logarithmic Rule to find the derivative of a function

GETTING STARTED   Despite the infamous "Got Milk?" advertising campaign, the per capita milk beverage consumption decreased from nearly 28 gallons per year in 1980 to less than 24 gallons per year in 1999. The annual per capita milk beverage consumption may be modeled by

$$M(t) = 27.76(0.9914)^t \text{ gallons}$$

where t is the number of years since 1980. Does the four-gallon decrease over the 19-year period indicate that consumer consumption of milk will continue to drop? According to the model, at what rate was consumption decreasing in 2000? Questions such as these may be answered using the Exponential Rule for derivatives.

In this section, we will introduce the Exponential Rule and the Logarithmic Rule for derivatives. These rules will allow us to calculate the exact derivative of these types of functions quickly and accurately instead of relying upon a numerical estimate as we have done previously.

Exponential Rule

Recall that an exponential function may be written in the form $y = ab^x$, where a and b are constants and b is positive and not equal to 1. Since the variable of an exponential function is in the exponent, the Power Rule cannot be used to find the derivative of this type of function.

EXPONENTIAL RULE

The derivative of a function $f(x) = b^x$ is given by

$$f'(x) = (\ln b)b^x$$

This rule is not entirely obvious. To better understand its origin, let's consider the limit definition of the derivative of an exponential function.

$$f'(x) = \lim_{h \to 0} \frac{f(x + h) - f(x)}{h}$$

$$= \lim_{h \to 0} \frac{b^{x+h} - b^x}{h} \qquad \text{Since } f(x) = b^x$$

$$= \lim_{h \to 0} \frac{b^x b^h - b^x}{h} \qquad \text{Algebraic rules of exponents}$$

$$= \lim_{h \to 0} \frac{b^x(b^h - 1)}{h}$$

$$= b^x \lim_{h \to 0} \frac{(b^h - 1)}{h} \qquad \text{Since } b^x \text{ is independent of the value of } h$$

Although we cannot simplify $\lim\limits_{h \to 0} \frac{(b^h - 1)}{h}$ algebraically, it turns out that $\lim\limits_{h \to 0} \frac{(b^h - 1)}{h} = \ln(b)$. Consequently,

$$f'(x) = b^x \lim_{h \to 0} \frac{(b^h - 1)}{h}$$

$$= (\ln b)b^x$$

Exponential functions frequently use the number e as the base. Recall that $e \approx 2.7182818$.

EXAMPLE 1

Finding the Derivative of an Exponential Function

Differentiate $y = e^x$.

SOLUTION

$$\frac{dy}{dx} = (\ln e)e^x$$

$$= 1 \cdot e^x \qquad \text{Since } \ln e = 1$$

$$= e^x$$

The derivative of e^x is e^x! This quirky result makes the function $y = e^x$ one of the most popular functions in calculus. For the function $f(x) = e^x$, the instantaneous rate of change in the function at a point $(a, f(a))$ is the same as $f(a)$. In other words, the slope of the tangent line to the graph of $f(x) = e^x$ at $x = a$ is $f(a)$.

EXPONENTIAL RULE: SPECIAL CASE

The derivative of a function $f(x) = e^x$ is given by

$$f'(x) = e^x$$

EXAMPLE 2 **Finding the Derivative of an Exponential Function**

Find the derivative of $g(x) = 2^x$.

SOLUTION

$$g'(x) = (\ln 2)2^x$$

EXAMPLE 3 **Finding the Derivative of an Exponential Function**

Calculate $\dfrac{d}{dx}(3 \cdot 4^x)$.

SOLUTION

$$\frac{d}{dx}(3 \cdot 4^x) = 3 \cdot \frac{d}{dx}(4^x) \qquad \text{Constant Multiple Rule}$$

$$= 3(\ln 4)4^x \qquad \text{Exponential Rule}$$

EXAMPLE 4 **Finding a Derivative by Using the Exponential and Chain Rules**

Find y' given $y = 3^{x^2+x}$.

SOLUTION This function is different because there is a function in the exponent rather than a single variable. Consequently, we'll have to use the Chain Rule in conjunction with the Exponential Rule to find the derivative.

$$y' = (\ln 3)3^{x^2+x} \cdot \frac{d}{dx}(x^2 + x) \qquad \text{Exponential and Chain Rules}$$

$$= (\ln 3)3^{x^2+x} \cdot (2x + 1) \qquad \text{Power Rule}$$

$$= (2x + 1)(\ln 3)3^{x^2+x}$$

EXAMPLE 5

Finding a Derivative by Using the Exponential and Chain Rules

Differentiate $y = e^{5x-6}$.

SOLUTION

$$\frac{dy}{dx} = (\ln e)e^{5x-6} \cdot \frac{d}{dx}(5x-6) \qquad \text{Exponential and Chain Rules}$$

$$= (1)e^{5x-6} \cdot (5) \qquad\qquad \text{Constant, Constant Multiple, and}$$
$$\qquad\qquad\qquad\qquad\qquad\qquad \text{Power Rules}$$

$$= 5e^{5x-6}$$

In general, a function of the form $y = e^u$, where u is some function of x, will have the derivative $y' = e^u \cdot u'$. This is referred to as the Generalized Exponential Rule; it is a special application of the Chain Rule together with the Exponential Rule.

GENERAL EXPONENTIAL RULE

Let $u = g(x)$. The derivative of a function $f(x) = b^u$ is given by

$$f'(x) = (\ln b)b^u u'$$

If $b = e$, the rule simplifies to

$$f'(x) = e^u u'$$

EXAMPLE 6

Finding a Derivative by Using Multiple Derivative Rules

Find the derivative of $P(t) = 2te^{3t}$.

SOLUTION Although this function looks relatively simple, there is a lot going on. To find the derivative, we must use the Constant Multiple Rule, the Power Rule, the Product Rule, and the Generalized Exponential Rule.

$$\frac{dP}{dt} = \frac{d}{dt}(2t) \cdot e^{3t} + \frac{d}{dt}(e^{3t}) \cdot (2t) \qquad \text{Product Rule}$$

$$= 2e^{3t} + (e^{3t} \cdot 3) \cdot (2t) \qquad\qquad \text{Constant Multiple, Generalized}$$
$$\qquad\qquad\qquad\qquad\qquad\qquad\qquad \text{Exponential, and Power Rules}$$

$$= 2e^{3t} + 6te^{3t}$$

$$= e^{3t}(2 + 6t)$$

$$= 2e^{3t}(3t + 1)$$

EXAMPLE 7

Finding a Derivative by Using the General Exponential Rule

The annual per capita milk beverage consumption may be modeled by

$$M(t) = 27.76(0.9914)^t \text{ gallons}$$

where t is the number of years since 1980. According to the model, at what rate was consumption decreasing in 2000?

SOLUTION

$$\frac{dM}{dt} = 27.76[\ln (0.9914)](0.9914)^t$$

$$\approx -0.2398(0.9914)^t \text{ gallons per year} \qquad \text{Evaluate ln (0.9914) and multtiply by 27.76}$$

In 2000, $t = 20$, so

$$\frac{dM}{dt}\bigg|_{t=20} = -0.2398(0.9914)^{20} \text{ gallons per year}$$

$$\approx -0.2398(0.8414)$$

$$\approx -0.2018$$

In 2000, the per capita milk beverage consumption was decreasing by about two-tenths (0.2) of a gallon per year.

It is important to note that although the per capita consumption of milk was decreasing, overall milk beverage sales still could have increased as a result of an increase in the U.S. population.

Logarithmic Rule

What is the derivative of a function of the form $y = \log_b x$? The answer is not obvious. We will state the rule here and show how the formula was determined when we discuss implicit differentiation in a subsequent section.

LOGARITHMIC RULE

The derivative of a function $f(x) = \log_b x$ is given by

$$f'(x) = \frac{1}{\ln b} \cdot \frac{1}{x}$$

EXAMPLE 8

Finding the Derivative of a Logarithmic Function

Differentiate $y = \ln x$.

SOLUTION Recall that $\ln x = \log_e x$.

$$y' = \frac{1}{\ln e} \cdot \frac{1}{x}$$

$$= \frac{1}{1} \cdot \frac{1}{x}$$

$$= \frac{1}{x}$$

EXAMPLE 9

Finding the Derivative of a Logarithmic Function

Find the derivative of $y = 3 \log t$.

SOLUTION Recall that $\log t$ means $\log_{10} t$.

$$\frac{dy}{dt} = \frac{d}{dt}(3 \log t)$$

$$= 3 \cdot \frac{d}{dt}(\log_{10} t) \qquad \text{Constant Multiple Rule}$$

$$= 3 \cdot \frac{1}{\ln 10} \cdot \frac{1}{t} \qquad \text{Logarithmic Rule}$$

$$= \frac{3}{\ln 10} \cdot \frac{1}{t}$$

GENERALIZED LOGARITHMIC RULE

The derivative of a function $f(x) = \ln u$ where u is a function of x is given by

$$f'(x) = \frac{1}{u}u'.$$

The Generalized Logarithmic Rule is simply a combination of the Logarithmic Rule and the Chain Rule.

EXAMPLE 10

Finding the Derivative by Using the Generalized Logarithmic Rule

Find $\frac{dy}{dx}$ for $y = \log(x^2)$.

SOLUTION Since y is the composition of $f(x) = \log x$ and $g(x) = x^2$, we must use the Chain Rule in conjunction with the Logarithmic Rule.

$$\frac{dy}{dx} = \left(\frac{1}{\ln 10} \cdot \frac{1}{x^2}\right) \cdot \frac{d}{dx}(x^2) \qquad \text{Generalized Logarithmic Rule}$$

$$= \left(\frac{1}{\ln 10} \cdot \frac{1}{x^2}\right) \cdot (2x) \qquad \text{Power Rule}$$

$$= \frac{1}{\ln 10} \cdot \frac{2x}{x^2}$$

$$= \frac{2}{(\ln 10)x}$$

EXAMPLE 11 **Finding a Derivative by Using Multiple Rules**

Differentiate $P(t) = e^{2t} \ln(3t)$.

SOLUTION We must use the Product Rule, the Chain Rule, the Exponential Rule, the Constant Multiple Rule, and the Power Rule in finding the derivative.

$$P'(t) = \frac{d}{dt}(e^{2t})\ln(3t) + \frac{d}{dt}[\ln(3t)]e^{2t} \qquad \text{Product Rule}$$

$$= (e^{2t} \cdot 2)\ln(3t) + \left(\frac{1}{3t} \cdot 3\right)e^{2t} \qquad \text{Generalized Exponential, Logarithmic,} \\ \text{Constant Multiple, and Power Rules}$$

$$= 2e^{2t}\ln(3t) + \left(\frac{3}{3t}\right)e^{2t}$$

$$= e^{2t}\left[2\ln(3t) + \frac{1}{t}\right]$$

EXAMPLE 12 **Applying the Logarithmic Rule in a Real-World Context**

The year in which the per capita consumption of bottled water will reach w gallons may be modeled by

$$T(w) = -9.249 + 9.839(\ln w) \text{ years since 1980}$$

(**Source:** *Statistical Abstract of the United States, 2001*, Table 204, p 130.) Calculate and interpret the meaning of $T'(20)$.

SOLUTION

$$T'(w) = 9.839 \cdot \frac{1}{w} \qquad \text{Constant and} \\ \text{Logarithmic Rules}$$

UNITS

$$T'(20) = \frac{9.839}{20} \frac{\text{years since 1980}}{\text{gallon}}$$

$$= 0.4920 \text{ year per gallon}$$

When the annual per capita consumption of bottled water is 20 gallons, it will take about 0.4920 year (roughly 6 months) for the annual per capita consumption to increase to 21 gallons.

EXAMPLE 13 **Applying the Logarithmic Rule in a Real-World Context**

Based on data from 1989 to 1999, the percentage of auto fatalities that are not alcohol-related may be modeled by

$$F(t) = 49.96 + 5.112 \ln t \text{ percent}$$

where t is the number of years since 1989. (**Source:** *Statistical Abstract of the United States, 2001*, Table 1099, p 688.) According to the model, how quickly was the percentage of non-alcohol-related fatalities increasing in 2000?

SOLUTION

UNITS

$$\frac{dF}{dt} = \frac{d}{dt}[49.96 + 5.112(\ln t)]$$

$$= 0 + 5.112 \cdot \frac{1}{t} \qquad \text{Constant, Constant Multiple, and Logarithmic Rules}$$

$$= \frac{5.112}{t} \frac{\text{percentage points}}{\text{year}}$$

In 2000, $t = 11$, so

$$\frac{dF}{dt}\bigg|_{t=11} = \frac{5.112}{11} \frac{\text{percentage points}}{\text{year}}$$

$$\approx 0.4647 \text{ percentage point per year}$$

In 2000, the percentage of *non-alcohol-related* auto fatalities was increasing by 0.46 percentage point per year. This means that *alcohol-related* auto fatalities were declining by about 0.46 percentage point per year in 2000, a good sign.

4.4 Summary

In this section, you learned how to use the Exponential Rule and the Logarithmic Rule for derivatives. You discovered that these rules allow you to calculate the exact derivative of exponential and logarithmic functions instead of having to rely on a numerical estimate.

4.4 Exercises

In Exercises 1–10, use the Exponential Rule or the Logarithmic Rule, as appropriate, to find the derivative of the function.

1. $y = 4^x$

2. $g(t) = 3(5)^t$

3. $s(t) = 4 \ln t$

4. $f(t) = 3 \log t$

5. $y = 5e^x + \ln x$

6. $P(n) = 4n^2(2^n)$

7. $f(x) = 5e^x \ln x$

8. $g(x) = e^{2x-1}$

9. $y = 2^{x^2+5x}$

10. $h(x) = \ln(x^3 + 3x)$

In Exercises 11–20, determine the slope of the function at the indicated domain value.

11. $S(n) = \log(n^2 - 4); n = 3$

12. $P(x) = xe^x; x = 0$

13. $w(t) = \log(2^t); t = -1$

14. $f(x) = 4 \log(3^x); x = 0$

15. $H(t) = 3^{\ln t}; t = e$

16. $y = 2^{\log 5t}; t = 0.2$

17. $f(t) = 3 \log(5^t); t = 0$

18. $y = \log(5x) + 4^{2x+5}; x = 1$

19. $C(n) = 3ne^{n^2+2n-1}; n = 1$

20. $g(t) = 5t(e^{t-2}); t = 3$

In Exercises 21–25, determine the equation of the tangent line at the indicated domain value.

21. $f(t) = 2^{3 \ln t}; t = 1$

22. $y = 5x^2(2^{3x}); x = 0$

23. $g(x) = [3 \log(4x)](3^x); x = 1$

24. $h(x) = \ln(3e^{5x}); x = -1$

25. $f(x) = \frac{\ln x}{x}; x = e$

In Exercises 26–30, answer the questions by using the Exponential Rule or the Logarithmic Rule, as appropriate.

26. **M** **Teacher Salaries**   Based on data from 1990 to 2000, the average salary of a public elementary or secondary school teacher may be modeled by

$$S(t) = 32.05(1.027)^t \text{ thousand dollars}$$

where t is the number of years since 1990.

(**Source**: Modeled from *Statistical Abstract of the United States, 2001*, Table 237, p. 151.)

At what rate (in thousands of dollars per year) were salaries increasing in 2000?

27. **M** **College Tuition**   Based on data from 1994 to 2001, the quarterly tuition of a resident student at Green River Community College may be modeled by

$$E(t) = 430.6(1.042)^t \text{ dollars}$$

where t is the number of years since 1994.

(**Source**: Modeled from Green River Community College data.)

How quickly was tuition increasing in 2001?

28. **M** **NBA Player Salaries**   Based on data from 1980 to 1998, the average salary of a National Basketball Association (NBA) player may be modeled by

$$P(t) = 161(1.17)^t \text{ thousand dollars}$$

where t is the number of years since 1980.

(**Source**: Modeled from *Statistical Abstract of the United States, 2001*, Table 1324, p. 829.)

Calculate and interpret the meaning of $P'(18)$.

29. **M** **VCR Usage**   Based on data from 1985 to 1999, the percentage of homes with a VCR may be modeled by

$$V(t) = 22.24 + 24.47 \ln t \text{ percent}$$

where t is the number of years since 1984.

(**Source**: Modeled from *Statistical Abstract of the United States, 2001*, Table 1126, p. 705.)

Calculate and interpret the meaning of $V'(15)$.

30. **M** **Army Personnel**   Based on data from 1980 to 2000, the number of active-duty Army personnel may be modeled by

$$A(t) = \frac{302.2}{1 + 0.0001047e^{0.7731t}} + 480 \text{ thousand people}$$

where t is the number of years since 1980.

(**Source**: Modeled from *Statistical Abstract of the United States, 2001*, Table 500, p. 329.)

At what rate was the number of active-duty army personnel changing in 2000?

Exercises 31–35 are intended to challenge your understanding of the Exponential and Logarithmic Rules.

31. Differentiate $f(x) = 2^{\ln(x^2)}$.

32. Find $\dfrac{dy}{dx}$ given $y = 2^x \ln(2x)$.

33. Find $g'(t)$ given $g(t) = 2^{2t} \ln(2^{2t})$.

34. Differentiate $h(t) = 3^{2^{4t}}$.

35. Find $\dfrac{dx}{dt}$ given $x(t) = \ln[\ln(3^{2^t})]$.

4.5 Implicit Differentiation

- Use implicit differentiation to differentiate functions and nonfunctions

GETTING STARTED   The intent of this section is to give you the skills necessary to work the related-rate problems in the next chapter. We will show you how to use *implicit differentiation* to find the derivative of functions and nonfunctions. Knowing how to do implicit differentiation is a key skill that is necessary for our discussion of related rates.

The equation of a circle may be written as $x^2 + y^2 = r^2$, where r is the radius of the circle. A circle is not a function because it fails the Vertical Line Test. However, we can draw tangent lines to the graph of a circle. Since the derivative is the slope of the tangent line, we should be able to find the derivative at any point on the circle. However, since most x values have two different y values, the derivative will be a function of both x and y.

Consider $x^2 + y^2 = 1$ and its associated graph Figure 4.3. What is the slope of each of the tangent lines shown? In other words, what is the derivative at the indicated points?

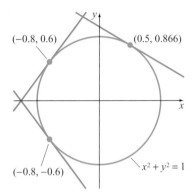

FIGURE 4.3

We will find the derivative by using **implicit differentiation.** We begin by taking the derivative of both sides of the equation with respect to x. Since y is a function of x, $\dfrac{d}{dx}(y) = \dfrac{dy}{dx}$. This fact will become important as we move through the problem.

$$x^2 + y^2 = 1$$

$$\frac{d}{dx}(x^2 + y^2) = \frac{d}{dx}(1)$$

$$\frac{d}{dx}(x^2) + \frac{d}{dx}(y^2) = 0 \qquad \text{Sum and Difference and Constant Rules}$$

$$2x + 2y \cdot \frac{d}{dx}(y) = 0 \qquad \text{Power and Chain Rules}$$

$$2x + 2y \cdot \frac{dy}{dx} = 0$$

$$2y\frac{dy}{dx} = -2x$$

$$\frac{dy}{dx} = \frac{-2x}{2y}$$

$$\frac{dy}{dx} = -\frac{x}{y}$$

Note that the derivative is a function of both x and y. We can now calculate the slope of the tangent line at each of the points shown.

At $(-0.8, 0.6)$, $\dfrac{dy}{dx} = -\dfrac{-0.8}{0.6} \approx 1.333$.

At $(-0.8, -0.6)$, $\dfrac{dy}{dx} = -\dfrac{-0.8}{-0.6} \approx -1.333$.

At $(0.5, 0.866)$, $\dfrac{dy}{dx} = -\dfrac{0.5}{0.866} \approx -0.5774$.

HOW TO **Steps of Implicit Differentiation**

1. Differentiate both sides of the equation with respect to x.
$$\left(\text{Recall } \frac{d}{dx}(y) = \frac{dy}{dx}.\right)$$

2. Algebraically isolate the $\dfrac{dy}{dx}$ term.

EXAMPLE 1   **Finding $\dfrac{dy}{dx}$ by Using Implicit Differentiation**

Differentiate $xy = 1$ using implicit differentiation.

SOLUTION

$$\frac{d}{dx}(xy) = \frac{d}{dx}(1)$$

$$\frac{d}{dx}(x) \cdot y + \frac{d}{dx}(y) \cdot x = 0 \qquad \text{Product Rule}$$

$$1 \cdot y + \frac{dy}{dx} \cdot x = 0 \qquad \text{Power Rule}$$

$$y + x\frac{dy}{dx} = 0$$

$$x\frac{dy}{dx} = -y$$

$$\frac{dy}{dx} = -\frac{y}{x}$$

If we first solve the equation for y and then differentiate, we get a very different-looking result.

$$xy = 1$$

$$y = \frac{1}{x}$$

$$\frac{d}{dx}(y) = \frac{d}{dx}\left(\frac{1}{x}\right)$$

$$\frac{dy}{dx} = \frac{d}{dx}(x^{-1})$$

$$= -x^{-2} \qquad \text{Power Rule}$$

$$= -\frac{1}{x^2}$$

Is this answer equivalent to our implicit differentiation result? Let's see.

$$\frac{dy}{dx} = -\frac{y}{x}$$

$$= -\frac{1}{x} y$$

$$= -\frac{1}{x} \cdot \frac{1}{x} \qquad \text{Since } y = \frac{1}{x}$$

$$= -\frac{1}{x^2}$$

The two equations are equivalent. When using implicit differentiation, it is common for correct answers to look very different from each other.

EXAMPLE 2   Finding $\dfrac{dy}{dx}$ by Using Implicit Differentiation

Differentiate $(x + y)^2 = 9$.

SOLUTION

$$\frac{d}{dx}(x + y)^2 = \frac{d}{dx}(9)$$

$$2(x + y) \cdot \frac{d}{dx}(x + y) = 0 \qquad \text{Constant and Chain Rules}$$

$$2(x + y) \cdot \left(1 + \frac{dy}{dx}\right) = 0 \qquad \text{Power Rule}$$

$$\left(1 + \frac{dy}{dx}\right) = \frac{0}{2(x + y)}$$

$$1 + \frac{dy}{dx} = 0$$

$$\frac{dy}{dx} = -1$$

For all values of x and y on the graph of $(x + y)^2 = 9$, the slope of the tangent line is -1.

Let's solve the original equation for y.

$$(x + y)^2 = 9$$

$$\sqrt{(x + y)^2} = \sqrt{9}$$

$$(x + y) = \pm 3$$

$$y = -x \pm 3$$

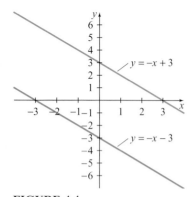

FIGURE 4.4

When we graph $y = -x - 3$ and $y = -x + 3$ simultaneously, we see a pair of parallel lines with slope -1 (Figure 4.4). This is consistent with the implicit differentiation result.

EXAMPLE **3**   Finding $\dfrac{dy}{dx}$ by Using Implicit Differentiation

Differentiate $x = y^2$.

SOLUTION

$$\frac{d}{dx}(x) = \frac{d}{dx}(y^2)$$

$$1 = 2y \cdot \frac{d}{dx}(y) \qquad \text{Power and Chain Rules}$$

$$1 = 2y \cdot \frac{dy}{dx}$$

$$\frac{dy}{dx} = \frac{1}{2y}$$

EXAMPLE **4**   Finding $\dfrac{dy}{dx}$ by Using Implicit Differentiation

Find the derivative of $y^2 + 2y + 1 = x$ with respect to x.

SOLUTION

$$\frac{d}{dx}(y^2 + 2y + 1) = \frac{d}{dx}(x)$$

$$2y \cdot \frac{d}{dx}(y) + 2 \cdot \frac{d}{dx}(y) + 0 = 1 \qquad \text{Power, Chain, and Constant Rules}$$

$$2y \cdot \frac{dy}{dx} + 2 \cdot \frac{dy}{dx} = 1$$

$$\frac{dy}{dx}(2y + 2) = 1$$

$$\frac{dy}{dx} = \frac{1}{(2y + 2)}$$

EXAMPLE 5   **Proving** $\dfrac{d}{dx}(\log_b(x)) = \dfrac{1}{\ln(b) \cdot x}$

Prove that the derivative of $y = \log_b(x)$ is $y' = \dfrac{1}{\ln(b) \cdot x}$.

SOLUTION We will do some clever manipulations with the function and then use the Exponential Rule to find the derivative.

Recall that according to the definition of logarithms, the following two equations are equivalent:

$$y = \log_b(x) \qquad \text{and}$$
$$b^y = x$$

Differentiating the second equation with respect to x yields

$$\frac{d}{dx}(b^y) = \frac{d}{dx}(x)$$

$$(\ln b)b^y \cdot \frac{d}{dx}(y) = 1 \qquad \text{Chain, Exponential, and Power Rules}$$

$$(\ln b)b^y \cdot \frac{dy}{dx} = 1 \qquad \text{Chain Rule}$$

$$\frac{dy}{dx} = \frac{1}{(\ln b)b^y}$$

$$= \frac{1}{(\ln b)x} \qquad \text{Since } b^y = x$$

EXAMPLE 6   **Finding** $\dfrac{dy}{dx}$ **by Using Implicit Differentiation**

Find $\dfrac{dy}{dx}$ given $3xy^2 = 4xy$.

SOLUTION

$$\frac{d}{dx}(3xy^2) = \frac{d}{dx}(4xy)$$

$$\frac{d}{dx}(3x) \cdot y^2 + \frac{d}{dx}(y^2) \cdot 3x = \frac{d}{dx}(4x) \cdot y + \frac{d}{dx}(y) \cdot 4x \qquad \text{Product Rule}$$

$$(3)y^2 + \left(2y\frac{dy}{dx}\right) \cdot 3x = (4)y + \frac{dy}{dx} \cdot 4x \qquad \begin{array}{l}\text{Constant Multiple, Power, and Chain} \\ \text{Rules}\end{array}$$

$$3y^2 + 6xy\frac{dy}{dx} = 4y + 4x\frac{dy}{dx}$$

$$6xy\frac{dy}{dx} - 4x\frac{dy}{dx} = 4y - 3y^2$$

$$\frac{dy}{dx}(6xy - 4x) = 4y - 3y^2$$

$$\frac{dy}{dx} = \frac{4y - 3y^2}{6xy - 4x}$$

4.5 Summary

In this section, you learned how to use implicit differentiation to find the derivative of functions and nonfunctions. Knowing how to do implicit differentiation will help you understand related rates in the next chapter.

4.5 Exercises

In Exercises 1–10, use implicit differentiation to find $\dfrac{dy}{dx}$. *Then evaluate the derivative function at the designated point.*

1. $2x^2 + y^2 = 32;\ (4, 0)$ **2.** $x^2 - y^2 = 9;\ (5, 4)$

3. $x^2y - 3 = y^2;\ (2, 1)$ **4.** $4y^2 = x^2;\ (4, 2)$

5. $xy = 6;\ (2, 3)$

6. $x^2y - xy^2 = 20;\ (5, 4)$

7. $9x - y^3 = 0;\ (3, 3)$

8. $xy^2 + xy + y = -1;\ (3, -1)$

9. $20y - x^2y^2 = 75;\ (1, 5)$

10. $xy^3 - x^3y = 30;\ (-2, -3)$

In Exercises 11–20, use implicit differentiation to find $\dfrac{dy}{dx}$.

11. $y^2 = x$ **12.** $xe^y = y$

13. $ye^y = 2$ **14.** $2^y = \ln x$

15. $\ln y = xy$ **16.** $x^2y - y^2 + y = x$

17. $2xy - y^2 = 9$ **18.** $x(x^2 - y^2) = y$

19. $\ln(xy^2) = y$ **20.** $ye^x = xy^2$

In Exercises 21–25, determine at what points, if any, $\dfrac{dy}{dx} = 0.$

21. $x^2 + y^2 = 1$ **22.** $3x^2 = y \ln y$

23. $x^2 + 2y^2 = y$ **24.** $x^3 - y^2 = xy$

25. $x^2 + xy = y^2 - 5$

Exercises 26–30 are intended to challenge your understanding of implicit differentiation.

26. Given $x^2 = y^2$, calculate $\dfrac{dy}{dx}$ and $\dfrac{dx}{dy}$. What is the relationship between $\dfrac{dy}{dx}$ and $\dfrac{dx}{dy}$?

27. Given $V = \pi r^2 h$, write $\dfrac{dr}{dt}$ in terms of $\dfrac{dV}{dt}, \dfrac{dh}{dt}, h,$ and r. (*Hint:* Differentiate both sides with respect to t.)

28. Find $\dfrac{dy}{dx}$ given $ye^x = xe^y$.

29. Given the equation $ye^x = xe^y$, find the solution to the equation $\dfrac{dy}{dx} = 0$, if it exists.

30. Find $\dfrac{dy}{dx}$, if possible, given $e^{xy} = \ln(xy)$.

Chapter 4 Review Exercises

Section 4.1 *In Exercises 1–4, use the Constant Rule, the Power Rule, the Constant Multiple Rule, and the Sum and Difference Rule (as appropriate) to find the derivative of the function.*

1. $f(x) = 9$

2. $g(x) = -12x + 14$

3. $v(t) = 7t^3 - 4t^{-3}$

4. $s(t) = -2t^4 + t^2 - t + 17$

In Exercises 5–6, answer the question by evaluating the derivative.

5. **M** **Military Personnel** Based on data from 1990 to 1999, the number of active-duty military personnel may be modeled by

$$A = 2.5626t^2 - 58.90t + 1321$$

thousand people, where t is the number of years since 1990. (**Source:** Modeled from *Statistical Abstract of the United States, 2001*, Table 499, p. 328.)

   According to the model, at what rate was the number of active-duty military personnel changing in 1999 and in 2000?

6. **M** **Prison Rate** Based on data from 1980 to 1998, the federal and state prison rate may be modeled by

$$R = 0.3750t^2 + 11.70t + 138.7$$

prisoners per 100,000 people, where t is the number of years since 1980. (**Source:** Modeled from *Statistical Abstract of the United States, 2001*, Table 332, p. 200.)

   Was the prison rate increasing more rapidly in 1988 or in 1998?

Section 4.2 *In Exercises 7–10, use the Product Rule to find the derivative of the function. Simplify the result.*

7. $f(x) = 3x^2(2x - 4)$

8. $g(x) = -15x(5x^2 - 2)$

9. $f(t) = (2t^2 - 4)(-t + 4)$

10. $s(t) = t^3(3t^2 + 7t)(t - 3)$

In Exercises 11–12, use the Product Rule in answering the questions from the real-life scenarios. You may find it helpful to use your calculator.

11. **M** **Prison Rate** Based on data from 1980 to 1998, the federal and state prison rate may be modeled by

$$R = 0.3750t^2 + 11.70t + 138.7$$

prisoners per 100,000 people, where t is the number of years since 1980. (**Source:** Modeled from *Statistical Abstract of the United States, 2001*, Table 332, p. 200.)

   The number of people in the United States may be modeled by

$$P = 0.525t^2 + 1.695t + 226.5$$

hundred thousand people, where t is the number of years since 1980. (**Source:** Modeled from www.census.gov.)

   In 2000, at what rate was the number of prisoners increasing (in prisoners per year)?

12. **M** **Milk Consumption** Based on data from 1980 to 1999, the per capita milk beverage consumption in the United States may be modeled by

$$M = -0.219t + 27.7$$

gallons per person, where t is the number of years since 1980. (**Source:** Modeled from *Statistical Abstract of the United States, 2001*, Table 202, p. 129.)

   The number of people in the United States may be modeled by

$$P = 0.525t^2 + 1.695t + 226.5$$

hundred thousand people, where t is the number of years since 1980. (**Source:** Modeled from www.census.gov.)

   In 1999, at what rate (in thousands of gallons per year) was milk beverage consumption changing?

Section 4.3 *In Exercises 13–14, find the function $f(g(x))$. Then simplify the result.*

13. $f(x) = -3x + 4$, $g(x) = 2x^2 - x$

14. $f(x) = -4x^2 + x$, $g(x) = 2x + 3$

In Exercises 15–16, use the Chain Rule to find the derivative of the function. Then write the result in factored form.

15. $f(x) = 2(9x + 5)^3$ **16.** $g(x) = 3(x^2 - 2x)^2$

In Exercises 17–18, use a composition of functions and the Chain Rule to answer the questions.

17. **M** **Advertising Expenditures** Based on data from 1990 to 2000, the amount of money spent on advertising on cable television may be modeled by

$$C(m) = 0.00009027m^2 - 0.0261m - 1130$$

million dollars, where m is the amount of money (in millions of dollars) spent on advertising in magazines.

The amount of money spent on advertising in magazines may be modeled by

$$m(t) = 42.29t^2 + 158.3t + 6575$$

million dollars, where t is the number of years since 1990. (**Source:** Modeled from *Statistical Abstract of the United States, 2001*, Table 1272, p. 777.)

In 2000, at what rate was cable television advertising spending changing?

18. **M** **Advertising Expenditures** Based on data from 1990 to 2000, the amount of money spent on advertising on radio may be modeled by

$$R(b) = 12.92b - 4984$$

million dollars, where b is the amount of money (in millions of dollars) spent on advertising on billboards.

The amount of money spent on advertising on billboards may be modeled by

$$b(t) = 9.396t^2 - 11.58t + 1063$$

million dollars, where t is the number of years since 1990. (**Source:** Modeled from *Statistical Abstract of the United States, 2001*, Table 1272, p. 777.)

In 2000, at what rate was radio advertising spending changing?

Section 4.4 *In Exercises 19–22, use the Exponential Rule or the Logarithmic Rule, as appropriate, to find the derivative of the function.*

19. $y = 9^x$ **20.** $g(t) = 20(0.5)^t$

21. $f(x) = \ln(x^2 + 3)$ **22.** $s(t) = \ln(2^{3t-4})$

In Exercises 23–24, answer the questions by using the Exponential Rule or the Logarithmic Rule, as appropriate.

23. **M** **Hawaii Population** The projected population of Hawaii may be modeled by

$$P = 1175(1.014)^t$$

thousand people, where t is the number of years since 1995. (**Source:** Modeled from www.census.gov.)

In 2010, at what rate is the population of Hawaii expected to be changing?

24. **M** **Church Growth** Based on data from 1990 and 2001, the number of Lutherans in the United States may be modeled by

$$L = 9110(1.005)^t$$

thousand people, where t is the number of years since 1990. (**Source:** Modeled from American Religious Identification Survey data.)

In 2001, at what rate was the number of Lutherans in the United States increasing?

Section 4.5 *In Exercises 25–29, use implicit differentiation to find $\dfrac{dy}{dx}$. Then evaluate the derivative function at the designated point.*

25. $xe^{3y} = 7$; $(7, 0)$

26. $\log(xy) = 2$; $(4, 25)$

27. $x^2y^2 - 1 = 0$; $(0.5, 2)$

28. $2xy^2 - y = 10$; $(1, -2)$

29. $4xy - 2^{xy} = 4$; $(-3, -1)$

Make It Real

What to do

1. Collect a minimum of five data points of something that is reported on a national basis in the United States each year. (For example, the total amount of vegetables and melons consumed in a given year.)

2. For each year identified in (1), record the U.S. population.

3. Model each of the data sets in (1) and (2) with a polynomial, exponential, or logarithmic function.

4. Write a third model defined as $g(t) = \dfrac{\text{first model}}{\text{population model}}$.

5. Find the derivative of the third model.

6. Evaluate and interpret the derivative at two different points. If the results don't make sense in their real-world context, choose a new data set for the first model.

Example

Years (Since 2000)	Vegetable and Melon Consumption (million pounds)
0	48,293
1	48,311
2	49,064
3	49,704
4	51,231

Source: *Statistical Abstract of the United States,* 2006, Table 830.

$v(t) = 207.50t^2 - 103.10t + 48,282$ million pounds t years after 2000

Years (Since 2000)	People (thousands)
0	282,402
1	285,329
2	288,173
3	291,028
4	293,907

Source: *Statistical Abstract of the United States,* 2006, Table 830.

$p(t) = 2870.9t + 282,426$ thousand people t years after 2000

(Continued)

259

The third model is

$$g(t) = \frac{v(t)}{p(t)} \frac{\text{million pounds}}{\text{thousand people}}$$

$$= \frac{207.50t^2 - 103.10t + 48{,}282}{2870.9t + 282{,}426} \text{ thousand pounds per person}$$

The derivative of the third model is found using the quotient rule.

$$g'(t) = \frac{(415t - 103.10)(2870.9t + 282{,}426) - (2870.9)(207.50t^2 - 103.10t + 48{,}282)}{(2870.9t + 282{,}426)^2}$$

thousand pounds per person per year

We evaluate the derivative at $t = 0$ and $t = 5$.

$$g'(0) = \frac{(415(0) - 103.10)(2870.9(0) + 282{,}426) - (2870.9)(207.50(0)^2 - 103.10(0) + 48{,}282)}{(2870.9(0) + 282{,}426)^2}$$

$$= -0.0021028 \text{ thousand pounds per person per year}$$

$$\approx -2.1 \text{ pounds per person per year}$$

In 2000, the consumption of vegetables and melons was decreasing at a rate of 2.1 pounds per person per year.

$$g'(5) = \frac{(415(5) - 103.10)(2870.9(5) + 282{,}426) - (2870.9)(207.50(5)^2 - 103.10(5) + 48{,}282)}{(2870.9(5) + 282{,}426)^2}$$

$$= 0.0049183 \text{ thousand pounds per person per year}$$

$$\approx 4.9 \text{ pounds per person per year}$$

In 2005, the consumption of vegetables and melons was increasing at a rate of 4.9 pounds per person per year.

Chapter 5

Derivative Applications

Businesses survive by being profitable. A savvy business owner effectively analyzes the factors that contribute to the financial success or failure of her business. Using mathematical models, she may forecast what prices, production levels, shipment schedules, and other such elements will result in maximum profits. Although no mathematical model is perfect at forecasting future results, a model can help a business owner make informed decisions.

5.1 Maxima and Minima
- Find relative and absolute extrema
- Use the First Derivative Test to find relative extrema

5.2 Applications of Maxima and Minima
- Analyze and interpret revenue, cost, and profit functions

5.3 Concavity and the Second Derivative
- Use the Second Derivative Test to find relative extrema
- Determine the concavity of the graph of a function

5.4 Related Rates
- Solve related-rate problems

5.1 Maxima and Minima

- Find relative and absolute extrema
- Use the First Derivative Test to find relative extrema

GETTING **STARTED** Based on data from 1984 to 2001, the rate at which tuition and fees at community colleges in Washington state are increasing may be modeled by

$$R(t) = -0.7473t^2 + 14.89t + 9.024 \text{ dollars per year}$$

where *t* is the number of years since the 1984–1985 school year. In what year were tuition and fees at Washington state community colleges increasing the fastest? (**Source:** Modeled from Washington State Higher Education Coordinating Board, Higher Education Statistics, September 2001.) Questions such as these may be answered using the concepts of relative and absolute extrema.

In this section, we will informally discuss continuity. In addition, we will discuss how to find *relative extrema* and *absolute extrema*.

Continuity

The notion of continuity is best understood graphically. Loosely speaking, a function is said to be **continuous** if its graph can be drawn by a pencil without lifting the pencil from the page. If there is a break in the graph, the graph is said to be **discontinuous.** This loose definition of continuity will be sufficient for our purposes. (The formal definition of continuity, which relies heavily on the concept of limits, may be obtained from a traditional calculus text.)

Many functions are continuous. For example, linear, polynomial, exponential, and logarithmic functions are all continuous. Frequently, functions that are discontinuous have domain restrictions or are defined piecewise. Consider the piecewise function $f(x) = \begin{cases} 2x + 6 & \text{if } x < 1 \\ 3x + 2 & \text{if } x \geq 1 \end{cases}$ shown in Figure 5.1. Since there is a break in the graph at $x = 1$, the graph is said to be discontinuous.

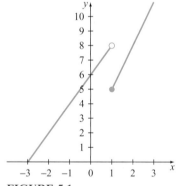

FIGURE 5.1

Relative and Absolute Extrema

An **extremum** is a maximum or minimum value of a function. The plural of extremum is **extrema.**

RELATIVE EXTREMA

- A **relative maximum** of a continuous function f occurs at a point $(c, f(c))$ if $f(x) \leq f(c)$ for all x in some interval (a, b) containing c. (That is, $a < c < b$.)
- A **relative minimum** of a continuous function f occurs at a point $(c, f(c))$ if $f(x) \geq f(c)$ for all x in some interval (a, b) containing c. (That is, $a < c < b$.)

According to the definition, a relative extremum may not occur at an endpoint. We will explore the concept of relative extrema by looking at the graph of

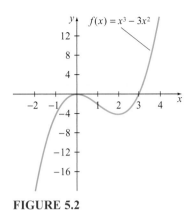

FIGURE 5.2

$f(x) = x^3 - 3x^2$ on the domain $[-2, 4]$ (Figure 5.2).

Graphically, we are looking for the peaks and valleys of the graph. In this case, a peak occurs at $(0, 0)$ and a valley occurs at $(2, -4)$. A relative maximum occurs at the point $(0, 0)$, since $y = 0$ is larger than all y values of the function *nearby* $x = 0$. (When we use the term *nearby*, we mean an arbitrarily small open interval (a, b) surrounding $x = 0$.) A relative minimum occurs at the point $(2, -4)$, since $y = -4$ is smaller than all y values of the function nearby $x = 2$.

ABSOLUTE EXTREMA

- An **absolute maximum** of a function f occurs at a point $(c, f(c))$ if $f(x) \leq f(c)$ for all x in the domain of f.
- An **absolute minimum** of a function f occurs at a point $(c, f(c))$ if $f(x) \geq f(c)$ for all x in the domain of f.

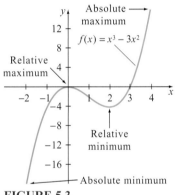

FIGURE 5.3

It is important to note that unlike relative extrema, *absolute extrema may occur at endpoints.*

To find the absolute maximum of the function $f(x) = x^3 - 3x^2$ on the domain $[-2, 4]$, we look for the largest y value. Since f is continuous (there are no breaks in the graph of f), we need only check the relative maxima and the y values of the endpoints. The graph of f has a relative maximum at $(0, 0)$ and endpoints at $(-2, -20)$ and $(4, 16)$. Since $(4, 16)$ has the largest y value, $y = 16$ is the absolute maximum.

To find the absolute minimum, we are looking for the smallest y value. Since f is continuous, we need only check the relative minima and the endpoints. The graph of f has a relative minimum at $(2, -4)$ and endpoints at $(-2, -20)$ and $(4, 16)$. Since $(-2, -20)$ has the smallest (most negative) y value, $y = -20$ is the absolute minimum (Figure 5.3).

Precisely determining relative extrema is difficult to do graphically, especially when extrema occur at irrational points such as $\left(\pi, \sqrt{2}\right)$. Fortunately, by using the derivative, we can quickly find relative extrema.

The derivative of $f(x) = x^3 - 3x^2$ is $f'(x) = 3x^2 - 6x$. Recall that the derivative at a point represents the slope of the tangent line (or slope of the graph) at that point. If the slope is positive, the graph is increasing. If the slope is negative, the graph is decreasing. If the slope is zero, the graph is neither increasing nor decreasing but remains flat. In other words, the graph has a horizontal tangent line if and only if $f'(x) = 0$. For $f(x) = x^3 - 3x^2$, we have the results shown in Table 5.1.

TABLE 5.1

x	$f(x)$	$f'(x)$	Slope of Tangent Line	Graph of f
-2	-20	24	Positive	Increasing
-1	-4	9	Positive	Increasing
0	0	0	Zero	Flat
1	-2	-3	Negative	Decreasing
2	-4	0	Zero	Flat
3	0	9	Positive	Increasing
4	16	24	Positive	Increasing

Observe that $f'(x) = 0$ where the relative extrema occurred $[(0, 0)$ and $(2, -4)]$.

THE VALUE OF THE DERIVATIVE AT A RELATIVE EXTREMUM

Let f be a function with a continuous graph. If a relative extremum of f occurs at $(c, f(c))$, then $f'(c) = 0$ or $f'(c)$ is undefined.

The converse is not true. That is, if $f'(c) = 0$ or $f'(c)$ is undefined, then a relative extremum is not guaranteed to occur at $(c, f(c))$. We will demonstrate this in Example 2.

EXAMPLE 1

Determining from a Graph Where $f'(x) = 0$

The graph of $f(x) = x^3 - 12x$ on the interval $[-3, 3]$ is shown (Figure 5.4). Determine the points on the graph where $f'(x) = 0$.

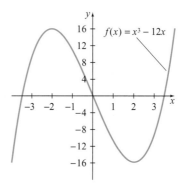

FIGURE 5.4

SOLUTION Since the graph is a smooth curve (no sharp points), the derivative will equal zero at the relative extrema. It appears that a relative maximum occurs when $x = -2$ and a relative minimum occurs when $x = 2$. We will now use the derivative to confirm our observation.

$$f(x) = x^3 - 12x$$
$$f'(x) = 3x^2 - 12$$
$$f'(2) = 3(2)^2 - 12$$
$$= 0$$
$$f'(-2) = 3(-2)^2 - 12$$
$$= 0$$

The results confirm our graphical estimate. At the point $(-2, 16)$ and the point $(2, -16)$, $f'(x) = 0$.

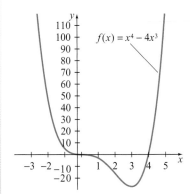

FIGURE 5.5

EXAMPLE 2

Determining from a Graph Where $f'(x) = 0$

The graph of $f(x) = x^4 - 4x^3$ on the interval $[-3, 5]$ is shown in Figure 5.5. Determine the points on the graph where $f'(x) = 0$.

SOLUTION   The graph is a continuous, smooth curve. We are looking for places on the graph where the function has a horizontal tangent line $[f'(x) = 0]$. It looks like the graph has a horizontal tangent line at $x = 0$ and $x = 3$. We will confirm this result by evaluating $f'(x)$ at $x = 0$ and $x = 3$.

$$f(x) = x^4 - 4x^3$$
$$f'(x) = 4x^3 - 12x^2$$
$$= 4x^2(x - 3)$$
$$f'(0) = 4(0)^2(0 - 3)$$
$$= 0$$
$$f'(3) = 4(3)^2(3 - 3)$$
$$= 0$$

So $f'(x) = 0$ at $(0, 0)$ and $(3, -27)$.

As discussed previously, the fact that $f'(c) = 0$ does not guarantee that a relative extremum occurs at $(c, f(c))$. In this example, a relative minimum occurred at $(3, -27)$, but a relative extremum did **not** occur at $(0, 0)$.

In Example 3, we will use the absolute value function. Recall that the absolute value function is formally defined as a piecewise function.

$$|x| = \begin{cases} x & \text{if } x \geq 0 \\ -x & \text{if } x < 0 \end{cases}$$

The graph of the absolute value function looks like a "V" (Figure 5.6).

Since the graph is decreasing for $x < 0$, the slope of the tangent line of f is negative for $x < 0$. Since the graph is increasing for $x > 0$, the slope of the tangent line of f is positive for $x > 0$. But what is the slope of the tangent line at $x = 0$? Recall that the derivative of a function f at $x = a$ is formally defined as

$$f'(a) = \lim_{h \to 0} \frac{f(a + h) - f(a)}{h}$$

For $f(x) = |x|$, the derivative of the function at $x = 0$ is

$$f'(0) = \lim_{h \to 0} \frac{|0 + h| - |0|}{h}$$
$$= \lim_{h \to 0} \frac{|h|}{h}$$

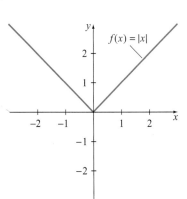

FIGURE 5.6

As h gets close to 0, what value does $\frac{|h|}{h}$ approach? We first observe that $h \neq 0$ because division by 0 is not defined. From the piecewise definition of the absolute value function, we know

$$\frac{|h|}{h} = \begin{cases} \dfrac{h}{h} & \text{if } h > 0 \\ \text{undefined} & \text{if } h = 0 \\ \dfrac{-h}{h} & \text{if } h < 0 \end{cases}$$

Simplifying the equation yields

$$\frac{|h|}{h} = \begin{cases} 1 & \text{if } h > 0 \\ \text{undefined} & \text{if } h = 0 \\ -1 & \text{if } h < 0 \end{cases}$$

For positive values of h, $\frac{|h|}{h} = 1$. For negative values of h, $\frac{|h|}{h} = -1$. Since $\frac{|h|}{h}$ doesn't approach a constant value as h nears 0, $\lim\limits_{h \to 0} \frac{|h|}{h}$ does not exist. That is, for $f(x) = |x|$, $f'(0)$ is undefined. In other words, the derivative of the absolute value function does not exist at $x = 0$. In general, the derivative of any function does not exist wherever the graph of a function has a "sharp point."

EXAMPLE   3

Finding the Relative Extrema of a Function

The graph of $f(x) = -|x| + 2$ is shown in Figure 5.7. Find the relative extrema of f on the interval $[-3, 3]$.

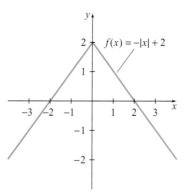

FIGURE 5.7

SOLUTION   From the graph, we can see that the function has a relative maximum at $(0, 2)$. Therefore, $f'(x) = 0$ or $f'(x)$ is undefined at that point. At any sharp point on a graph, the derivative is undefined. Therefore, $f'(0)$ is undefined. Does $f'(x) = 0$ anywhere? From the graph, we see that f is increasing for all $x < 0$. Therefore, $f'(x)$ is positive for $x < 0$. Similarly, f is decreasing for all $x > 0$. Therefore, $f'(x)$ is negative for $x > 0$. Hence, $f'(x) \neq 0$ anywhere. Thus the only relative extremum occurs at $(0, 2)$.

Critical Values

In Example 2, we saw that relative extrema may occur where $f'(x) = 0$. In Example 3, we saw that relative extrema may also occur where $f'(x)$ is undefined. These two situations represent two different types of *critical values* of a function.

CRITICAL VALUES

Let f be a continuous function on the interval (a, b) with $a < c < b$. If

- $f'(c) = 0$ or
- $f'(c)$ is undefined

then c is called a **critical value** of f.

The two types of critical values may be further classified as **stationary values** $[f'(c) = 0]$ and **singular values** $[f'(c)$ is undefined$]$. The corresponding point $(c, f(c))$ is called a **critical point, stationary point,** or **singular point** of f, respectively.

EXISTENCE OF RELATIVE EXTREMA

Let f be a continuous function on the interval (a, b). All relative extrema of f occur at critical values of f.

EXAMPLE   4

Finding Critical Values of a Function

Find the critical values of $f(x) = x^3 - 3x$ on the interval $[-2, 4]$.

SOLUTION   To find the stationary values, we set the derivative equal to zero and solve for x.

$$f'(x) = 3x^2 - 3$$
$$0 = 3x^2 - 3 \qquad \text{Set } f'(x) = 0$$
$$0 = 3(x^2 - 1)$$
$$0 = 3(x - 1)(x + 1)$$
$$x - 1 = 0 \qquad x + 1 = 0 \qquad \text{Set each factor equal to 0 and solve for } x$$
$$x = 1 \qquad\quad x = -1$$

The stationary values are $x = 1$ and $x = -1$. Since the derivative is defined for all values of x, there are no singular values.

EXAMPLE   5

Finding Critical Points of a Function

Find the critical points of $f(x) = x^{1/3}$ on the interval $[-2, 2]$.

SOLUTION

$$f'(x) = \frac{1}{3}x^{-2/3}$$

$$= \frac{1}{3x^{2/3}}$$

$f'(x) \neq 0$ for all values of x, since a fraction can equal zero only if its numerator is zero. Therefore, f does not have any stationary points.

A rational expression $\dfrac{g(x)}{h(x)}$ is undefined when $h(x)$ is equal to zero. What value of x makes the denominator of $f'(x)$ equal zero?

$0 = 3x^{2/3}$	Set the denominator equal to 0
$0 = x^{2/3}$	Divide by 3
$(0)^3 = (x^{2/3})^3$	Cube both sides of the equation
$0 = x^2$	$(x^{2/3})^3 = x^{2/3 \cdot 3} = x^2$
$x = 0$	Take the square root

Therefore, $f'(0)$ is undefined. A singular point may occur at $x = 0$. We need only confirm that the function f is defined at $x = 0$. Since $f(0) = 0$, $x = 0$ is in the domain of f and is indeed a singular value of f. The point $(0, 0)$ is a singular point of f.

The First Derivative Test

Is there a nongraphical way to tell whether a relative maximum or a relative minimum occurs at a critical value? Fortunately, yes. Recall that when a function is increasing, its derivative is positive. When a function is decreasing, its derivative is negative. If the derivative changes sign at a critical value, a relative extremum occurs at that value. The First Derivative Test will allow us to determine where relative extrema occur for *nonconstant* functions.

If f is a constant function $[f(x) = k$ for some constant $k]$, then $f'(x) = 0$ for all x and all points of the function are critical points. Furthermore, since $k \leq f(x) \leq k$ for all x, both relative maxima and relative minima occur at all points of f. Since most real-life data models are not constant functions, we will focus our discussion on *nonconstant* functions.

FIRST DERIVATIVE TEST

Let $(c, f(c))$ be a critical point of a non-constant function f.

- If f' changes from positive to negative at $x = c$, then a relative maximum occurs at $(c, f(c))$.
- If f' changes from negative to positive at $x = c$, then a relative minimum occurs at $(c, f(c))$.
- If f' doesn't change sign at $x = c$, then a relative extremum does not occur at $(c, f(c))$.

It is often helpful to develop a *sign chart* when determining where relative extrema occur. We will demonstrate this with an example and then detail the specific steps of the process.

EXAMPLE 6

Finding Relative and Absolute Extrema

Find the relative and absolute extrema of $f(x) = x^4 - 4x^3$ on the interval $[-1, 5]$.

SOLUTION We will first find the relative extrema by using the First Derivative Test.

$$f'(x) = 4x^3 - 12x^2$$
$$= 4x^2(x - 3)$$

$4x^2 = 0 \qquad x - 3 = 0$ Set each factor equal to zero
$x = 0 \qquad\qquad x = 3$ Solve for x

Setting the derivative equal to zero, we see that $x = 0$ and $x = 3$ are critical values of f. We begin by drawing a number line with the critical values clearly marked.

We will evaluate $f'(x)$ at an x value in each of the following three intervals: $[-1, 0)$, $(0, 3)$, and $(3, 5]$. (These intervals may be equivalently expressed as $-1 \le x < 0$, $0 < x < 3$, and $3 < x \le 5$, respectively.) We are not concerned with the actual value of the derivative. We simply want to know if the derivative is positive or negative.

From the interval $[-1, 0)$, we pick $x = -1$.

$f'(-1) = 4(-1)^2(-1-3)$ The product is equivalent to $(4)(1)(-4)$
$f'(-1) < 0$ Since there are an odd number of negative factors

On the interval $[-1, 0)$, the derivative is negative. We update the chart to reflect our finding.

From the interval $(0, 3)$, we pick $x = 1$.

$f'(1) = 4(1)^2(1-3)$ The product is equivalent to $(4)(1)(-2)$
$f'(-1) < 0$ Since there are an odd number of negative factors

On the interval $(0, 3)$, the derivative is negative. We update the chart to reflect our finding.

From the interval $(3, 5]$, we pick $x = 4$.

$f'(4) = 4(4)^2(4-3)$ The product is equivalent to $(4)(16)(1)$
$f'(4) > 0$ Since there are no negative factors

On the interval $(3, 5]$, the derivative is positive. We update the chart to reflect our finding.

At the critical value $x = 0$, the derivative did not change sign, so a relative extremum does not occur there. Graphically speaking, the graph of f is decreasing through the whole interval $[-1, 3)$. At the critical value $x = 3$, the derivative changed sign from negative to positive, so a relative minimum occurs there.

Graphically speaking, the graph of f changed from decreasing to increasing at $x = 3$.

To find absolute extrema, we consider the relative extrema and evaluate the function f at the endpoints (see Table 5.2).

TABLE 5.2

x	$f(x)$	
-1	5	
3	-27	Relative minimum, absolute minimum
5	125	Absolute maximum

The largest value of $f(x)$ in the table is the absolute maximum, and the smallest value is the absolute minimum. In this case, the relative and absolute minimum occurs at $(3, -27)$ and is equal to -27. The absolute maximum occurs at $(5, 125)$ and is equal to 125. There is no relative maximum. (Recall from our definition of a relative extremum that a relative extremum may not occur at an endpoint. This definition allows us to use critical values to find all relative extrema.)

HOW TO   **Constructing a Derivative Sign Chart**

1. Label a number line with the critical values of the function f.
2. Write f' next to the number line.
3. Evaluate f' at a value in each of the number line intervals.
4. Record a "+" if the derivative is positive and a "−" if the derivative is negative.
5. Use the First Derivative Test to determine where relative maxima and minima occur and record the results on the chart.

EXAMPLE 7   **Finding Relative and Absolute Extrema**

Find the absolute and relative extrema of $g(x) = x^3 - 2x^2 - 5x + 6$ on the interval $[-3, 4]$.

SOLUTION

$$g'(x) = 3x^2 - 4x - 5$$

We set $g'(x) = 0$.

$$3x^2 - 4x - 5 = 0$$

This is a quadratic equation in the form $ax^2 + bx + c = 0$. Since the derivative doesn't factor, we must use the Quadratic Formula to find the stationary points (or use a calculator or computer).

$$x = \frac{-b \pm \sqrt{b^2 - 4ac}}{2a}$$

$$= \frac{-(-4) \pm \sqrt{(-4)^2 - 4(3)(-5)}}{2(3)}$$

$$= \frac{4 \pm \sqrt{76}}{6}$$

$$= \frac{4 \pm 2\sqrt{19}}{6}$$

$$= \frac{2}{3} \pm \frac{\sqrt{19}}{3}$$

Since $\frac{2}{3} + \frac{\sqrt{19}}{3} \approx 2.120$ and $\frac{2}{3} - \frac{\sqrt{19}}{3} \approx -0.7863$, the stationary values are $x \approx -0.7863$ and $x \approx 2.120$. The derivative is defined for all values of x, so there are no singular values. We now construct a sign chart.

$$g' \xleftarrow{\quad\quad\underset{-0.7863}{|}\quad\quad\quad\underset{2.120}{|}\quad\quad} $$

To determine the sign of the derivative $g'(x) = 3x^2 - 4x - 5$ in each interval, we evaluate the derivative at $x = -1$, $x = 0$, and $x = 3$.

$$g'(-1) = 3(-1)^2 - 4(-1) - 5$$
$$= 3 + 4 - 5$$
$$g'(-1) > 0$$
$$g'(0) = 3(0)^2 - 4(0) - 5$$
$$= 0 - 0 - 5$$
$$g'(0) < 0$$
$$g'(3) = 3(3)^2 - 4(3) - 5$$
$$= 27 - 12 - 5$$
$$g'(3) > 0$$

We record the sign of the derivative in each interval on the sign chart.

$$g' \xleftarrow{\quad \overset{+\ \ \text{max}}{}\quad\underset{-0.7863}{|}\quad \overset{-}{}\quad \overset{\text{min}\ +}{}\underset{2.120}{|}\quad}$$

A relative maximum occurs at $x = \frac{2}{3} - \frac{\sqrt{19}}{3} \approx -0.7863$ and a relative minimum occurs at $x = \frac{2}{3} + \frac{\sqrt{19}}{3} \approx 2.120$.

To find the absolute extrema, we consider the relative extrema and evaluate the function g at the endpoints, as shown in Table 5.3.

TABLE 5.3

x	$g(x)$	
-3	-24	Absolute minimum
-0.7863	8.209	Relative maximum
2.120	-4.061	Relative minimum
4	18	Absolute maximum

The absolute minimum occurs at $(-3, -24)$, although there is a relative minimum at $(2.120, -4.061)$. The absolute maximum occurs at $(4, 18)$, although there is a relative maximum at $(-0.7863, 8.209)$.

It is often helpful to use a graphing calculator to calculate values of the derivative, especially when evaluating the derivative at other than whole-number values. The following Technology Tip will detail how to use your calculator to evaluate the derivative at a point quickly.

TECHNOLOGY TIP

Evaluating the Derivative at a Point

1. Enter the function $f(x)$ as Y1 by pressing the Y= button and typing the equation.

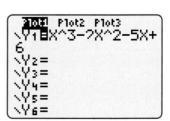

2. Graph the function over the specified domain. (You may need to press WINDOW and adjust the Xmin and Xmax settings. Press Zoom and 0:ZoomFit then ENTER to automatically adjust the y values so that the entire graph will appear on screen.)

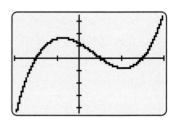

3. Press 2nd then TRACE to bring up the CALCULATE menu. Select item 6:dy/dx and press ENTER.

(Continued)

4. Type in the *x* value of the point where you want to evaluate the derivative or use the blue arrows to select the point graphically. Then press (ENTER).

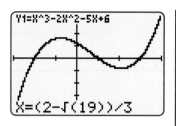

5. The value of the derivative is displayed at the bottom of the screen. (You can assume that a value like "1E-6" is equal to zero. 1E-6 = 1 × 10⁻⁶ = 0.000001. The error occurs because the calculator uses numerical methods to differentiate.)

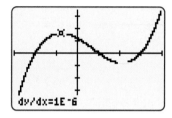

EXAMPLE 8

Interpreting Extrema in a Real-World Context

Based on data from 1980 to 1999, the per capita consumption of breakfast cereal may be modeled by

$$C(t) = -0.004718t^3 + 0.1165t^2 - 0.3585t + 12.17 \text{ pounds per person}$$

where *t* is the number of years since 1980. (**Source:** *Statistical Abstract of the United States, 2001*, Table 202, p. 129.) Find the relative and absolute extrema of $C(t)$ on the interval $[0, 19]$ and interpret the meaning of the results.

SOLUTION We must first find the critical values of $C(t)$.

$$C'(t) = -0.014154t^2 + 0.2330t - 0.3585$$
$$0 = -0.014154t^2 + 0.2330t - 0.3585 \qquad \text{Set } C'(t) \text{ equal to } 0$$

Recall that the quadratic formula states that the solutions to a quadratic equation of the form $at^2 + bt + c = 0$ are given by

$$t = \frac{-b \pm \sqrt{b^2 - 4ac}}{2a}$$

$$= \frac{-(0.2330) \pm \sqrt{(0.2330)^2 - 4(-0.014154)(-0.3585)}}{2(-0.014154)}$$

$$= \frac{-0.2330 \pm 0.1844}{-0.028308}$$

$$t = \frac{-0.2330 + 0.1844}{-0.028308} \qquad t = \frac{-0.2330 - 0.1844}{-0.028308}$$

$$\approx 1.7 \qquad\qquad \approx 14.7$$

The critical values are $t \approx 1.7$ and $t \approx 14.7$. Using these values together with the interval test values of $t = 0$, $t = 10$, and $t = 18$, we construct the first derivative sign chart.

C has a relative minimum at $t \approx 1.7$ and a relative maximum at $t \approx 14.7$. Evaluating $C(t) = -0.004718t^3 + 0.1165t^2 - 0.3585t + 12.17$ at the critical values and the endpoints, we construct a table of values, as shown in Table 5.4.

$C'(t)$ ◄————————————————►
$\quad$ − $\quad$ min of C $\quad$ + $\quad$ max of C $\quad$ −
$\qquad\qquad$ 1.7 $\qquad\qquad$ 14.7

TABLE 5.4

t	$C(t)$	
0	12.2	
1.7	11.9	Relative and absolute minimum
14.7	17.1	Relative and absolute maximum
19	15.1	

According to the model, the per capita consumption of breakfast cereal was decreasing between the end of 1980 ($t = 0$) and mid-September 1982 ($t = 1.7$). It reached a relative low of 11.9 pounds per person in mid-September 1982. From mid-September 1982 to mid-September 1995, the per capita consumption of cereal was increasing. It reached a relative high of 17.1 pounds per person in mid-September 1995. Between mid-September 1995 and the end of 1999, the per capita consumption of cereal decreased. The highest per capita consumption of cereal between 1980 and 1999 was 17.1 pounds per person in the one-year period ending in mid-September 1995, and the lowest per capita consumption of cereal between 1980 and 1999 was 11.9 pounds per person in the one-year period ending in mid-September 1982.

EXAMPLE 9

Interpreting Extrema in a Real-World Context

Based on data from 1980 to 1999, the annual per capita bottled water consumption may be modeled by

$$W(t) = 2.593(1.106)^t \text{ gallons}$$

where t is the number of years since the end of 1980. (**Source:** *Statistical Abstract of the United States, 2001*, Table 204, p. 130.) In what year between 1980 and 1999 was per capita bottled water consumption the greatest?

SOLUTION We are asked to find the absolute maximum of $W(t) = 2.593(1.106)^t$ on the interval $[0, 19]$. We first observe that W is an exponential function. Recall that the derivative of an exponential function $y = ab^x$ is $y' = a\ln(b)b^x$. Since $a \neq 0$ and $\ln(b) \neq 0$, $y' = 0$ only when $b^x = 0$. But no value of x can make $b^x = 0$. Similarly, no value of x can make b^x undefined. Therefore, an exponential function $y = ab^x$ doesn't have any critical values. Thus exponential functions do not have any relative extrema. So to determine when the per capita bottled water consumption was the greatest, we only need to evaluate $W(t) = 2.593(1.106)^t$ at the endpoints, as shown in Table 5.5.

TABLE 5.5

t	$W(t)$	
0	2.6	Absolute minimum
19	17.6	Absolute maximum

According to the model, per capita bottled water consumption reached a maximum of 17.6 gallons at the end of 1999.

EXAMPLE 10   **Forecasting Maximum Revenue**

Based on data from 1985 to 1999, the number of movie theater admissions purchased may be modeled by

$$A(p) = -1036.5p^3 + 12{,}915p^2 - 53{,}497p + 74{,}871 \text{ million admissions}$$

where p is the price of a movie ticket in constant 1990 dollars. (That is, prices have been adjusted for inflation.) (**Source:** *Modeled from Statistical Abstract of the United States, 2001*, Table 1244, p. 761 and Table 694, p. 455.)

Assuming that prices (in constant 1990 dollars) will continue to remain between $3.45 and $4.76, at what price will the revenue from movie ticket sales be the greatest?

SOLUTION   Revenue is the product of the price and the number of admissions sold.

$$\begin{aligned} R(p) &= pA \\ &= p(-1{,}036.5p^3 + 12{,}915p^2 - 53{,}497p + 74{,}871) \\ &= -1036.5p^4 + 12{,}915p^3 - 53{,}497p^2 + 74{,}871p \text{ million dollars of revenue} \end{aligned}$$

The change in revenue is the derivative of the revenue function.

 UNITS

$$R'(p) = -4146p^3 + 38{,}745p^2 - 106{,}994p + 74{,}871 \quad \frac{\text{revenue in millions of dollars}}{\text{price in dollars}}$$

The graph of $R'(p)$ on the interval $[3.45, 4.76]$ is shown in Figure 5.8.

Using the Technology Tip following this example, we determined that the horizontal intercepts of $R'(p)$ occur at $p = 3.77$ and $p = 4.52$. In other words, $R'(3.77) = 0$ and $R'(4.52) = 0$. That is, the horizontal intercepts of $R'(p)$ are the critical points of $R(p)$.

From the graph of $R'(p)$, we can readily construct the sign chart of $R'(p)$. When the graph of $R'(p)$ is below the horizontal axis, $R'(p) < 0$. When the graph of $R'(p)$ is above the horizontal axis, $R'(p) > 0$.

A relative minimum occurs at $p = 3.77$, and a relative maximum occurs at $p = 4.52$. We evaluate the function R at the endpoints and the relative extrema, as shown in Table 5.6.

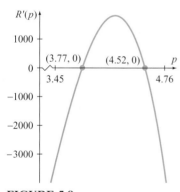

FIGURE 5.8

TABLE 5.6

p	$R(p)$	
3.45	5,053	
3.77	4,556	Relative and absolute minimum
4.52	5,456	Relative and absolute maximum
4.76	5,053	

According to the model, the maximum revenue of $5,456 million is achieved when the ticket price is $4.52. Recall that the price and revenue are in constant 1990 dollars. A movie theater would need to adjust these figures for inflation if it were to use the model as a guide in setting ticket prices today.

TECHNOLOGY **TIP**

Finding the *x*-intercepts of a Function

1. Enter the function $f(x)$ as Y1 by pressing the [Y=] button and typing the equation.

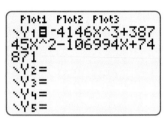

2. Graph the function over the specified domain. (Press [WINDOW] and adjust the Xmin and Xmax settings. Press [ZOOM] then 0:ZoomFit then [ENTER] to automatically adjust the *y* values so that the entire graph will fit on the screen.)

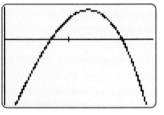

3. Press [2nd] then [TRACE] to bring up the CALCULATE menu. Select item 2:zero and press [ENTER].

4. The calculator asks Left Bound?. Enter an *x* value that is smaller than the *x*-intercept. Or, if you prefer, use the blue arrows to select a point to the left of the *x*-intercept visually.

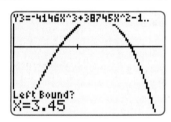

5. The calculator asks Right Bound?. Enter an *x* value that is larger than the *x*-intercept. Or, if you prefer, use the blue arrows to select a point to the right of the *x*-intercept visually.

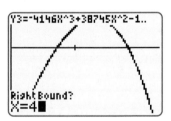

6. The calculator asks Guess?. You may enter a guess or simply press [ENTER]. The calculator returns the coordinates of the *x*-intercept. Repeat Steps 3 through 6 to find additional *x*-intercepts.

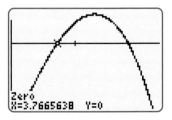

5.1 Summary

In the section, you learned how to find the relative and absolute extrema of a function. You discovered that the First Derivative Test is an excellent tool for finding the location of relative extrema algebraically.

5.1 Exercises

In Exercises 1–10, determine the points on the graph where $f'(x) = 0$ or $f'(x)$ is undefined.

1. $f(x) = x^2 - 2x$

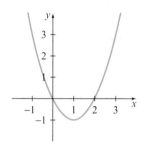

2. $f(x) = |x - 1| + 2x$

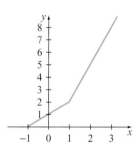

3. $f(x) = x^3 + 1$

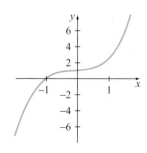

4. $f(x) = -x^3 + 3x^2 - 3x + 5$

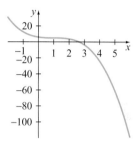

5. $f(x) = -x^2 + 4x - 4$

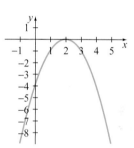

6. $f(x) = x^3 - 9x^2 + 24x + 3$

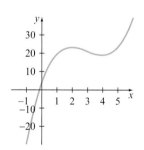

7. $f(x) = -|x - 2| + 4$

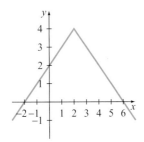

8. $f(x) = |x + 1| - |2x|$

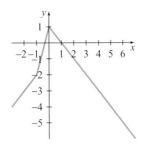

9. $f(x) = x^5 - 5x^4 + 200$

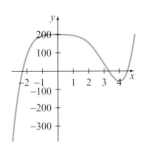

10. $f(x) = 5$

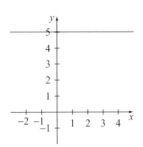

In Exercises 11–20, find the critical values of the function. Then classify the critical values as stationary values or singular values.

11. $f(x) = x^2 - 4x$ **12.** $g(t) = t^3 - 1$

13. $h(t) = t^3 - 6t^2$ **14.** $f(x) = x^{1/5} + 4$

15. $g(x) = 3x^{1/3} - x$

16. $R(p) = p^4 - 2p^2 + 1$

17. $h(x) = \dfrac{x}{x^2 + 1}$ **18.** $g(t) = \dfrac{1}{t^2}$

19. $C(p) = 4p^3 - 12p + 7$

20. $f(x) = -x + \sqrt{x}$

In Exercises 21–35, construct a sign chart for the derivative. Then determine the relative and absolute extrema of the function on the specified domain. Use the First Derivative Test as appropriate. (Note: Exercises 21–30 use the same functions as Exercises 11–20.)

21. $f(x) = x^2 - 4x; [-1, 5]$

22. $g(t) = t^3 - 1; [-2, 2]$

23. $h(t) = t^3 - 6t^2; [-1, 7]$

24. $f(x) = x^{1/5} + 4; [-1, 1]$

25. $g(x) = 3x^{1/3} - x; [-3, 3]$

26. $R(p) = p^4 - 2p^2 + 1; [0, 4]$

27. $h(x) = \dfrac{x}{x^2 + 1}; [-2, 2]$

28. $g(t) = \dfrac{1}{t^2}; [-2, 2]$

29. $C(p) = 4p^3 - 12p + 7; [-3, 3]$

30. $f(x) = -x + \sqrt{x}; [0, 2]$

31. $f(x) = x^3 - 4x; [-3, 3]$

32. $g(t) = e^t - t; [-2, 2]$

33. $h(t) = t \ln(t); [0.1, 3]$

34. $p(n) = n(2^n); [-2, 1]$

35. $g(x) = e^x - \ln x; [0.1, 3]$

In Exercises 36–40, the number of tickets sold, $N(p)$, is given as a function of the ticket price p. Determine what ticket price will maximize the revenue from ticket sales.

36. $N(p) = -200p + 1000$

37. $N(p) = -60p + 3000$

38. $N(p) = 2500e^{-0.1p}$

39. $N(p) = 625e^{-0.025p}$

40. $N(p) = 8500e^{-0.04p}$

In Exercises 41–45, use the methods discussed in this section to answer the questions.

 41. **Margarine Consumption** Based on data from 1980 to 1999, the per capita consumption of margarine may be modeled by

$$M(t) = 0.0002868t^4 - 0.01149t^3 + 0.1329t^2 - 0.5098t + 11.30$$

pounds per year, where t is the number of years since 1980. (**Source:** Modeled from *Statistical Abstract of the United States, 2001*, Table 202, p. 129.)

Between 1980 and 1999, when was the per capita consumption of margarine increasing?

42. **Gaming Software Sales** Based on data from 1990 to 1999, electronic gaming *software* sales may be modeled by

$$S(t) = 39.79t^2 - 48.41t + 2378$$

million dollars, where t is the number of years since 1990. (**Source:** Modeled from *Statistical Abstract of the United States, 2001*, Table 1005, p. 634.)

In what year (between 1990 and 1999) did gaming *software* sales reach their highest point and in what year did they reach their lowest point?

 43. **Gaming Hardware Sales** Based on data from 1990 to 1999, electronic gaming *hardware* sales may be modeled by

$$S(t) = 8.451t^3 - 111.9t^2 + 470.0t + 965.0$$

million dollars, where t is the number of years since 1990. (**Source:** Modeled from *Statistical Abstract of the United States, 2001*, Table 1005, p. 634.)

In what year (between 1990 and 1999) did gaming *hardware* sales reach their highest point, and in what year did they reach their lowest point?

44. **Lumber Industry Earnings** Based on data from 1995 to 1999, the average earnings of a lumber and wood products industry employee may be modeled by

$$S(n) = -0.7685n^2 + 1295n - 516,596$$

dollars, where n is the number of employees in thousands. (**Source:** Modeled from *Statistical Abstract of the United States, 2001*, Table 979, p. 622.)

Assuming that the employee union wants to maximize the average wage of its members, what is the optimal number of employees to have working in the lumber and wood products industry (according to the model)?

45. **DVD Market Value** Based on data from 1998 to 2001, the total dollar value of DVDs shipped by manufacturers may be modeled by

$$V(x) = -0.2173x^2 + 25.84x - 0.02345$$

million dollars, where x is the number of DVDs shipped in millions. (**Source:** Modeled from Recording Industry Association of America data.)

According to the model, how many DVDs should be shipped in order to maximize the value of the manufacturers' DVD shipments?

Exercises 46–50 are intended to challenge your understanding of extrema.

46. A continuous function f has the property that $f'(2) > 0$, $f'(3) = 0$, and $f'(4) < 0$. Does f have a relative maximum at $x = 3$? Justify your answer.

47. Give an example of a nonlinear function that does not have any critical points.

48. Give an example of a continuous function that has a singular point but no relative extrema.

49. The graph of a polynomial function of degree n changes from increasing to decreasing (or vice versa) at most $n - 1$ times. What is the maximum number of stationary points the function can have? Defend your conclusions.

50. A continuous function f' has two x-intercepts: $x = 2$ and $x = 5$. The function f has exactly two relative extrema. (*Note:* $\dfrac{d}{dx}(f) = f'$.)

If $f'(3) = 1$, determine the x values of the relative maximum and relative minimum of f.

5.2 Applications of Maxima and Minima

- Analyze and interpret revenue, cost, and profit functions

GETTING STARTED   The equation that expresses the price of an item as a function of the number of items sold is referred to as a **demand equation.** Based on data from 1992 to 2000, the wholesale demand equation for audio compact discs may be modeled by

$$p = -0.000002549x^2 + 0.001036x + 13.12$$

where x is the number of CDs shipped by manufacturers (in millions) and p is the price per CD (in constant 1992 dollars). The financial data have been adjusted for inflation. (**Source:** Modeled from Recording Industry Association of America data.)

How many CDs should be sold in order to maximize revenue? Questions such as these can be answered using the notion of the derivative.

In this section, we will further illustrate applications of maxima and minima. Specifically, we will look at the business concepts of revenue, cost, and profit as well as marginal revenue, marginal cost, and marginal profit. We will also explore area and volume problems.

Revenue, Cost, and Profit

A company's **revenue** is the total amount of money it brings in. The company's **cost** is the total amount of money it spends; this includes both fixed costs and variable costs. **Fixed costs** are those costs that the company incurs regardless of production. **Variable costs** are typically expenses that vary based on the level of production. **Profit** is the difference between the company's revenue and its cost. The **break-even point** is the production level that results in revenue equaling cost.

Marginal revenue is an approximation of the additional revenue generated if one more unit is produced and sold. **Marginal cost** is the approximate cost incurred by producing one more unit, and **marginal profit** is the approximate profit resulting from the production and sale of one more unit. In order to maximize profits, product manufacturers should continue to increase production as long as the marginal profit is positive. If they continue production when the marginal profit turns negative, they will actually lose money by producing more items.

Marginal profit is determined by taking the derivative of the profit function. This relationship between the derivative and marginal profit is seen by looking at the difference quotient of the profit function $P(x)$. Suppose we wanted to estimate the additional profit earned by selling one more item after 100 items had been sold. We know that $P(101) - P(100)$ is the additional profit earned by selling the 101st item. Note that

$$P(101) - P(100) = \frac{P(101) - P(100)}{101 - 100}$$

$$\approx \lim_{h \to 0} \frac{P(100 + h) - P(100)}{(100 + h) - 100}$$

$$\approx \lim_{h \to 0} \frac{P(100 + h) - P(100)}{h}$$

$$\approx P'(100)$$

It may seem illogical to evaluate the derivative at $x = 100$ when we could calculate the exact amount of the additional profit earned by producing and selling the 101st item by evaluating $P(101) - P(100)$. However, if P is a complicated function, it is often easier to evaluate the derivative at a single value instead of calculating P at two different values.

In a similar manner, we can see that marginal cost is the derivative of the cost function and marginal revenue is the derivative of the revenue function. We'll begin by working a straightforward example to demonstrate the concepts of marginal revenue and marginal profit.

EXAMPLE **1**

Forecasting Maximum Profit

The demand equation for a certain brand of squirt gun is given by

$$p = -x + 15$$

where x is the number of squirt guns sold (in thousands) and p is the price per squirt gun (in dollars).

The cost to produce x thousand squirt guns is

$$C(x) = 5x \text{ thousand dollars}$$

(a) If the company is currently producing 7000 squirt guns per year, should it increase or decrease squirt gun production? Explain using the concepts of marginal revenue and marginal profit.

(b) Determine the production level that will maximize profit.

SOLUTION

(a) The revenue function $R(x)$ is the product of the price p and the number of squirt guns sold.

$$R(x) = \left(p\,\frac{\text{dollars}}{\text{squirt gun}}\right) \cdot (x \text{ thousand squirt guns})$$

$$= (-x + 15)x\,\frac{(\text{dollars})(\text{thousand squirt guns})}{\text{squirt gun}}$$

$$= -x^2 + 15x \text{ thousand dollars}$$

The marginal revenue is the derivative of the revenue function.

$$R'(x) = -2x + 15\,\frac{\text{thousand dollars}}{\text{thousand squirt guns}}$$

Evaluating the marginal revenue at a production level of 7 thousand squirt guns yields

$$R'(7) = -2(7) + 15$$

$$= 1\,\frac{\text{thousand dollars}}{\text{thousand squirt guns}}$$

Increasing production from 7 thousand squirt guns to 8 thousand squirt guns will increase revenue by about $1000.

Profit is the difference between revenue and cost. The profit function is

$$P(x) = R(x) - C(x)$$
$$= (-x^2 + 15x) - (5x)$$
$$= -x^2 + 10x \text{ thousand dollars}$$

The marginal profit function is

$$P'(x) = -2x + 10 \frac{\text{thousand dollars}}{\text{thousand squirt guns}}$$

Evaluating the marginal profit at a production level of 7 thousand squirt guns yields

$$P'(7) = -2(7) + 10$$
$$= -4 \text{ thousand dollars per thousand squirt guns}$$

Increasing production from 7 thousand squirt guns to 8 thousand squirt guns will decrease profit by about $4000.

Although marginal revenue is positive at a production level of 7 thousand squirt guns, marginal profit is negative at that production level. This means that any additional revenue brought in won't be enough to cover the cost of producing the extra squirt guns. The company should reduce production.

(b) The company wants to maximize profit. When profit is maximized, the marginal profit will be equal to 0. Recall that the profit function is

$$P(x) = -x^2 + 10x \text{ thousand dollars}$$

and the marginal profit function is the derivative

$$P'(x) = -2x + 10 \text{ thousand dollars per thousand squirt guns}$$

Setting the marginal profit to zero yields

$$0 = -2x + 10$$
$$2x = 10$$
$$x = 5 \text{ thousand squirt guns}$$

Does a maximum or a minimum of the profit function occur at the critical value $x = 5$? Observe that the graph of $P(x) = -x^2 + 10x$ is a concave down parabola. A concave down parabola has one relative maximum and no relative minima. The relative maximum is also an absolute maximum. Consequently, an absolute maximum of the profit function $P(x) = -x^2 + 10x$ occurs at the critical value $x = 5$. At a production level of five thousand squirt guns, the profit will be maximized. The maximum profit is

$$P(5) = -(5)^2 + 10(5)$$
$$= 25 \text{ thousand dollars}$$

We can verify the accuracy of our result by evaluating the profit function at other production levels on either side of $x = 5$.

$$P(4) = -(4)^2 + 10(4)$$
$$= 24 \text{ thousand dollars}$$

$$P(7) = -(7)^2 + 10(7)$$
$$= 21 \text{ thousand dollars}$$

Increasing or decreasing the production level from 5000 squirt guns will decrease profit.

EXAMPLE 2

Forecasting Maximum Revenue

Based on data from 1992 to 2000, the wholesale demand equation for audio compact disks may be modeled by

$$p = -0.000002549x^2 + 0.001036x + 13.12$$

where x is the number of CDs shipped by manufacturers (in millions) and p is the price per CD (in constant 1992 dollars). The financial data have been adjusted for inflation. (**Source:** Modeled from Recording Industry Association of America data.)

(a) What is the revenue function for the wholesale shipment of CDs?
(b) At a production level of 1 billion CDs, what is the marginal revenue?
(c) At a production level of 2 billion CDs, what is the marginal revenue?
(d) How many CDs should be shipped in order to maximize revenue?

SOLUTION

(a) Revenue is the product of an item's price and the number of items sold.

$$R(x) = (-0.000002549x^2 + 0.001036x + 13.12)x$$
$$= -0.000002549x^3 + 0.001036x^2 + 13.12x \text{ million dollars}$$

(b) Marginal revenue is the derivative of the revenue function.

UNITS

$$R'(x) = \frac{d}{dx}(-0.000002549x^3 + 0.001036x^2 + 13.12x)$$

$$= -0.000007647x^2 + 0.002072x + 13.12 \frac{\text{millions of dollars}}{\text{millions of CDs}}$$

$$R'(1000) = -0.000007647(1000)^2 + 0.002072(1000) + 13.12$$
$$= 7.55 \text{ million dollars per million CDs}$$

Increasing production from 1000 million to 1001 million CDs will increase revenue by about 7.55 million dollars.

(c) Evaluating marginal revenue at a production level of 2000 million CDs yields

$$R'(2000) = -0.000007647(2000)^2 + 0.002072(2000) + 13.12$$
$$= -13.32 \text{ million dollars per million CDs}$$

Increasing production from 2000 million to 2001 million CDs will decrease revenue by about 13.32 million dollars.

(d) We will use a sign chart and the First Derivative Test to determine the production level that will maximize revenue.

$$R'(x) = -0.000007647x^2 + 0.002072x + 13.12$$

The derivative is defined for all values of x, so there are no singular values. We'll find the stationary values by setting the derivative equal to zero and solving the equation with the Quadratic Formula.

$$0 = -0.000007647x^2 + 0.002072x + 13.12$$

$$0 = -7.647x^2 + 2072x + 13,120,000 \qquad \text{Multiply equation by 1,000,000}$$

$$x = \frac{-b \pm \sqrt{b^2 - 4ac}}{2a}$$

$$= \frac{-2072 \pm \sqrt{(2072)^2 - 4(-7.647)(13,120,000)}}{2(-7.647)}$$

$$= \frac{-2072 \pm \sqrt{405,607,744}}{2(-7.647)}$$

$$= \frac{-2072 \pm 20,140}{-15.294}$$

$$\approx -1181; 1452$$

The revenue function has two critical values: $x \approx -1181$ and $x \approx 1452$. In the context of the problem, it doesn't make sense to talk about a production level of -1181 million CDs. Consequently, we'll throw out that critical value. We're only interested in what happens when 1452 million CDs are produced. Using the data from parts (b) and (c) of this problem, we can construct the sign chart.

A relative maximum exists at $x = 1452$. Is this an absolute maximum? The domain of the revenue function is not explicitly stated; however, it is implied. Clearly, $x \geq 0$, since it doesn't make sense to talk about a negative quantity of CDs being shipped. To determine the maximum domain value, we analyze the graph of the cubic function $R(x) = -0.000002549x^3 + 0.001036x^2 + 13.12x$ (Figure 5.9).

Using a graphing calculator, we determined that the revenue function has an x-intercept at $x = 2481.03$. As seen on the graph, the revenue function R is negative for values of x greater than 2481.03. Since it doesn't make sense to have negative revenue, $x \leq 2481.03$. Therefore, the implied domain of the function is $[0, 2481.03]$. Redrawing $R(x)$ on the implied domain yields Figure 5.10.

From the graph and equation of $R(x)$, we confirm the results of our earlier analysis, as shown in Table 5.7.

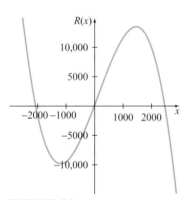

FIGURE 5.9

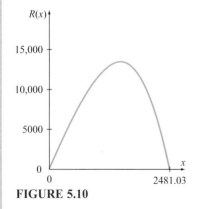

FIGURE 5.10

TABLE 5.7

x	$R(x)$	
0	0	Absolute minimum
1452	13,431	Relative and absolute maximum
2481.03	0	Absolute minimum

According to the model, 1452 million CDs should be shipped in order to maximize revenue. The maximum revenue is forecasted to be 13,431 million dollars (in constant 1992 dollars).

EXAMPLE **3**

Forecasting Maximum Profit

Based on data from 1992 to 2000, the wholesale demand for cassette tapes may be modeled by

$$p = 8.954 - \frac{156.2}{x} \text{ dollars per cassette (in constant 1992 dollars)}$$

where x is the number of cassettes shipped by manufacturers (in millions). The financial data have been adjusted for inflation. In order for the price to be positive, x must equal at least 17.5 million cassettes. (**Source:** Modeled from Recording Industry Association of America data.)

The cassette cost function for cassette tape retailers is the same as the revenue function for cassette tape manufacturers, since retailers purchase their cassettes from the manufacturers.

(a) Find the product cost function for cassette tape retailers.
(b) Assuming that the average retail price of a cassette is $p = 11 - \frac{x}{400}$, find the revenue function for cassette tape retailers.
(c) Find the profit function for cassette tape retailers.
(d) Determine what quantity of cassettes will maximize the profit of cassette tape retailers.

SOLUTION

(a) The cost function for cassette tape retailers is

$$C(x) = \left(8.954 - \frac{156.2}{x}\right)x$$

$$= 8.954x - 156.2 \text{ million dollars}$$

with the domain $[17.5, \infty)$.

(b) The revenue function for cassette tape retailers is

$$R(x) = \left(11 - \frac{x}{400}\right)x$$

$$= 11x - \frac{x^2}{400} \text{ million dollars}$$

(c) The profit function for cassette tape retailers is

$$P(x) = R(x) - C(x)$$

$$= \left(11x - \frac{x^2}{400}\right) - (8.954x - 156.2)$$

$$= 2.046x - \frac{x^2}{400} + 156.2 \text{ with domain } [17.5, \infty)$$

(d) The marginal profit function is

$$P'(x) = 2.046 - \frac{2x}{400}$$

$$= 2.046 - \frac{x}{200} \text{ with domain } [17.5, \infty)$$

We want to know when the marginal profit is zero.

$$0 = 2.046 - \frac{x}{200}$$

$$\frac{x}{200} = 2.046$$

$$x = 409.2$$

Evaluating the derivative on either side of the critical value $x = 409.2$, we get

$$P'(0) = 2.046$$

$$P'(500) = 2.046 - \frac{500}{200}$$

$$= -0.454$$

The profit function has a relative maximum when 409.2 million cassettes are sold, as shown in the sign chart and Table 5.8.

TABLE 5.8

x	$P(x)$
17.5	191.2
409.2	574.8
1,000	−297.8

Relative and absolute maximum

We do not have a right-hand endpoint, since the domain of the function is $[17.5, \infty)$. However, we can calculate the profit at a large value of x greater than 409.2 and compare it to $P(409.2)$. Additionally, since the graph of the profit function is a concave down parabola, we know that there will be an absolute maximum at the vertex. The vertex is $(409.2, 574.8)$. According to the model, the maximum profit of $574.8 million occurs when 409.2 million cassettes are sold.

EXAMPLE 4

Minimizing Inventory Costs

Some authors turn to self-publishing when they are unable to convince a publisher to produce their work. Authors such as Deepak Chopra, Virginia Woolf, and James Joyce have self-published. (**Source:** The Self Publishing Manual by Dan Poynter.)

An author anticipates that she will sell 10,000 copies of her book annually. The setup cost for each print run is $300. The storage cost for each printed book is $1.50 per year. Assuming that customer demand for the book is constant for the 250 business days of a year, determine the number and size of each print run that will minimize her inventory cost. (Inventory cost is the sum of the setup and storage costs.)

SOLUTION The author expects to sell 10,000 copies in the 250 business days of the year. The number of books sold per day is $\frac{10,000}{250} = 40$. She could print

40 books a day; however, her annual setup cost would be $75,000! She could print all 10,000 books at once; however, her annual storage cost would be $7500. We seek to determine the size of the print run that will minimize her annual inventory cost.

Let x be the number of books in a print run. Since the demand is 40 books per day, the number of books in storage will decrease linearly from x books to 0 books at a rate of 40 books per day. Because the number of books in storage decreases linearly, the average number of books in storage is equal to the average of the starting and ending quantities. That is, the average number of books in storage is given by $\frac{x + 0}{2} = \frac{x}{2}$. On the same day that the number of books from the first print run reaches zero, the x books from the second print run are delivered.

For example, if 400 books were produced in a print run, then the number of books in inventory would be as shown in Figure 5.11.

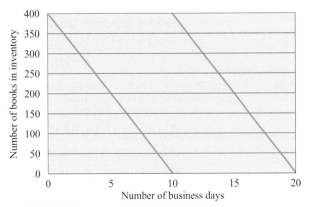

FIGURE 5.11

It takes 10 days to sell all of the books from a print run. The average number of books in inventory over each 10-day period is $\frac{400 + 0}{2} = 200$ books. Since this value is the same for every 10-day period, the average inventory over the course of the year is 200 books when 400 books are produced per print run.

Since the average number of books in storage over the course of the year is $\frac{x}{2}$ and the storage cost is $1.50 per book per year, the annual storage cost is

$$S(x) = 1.5\left(\frac{x}{2}\right)$$

$$= 0.75x \text{ dollars}$$

Since she needs to produce 10,000 books annually, the total number of print runs is equal to $\frac{10,000}{x}$, where x is the number of books in a print run. Since the setup cost of each print run is $300, her annual setup cost is

$$C(x) = 300\left(\frac{10,000}{x}\right)$$

$$= 3,000,000x^{-1} \text{ dollars}$$

The author's annual inventory cost is the sum of the storage and the setup costs.

$$I(x) = S(x) + C(x)$$
$$= 0.75x + 3,000,000x^{-1}$$

Differentiating I and setting it equal to 0 yields

$$0 = 0.75 - 3,000,000x^{-2}$$
$$\frac{3,000,000}{x^2} = 0.75$$
$$3,000,000 = 0.75x^2$$
$$4,000,000 = x^2$$
$$x = 2000$$

The critical value of $I(x)$ is $x = 2000$. We construct a sign chart for $I'(x)$.

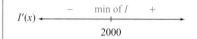

$I'(x)$ —   min of I +   2000

The inventory cost function is minimized when 2000 books are ordered. The author will need to make five 2000-book orders during the year. Her minimum annual inventory cost is given by

$$I(2000) = 0.75(2000) + 3,000,000(2000)^{-1}$$
$$= \$3000$$

Area and Volume

Shipping companies such as United Parcel Service (UPS) and the U.S. Postal Service (USPS) classify packages by size and weight. USPS advises customers that Parcel Post™ is the best value when shipping a single package or a small number of packages. A Parcel Post package "can weigh up to 70 pounds and measure up to 130 inches in length and girth combined." (**Source:** U.S. Postal Service.) The length of a package is the length of the longest side, and the girth of a package is the distance around its thickest part. Businesses are often interested in fitting the maximum amount of their product into a box whose dimensions meet the postal service guidelines. They can save a substantial amount of money in shipping costs by conforming to the postal service standard.

EXAMPLE 5

Maximizing Volume

A compact disc in a plastic jewel case weighs 3.75 ounces and has dimensions 5.6" × 4.9" × 0.41". A box 22 inches long can fit four CDs placed end to end with an additional 2.4" of space to ensure ease of packing. The box will be shipped Parcel Post.

(a) What are the dimensions of the 22-inch rectangular box that conforms to the Parcel Post criteria that has maximum volume?

(b) How many CDs could be packed into the box without exceeding the 70-pound weight restriction?

SOLUTION To help us visualize what is going on in the problem, we draw a picture of the box (Figure 5.12).

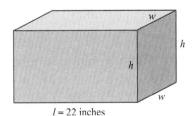

$l = 22$ inches

FIGURE 5.12

(a) The volume of a box is $V = lwh$. We want to maximize $V = 22wh$; however, we can't take the derivative of V because it is a function of two variables, w and h. We must find a *secondary equation* that will allow us to write one of the variables in terms of the other.

The girth of the package is

$$g = w + h + w + h$$
$$= 2w + 2h$$

A Parcel Post package can measure up to 130 inches in length and girth combined ($l + g = 130$).

$$l + g = 130$$
$$22 + (2w + 2h) = 130$$
$$2w + 2h = 108$$
$$w + h = 54$$
$$w = 54 - h$$

Now that we have w in terms of h, we may write V in terms of a single variable.

$$V = 22wh$$
$$= 22(54 - h)h$$
$$= 1188h - 22h^2$$

Do we have any domain restrictions on h? Yes. Clearly, $h \geq 0$ and $w \geq 0$, since neither the height nor the width of the package can be negative. Recall that $w = 54 - h$. Thus we may write $w \geq 0$ in terms of h.

$$w \geq 0$$
$$54 - h \geq 0$$
$$54 \geq h$$
$$h \leq 54$$

The domain of the volume function $V = 1188h - 22h^2$ is $[0, 54]$. We must maximize V on this interval.

$$V'(h) = 1188 - 44h$$
$$0 = 1188 - 44h \qquad \text{Set } V'(h) = 0$$
$$44h = 1188$$
$$h = 27$$

The critical value of the volume function is $h = 27$. Since V is a quadratic equation with a negative coefficient on h^2, the graph of V is a concave down parabola, and a relative and absolute maximum occurs at $h = 27$ (see Table 5.9).

TABLE 5.9

h	$V(h)$	
0	0	
27	16,038	Relative and absolute maximum
54	0	

We need to find the value of w in order to determine the dimensions of the box.

$$w = 54 - h$$
$$= 54 - 27$$
$$= 27$$

A rectangular box of size $22'' \times 27'' \times 27''$ will have a volume of 16,038 cubic inches. This is the maximum volume of a box 22 inches long that meets the Parcel Post criteria.

(b) We'll first determine how many CDs can fit across the width of the box.

UNITS

$$\frac{\text{Width of box}}{\text{Width of CD}} = \frac{27 \text{ inches}}{5.6 \text{ inches per CD}}$$
$$= 4.8 \text{ CDs}$$

Since it doesn't make sense to talk about a fraction of a CD, we'll round this figure to 4 CDs. The width of a row 4 CDs wide is $5.6(4) = 22.4$ inches. There is an additional 4.6 inches of space available for packing material.

We'll then determine how many CDs can fit across the length of the box.

UNITS

$$\frac{\text{Length of box}}{\text{Length of CD}} = \frac{22 \text{ inches}}{4.9 \text{ inches per CD}}$$
$$= 4.5 \text{ CDs}$$

Since it doesn't make sense to talk about a fraction of a CD, we'll round this figure to 4 CDs. The length of a row 4 CDs long is $4.9(4) = 19.6$ inches. There is an additional 2.4 inch of space available for packing material. (This confirms the information given at the start of the problem.)

We'll next determine how high we can stack the CDs.

UNITS

$$\frac{\text{Height of box}}{\text{Height of CD}} = \frac{27 \text{ inches}}{0.41 \text{ inches per CD}}$$
$$\approx 65 \text{ CDs}$$

We can stack the CDs 65 units high. Visually, a stack 4 CDs wide, 4 CDs long, and 24 CDs high in the box would look like Figure 5.13.

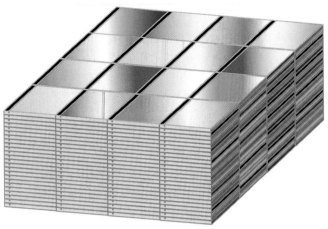

FIGURE 5.13

A total of $4 \cdot 4 \cdot 65 = 1040$ CDs can fit in the box.

The total weight of the CDs will be $1040 \cdot 3.75 = 3900$ ounces. There are 16 ounces in a pound. The weight of the CDs is

$$3900 \text{ ounces} \cdot \frac{1 \text{ pound}}{16 \text{ ounces}} \approx 244 \text{ pounds}$$

Since the weight of the package exceeds the 70-pound limit, we will need to put fewer CDs in the box. Since 70 pounds is equivalent to 1120 ounces, the combined weight of the CDs and packing material can be at most 1120 ounces.

$$\frac{1120 \text{ ounces}}{3.75 \dfrac{\text{ounces}}{\text{CD}}} = 298\frac{2}{3} \text{ CDs}$$

We can pack at most 298 CDs into the box. Depending upon the weight of the packing materials, we may have to remove additional CDs to remain within the weight limit.

EXAMPLE 6

Minimizing Use of Resources

Moving companies are hired by businesses and private individuals to pack, load, move, and unload household goods. As part of their service, movers bring cardboard boxes that can be easily constructed and easily broken down. Boxes are purchased by the moving company from a box manufacturer.

In an effort to cut costs, a box manufacturer wants to design a closable box with a square base that has a volume of 8 cubic feet and uses the least amount of cardboard. The box must be at least 24 inches in height and will be constructed according to the design shown in Figure 5.14. What are the dimensions of the box that uses the least material?

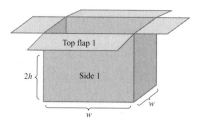

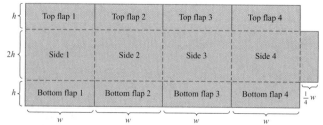

FIGURE 5.14

SOLUTION   The surface area of the cardboard material is

$$A = (h + 2h + h)(w + w + w + w) + \frac{1}{4}w(2h)$$

$$= (4h)(4w) + \frac{1}{2}hw$$

$$= 16hw + \frac{1}{2}hw$$

$$= 16.5hw \text{ square feet}$$

Since we want to minimize the amount of cardboard used, we must minimize the surface area. We are unable to find A' because it is a function of two variables. We must use a secondary equation to write one variable in terms of the other.

The volume of the box is

$$V = w \cdot w \cdot 2h$$

$$= 2hw^2$$

Since the box has a volume of 8 cubic feet, we have

$$8 = 2hw^2$$

$$h = \frac{8}{2w^2}$$

$$= 4w^{-2}$$

Substituting this result into the surface area equation, we get

$$A = 16.5hw$$

$$= 16.5(4w^{-2})w$$

$$= 66w^{-1}$$

Do we have any restrictions on w? Clearly, $w > 0$, since the box must have a positive width. (If $w = 0$, the surface area function is undefined.) Additionally, the box is required to have a height of at least 24 inches (2 feet). That is,

$$2h \geq 2$$

$$h \geq 1$$

Since $h = 4w^{-2}$, we can write $h \geq 1$ in terms of w.

$$h \geq 1$$
$$4w^{-2} \geq 1$$
$$4 \geq w^2$$
$$4 - w^2 \geq 0$$

Graphically speaking, we want to know when the graph of $f(w) = 4 - w^2$ is on or above the horizontal axis. [This is when $f(w) \geq 0$.] Rewriting $f(w)$ in the standard form of a quadratic equation $[f(w) = aw^2 + bw + c]$ yields $f(w) = -w^2 + 4$. Since $a < 0$, the parabola is concave down. Since $c = 4$, the graph has a vertical intercept of $(0, 4)$. Since $f(-2) = 0$ and $f(2) = 0$, $(-2, 0)$ and $(2, 0)$ are the horizontal intercepts of $f(w) = -w^2 + 4$ (Figure 5.15).

From the graph, we see that when $-2 \leq w \leq 2$, $f(w) = -w^2 + 4 \geq 0$. Recall that since w is the width, we also required $w > 0$. Thus the domain of the surface area function is $(0, 2]$.

To minimize the surface area, we will differentiate A.

$$A(w) = 66w^{-1}$$
$$A'(w) = -66w^{-2}$$
$$= -\frac{66}{w^2}$$

$A'(w)$ is negative for all values of w in the domain of A, so there are no critical values. Since the derivative is negative, the surface area function is decreasing over its entire domain. Consequently, the smallest area will occur at the rightmost endpoint, $w = 2$. This may be easily confirmed by looking at a table of values for A (Table 5.10).

TABLE 5.10

w	$A(w)$
0	Undefined
0.5	132
1.0	66
1.5	44
2.0	33

Absolute minimum

When $w = 2$, $h = 1$, since

$$h = \frac{4}{w^2}$$
$$= \frac{4}{(2)^2}$$
$$= \frac{4}{4}$$
$$= 1$$

The box has dimensions $w \times w \times 2h$, so the optimal dimensions are $2' \times 2' \times 2'$.

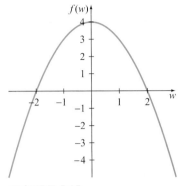

FIGURE 5.15

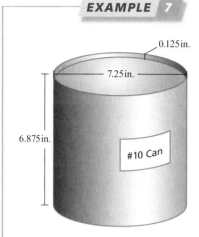

0.125 in.

7.25 in.

6.875 in.

#10 Can

FIGURE 5.16

EXAMPLE **7**

Minimizing Use of Resources

Many bulk food commodities are packaged in #10 cans. These cylindrical metal cans are 6.875 inches tall and have a 7.25-inch diameter. There is a 0.125-inch lip on the top and on the bottom of the can, which reduces the height of the interior of the can (from lid to lid) to 6.625 inches (Figure 5.16). The volume of a cylindrical can is $V = \pi r^2 h$, where r is the radius of the can and h is the interior height of the can.

A cylindrical can may be constructed from a flat sheet of metal of length $2\pi r$ and two circular lids each with surface area $A = \pi r^2$. The height of the sheet of metal is $h + 0.5$ (Figure 5.17).

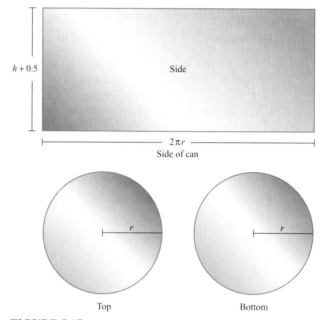

$h + 0.5$

Side

$2\pi r$

Side of can

r

r

Top

Bottom

FIGURE 5.17

The additional 0.5 inch of metal is required to create the lip at the top and the bottom of the can. (The metal initially extends upward about 0.25 inch but is folded in half when the can is sealed, creating the 0.125-inch top lip. We assume that the 0.125-inch bottom lip was created in a similar manner.)

(a) What is the volume of a #10 can?
(b) How much metal is required to construct a #10 can?
(c) Is it possible to construct a can with the same volume as a #10 can that requires less metal? If so, what is the minimum amount of metal required?

SOLUTION

(a) The diameter is 7.25 inches. Since the radius r is equal to half of the diameter,

$$r = 0.5(7.25)$$
$$= 3.625$$

The volume V of a cylinder with radius r and height h is given by $V = \pi r^2 h$.

$$V = \pi r^2 h$$
$$= \pi (3.625)^2 (6.625)$$
$$= 273.5 \text{ cubic inches}$$

The volume of the can is 273.5 cubic inches.

(b) The amount of metal required to construct the sides of the can is

$$S = 2\pi r(h + 0.5)$$
$$= 2\pi(3.625)(6.625 + 0.5)$$
$$= 162.3 \text{ square inches}$$

The amount of metal required to construct each lid is

$$L = \pi r^2$$
$$= \pi(3.625)^2$$
$$= 41.28 \text{ square inches}$$

The total amount of metal required is

$$A = 2\pi r(h + 0.5) + \pi r^2 + \pi r^2$$
$$= 162.3 + 41.28 + 41.28$$
$$= 244.9 \text{ square inches}$$

(c) The surface area of a cylindrical can with a 0.125-inch lip at the top and bottom of the can is

$$A = 2\pi r(h + 0.5) + \pi r^2 + \pi r^2$$
$$= 2\pi rh + \pi r + 2\pi r^2$$
$$= \pi r(2h + 1 + 2r)$$

In order to minimize A, we must find A'. However, A is presently a function of two variables. We must use a secondary equation to solve for one variable in terms of the other.

The volume of the can is $V = \pi r^2 h$. We want the can to have the same volume as a #10 can. So

$$V = \pi r^2 h$$
$$273.5 = \pi r^2 h \qquad \text{The volume of a \#10 can is 273.5 cubic inches}$$
$$h = \frac{273.5}{\pi r^2}$$

We may now rewrite the surface area equation in terms of a single variable, r.

$$A = \pi r(2h + 1 + 2r)$$
$$= \pi r\left[2\left(\frac{273.5}{\pi r^2}\right) + 1 + 2r\right]$$
$$= \pi r\left(\frac{547}{\pi r^2} + 1 + 2r\right)$$
$$= \frac{547}{r} + \pi r + 2\pi r^2$$
$$= 547r^{-1} + \pi r + 2\pi r^2$$
$$A'(r) = -547r^{-2} + \pi + 4\pi r$$
$$0 = -\frac{547}{r^2} + \pi + 4\pi r$$

We will use the graphing calculator to find the x-intercept(s) of the graph of $A'(r)$ (Figure 5.18).

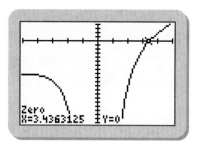

FIGURE 5.18

The only critical value is $r = 3.436$. We can visually see that $A'(r) < 0$ for values of $r < 3.436$. Likewise, $A'(r) > 0$ for values of $r > 3.436$. Since the sign of $A'(r)$ changes from negative to positive at $r = 3.436$, a relative minimum of A occurs at $r = 3.436$. The corresponding value of h is

$$h = \frac{273.5}{\pi(3.436)^2}$$

$$= 7.374 \text{ inches}$$

Using these values of h and r, we calculate the surface area.

$$A(r) = 547r^{-1} + \pi r + 2\pi r^2$$
$$A(3.436) = 547(3.436)^{-1} + \pi(3.436) + 2\pi(3.436)^2$$
$$= 244.2 \text{ square inches}$$

Remarkably, the #10 can very nearly minimizes the amount of metal required (244.9 square inches versus the minimal 244.2 square inches).

EXAMPLE 8   Minimizing Landscaping Costs

A landscape designer offers an Economy garden package to her clients. The rectangular garden has flowers along the front of the garden and shrubs along the back and sides (Figure 5.19).

FIGURE 5.19

The designer charges $25 per linear foot for the shrubs and $6 per linear foot for the flowers. A client wants to install a 400-square-foot garden. What will be the dimensions of the least expensive garden?

SOLUTION We begin by labeling the figure. We let l represent the length of the garden and w represent the width of the garden (Figure 5.20). The area of the garden is given by

$$A = lw$$
$$400 = lw$$

FIGURE 5.20

Assuming that the shrubs extend across the back of the garden and fully down both sides, the cost of the shrubs (in dollars) is

$$S = 25(w + l + w)$$
$$= 25l + 50w$$

Assuming that the flowers extend across the front of the garden, the cost of the flowers (in dollars) is

$$F = 6l$$

The combined cost for the garden is

$$C = (25l + 50w) + 6l$$
$$= 31l + 50w$$

We want to minimize this cost. We can't differentiate C because it contains two input variables, l and w. However, recall that

$$lw = 400$$
$$l = \frac{400}{w}$$

Therefore, the cost function may be rewritten as

$$C = 31l + 50w$$

$$= 31\left(\frac{400}{w}\right) + 50w$$

$$= 12{,}400w^{-1} + 50w$$

Differentiating the cost function and setting it equal to zero yields

$$C'(w) = -12{,}400w^{-2} + 50$$

$$0 = -12{,}400w^{-2} + 50 \qquad \text{Set } C'(w) = 0$$

$$\frac{12{,}400}{w^2} = 50$$

$$12{,}400 = 50w^2$$

$$w^2 = 248$$

$$w \approx 15.75 \qquad\qquad \text{Since } w > 0, \text{ we omit the solution}$$
$$w \approx -15.75$$

Applying the First Derivative Test, we determine that a relative minimum occurs at $w = 15.75$. (*Note:* $C'(15) < 0$ and $C'(16) > 0$.) To find the corresponding value of l, we substitute the value of w into the area equation.

$$l = \frac{400}{w}$$

$$= \frac{400}{15.75}$$

$$= 25.40$$

The dimensions that will minimize the cost are 25.40 feet by 15.75 feet. The minimum cost is

$$C = 31l + 50w$$

$$= 31(25.40) + 50(15.75)$$

$$= \$1574.90$$

The shape of the garden (drawn to scale) is shown in Figure 5.21. The client may want to change the dimensions; however, any change in dimensions will result in a higher cost.

FIGURE 5.21 25.4 feet $\times$ 15.75 feet

5.2 Summary

In this section, you learned how to find marginal revenue, marginal cost, and marginal profit. You also learned how to use the derivative to optimize specified areas and volumes.

5.2 Exercises

In Exercises 1–30, use the derivative techniques demonstrated in this section to determine the answer to each question.

1. **M** **Company Profit** In 2001, the Kellogg Company introduced a new breakfast cereal: Special K® Red Berries cereal. It quickly achieved an impressive 1 percent market share during its first six months on the market and helped to boost corporate sales. Based on data from 1999 to 2001, the net sales of the Kellogg Company may be modeled by

$$R(t) = 964.1t^2 - 993.6t + 6984$$

million dollars, where t is the number of years since 1999. The cost of goods sold may be modeled by

$$C(t) = 399.8t^2 - 379.9t + 3325.1$$

million dollars, where t is the number of years since 1999. (**Source:** Modeled from Kellogg Company 2001 Annual Report, pp. 7, 27.)

(a) Find the equation for the gross profit. (This is net sales minus the cost of goods sold.)

(b) Between 1999 and 2001, in which year was gross profit minimized?

2. **Carton Design** A carton of Peer Foods Sliced Canadian Style Bacon holds four 3-pound packages of sliced Canadian bacon. The dimensions of a carton are $11'' \times 19.75'' \times 4''$. (**Source:** www.peerfoods.com.)

(a) What is the volume of the carton?

(b) Is it possible to construct a carton with the same volume but less surface area? If so, find the dimensions of the carton with a square base with minimal surface area.

(c) What other issues should Peer Foods consider before moving to a different sized carton?

3. **Box Design** Peer Foods sells its bacon in two different case sizes. The Hotel Pack case has dimensions

$10.1875'' \times 8.5'' \times 7.5625''$. The Single Slice Pack case has dimensions $19.75'' \times 11'' \times 4''$. (**Source:** www.peerfoods.com.)

(a) Calculate the surface area and volume of each of the cases.

(b) Find the dimensions of the box with a square base that has the same surface area as the Hotel Pack case but has the greatest possible volume.

(c) Find the dimensions of the box with a square base that has the same surface area as the Single Slice Pack case but has the greatest possible volume.

(d) Based on the results of (b) and (c), what conclusion can you draw regarding the dimensions of a box that has the maximum possible volume for its surface area? (*Hint:* Divide the height of each box by its width and look for a pattern.)

4. **M** **Company Revenue** Uline Shipping Supply Specialists sell $40'' \times 48''$ hardwood pallets. The price they charge per pallet is reduced as the order quantity increases. The price of hardwood pallets is given by the piecewise function

$$p = \begin{cases} -5.5x + 124.5 & 1 \le x \le 2 \\ 105 & 3 \le x \le 10 \end{cases}$$

dollars, where x is the number of groups of 10 pallets. (**Source:** www.uline.com.)

(a) Write the equation for revenue as a function of the number of 10-pallet groups sold.

(b) Determine the marginal revenue when 10, 20, 30, and 40 pallets are sold.

(c) Explain the real-world meaning of the results of (b).

5. **M** **Company Revenue** In its 2001 Annual Report, the Coca-Cola Company reported that "Our worldwide unit case volume increased 4 percent in 2001, on top of a 4 percent increase in 2000. The increase in unit case volume reflects consistent performance across certain key operations despite difficult global economic conditions. Our business system sold 17.8 billion unit cases in 2001." (**Source:** Coca-Cola Company 2001 Annual Report, p. 46.) A unit case is equivalent to 24 eight-ounce servings of finished beverage.

(a) Find a model for the number of unit cases sold, s, as a function of t, where t is the number of years since 1999.

(b) Based on data from 1999 to 2001, the net operating revenue of the Coca-Cola Company may be modeled by

$$R(s) = -0.4029s^2 + 14.44s - 109.2$$

billion dollars, where s is the number of unit cases sold (in billions). Find the marginal revenue function.

(c) According to the model, at what unit case production level will revenue be maximized?

(d) Rewrite the revenue function as a function of t (years since 1999) instead of as a function of s (billions of unit cases sold). (*Hint:* Use a composition of functions.)

(e) According to the model, in what year will revenue be maximized?

6. **M**   **Company Profit**   Based on data from 1999 to 2001, the net operating revenues of the Coca-Cola Company may be modeled by

$$R(t) = -201t^2 + 806t + 19{,}284$$

million dollars and the cost of goods sold may be modeled by

$$C(t) = -177.5t^2 + 372.5t + 6009$$

million dollars, where t is the number of years since 1999.

(a) Find the gross profit function.

(b) According to the model, in what year is the gross profit projected to reach a maximum?

(c) Does the result of part (b) seem reasonable? Explain.

7. **M**   **Company Costs**   Frito-Lay North America, a subsidiary of PepsiCo, produces Doritos, Cheetos, Fritos corn chips, and a variety of other salty, sweet, or grain-based snacks. Based on data from 1999 to 2001, the net sales (revenue) of Frito-Lay North America may be modeled by

$$R(t) = -168t^2 + 907t + 8232$$

million dollars and the operating profit (earnings before interest and taxes) may be modeled by

$$P(t) = -47.5t^2 + 283.5t + 1679$$

million dollars, where t is the number of years since 1999. (**Source:** Modeled from 2001 PepsiCo Annual Report, pp. 23, 44.)

(a) In what year are net sales projected to reach a maximum?

(b) Find the cost function for Frito-Lay North America.

(c) According to the model, in what year are costs expected to reach a maximum?

(d) Compare the results of (a) and (c). Do the results seem reasonable?

8. **M**   **Company Revenue Versus Profit**   Based on data from 1999 to 2001, the net sales (revenue) of Pepsi-Cola North America may be modeled by

$$R(t) = -65.5t^2 - 749.5t + 2605$$

million dollars and the operating profit may be modeled by

$$P(t) = 6t^2 + 76t + 751$$

million dollars, where t is the number of years since 1999. (**Source:** Modeled from PepsiCo 2001 Annual Report, p. 44.)

(a) Find the marginal revenue and marginal profit functions.

(b) At what rate was revenue increasing in 2001?

(c) At what rate was profit increasing in 2001?

(d) Project the rate of change in revenue and in profit in 2014. Does this projection seem reasonable? Explain.

9. **M**   **Employee Wages**   Based on data from 1991 to 2001, the average wage of a Ford Motor Company employee may be modeled by

$$W(t) = -0.003931t^4 + 0.1005t^3 - 0.8295t^2 + 3.188t + 16.48$$

dollars per hour, where t is the number of years since 1990. (**Source:** Modeled from Ford Motor Company 2001 Annual Report, p. 71.)

Was the average wage changing more rapidly in 1995 or 2000?

10. **Pool Size Optimization**   A pool builder makes two types of economy-priced pools: square and circular. He estimates the price of the job by multiplying the perimeter or circumference of the pool (in linear feet) by a fixed price per linear foot. A homeowner wants her pool to have the

maximum amount of water surface area for the lowest possible price.

(a) Should the homeowner have a square or a circular pool built? Explain. (The area of a square with a perimeter p is $S = \frac{1}{16}p^2$. The area of a circle with circumference c is $A = \frac{1}{4\pi}c^2$.)

(b) Shortly before construction, the homeowner decides to increase the 100-foot distance around the pool by 1 foot. For both shapes of pool, use the derivative to approximate how much the surface area of the pool will increase by increasing the perimeter (circumference) by 1 foot.

(c) Does your result in (b) confirm your conclusion in (a)? Explain.

11. **Apple Farming** Historically, many apple farmers spaced trees 40 feet by 40 feet apart (27 per acre). Trees typically took 25 years to reach their maximum production of 500 bushels per acre. (A bushel is about 44 pounds.) In recent years, agriculturists have created dwarf and semi-dwarf varieties that allow trees to be spaced 10 feet by 10 feet apart. (Source: USDA.)

Suppose that a farmer has an apple orchard with 40 trees per acre. The orchard yields 10 bushels per tree. The farmer estimates that for each additional tree planted (per acre), the average yield per tree is reduced by 0.1 bushel.

If $y = f(x)$ is the total number of bushels of apples produced per acre when an additional x trees per acre are planted, determine how many additional trees should be planted in order to maximize the number of bushels of apples produced.

12. **Apple Farming** Suppose that a farmer has an apple orchard with 30 trees per acre. The orchard yields 12 bushels per tree. The farmer estimates that for each additional tree planted (per acre), the average yield per tree is reduced by 0.1 bushel.

If $y = f(x)$ is the total number of bushels of apples produced per acre when an additional x trees per acre are planted, determine how many additional trees should be planted in order to maximize the number of bushels of apples produced.

13. **Apple Supplier Prices** A fruit farmer sells apples to a grocery store chain. The amount of apples the store buys depends linearly upon the price per pound the farmer charges. The farmer estimates that for every $0.02 per pound increase in the price, the store will reduce its order by 50 pounds. The store presently orders 500 pounds per week and pays $0.18 per pound. What price should the farmer charge in order to maximize her revenue from apple sales?

14. **Apple Retailer Prices** A grocery store has priced apples at $0.70 per pound and sells 1000 pounds per week. The amount of apples the store sells depends linearly upon the price per pound the store charges. The store manager estimates that for every $0.04 per pound increase in the price, the store will reduce its sales by 100 pounds. What price should the store charge for its apples in order to maximize revenue from apple sales?

15. **Company Sales** Based on data from 1993 to 2002, the annual sales of Starbucks Corporation may be modeled by

$$S(t) = 29.23t^2 + 79.33t + 177.4$$

million dollars, where t is the number of years since 1993.

According to the model, how much more rapidly were sales increasing in 2001 than in 1999?

16. **Company Sales** Based on data from 1991 to 2001, the franchised sales of McDonalds Corporation may be modeled by

$$S(t) = -41.293t^2 + 1729.0t + 11,139$$

million dollars, where t is the number of years since 1990.

Calculate the instantaneous rate of change of sales in 1999 and 2001. Explain the financial significance of the result.

17. **Landscape Design** A landscaping company offers its clients the Rectangular Garden package at a reduced rate.

The Rectangular Garden has shrubs along the back and sides of the garden and flowers along the front of the garden. (See figure.)

The designer charges $25 per linear foot for the shrubs and $8 per linear foot for the flowers. A client wants to install a 500-square-foot garden. What will be the dimensions of the least expensive garden?

18. Landscape Design A landscaping company offers its clients the Rectangular Garden package at a reduced rate.

The Rectangular Garden has shrubs along the back and sides of the garden and flowers along the front of the garden. (See figure.)

The designer charges $20 per linear foot for the shrubs and $5 per linear foot for the flowers. A client wants to install a 250-square-foot garden. What will be the dimensions of the least expensive garden?

19. Inventory Cost An author anticipates that she will sell 6,000 copies of her book annually. The setup cost for each print run is $300. The storage cost for each printed book is $2.50 per year. Assuming that customer demand for the book is constant for the 250 business days of a year, determine the size of each print run that will minimize her inventory cost.

20. Inventory Cost An author anticipates that she will sell 5,000 copies of her book annually. The setup cost for each print run is $300. The storage cost for each printed book is $1.92 per year. Assuming that customer demand for the book is constant for the 250 business days of a year, determine the size of each print run that will minimize her inventory cost.

21. Inventory Cost An author anticipates that he will sell 500 copies of his book annually. The setup cost for each print run is $300. The storage cost for each printed book is $1.20 per year. Assuming that customer demand for the book is constant for the 250 business days of a year, determine the size of each print run that will minimize his inventory cost.

22. Inventory Cost An author anticipates that he will sell 12,000 copies of his book annually. The setup cost for each print run is $300. The storage

cost for each printed book is $1.25 per year. Assuming that customer demand for the book is constant for the 250 business days of a year, determine the size of each print run that will minimize his inventory cost.

23. Fleet Vehicle Sales A business that owns or leases 10 or more vehicles may qualify for auto manufacturer fleet purchase incentives. Incentives vary from $300 to $7000 on new vehicles. In August 2003, the manufacturer's fleet incentive for a 2004 Chevrolet Cavalier was $1400. (**Source:** www.fleet-central.com.)

An auto dealer offers an additional discount to fleet buyers who purchase one or more new Cavaliers. To encourage sales, the dealer reduces the after-incentive price of each car by $x\%$, where x is the total number of cars purchased. Assuming that the pre-incentive price of a 2004 Chevrolet Cavalier is $14,400, how many vehicles would the dealer need to sell in order to maximize revenue?

24. Fleet Vehicle Sales In August 2003, the manufacturer's fleet incentive for a 2004 Cadillac Seville was $2000. (**Source:** www.fleet-central.com.)

An auto dealer offers an additional discount to fleet buyers who purchase one or more new Cadillac Sevilles. To encourage sales, the dealer reduces the after-incentive price of each car by $2x\%$, where x is the total number of cars purchased. Assuming that the pre-incentive price of a 2004 Cadillac Seville is $41,000, how many vehicles would the dealer need to sell in order to maximize revenue?

25. Fleet Vehicle Sales In August 2003, the manufacturer's fleet incentive for a 2004 Buick Century was $1800. (**Source:** www.fleet-central.com.)

An auto dealer offers an additional discount to fleet buyers who purchase one or more new Buick Century vehicles. To encourage sales, the dealer reduces the after-incentive price of each car by $200x$ dollars, where x is the total number of cars purchased. Assuming that the pre-incentive price of a 2004 Buick Century is $19,000, how many vehicles would the dealer need to sell in order to maximize revenue?

26. Fleet Vehicle Sales In August 2003, the manufacturer's fleet incentive for a 2004 Pontiac Bonneville was $900. (**Source:** www.fleet-central.com.)

An auto dealer offers an additional discount to fleet buyers who purchases one or more new Pontiac Bonneville vehicles. To encourage sales, the dealer reduces the after-incentive price of each car by $250x$ dollars, where x is the total number of cars purchased. Assuming that the pre-incentive price of a 2004 Pontiac Bonneville is \$24,000, how many vehicles would the dealer need to sell in order to maximize revenue?

27. Pricing Analysis   Based on the results of Exercises 23 to 26, what additional restrictions (if any) should the dealer place on his advertised discount in order to ensure that selling additional vehicles won't reduce his revenue?

28. Profit Lost Due to Waste   A fruit vendor purchases x pounds of fruit from her supplier. She estimates that $0.2x\%$ of each pound of produce she buys spoils before it is purchased by a consumer. For each pound of fruit she sells, she makes a profit of \$0.40. How many pounds of fruit should she buy in order to maximize profit?

29. Profit Lost Due to Waste   A fruit vendor purchases x pounds of fruit from his supplier. He estimates that $0.2x\%$ of each pound of produce he buys spoils before it is purchased by a consumer. For each pound of fruit he sells, he makes a profit of \$0.50. How many pounds of fruit should he buy in order to maximize profit?

30. Profit Analysis   For many commodities, the marginal profit fluctuates. Based on the results of Exercises 28 and 29, what effect does the profit per pound have on the quantity of fruit that maximizes profit?

Exercises 31–36 are intended to challenge your skill in finding and interpreting the meaning of extrema. Each of the questions deals with a parametric function model of the position of a roller coaster train over a six-second interval. The x value of the model represents the horizontal position of a roller coaster train (in meters) at time t seconds. The y value of the model represents the vertical position of the roller coaster train (in meters) at time t seconds. The model is

$$x = t^3 - 9t^2 + 23t$$
$$y = 6t - t^2$$

31. Calculate and interpret the meaning of $\dfrac{dx}{dt}$ and $\dfrac{dy}{dt}$.

32. Determine when $\dfrac{dx}{dt} = 0$ and $\dfrac{dy}{dt} = 0$.

33. Given that the domain of the function is $0 \le t \le 6$, find the relative and absolute extrema of x and y.

34. Over the six-second interval, what was the slowest horizontal speed of the train?

35. Over the six-second interval, what was the slowest vertical speed of the train?

36. Do you think that the roller coaster would be safe to ride? Defend your conclusion.

5.3 Concavity and the Second Derivative

- Use the Second Derivative Test to find relative extrema
- Determine the concavity of the graph of a function

GETTING STARTED   Based on data from 1992 to 1998, the average annual earnings of a prepackaged software retailer may be modeled by

$$W(t) = 49.62t^4 - 637.12t^3 + 2895t^2 - 3768t + 32,150 \text{ dollars}$$

where t is the number of years since the end of 1992. (**Source:** Modeled from *Statistical Abstract of the United States, 2001*, Table 1123, p. 703.)

In what year were annual earnings increasing the fastest, and in what year were they increasing the slowest? Did the rate of increase in earnings continually grow, or did the rate of increase in earnings shrink at some point? Questions such as these may be answered using the *second derivative*.

In this section, we will introduce the second derivative and discuss the graphical concepts of concavity and inflection points. We will use these concepts to investigate the relationship between position, velocity, and acceleration. We will also introduce the Second Derivative Test as an alternative means of finding relative extrema. We will conclude with curve-sketching techniques.

Concavity

The term **concavity** refers to the curvature of a graph. A graph is said to be **concave up** if it curves upward and **concave down** if it curves downward. A simple rhyme is helpful in remembering the meaning of the terms.

Concave up is like a cup.

Concave down is like a frown.

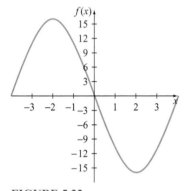

FIGURE 5.22

We will explore the concept of concavity by looking at the graph of the function $f(x) = x^3 - 12x$ on the interval $[-3, 3]$ (Figure 5.22).

From $x = -3$ to $x = 0$, the graph is curved downward (looks like a frown). At $x = 0$, the concavity appears to change. From $x = 0$ to $x = 3$, the graph is curved upward (looks like a cup). Is there a way to determine the concavity of a graph algebraically? Let's see. We'll begin by finding the derivative of the function and then the derivative of the derivative. The derivative of the derivative is called the **second derivative** and is commonly denoted by y'' (read "y double prime") or $f''(x)$ (read "f double prime of x").

$$f(x) = x^3 - 12x$$
$$f'(x) = 3x^2 - 12$$
$$f''(x) = 6x$$

We will generate a table of values for each of the functions (see Table 5.11).

TABLE 5.11

x	$f(x)$	$f'(x)$	$f''(x)$
-3	9	15	-18
-2	16	0	-12
-1	11	-9	-6
0	0	-12	0
1	-11	-9	6
2	-16	0	12
3	-9	15	18

We can learn much about the graph of the function from this table.

The graph of the function has x-intercepts when $f(x) = 0$. The x-intercept of the function on the domain $[-3, 3]$ occurs at $x = 0$, since $f(0) = 0$.

The graph of the function has relative extrema when $f'(x) = 0$ and the derivative changes sign. A relative maximum occurs at $x = -2$, since $f'(-2) = 0$ and the derivative changes from positive to negative at $x = -2$. A relative minimum occurs at $x = 2$, since $f'(2) = 0$ and the derivative changes from negative to positive at $x = 2$.

We observe that on the interval $[-3, 0)$, $f''(x) < 0$, and on the interval $(0, 3]$, $f''(x) > 0$. At $x = 0$, $f''(x) = 0$. Observe that the graph of f is concave down when $f''(x) < 0$ and is concave up when $f''(x) > 0$. The graph changes concavity from down to up at $x = 0$. The point on the graph where a function changes concavity is called an **inflection point.**

CONCAVITY AND INFLECTION POINTS OF THE GRAPH OF A FUNCTION

1. A continuous function f is *concave up* at a point $(c, f(c))$ if $f''(c) > 0$.
2. A continuous function f is *concave down* at a point $(c, f(c))$ if $f''(c) < 0$.
3. A continuous function f has an *inflection point* at $(c, f(c))$ if
   - $f''(c) = 0$ or $f''(c)$ is undefined **and**
   - $f''(c)$ changes sign at $x = c$

By referring back to the earlier rhyme, we can come up with a clever strategy to remember how to determine the concavity of a graph.

When we're positive, we are up (like the cup).

When we're negative, we're down and we frown.

Admittedly, the rhyme is a bit silly; however, its sheer wackiness will make the concept easier to remember.

A common error made by learners in their search for inflection points is to assume that if $f''(c) = 0$, then $(c, f(c))$ is an inflection point. Although $f''(c) = 0$ is a necessary condition for an inflection point, it is not a sufficient condition. Consider the function $f(x) = x^4$ with its associated second derivative $f''(x) = 12x^2$. Although $f''(0) = 0$, the point $(0, 0)$ is not an inflection point of f since the graph of f is concave up on both sides of the point (Figure 5.23).

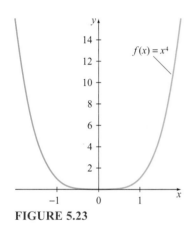

FIGURE 5.23

EXAMPLE   1

Determining the Concavity of a Graph

The graph of $f(x) = x^4 - 4x^3$ on the interval $[-3, 5]$ is shown in Figure 5.24. Determine where the graph is concave up and where the graph is concave down. Identify all inflection points.

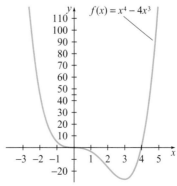

FIGURE 5.24

SOLUTION The graph appears to be concave up from $[-3, 0]$, concave down from $[0, 2]$, and concave up again from $[2, 5]$. It is difficult to determine the exact location of inflection points visually, so we will proceed algebraically to verify our visual conclusions.

$$f(x) = x^4 - 4x^3$$
$$f'(x) = 4x^3 - 12x^2$$
$$f''(x) = 12x^2 - 24x$$

In order to identify potential inflection points, we must set $f''(x)$ equal to zero.

$$0 = 12x^2 - 24x$$
$$0 = 12x(x - 2)$$

$$12x = 0 \qquad \text{or} \qquad x - 2 = 0$$
$$x = 0 \qquad\qquad\qquad x = 2$$

Inflection points may occur at $x = 0$ or $x = 2$; however, we must verify that $f''(x)$ changes sign at each of these points. We will do this using a sign chart for f''. We evaluate $f''(x)$ using points on either side of the potential inflection points.

$$f''(x) = 12x\,(x - 2)$$
$$f''(-1) = 12(-1)(-1 - 2) > 0$$
$$f''(1) = 12(1)(1 - 2) < 0$$
$$f''(3) = 12(3)(3 - 2) > 0$$

We update the sign chart for f'' with the results.

Since $f''(x)$ changes sign at $x = 0$ and $x = 2$, inflection points occur at each of these points. Since $f''(x) > 0$ on $[-3, 0)$ and on $(2, 5]$, it is concave up on those intervals. Since $f''(x) < 0$ on $(0, 2)$, it is concave down on that interval. Our algebraic analysis confirms our graphical estimation.

f''
$\quad +\qquad\quad -\qquad\quad +$
$\qquad 0 \qquad\qquad 2$

EXAMPLE 2

Determining the Concavity of a Graph

Determine the concavity of the graph $f(x) = x^3 + 3x^2 + 3x + 1$ on the interval $[-2, 2]$. Identify the location of any inflection points.

SOLUTION

$$f(x) = x^3 + 3x^2 + 3x + 1$$
$$f'(x) = 3x^2 + 6x + 3$$
$$f''(x) = 6x + 6$$

We set $f''(x) = 0$ and solve.

$$0 = 6x + 6$$
$$-6x = 6$$
$$x = -1$$

An inflection point may occur at $x = -1$. We'll evaluate the second derivative on either side of $x = -1$ and construct a sign chart for f''.

$$f''(-2) = 6(-2) + 6$$
$$= -6$$

f''
$\quad -\qquad\qquad +$
$\qquad -1$

$$f''(0) = 6(0) + 6$$
$$= 6$$

$f(x)$ is concave down on the interval $[-2, -1)$ and concave up on the interval $(-1, 2]$. An inflection point occurs at the point $(-1, 0)$ (Figure 5.25).

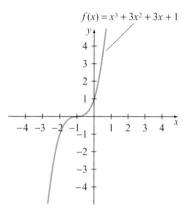

FIGURE 5.25

EXAMPLE 3

Using the Second Derivative to Determine Maximum and Minimum Rates of Change

Based on data from 1992 to 1998, the average annual earnings of a prepackaged software retailer may be modeled by

$$W(t) = 49.62t^4 - 637.12t^3 + 2895t^2 - 3768t + 32{,}150 \text{ dollars}$$

where t is the number of years since 1992. (**Source:** Modeled from *Statistical Abstract of the United States, 2001*, Table 1123, p. 703.)

In what year were annual earnings increasing the fastest, and in what year were they increasing the slowest? Did the rate of increase in earnings continually grow, or did the rate of increase in earnings shrink at some point?

SOLUTION The rate of change of the earnings function is $W'(t)$.

$$W'(t) = 198.48t^3 - 1911.36t^2 + 5790t - 3768 \frac{\text{dollars}}{\text{year}}$$

This function tells us how fast or how slowly earnings are increasing. We want to determine the extrema of $W'(t)$. To do this, we differentiate $W'(t)$.

UNITS

$$W''(t) = 595.44t^2 - 3822.72t + 5790 \frac{\text{dollars per year}}{\text{year}}$$

We set $W''(t)$ equal to zero and solve using the quadratic formula.

$$t = \frac{-b \pm \sqrt{b^2 - 4ac}}{2a}$$

$$= \frac{3822.72 \pm \sqrt{(-3822.72)^2 - 4(595.44)(5790)}}{2(595.44)}$$

$$= \frac{3822.72 \pm 907.08}{1190.88}$$

$$= 2.448; \ 3.972$$

We evaluate $W''(t)$ on either side of the critical points of $W'(t)$ and construct a sign chart for $W''(t)$.

The relative extrema of $W'(t)$ are in fact the inflection points of $W(t)$. Inflection points tell us when the growth rate of $W(t)$ is accelerating or slowing down. Let's interpret the financial meaning of our results.

$$W'(2.448) = 1863 \text{ dollars per year}$$

$$W'(3.972) = 1513 \text{ dollars per year}$$

Let's look again at the sign chart of W''.

Recall that W' is the growth rate of employee earnings (in dollars per year). Between the end of 1992 ($t = 0$) and the middle of 1995 ($t = 2.448$), the growth rate of employee earnings was increasing. However, between the middle of 1995 and the end of 1996 ($t = 3.972$), the growth rate of employee earnings was decreasing. That is, although earnings were increasing, they were increasing at a slower rate. In the middle of 1995 ($t = 2.448$), the earnings growth rate was $1863 per year, but by the end of 1996 ($t = 3.972$), the earnings growth rate had fallen to $1513 per year. Between the end of 1996 and the end of 1998, the growth rate of employee earnings was again increasing. We can verify our conclusions by generating a table of values for W, W', and W''. Tables such as Table 5.12 may be quickly generated using the Technology Tip following this example.

TABLE 5.12

Years Since 1992 (t)	$W(t)$ (dollars)	$W'(t)$ (dollars per year)	$W''(t)$ (dollars per year per year)
0	32,150	−3,768	5,790
1	30,690	309	2,563
2	31,891	1,754	526
3	33,718	1,759	−319
4	35,325	1,513	26
5	37,058	2,208	1,562
6	40,452	5,035	4,290

In what year were annual earnings increasing the fastest, and in what year were they increasing the slowest? Earnings were diminishing at a rate of $3768 per year in 1992. This was the slowest rate of growth between 1992 and 1998. Earnings were increasing at a rate of $5035 per year in 1998. This is the fastest rate of growth between 1992 and 1998.

Did the rate of increase in earnings grow continually, or did the rate of increase in earnings shrink at some point? Notice that the values of $W'(t)$ increased every year except between 1995 ($t = 3$) and 1996 ($t = 4$). The fact that $W''(3)$ was negative told us that the rate of increase in earnings was decreasing in 1995.

TECHNOLOGY `TIP`

Creating a Table of Values

1. Use the [Y=] editor to enter the function(s) to be evaluated. In this case we enter W, W', and W''.

```
Plot1 Plot2 Plot3
\Y1■49.62X^4-637
.12X^3+2895X^2-3
768X+32150
\Y2■198.48X^3-19
11.36X^2+5790X-3
768
\Y3■595.44X^2-38
```

2. Press [2nd] [WINDOW] to open Table Setup. `TblStart=` is the x value where we want the table to begin, and ΔTbl is the distance between consecutive x values. We will start the table at $x = 0$ and space values one unit apart.

```
TABLE SETUP
 TblStart=0
 △Tbl=1
Indpnt: Auto  Ask
Depend: Auto  Ask
```

3. Press [2nd] [GRAPH] to display the table.

X	Y₁	Y₂
0	32150	⁻3768
1	30690	309.12
2	31891	1754.4
3	33718	1758.72
4	35325	1513
5	37058	2208
6	40452	5034.7

`Y₂=1758.72`

4. To see additional function values, use the blue arrows to navigate between rows and columns.

X	Y₂	Y₃
2	1754.4	526.32
3	1758.7	⁻319.7
4	1513	26.16
5	2208	1562.4
6	5034.7	4289.5
7	11184	8207.5
8	21847	13316.4

`Y₃=13316.4`

5. If you prefer to select the x values of the function(s) manually, press [2nd] [WINDOW] and select `Indpnt:Ask`.

```
TABLE SETUP
 TblStart=0
 △Tbl=1
Indpnt: Auto  Ask
Depend: Auto  Ask
```

6. Press [2nd] [GRAPH] to bring up the table, then enter the desired x values.

X	Y₁	Y₂
0	32150	⁻3768
2.448	32710	1863.4
3	33718	1758.7
3.972	35283	1512.6
5	37058	2208

`X=`

The Second Derivative Test

As we've seen, the second derivative may be used to determine the concavity of the graph and find the location of inflection points. The second derivative may also be used to find relative extrema. Consider the function $f(x) = x^3 - 12x$. Differentiating the function yields

$$f'(x) = 3x^2 - 12$$
$$0 = 3(x^2 - 4) \qquad \text{Set } f'(x) = 0 \text{ to find critical values}$$
$$0 = 3(x - 2)(x + 2)$$
$$x - 2 = 0 \qquad\qquad x + 2 = 0$$
$$\text{or}$$
$$x = 2 \qquad\qquad x = -2$$

The function f has critical values $x = -2$ and $x = 2$, since $f'(-2) = 0$ and $f'(2) = 0$. Graphically speaking, the graph of $f(x) = x^3 - 12x$ is flat (has a horizontal tangent line) at $x = -2$ and $x = 2$.

The second derivative of $f(x) = x^3 - 12x$ is $f''(x) = 6x$. We evaluate f'' at each of the critical values of f.

$$f''(x) = 6x$$
$$f''(-2) = 6(-2)$$
$$= -12$$
$$f''(2) = 6(2)$$
$$= 12$$

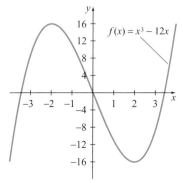

FIGURE 5.26

Since $f''(-2) < 0$, the graph of f is concave down at $x = -2$. However, since f also has a horizontal tangent line at $x = -2$, a relative maximum occurs there.

Since $f''(2) > 0$, the graph of f is concave up at $x = 2$. However, since f also has a horizontal tangent line at $x = 2$, a relative minimum occurs there (Figure 5.26).

These results are summarized in the Second Derivative Test.

THE SECOND DERIVATIVE TEST

Let f be a continuous function with $f'(c) = 0$. If

- $f''(c) > 0$, then a relative minimum of f occurs at $x = c$.
- $f''(c) < 0$, then a relative maximum of f occurs at $x = c$.
- $f''(c) = 0$, then the test is inconclusive.

EXAMPLE 4   **Finding Relative Extrema with the Second Derivative Test**

Use the Second Derivative Test to find the relative extrema of the function $f(x) = x^5 - 5x^4$.

SOLUTION

$$f'(x) = 5x^4 - 20x^3$$
$$0 = 5x^3(x - 4) \qquad \text{Set } f'(x) = 0 \text{ and find critical values}$$

$$5x^3 = 0 \qquad \text{or} \qquad x - 4 = 0$$
$$x = 0 \qquad\qquad\qquad x = 4$$

The critical values are $x = 0$ and $x = 4$. To apply the Second Derivative Test, we calculate $f''(x)$ and evaluate it at the critical values, $x = 0$ and $x = 4$. Writing the second derivative in factored form will make it easier to evaluate.

$$f''(x) = 20x^3 - 60x^2$$
$$= 20x^2(x - 3)$$
$$f''(0) = 20(0)^2(0 - 3)$$
$$= 0$$
$$f''(4) = 20(4)^2(4 - 3)$$
$$= 320$$

Since $f''(4) > 0$, a relative minimum occurs at $x = 4$. The relative minimum is $(4, -256)$.

At $x = 0$, the Second Derivative Test is inconclusive. What is happening at $x = 0$? Let's evaluate $f'(x)$ at points on either side of the critical value $x = 0$ to see if the slope of f changes sign at $x = 0$. (This is the First Derivative Test.)

$$f'(-1) = 5(-1)^3(-1 - 4)$$
$$= 25$$

Since $f'(-1) > 0$, the graph of f is increasing at $x = -1$.

$$f'(1) = 5(1)^3(1 - 4)$$
$$= -15$$

Since $f'(1) < 0$, the graph of f is decreasing at $x = 1$. Since f' changes from positive to negative at $x = 0$, a relative maximum occurs at $x = 0$. The relative maximum is $(0, 0)$. The graph of the function (Figure 5.27) confirms our conclusion.

Although calculations are often simpler when we use the Second Derivative Test, we may have to revert back to the First Derivative Test if the Second Derivative Test yields an inconclusive result.

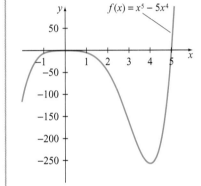

FIGURE 5.27

Sometimes we want to maximize the value of the first derivative or the second derivative instead of the value of the original function. In Example 5, we will maximize the derivative using the First Derivative Test. The notation gets a bit tricky because the critical points of f' are found by setting f'' equal to zero.

EXAMPLE 5

Using the Second Derivative in a Real-World Context

Based on data from 1991 to 2001, the average wage of a Ford Motor Company employee may be modeled by

$$W(t) = -0.003931t^4 + 0.1005t^3 - 0.8295t^2 + 3.188t + 16.48 \text{ dollars per hour}$$

where t is the number of years since 1990. (**Source:** Modeled from Ford Motor Company 2001 Annual Report, p. 71.) According to the model, did the rate of increase in employee hourly wages (in dollars) ever diminish between 1991 and 2001? What was the minimum and what was the maximum *annual increase* in the average hourly wage?

SOLUTION   The rate of increase in hourly wages is given by

$$W'(t) = -0.015724t^3 + 0.3015t^2 - 1.659t + 3.188 \frac{\text{dollars}}{\text{year}}$$

We want to know if W' ever decreased between 1991 and 2001. That is, we want to know when $W'' < 0$.

$$W''(t) = -0.047172t^2 + 0.603t - 1.659 \frac{\text{dollars per year}}{\text{year}}$$

We set $W''(t) = 0$ and solve using the Quadratic Formula.

$$0 = -0.047172t^2 + 0.603t - 1.659$$

$$t = \frac{-0.603 \pm \sqrt{(0.603)^2 - 4(-0.047172)(-1.659)}}{2(-0.047172)}$$

$$= \frac{-0.603 \pm 0.2249}{-0.094344}$$

$$= 4.008, \ 8.775$$

The critical values of W' are $t = 4.008$ and $t = 8.775$. We'll evaluate $W''(t)$ on either side of the critical values of W'.

$$W''(0) = -1.659$$

$$W''(5) = 0.1767$$

$$W''(10) = -0.3462$$

From 1991 ($t = 1$) to early 1995 ($t = 4.008$), the annual rate of increase in hourly wages was decreasing (since $W'' < 0$). Although the hourly wage continued to rise during those years, each annual increase was less than that of the year before.

From early 1995 ($t = 4.008$) to late 1999 ($t = 8.775$), the annual rate of increase in hourly wages was increasing (since $W'' > 0$). The hourly wage continued to rise during those years, and each annual increase was more than that of the year before.

From late 1999 ($t = 8.775$) to the end of 2001 ($t = 11$), the annual rate of increase in hourly wages was decreasing (since $W'' < 0$). Although the hourly wage continued to rise during those years, each annual increase was less than that of the year before.

The maximum annual increase in hourly wage will occur when W' reaches its maximum value. From the sign chart for W'', we see that a relative maximum of W' occurs when $t = 8.775$. We'll need to check the endpoints to see if this is an absolute maximum.

The minimum annual increase in hourly wage will occur when W' reaches its minimum value. From the sign chart for W'', we see that a relative minimum of W' occurs when $t = 4.008$. We'll need to check the endpoints to see if this is an absolute minimum. (See Table 5.13.)

TABLE 5.13

Years Since 1990 (t)	Dollars per Year [$W'(t)$]	
1	1.81	Absolute maximum of $W'(t)$
4.008	0.37	Relative and absolute minimum of $W'(t)$
8.775	1.22	Relative maximum of $W'(t)$
11	0.49	

According to the model, the average hourly wage was increasing most slowly in early 1995 ($t = 4.008$). At that time, it was increasing at a rate of $0.37 per year.

According to the model, the average hourly wage was increasing most rapidly at the end of 1991 ($t = 1$). At that time, it was increasing at a rate of $1.81 per year.

Point of Diminishing Returns

In business, we are often interested in knowing when the rate of change in revenue or profit from the sale of an item reaches a maximum value. The point at which this occurs is called the **point of diminishing returns.** In terms of calculus, we want to know when the derivative $R'(x)$ of a revenue function $R(x)$ attains a maximum.

EXAMPLE 6

Finding the Point of Diminishing Returns

Based on data from 1990 to 2000, the revenue from sales of newspaper advertising may be modeled by

$$A(t) = -32.08t^3 + 656.2t^2 - 1641t + 31{,}914 \text{ million dollars}$$

where t is the number of years since the end of 1990. (**Source:** Modeled from *Statistical Abstract of the United States, 2001*, Table 1272, p. 777.) Find the point of diminishing returns for newspaper advertising revenue and determine the rate at which advertising revenue is changing at that point.

SOLUTION   The rate of change in newspaper advertising is $A'(t)$.

$$A(t) = -32.08t^3 + 656.2t^2 - 1641t + 31{,}914$$

$$A'(t) = -96.24t^2 + 1312.4t - 1641 \text{ million dollars per year}$$

To determine when $A'(t)$ attains a maximum, we must find $A''(t)$.

$$A''(t) = -192.48t + 1312.4 \text{ million dollars per year per year}$$

To find the critical values of $A'(t)$, we set $A''(t) = 0$.

$$0 = -192.48t + 1312.4$$

$$192.48t = 1312.4$$

$$t = \frac{1312.4}{192.48}$$

$$t \approx 6.8$$

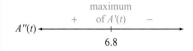

At $t = 6.8$, $A'(t)$ attains a relative maximum. Thus the point of diminishing returns occurs at $t = 6.8$. The point of diminishing returns is $(6.8, A(6.8)) = (6.8, 41{,}011)$. According to the model, revenue from newspaper advertising was \$41,011 million in late 1997 ($t = 6.8$).

To determine the rate of change in revenue at the point of diminishing returns, we evaluate $A'(t)$ at the critical value. (We also evaluate $A'(t)$ at the implied endpoints as one strategy for verifying the accuracy of our work; see Table 5.14.)

TABLE 5.14

t	$A'(t) = -96.24t^2 + 1{,}312.4t - 1641$	
0	$-1{,}641$	Absolute minimum of $A'(t)$
6.8	$2{,}833$	Relative and absolute maximum of $A'(t)$
10	$1{,}859$	

Newspaper advertising revenue was increasing most rapidly in late 1997 ($t = 6.8$). At that time, revenue was increasing at a rate of \$2833 million per year.

Position, Velocity, and Acceleration

The velocity of an object is the rate of change in its position over time. The acceleration of an object is the rate of change in its velocity over time. These relationships between position, velocity, and acceleration are nicely captured using the derivative concept.

RELATIONSHIP BETWEEN POSITION, VELOCITY, AND ACCELERATION

Let $s(t)$ be the function that describes the position of an object at time t. Then

- $v(t) = s'(t)$, where $v(t)$ is the velocity of the object at time t.
- $a(t) = v'(t) = s''(t)$, where $a(t)$ is the acceleration of the object at time t.

EXAMPLE 7   **Using a Position Function to Determine Velocity and Acceleration**

The author tracked his mileage as he drove through a residential area and into a cemetery. Every 15 seconds he recorded the mileage (accurate to 0.05 mile). Based on 1.5 minutes of data, his distance from a stoplight at the bottom of a hill is given by the position function

$$s(t) = -0.5333t^4 + 1.333t^3 - 0.7667t^2 + 0.3167t \text{ miles}$$

where t is in minutes.

Determine the velocity and acceleration functions. Then calculate the author's velocity and acceleration at 1 minute into his timed trip and at 1.25 minutes into his trip.

SOLUTION

$$v(t) = s'(t)$$

$$v(t) = \frac{d}{dt}(-0.5333t^4 + 1.333t^3 - 0.7667t^2 + 0.3167t)$$

$$v(t) = -2.1332t^3 + 3.999t^2 - 1.5334t + 0.3167 \, \frac{\text{miles}}{\text{minute}}$$

$$a(t) = v'(t)$$

$$a(t) = \frac{d}{dt}(-2.1332t^3 + 3.999t^2 - 1.5334t + 0.3167)$$

UNITS ➤

$$a(t) = -6.3996t^2 + 7.998t - 1.5334 \, \frac{\text{miles per minute}}{\text{minute}}$$

To calculate the velocity and acceleration one minute into the trip, we evaluate $v(1)$ and $a(1)$.

$$v(1) = -2.1332(1)^3 + 3.999(1)^2 - 1.5334(1) + 0.3167 \, \frac{\text{miles}}{\text{minute}}$$

$$= 0.6491 \, \frac{\text{miles}}{\text{minute}}$$

UNITS ➤

$$= 0.6491 \, \frac{\text{miles}}{\text{minute}} \cdot \frac{60 \, \text{minutes}}{1 \, \text{hour}}$$

$$= 38.946 \text{ miles per hour}$$

$$\approx 39 \text{ miles per hour}$$

$$a(1) = -6.3996(1)^2 + 7.998(1) - 1.5334 \, \frac{\text{miles per minute}}{\text{minute}}$$

$$= 0.065 \, \frac{\text{miles per minute}}{\text{minute}}$$

UNITS ➤

$$= 0.065 \, \frac{\text{miles}}{\text{minute}} \cdot \frac{1}{\text{minute}}$$

$$= 0.065 \, \frac{\text{miles}}{\text{minute}} \cdot \frac{1}{\text{minute}} \cdot \frac{60 \, \text{minutes}}{1 \, \text{hour}}$$

$$= 3.9 \, \frac{\text{miles}}{\text{hour}} \cdot \frac{1}{\text{minute}}$$

$$\approx 4 \text{ miles per hour per minute}$$

At 1 minute into his trip, the author was traveling at approximately 39 miles per hour and was accelerating at a rate of 4 miles per hour per minute. That is, if he maintained his current rate of acceleration for the next minute, his speed would increase by about 4 miles per hour.

To calculate the velocity and acceleration 1.25 minutes into the trip, we evaluate $v(1.25)$ and $a(1.25)$.

$$v(1.25) = -2.1332(1.25)^3 + 3.999(1.25)^2 - 1.5334(1.25) + 0.3167 \, \frac{\text{miles}}{\text{minute}}$$

$$= 0.4820 \, \frac{\text{miles}}{\text{minute}}$$

$$= 0.4820 \frac{\text{miles}}{\text{minute}} \cdot \frac{60 \text{ minutes}}{1 \text{ hour}}$$

$$\approx 29 \text{ miles per hour}$$

$$a(1.25) = -6.3996(1.25)^2 + 7.998(1.25) - 1.5334 \frac{\text{miles per minute}}{\text{minute}}$$

$$= -1.535 \frac{\text{miles per minute}}{\text{minute}}$$

UNITS

$$= -1.535 \frac{\text{miles}}{\text{minute}} \cdot \frac{1}{\text{minute}} \cdot \frac{60 \text{ minutes}}{1 \text{ hour}}$$

$$= -92.10 \frac{\text{miles}}{\text{hour}} \cdot \frac{1}{\text{minute}}$$

$$\approx -92 \text{ miles per hour per minute}$$

At 1 minute and 15 seconds into his trip, he was traveling at approximately 29 miles per hour and was decelerating at a rate of 92 miles per hour per minute. That is, if he maintained his current rate of deceleration for the next minute, his speed would decrease by about 92 miles per hour. The rapid rate of deceleration indicates that he is braking, probably in preparation for turning into the cemetery. (Since he was traveling at 29 mph, we know that he won't be able to maintain the high rate of deceleration for an entire minute. Otherwise, his velocity would turn negative, indicating that he was moving away from the cemetery and back toward the stop light.)

Curve Sketching

Although nowadays it is customary to graph functions using a graphing calculator, knowing how to graph functions by hand will greatly increase your understanding of calculus. If you know the x- and y-intercepts, the relative and absolute extrema, and the concavity and inflection points of a function, it is relatively easy to come up with a good sketch of the curve. We will demonstrate the curve-sketching process in Example 8.

EXAMPLE 8

Sketching a Polynomial Function

Sketch the graph of $f(x) = x^3 - 2x^2 - 4x + 8$ on the interval $[-3, 4]$.

SOLUTION We will set $f(x) = 0$ to find x-intercepts, $f'(x) = 0$ to find critical values, and $f''(x) = 0$ to determine where inflection points may occur.

x- and y-intercepts

$$f(x) = x^3 - 2x^2 - 4x + 8$$

$$0 = (x^3 - 2x^2) - (4x - 8) \qquad \text{Set } f(x) = 0 \text{ and group terms}$$

$$0 = x^2(x - 2) - 4(x - 2) \qquad \text{Factor each group}$$

$$0 = (x - 2)(x^2 - 4) \qquad \text{Factor } (x - 2) \text{ out of each term}$$

$$0 = (x - 2)(x - 2)(x + 2) \qquad \text{Factor } (x^2 - 4)$$

$$0 = (x - 2)^2(x + 2) \qquad \text{Group like terms}$$

$$x - 2 = 0 \qquad\qquad x + 2 = 0$$
$$\text{or}$$
$$x = 2 \qquad\qquad x = -2$$

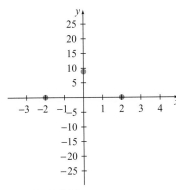

FIGURE 5.28

The graph of $f(x) = x^3 - 2x^2 - 4x + 8$ has x-intercepts at $(-2, 0)$ and $(2, 0)$. To find the y-intercept, we evaluate $f(x) = x^3 - 2x^2 - 4x + 8$ at $x = 0$.

$$f(0) = (0)^3 - 2(0)^2 - 4(0) + 8$$
$$= 8$$

The y-intercept is $(0, 8)$. We plot these points (Figure 5.28).

Relative and Absolute Extrema

To find critical values, we find $f'(x)$ and set it equal to zero.

$$f(x) = x^3 - 2x^2 - 4x + 8$$
$$f'(x) = 3x^2 - 4x - 4 \qquad \text{Differentiate } f(x)$$
$$0 = 3x^2 - 4x - 4 \qquad \text{Set } f'(x) = 0$$
$$0 = (3x + 2)(x - 2) \qquad \text{Factor}$$

$$3x + 2 = 0 \qquad \qquad x - 2 = 0$$
$$\text{or}$$
$$3x = -2 \qquad \qquad x = 2$$

$$x = -\frac{2}{3}$$

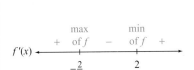

$f(x)$ has critical values at $x = -\frac{2}{3}$ and $x = 2$. We construct a sign chart for $f'(x)$ to determine where the relative extrema occur.

We evaluate $f(x)$ at the critical values and the endpoints (see Table 5.15).

TABLE 5.15

x	$f(x) = x^3 - 2x^2 - 4x + 8$	
-3	-25	Absolute minimum
$-\frac{2}{3}$	$9\frac{13}{27} \approx 9.5$	Relative maximum
2	0	Relative minimum
4	24	Absolute maximum

We plot the corresponding points on the graph of f (Figure 5.29).

From the sign chart of $f'(x)$, we can also determine the intervals on which f is increasing or decreasing. Since $f'(x) > 0$ on $\left[-3, -\frac{2}{3}\right)$ and $(2, 4]$, f is increasing on those intervals. Similarly, since $f'(x) < 0$ on the interval $\left[-\frac{2}{3}, 2\right)$, f is decreasing on that interval. We won't update the graph of f with this information yet, since we still don't know the concavity of the graph on these intervals.

Inflection Points

To find the inflection points of f, we set $f''(x) = 0$.

$$f'(x) = 3x^2 - 4x - 4$$
$$f''(x) = 6x - 4 \qquad \text{Differentiate } f'(x)$$
$$0 = 6x - 4 \qquad \text{Set } f''(x) = 0$$
$$4 = 6x$$
$$x = \frac{2}{3} \qquad \text{Simplify } \frac{4}{6} \text{ to } \frac{2}{3}$$

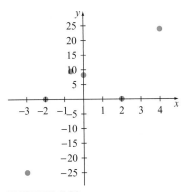

FIGURE 5.29

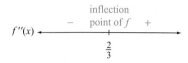

An inflection point *may* occur at $x = \frac{2}{3}$. We'll construct a sign chart for $f''(x)$ and see if the concavity of f changes at $x = \frac{2}{3}$.

Since $f''(x) < 0$ on the interval $\left[-3, \frac{2}{3}\right)$, the graph of f is concave down on that interval. Since $f''(x) > 0$ on the interval $\left(\frac{2}{3}, 4\right]$, the graph of f is concave up on that interval. An inflection point occurs at $x = \frac{2}{3}$.

TABLE 5.16

x	$f(x) = x^3 - 2x^2 - 4x + 8$	
$\frac{2}{3}$	$4\frac{20}{27} \approx 4.7$	Inflection point

We update the graph with the inflection point (Figure 5.30).

Referring to our earlier observations regarding the increasing/decreasing behavior and concavity of f, we finish the graph (Figure 5.31).

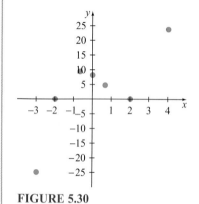

FIGURE 5.30

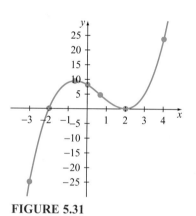

FIGURE 5.31

We summarize the curve-sketching steps in the following box.

HOW TO Curve Sketching

To graph a function $f(x)$ on an interval $[a, b]$, complete the following steps. After each step, graph the corresponding point(s).

1. Find the x-intercepts of $f(x)$ by setting $f(x) = 0$ and solving for x.
2. Find the y-intercept of $f(x)$ by evaluating $f(0)$.
3. Find the relative extrema and increasing/decreasing behavior of $f(x)$ by constructing a sign chart for $f'(x)$.

(Continued)

4. Find the absolute extrema of $f(x)$ by evaluating $f(x)$ at each critical value and at the endpoints $x = a$ and $x = b$.
5. Find the inflection points and concavity of $f(x)$ by constructing a sign chart for $f''(x)$.
6. Connect the points, paying attention to the increasing/decreasing behavior and concavity of $f(x)$.

EXAMPLE   9

Sketching a Function

Using the curve-sketching techniques of calculus, graph the function $f(x) = \frac{\ln(x)}{2x}$ on the domain $[1, 8]$.

SOLUTION

x- and y-intercepts

$$f(x) = \frac{\ln(x)}{2x}$$

$$0 = \frac{\ln(x)}{2x} \qquad \text{Set } f(x) = 0$$

$$0 = \ln(x) \qquad \text{Multiply both sides by } 2x$$

$$e^0 = x \qquad \text{Rewrite as an exponential function}$$

$$x = 1 \qquad \text{Since } e^0 = 1$$

The function has an x-intercept at $(1, 0)$. Since $x = 0$ is not in the domain of the function, the function does not have a y-intercept.

Relative Extrema and Increasing/Decreasing Behavior

$$f(x) = \frac{\ln(x)}{2x}$$

$$f'(x) = \frac{\left(\frac{1}{x}\right)(2x) - (2)[\ln(x)]}{(2x)^2} \qquad \text{Apply the Quotient Rule}$$

$$= \frac{2 - 2\ln(x)}{4x^2}$$

$$0 = \frac{2 - 2\ln(x)}{4x^2} \qquad \text{Set } f'(x) = 0$$

$$0 = 2 - 2\ln(x) \qquad \text{Set numerator equal to 0}$$

$$2 = 2\ln(x)$$

$$1 = \ln(x)$$

$$e^1 = e^{\ln(x)} \qquad \text{Exponentiate both sides}$$

$$e = x \qquad \text{Since } e^{\ln(x)} = x$$

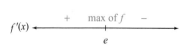

The critical value of the function is $x = e$. [Although the value $x = 0$ makes $f'(x)$ undefined, $x = 0$ is not a critical value because it is not in the domain of $f(x)$.]

We construct a sign chart to determine the relative extrema and the increasing/decreasing behavior of f.

The graph of f is increasing on the interval $[1, e)$ and decreasing on the interval $(e, 8]$. A relative maximum occurs at $x = e$. We evaluate f at the critical value and endpoints (see Table 5.17).

TABLE 5.17

x	$f(x) = \dfrac{\ln(x)}{2x}$	
1	0	Absolute minimum
e	$\dfrac{1}{2e} \approx 0.18$	Relative maximum, absolute maximum
8	$\dfrac{\ln(8)}{16} \approx 0.13$	

We create a graph of the points (Figure 5.32).

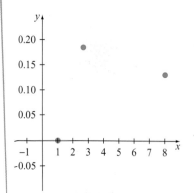

FIGURE 5.32

Inflection Points and Concavity

$$f'(x) = \frac{2 - 2\ln(x)}{4x^2}$$

$$f''(x) = \frac{\left(-\dfrac{2}{x}\right)(4x^2) - 8x[2 - 2\ln(x)]}{(4x^2)^2}$$ Apply the Quotient Rule

$$= \frac{-8x - 16x + 16x\ln(x)}{16x^4}$$

$$= \frac{-24x + 16x\ln(x)}{16x^4}$$

$$= \frac{-8x[3 - 2\ln(x)]}{16x^4}$$ Factor the numerator

$$0 = \frac{-8x[3 - 2\ln(x)]}{16x^4}$$ Set $f'(x) = 0$

$$0 = -8x[3 - 2\ln(x)]$$ Multiply both sides by $16x^4$

$$-8x = 0 \qquad\text{or}\qquad 3 - 2\ln(x) = 0$$
$$x = 0 \qquad\qquad 3 = 2\ln(x)$$
$$\frac{3}{2} = \ln(x)$$
$$x = e^{3/2}$$

Since $x = 0$ is not in the domain of the function, $x = 0$ is not a point of inflection. We construct a sign chart for $f''(x)$.

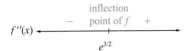

The graph of f is concave down on $[1, e^{3/2})$ and concave up on $(e^{3/2}, 8]$. An inflection point occurs at $x = e^{3/2} \approx 4.5$ (see Table 5.18).

TABLE 5.18

x	$f(x) = \dfrac{\ln (x)}{2x}$	
$e^{3/2}$	$\dfrac{3}{4e^{3/2}} \approx 0.17$	Inflection point

We update the graph with the inflection point (Figure 5.33).

Paying attention to the increasing/decreasing behavior and concavity of f, we finish the graph (Figure 5.34).

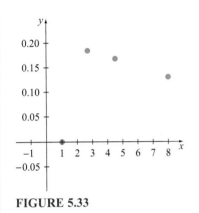

FIGURE 5.33

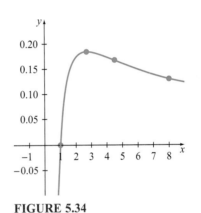

FIGURE 5.34

5.3 Summary

In this section, you learned how to use the second derivative in discussing the graphical concepts of concavity and inflection points. You saw how the Second Derivative Test may be used as an alternative means of finding relative extrema. You also discovered that velocity is the derivative of the position function and that acceleration is the second derivative of the position function. You also learned some curve sketching techniques.

5.3 Exercises

In Exercises 1–10, indicate on the graph where the function is concave up and where the function is concave down. Mark the graph where you think inflection points occur.

1. $f(x) = x^3 - 3x^2$

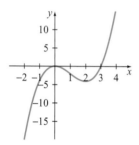

2. $y = \dfrac{\ln(x)}{x}$

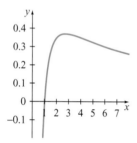

3. $f(x) = x^4 - 12x^2$

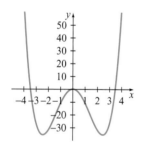

4. $g(x) = x^2$

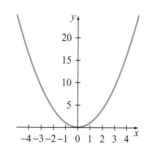

5. $y = x^4 - 8x^3 + 18x^2$

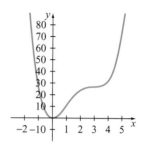

6. $f(x) = x^4 + 2x^3 - 12x$

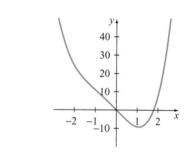

7. $y = -x^4 - 2x^3 + 12x^2$

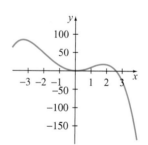

8. $f(x) = x^5 - 5x^4$

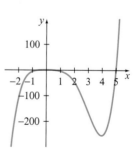

9. $f(x) = \dfrac{x^2}{x^2 + 1}$

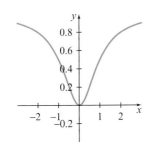

10. $g(x) = \dfrac{3}{x^2 + 1}$

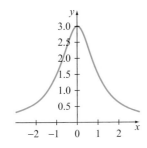

In Exercises 11–20, use the second derivative to find the inflection points of the function algebraically. You may find it helpful to refer to the graphs of the functions in Exercises 1–10.

11. $f(x) = x^3 - 3x^2$

12. $y = \dfrac{\ln(x)}{x}$

13. $f(x) = x^4 - 12x^2$

14. $g(x) = x^2$

15. $y = x^4 - 8x^3 + 18x^2$

16. $f(t) = t^4 + 2t^3 - 12t$

17. $y = -x^4 - 2x^3 + 12x^2$

18. $f(x) = x^5 - 5x^4$

19. $h(t) = \dfrac{t^2}{t^2 + 1}$

20. $g(x) = \dfrac{3}{x^2 + 1}$

In Exercises 21–30, find the stationary points of the function. Then use the Second Derivative Test to determine where relative extrema occur. If the Second Derivative Test is inconclusive, use the First Derivative Test. You may find it helpful to refer to the graphs of the functions in Exercises 1–10 and your work in Exercises 11–20.

21. $f(x) = x^3 - 3x^2$

22. $y = \dfrac{\ln(x)}{x}$

23. $f(r) = r^4 - 12r^2$

24. $g(x) = x^2$

25. $y = x^4 - 8x^3 + 18x^2$

26. $s(n) = n^4 + 2n^3 - 12n$

27. $y = -x^4 - 2x^3 + 12x^2$

28. $f(t) = t^5 - 5t^4$

29. $f(x) = \dfrac{x^2}{x^2 + 1}$

30. $g(x) = \dfrac{3}{x^2 + 1}$

In Exercises 31–40, use the curve-sketching techniques introduced in the section to graph each function by hand.

31. $f(x) = \dfrac{1}{3}x^3 - x^2$ on $[-2, 4]$

32. $g(x) = x^4 + 4x^3$ on $[-2, 5]$

33. $f(x) = -x^3 + 3x + 1$ on $[-2, 2]$

34. $f(x) = -x^4 + 2x^2$ on $[-2, 2]$

35. $g(x) = -x^3 + 12x$ on $[-4, 4]$

36. $f(x) = x^4 - 2x^2 + 1$ on $[-2, 2]$

37. $g(x) = \dfrac{1}{5}x^5 - x^4$ on $[-3, 5]$

38. $f(x) = \ln(x)/x$ on $[1, 10]$

39. $g(x) = 2^x - x \ln(2)$ on $[-2, 2]$

 Hint: This function doesn't have any *x*-intercepts.

40. $h(x) = 2^{-x} - 1$ on $[-2, 4]$

In Exercises 41–50, answer the questions by using the Second Derivative Test, as appropriate. In determining where absolute extrema occur, remember to check the endpoints.

41. **West Coast Sailor Wages**   Based on data from 1975 to 2000, the average monthly wage (in addition to room and board) of a sailor working on the West Coast may be modeled by

$$W(t) = 0.1446t^2 - 5.765t + 883.7$$

dollars, where t is the number of years since 1975. (**Source:** Modeled from *Statistical Abstract of the United States, 2001*, Table 1072, p. 674.)

What was the maximum and what was the minimum increase in monthly wages, and in what year (between 1975 and 2000) did each occur? (Round to the nearest year.) *Hint:* A decrease in wages is a negative increase.

42. **East Coast Sailor Wages**   Based on data from 1975 to 2000, the average monthly wage (in addition to room and board) of a sailor working on the East Coast may be modeled by

$$E(t) = 0.1643t^2 - 5.658t + 112.3$$

dollars, where t is the number of years since 1975. (**Source:** Modeled from *Statistical Abstract of the United States, 2001*, Table 1072, p. 674.)

What was the maximum and what was the minimum increase in monthly wages, and in what year (between 1975 and 2000) did each occur? (Round to the nearest year.)

43. **Salary Negotiations**   How could an East Coast sailors' labor union use the information in Exercises 41 and 42 in negotiating future wage increases?

44. **Homicide Rate**   Based on data from 1990 to 2000, the homicide rate may be modeled by

$$H(t) = 0.01129t^3 - 0.2002t^2 + 0.4826t + 9.390$$

deaths per 100,000 people, where t is the number of years since 1990. (**Source:** Modeled from *FBI Uniform Crime Reports, 2000.*)

Between 1990 and 2000, in what year did the homicide rate increase most rapidly, and in what year did it decrease most rapidly?

45. **Internet Usage**   Based on usage data from 1995 to 1999 and Census Bureau projections of future usage, the number of hours that the average person spends on the Internet annually may be modeled by

$$L(t) = -0.4514t^3 + 6.125t^2 + 6.171t + 1.994$$

hours, where t is the number of years since 1995. (**Source:** Modeled from *Statistical Abstract of the United States, 2001*, Table 1125, p. 704.)

In what year (between 1995 and 2004) was Internet usage expected to increase most slowly, and in what year was it expected to increase most rapidly?

46. Based on the results of Exercise 45, do you think that the rate of change in annual Internet usage will ever exceed 35 hours per year? Justify your answer.

47. **Manufacturing Employees**   Based on data from 1995 to 1999, the number of employees in the leather and leather products manufacturing industry may be modeled by

$$A(t) = -0.6667t^3 + 4.5t^2 - 14.83t + 106$$

thousand employees, where t is the number of years since 1995. (**Source:** Modeled from *Statistical Abstract of the United States, 2001*, Table 979, p. 622.)

In what year did the number of employees in the leather and leather products industry drop most rapidly? In what year did the number of employees increase most rapidly (or decrease most slowly)?

48. **Political Influence**   People for the Ethical Treatment of Animals (PETA) is a nonprofit organization that believes that animals are not ours to eat or wear. (**Source:** www.peta.org.)

Based on the results of Exercise 47, do you think PETA is having an influence on the leather and leather products industry? Justify your answer.

49. **Love Triangles**   Based on data from 1991 to 2000, the cumulative number of homicides since the *start* of 1991 and the *end* of year t that are due to a romantic triangle may be modeled by

$$R(t) = -20.45t^2 + 471.8t - 152.5$$

homicides, where t is the number of years since the end of 1990. (**Source:** Modeled from *Crime in the United States 2000*, Uniform Crime Report, FBI .)

In what year between 1991 and 2000 did the cumulative number of such homicides increase most rapidly? How many homicides due to a romantic triangle occurred that year?

50. **[M]** **Breakfast Cereal Consumption** Based on data from 1980 to 1999, the per capita consumption of breakfast cereal may be modeled by

$$C(t) = -0.004718t^3 + 0.1165t^2 - 0.3585t + 12.17$$

pounds, where t is the number of years since 1980. (**Source:** Modeled from *Statistical Abstract of the United States, 2001*, Table 202, p. 129.)

In what year (between 1980 and 1999) was the per capita breakfast cereal consumption increasing most rapidly, and in what year was it decreasing most rapidly?

Exercises 51–55 deal with position, velocity, and acceleration. The vertical position of a free-falling object on Earth may be modeled by $s(t) = -16t^2 + v_0t + s_0$ feet above the earth after being airborne for t seconds. v_0 is the velocity of the object, and s_0 is the position of the object at time $t = 0$. Use this information and the techniques shown in the section to answer the questions.

51. A rock is *thrown* upward into the air at a rate of 32 feet per second. After how many seconds of flight will it attain its maximum height above the ground?

52. When the rock in Exercise 51 attains its maximum height, what will be its velocity?

53. Given $s(t) = -16t^2 + v_0t + s_0$, calculate $s''(t)$ and interpret the meaning of the result.

54. A pebble is *dropped* into the ocean from the top of a 240-foot cliff. What will be the pebble's velocity when it hits the surface of the ocean?

55. When will the pebble in Exercise 54 attain its maximum speed?

In Exercises 56–57, find the point of diminishing returns for the given function.

56. **[M]** **Electronic Game Unit Sales** Based on data from 1996 to 2000, the annual sales of computer and video game units may be modeled by

$$S(t) = \frac{121.8}{1 + 17.23e^{-1.794t}} + 100$$

units, where t is the number of years since the end of 1996. (**Source:** Modeled from Interactive Digital Software Association State of the Industry Report, 2000–2001.)

Find the point of diminishing returns and interpret its real-life meaning.

57. **[M]** **Box Office Movie Sales** The movie *My Big Fat Greek Wedding* was first released on April 19, 2002. Based on the first 52 weeks of sales, the cumulative gross box office sales of the movie may be modeled by

$$S(t) = \frac{240,800,000}{1 + 125.4e^{-0.2086t}}$$

dollars, where t is in weeks with $t = 1$ corresponding to the first week the movie was released.

Find the point of diminishing returns and interpret its real-life meaning.

Exercises 58–63 are intended to challenge your skill in finding and interpreting the meaning of concavity and the second derivative. Each of the questions deals with a parametric function model of the position of a roller-coaster train over a six-second interval $[0, 6]$. The x value of the model represents the horizontal position of a roller-coaster train (in meters) at time t seconds. The y value of the model represents the vertical position of the roller-coaster train (in meters) at time t seconds. The model is

$$x = t^3 - 9t^2 + 23t$$
$$y = 6t - t^2$$

58. Find the second derivative of x and of y with respect to t. (These may be represented as x'' and y''.)

59. Determine when $x'' = 0$ and $y'' = 0$.

60. Given that the domain of the function is $0 \leq t \leq 6$, find the points of inflection of x and y.

61. Over the six-second interval $[0, 6]$, what was the quickest horizontal acceleration of the train?

62. Over the same six-second interval, what was the slowest horizontal acceleration of the train?

63. Do you think that the roller coaster would be safe to ride? Defend your conclusion.

5.4 Related Rates

■ Solve related-rate problems

GETTING **STARTED** On March 24, 1989, the oil tanker *Exxon Valdez* ran aground on Bligh Reef in Prince William Sound, Alaska. In the ensuing spill, 11 million gallons of crude oil flowed into the water. About 250,000 seabirds; 2,650 sea otters; 1,000 cormorants, and hundreds of loons, harbor seals, and bald eagles were killed. Although much larger spills have occurred internationally, the *Exxon Valdez* spill was one of the most environmentally destructive. Because of the location of the accident, approximately 1,300 miles of the 9,000 miles of shoreline in the region were contaminated. Despite cleanup efforts, there were still pockets of oil remaining a decade after the accident.

When an oil spill occurs, many questions immediately arise. How quickly is the oil spilling into the water? How rapidly is the area covered by the oil expanding? If the oil continues to spill at its current rate, how much area will be covered an hour from now? These questions may be addressed using the concept of *related rates*.

In this section, we will demonstrate how related rates can be used to measure how the rate of change in one variable affects the rate of change in another variable. We will use the implicit differentiation techniques covered in Section 4.5 extensively in this section.

The area of a circular oil spill with radius r feet is given by the formula $A = \pi r^2$. If the area of the circle is increasing at a rate of 100 square feet per minute, how quickly is the radius changing? Since the radius rate of change function is the derivative of the radius function, we are looking for $\dfrac{dr}{dt}$ with units $\dfrac{\text{feet}}{\text{minute}}$. The rate of change in the area is given by $\dfrac{dA}{dt}$. The units of $\dfrac{dA}{dt}$ are $\dfrac{\text{square feet}}{\text{minute}}$. Notice that both A and r are differentiated with respect to t. Using implicit differentiation, we have

$$A = \pi r^2$$

$$\frac{d}{dt}(A) = \frac{d}{dt}(\pi r^2)$$

$$\frac{dA}{dt} = \pi \frac{d}{dt}(r^2)$$

$$\frac{dA}{dt} = \pi \left(2r \frac{dr}{dt} \right)$$

$$\frac{dA}{dt} = 2\pi r \frac{dr}{dt}$$

$$\frac{dr}{dt} = \frac{1}{2\pi r} \frac{dA}{dt}$$

Thus the rate of change in the radius is the rate of change in the area divided by the product of 2π and the current radius. For this problem, we indicated that the area was increasing at a rate of 100 square feet/minute. That is, $\frac{dA}{dt} = 100$.

Therefore, the rate of change in the radius is

$$\frac{dr}{dt} = \frac{1}{2\pi r} \frac{dA}{dt}$$

$$= \frac{1}{2\pi r}(100)$$

$$= \frac{50}{\pi r}$$

When the radius is 1 foot, $\frac{dr}{dt} = \frac{50}{\pi(1)} \approx 15.9$ feet per minute. When the radius is 2 feet, $\frac{dr}{dt} = \frac{50}{\pi(2)} \approx 7.96$ feet per minute. When the radius is 1000 feet, $\frac{dr}{dt} = \frac{50}{\pi(1000)} \approx 0.01592$ feet per minute. Assuming that the shape of the spill remains circular, we can calculate the rate of change in the radius for any value of the radius.

Related rates are most often used when comparing the changes in two related quantities over time. In the next several examples, we will illustrate how related rates are used.

EXAMPLE 1   Interpreting the Meaning of $\frac{dV}{dt}, \frac{dr}{dt},$ and $\frac{dh}{dt}$

The volume of space occupied by a liquid in a cylindrical can is given by $V = \pi r^2 h$, where r is the radius of the can and h is the depth of the liquid in the can. Differentiate V with respect to t and interpret the meaning of $\frac{dV}{dt}, \frac{dr}{dt},$ and $\frac{dh}{dt}$.

Assume that the radius and height are measured in centimeters and that time is measured in seconds. Assuming that the radius of the can is fixed, simplify the result for $\frac{dV}{dt}$.

SOLUTION

$$\frac{d}{dt}(V) = \frac{d}{dt}(\pi r^2 h)$$

$$\frac{dV}{dt} = \pi \frac{d}{dt}(r^2 h) \qquad \text{Constant Multiple Rule}$$

$$= \pi \left(2r \frac{dr}{dt} \cdot h + \frac{dh}{dt} \cdot r^2 \right) \qquad \text{Product and Chain Rules}$$

$$= \pi r \left(2h \frac{dr}{dt} + r \frac{dh}{dt} \right)$$

$\dfrac{dr}{dt}$ is the rate of change in the radius in centimeters per second.

$\dfrac{dh}{dt}$ is the rate of change in depth of the liquid in centimeters per second.

$\dfrac{dV}{dt}$ is the rate of change in the volume of the liquid in the can in cubic centimeters per second. It is a function of the radius, the depth, and the rates of change in the radius and the depth. Notice that if the can has a fixed radius, then $\dfrac{dr}{dt} = 0$ and

$$\frac{dV}{dt} = \pi r \left[2h(0) + r \frac{dh}{dt} \right]$$

$$= \pi r^2 \frac{dh}{dt}$$

EXAMPLE 2

Using Related Rates in a Real-World Context

The volume of a cylindrical can is given by $V = \pi r^2 h$, where r is the radius and h is the height of the interior of the can from lid to lid. In Section 5.2, we learned that a #10 can has a radius of 3.625 inches (9.208 centimeters). Suppose that the can is being filled with rice and that the rice is being poured into the can at a constant rate of 20 cups per minute. At what rate is the height of the rice in the can changing? Will it take more or less than a minute to fill the can?

SOLUTION You may immediately ask, "Why do we care?" Admittedly, we may not care if we have to fill only one can. However, if we worked in a cannery and wanted to program a machine to fill thousands of cans, the problem would immediately become meaningful. Failure to program the machine with the correct fill rate would result in overfilled cans or insufficiently filled cans. Both of these would hamper the cannery's effort to create a uniform product with minimum waste.

Since we are filling #10 cans, the volume equation becomes

$$V = \pi (9.208)^2 h$$

$$= 84.79\pi h \text{ cubic centimeters}$$

and the rate of change in the volume is given by

$$\frac{dV}{dt} = 84.79\pi \frac{dh}{dt} \frac{\text{cubic centimeters}}{\text{minute}}$$

One of the challenges of the standard measurement system (inches, feet, miles, cups, gallons, etc.) is that conversion between quantities (e.g., cups to cubic inches) is difficult. For this reason, we will convert standard measurements to their metric equivalents. One cup is equivalent to 236.6 cubic centimeters. Since the volume is changing at a rate of 20 cups per minute, we have

UNITS

$$\frac{dV}{dt} = 20 \; \frac{\text{cups}}{\text{minute}} \cdot \frac{236.6 \text{ cubic centimeters}}{1 \text{ cup}}$$

$$= 4732 \; \frac{\text{cubic centimeters}}{\text{minute}}$$

We must find $\dfrac{dh}{dt}$.

$$\frac{dV}{dt} = 84.79\pi\frac{dh}{dt}$$

$$4732 = 266.4\frac{dh}{dt}$$

$$\frac{dh}{dt} = \frac{4732}{266.4}$$

$$= 17.76 \text{ centimeters per minute}$$

The height of the rice in the can is increasing by about 17.76 centimeters per minute. Since the interior height of a #10 can is 16.83 centimeters (6.625 inches), it will take less than a minute to fill the can. (To be precise, it will take 57 seconds.)

EXAMPLE 3

Using Related Rates in a Real-World Context

A circular above-ground swimming pool has a 12-foot radius and may be filled to a maximum depth of 4 feet. By measuring the amount of time it took to fill a bucket, the author determined that water leaves his outside faucet at a rate of 9 gallons per minute. How quickly is the depth of the water in the pool changing when the pool is being filled from a hose connected to the faucet? How long will it take to fill the pool?

SOLUTION   We will first convert the dimensions to metric measurements. We know that 1 foot ≈ 30.48 centimeters and 1 gallon ≈ 3785 cubic centimeters.

The radius of the pool is $12(30.48) = 365.76$ centimeters. The rate at which the water is leaving the faucet is $9(3785) = 34{,}065$ cubic centimeters per minute.

The pool is cylindrical, so its volume is $V = \pi r^2 h$, where r is the radius and h is the depth of the water in the pool. Since the radius of the pool will not change, we have

$$V = \pi(365.76)^2 h$$

$$= 420{,}283.45h \text{ cubic centimeters}$$

Differentiating with respect to t, we get

$$\frac{dV}{dt} = 420{,}283.45\frac{dh}{dt} \; \frac{\text{cubic centimeters}}{\text{minute}}$$

But we know that $\dfrac{dV}{dt}$ = 34,065 cubic centimeters per minute, so

$$34{,}065 = 420{,}283.45\dfrac{dh}{dt}$$

$$\dfrac{dh}{dt} = 0.08105\ \dfrac{\text{centimeters}}{\text{minute}}$$

$$= 0.08105\ \dfrac{\text{centimeters}}{\cancel{\text{minute}}} \cdot \dfrac{60\ \cancel{\text{minutes}}}{1\ \text{hour}}$$

$$= 4.863\ \text{centimeters per hour}$$

The pool depth is increasing at a rate of 4.863 centimeters per hour. Since 4 feet = 121.92 centimeters, the amount of time it will take to fill the pool is

UNITS

$$\dfrac{121.92\ \text{centimeters}}{4.863\ \text{centimeters per hour}} = 25.07\ \text{hours}$$

It will take about 25 hours to fill the pool.

EXAMPLE 4 **Calculating a Change in Wind Chill Temperature**

Wind chill temperature is the temperature in calm air that has the same chilling effect on a person as that of a particular combination of temperature and wind. For example, when the temperature is 35° Fahrenheit (°F) and the wind speed is 40 miles per hour, it feels like it is 3°F. The wind chill temperature may be calculated according to the following formula:

$$F = 91.4 - \left(0.474677 - 0.020425w + 0.303107\sqrt{w}\right)(91.4 - T)\ \text{degrees Fahrenheit}$$

where w is the wind speed (in mph) and T is the temperature (in °F). (**Source:** National Climatic Data Center.)

If the temperature is currently 20° and is dropping at a rate of 3° per hour, and if the wind is currently blowing at 16 mph and is increasing at a rate of 2 mph per hour, how quickly is the wind chill temperature changing?

SOLUTION We must first differentiate the wind chill equation with respect to time.

$$\dfrac{d}{dt}(F) = \dfrac{d}{dt}\left[91.4 - \left(0.474677 - 0.020425w + 0.303107\sqrt{w}\right)(91.4 - T)\right]$$

$$\dfrac{dF}{dt} = 0 - \left[-0.020425\dfrac{dw}{dt} + 0.5(0.303107w^{-0.5})\dfrac{dw}{dt}\right](91.4 - T)$$

$$+ \left(-\dfrac{dT}{dt}\right)\left[-\left(0.474677 - 0.020425w + 0.303107\sqrt{w}\right)\right] \qquad \text{Product Rule}$$

$$\dfrac{dF}{dt} = \left[0.020425\dfrac{dw}{dt} - 0.5(0.303107w^{-0.5})\dfrac{dw}{dt}\right](91.4 - T)$$

$$+ \left(\dfrac{dT}{dt}\right)\left(0.474677 - 0.020425w + 0.303107\sqrt{w}\right)$$

The current wind speed is 16 mph, so $w = 16$.

$\dfrac{dw}{dt}$ is the rate of change in the wind and equals 2 mph per hour.

The current temperature is $20°$, so $T = 20$.

$\dfrac{dT}{dt}$ is the rate of change in the temperature and equals -3 degrees per hour.

Substituting these values into the derivative equation, we get

$$\frac{dF}{dt} = \{0.020425(2) - 0.5[0.303107(16)^{-0.5}](2)\}[91.4 - (20)]$$

$$+ (-3)\left[0.474677 - 0.020425(16) + 0.303107\sqrt{16}\right]$$

$$= (-0.03493)(71.4) - 4.081$$

$$= -2.494 - 4.081$$

$$= -6.575$$

The wind chill index is dropping by about 6.6 degrees per hour. That is, although the temperature is dropping by only 3 degrees per hour, it feels like it is dropping by 6.6 degrees per hour because of the wind.

Because of the number of terms and the complexity of the coefficients, the type of calculation shown in Example 4 is time-consuming and prone to error. For this reason, we recommend using the following Technology Tip to speed up calculations and increase accuracy.

TECHNOLOGY TIP

Using the Equation Solver

1. Press MATH □:Solver... to bring up the Equation Solver. Immediately press the blue up arrow. Press CLEAR to delete any equation that may be visible.

```
EQUATION SOLVER
eqn:0=
```

2. Enter the equation you need to solve, using the alpha keys for variables. Since the calculator requires the equation to be set equal to zero, we must subtract $\dfrac{dF}{dt}$ from both sides of our equation. For our equation, we let $\dfrac{dF}{dt} = A$, $\dfrac{dT}{dt} = B$, and $\dfrac{dw}{dt} = C$.

```
EQUATION SOLVER
eqn:0=(.020425*C
-.5*(.303107*W^-
.5)*C)*(91.4-T)+
B*(.474677-.0204
25W+.303107*W^.5
)-A
```

(Continued)

3. Press [ENTER] to display the variable input menu. Edit the variables to reflect the known values from our problem.

```
(.020425*C-.5…=0
C=2
W=16
T=20
B=-3
A=█
bound={-1ε99,1…
left-rt=0
```

4. Move the cursor to the line containing A= and press [ALPHA] [ENTER] to solve the equation for A.

```
(.020425*C-.5…=0
C=2
W=16
T=20
B=-3
■A=-6.57468495
bound={-1ε99,1…
■left-rt=0
```

EXAMPLE 5

Calculating a Change in Wind Chill Temperature

If the temperature is currently 30° and dropping at a rate of 5° per hour, and the wind is currently blowing at 20 mph and is decreasing at a rate of 10 mph per hour, how quickly is the wind chill temperature changing? Use the Technology Tip and the equation for $\dfrac{dF}{dt}$ from Example 4 to answer the question.

SOLUTION

The current wind speed is 20 mph, so $w = 20$.

$C = \dfrac{dw}{dt}$ is the rate of change in the wind and equals -10 mph per hour.

The current temperature is 30°, so $T = 30$.

$B = \dfrac{dT}{dt}$ is the rate of change in the temperature and equals -5 degrees per hour.

We need to solve for $A = \dfrac{dF}{dt}$. Since we have already entered the equation (from Example 4) in our calculator, we'll use the Technology Tip to solve the problem (see Figure 5.35).

The wind chill temperature is increasing by $1.1579 \approx 1.2$ degrees per hour. Although the temperature is dropping 5 degrees per hour, it feels like it is warming by 1.2 degrees per hour because the wind speed is decreasing.

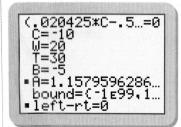

```
(.020425*C-.5…=0
C=-10
W=20
T=30
B=-5
■A=1.157959286…
bound={-1ε99,1…
■left-rt=0
```

FIGURE 5.35

5.4 Summary

In this section, you learned how related rates can be used to measure how the rate of change in one variable affects the rate of change in another variable. You discovered that technology can simplify the arithmetic computations of related-rate problems.

5.4 Exercises

In Exercises 1–10, differentiate the function with respect to t. Explain the physical meaning of each of the rates. (For each of the problems, A is area, V is volume, r is radius, h is height, l is length, w is width, b is base, and t is time. Use inches, square inches, cubic inches, and minutes as the units of length, area, volume, and time, respectively.)

1. $A = lw$ (area of a rectangle)

2. $A = 2\pi rh + 2\pi r^2$ (surface area of a cylinder including top and bottom)

3. $V = lwh$ (volume of a box)

4. $V = \frac{4}{3}\pi r^3$ (volume of a sphere)

5. $A = 2lw + 2wh + 2lh$ (surface area of a box)

6. $V = \frac{1}{3}\pi r^2 h$ (volume of a cone)

7. $A = 4\pi r^2$ (surface area of a sphere)

8. $A = \frac{1}{2}bh$ (area of a triangle)

9. $A = \pi r^2$ (area of a circle)

10. $A = 2\pi r^2$ (surface area of a hemisphere)

In Exercises 11–15, solve the related-rate problem using the techniques demonstrated in this section.

11. Swimming Pool Depth   A circular above-ground swimming pool has a 9-foot radius and may be filled to a maximum depth of 4 feet. The pool is being filled with a faucet that releases water at a rate of 10 gallons per minute. How quickly is the depth of the water in the pool changing when the pool is being filled from a hose connected to the faucet? How long will it take to fill the pool?

12. Water Pressure   A circular above-ground swimming pool has a 9-foot radius and may be filled to a maximum depth of 4 feet. If the depth of the water in the pool is rising at a rate of 0.5 inch per minute, at what rate is the water leaving the hose that is filling the pool?

13. Water Depth   A cylindrical can has a 5-inch radius and an 8-inch height and is to be filled with water. At what constant rate must the depth of the water increase if the can is to be filled in exactly 30 seconds?

14. Pizza Size   The owner of a pizzeria is contemplating changing the size of his large pizza. He intends to decrease the size of the large pizza from a 16-inch diameter to a 14-inch diameter by gradually reducing the diameter by $\frac{1}{2}$ inch each month. (He figures that consumers won't notice the change if it is gradual.) At what rate will the surface area of the pizza be changing two months into the reduction period?

15. Oil Spill Spread   The surface area of a circular oil spill is increasing at a rate of 100 square feet per hour. How quickly is the radius increasing when the radius is 157 feet?

In Exercises 16–20, solve the related-rate problems using the Technology Tip demonstrated in this section. The wind chill temperature may be estimated using the following model equation:

$$F = 91.4 - (0.474677 - 0.020425w + 0.303107\sqrt{w})$$
$$\cdot (91.4 - T) \text{ degrees Fahrenheit}$$

where w is the wind speed (in mph) and T is the temperature (in °F). (**Source:** National Climatic Data Center.)

16.   **Wind Chill**   If the temperature is currently 0°F and is dropping at a rate of 5°F per hour, and the wind is currently blowing at 18 mph and is increasing at a rate of 2 mph per hour, how quickly is the wind chill temperature changing?

17.   **Wind Chill**   If the temperature is currently −10°F and is warming at a rate of 8°F per hour, and the wind is currently blowing at 25 mph and is increasing at a rate of 3 mph per hour, how quickly is the wind chill temperature changing?

18.   **Wind Chill**   If the temperature is currently −2°F and is decreasing at a rate of 4°F per hour, and the wind is currently blowing at 55 mph and is decreasing at a rate of 5 mph per hour, how quickly is the wind chill temperature changing?

19.   **Wind Chill**   If the temperature is currently 40°F and is increasing at a rate of 4°F per hour, and the wind is currently blowing at 5 mph and is not increasing its speed, how quickly is the wind chill temperature changing?

20.   **Wind Chill**   If the temperature is currently 32°F and is not changing, and the wind is currently blowing at 25 mph and is increasing its speed by 4 mph per hour, how quickly is the wind chill temperature changing?

Chapter 5 Review Exercises

Section 5.1 *In Exercises 1–2, determine the points on the graph where $f'(x) = 0$ or $f'(x)$ is undefined.*

1. $f(x) = |x - 2| - (x - 2)^2 + 2$

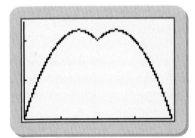

2. $f(x) = x^5 - 4x^3 + 10$

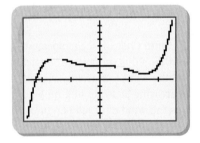

In Exercises 3–6, find the critical points of the function.

3. $f(x) = x^5 - 4x^3 + 10$

4. $g(x) = 5x^{1/5}$ **5.** $y = -4x^2 + 4$

6. $y = -2x + 4$

In Exercises 7–10, construct a sign chart for the derivative. Then determine the relative and absolute extrema of the function on the specified domain. Use the First Derivative Test as appropriate. (These functions are the same as those in Exercises 3–6.)

7. $f(x) = x^5 - 4x^3 + 10; [-2, 2]$

8. $g(x) = 5x^{1/5}; [-2, 2]$

9. $y = -4x^2 + 4; [-3, 3]$

10. $y = -2x + 4; [0, 4]$

In Exercises 11–12, use the concepts of relative and absolute extrema to find the answer to the question.

11. **M** *Exxon Valdez* **Oil Spill**   Based on data from the *fourth* to the fifty-sixth day after the accident, the spread of the oil from the site of the *Exxon Valdez* oil spill may be modeled by

$$D(t) = 0.004108t^3 - 0.4432t^2 + 21.07t - 42.07$$

miles from the spill site t days after the accident. Note that the model is invalid for $t < 4$. (**Source:** Modeled from www.oilspill.state.ak.us/facts/spillmap.html.)

According to the model, what was the greatest distance the oil had spread between the fourth and the fifty-sixth day?

12. **M** **Industry Work Force Size**   Based on data from 1995 to 1999, the number of employees in the printing and publishing industry may be modeled by

$$E(t) = -4.417t^3 + 24.93t^2 - 25.30t + 1450$$

thousand employees, where t is the number of years since 1995. (**Source:** Modeled from *Statistical Abstract of the United States, 2001*, Table 979, p. 622.)

According to the model, when did the industry have its maximum and minimum number of employees? How many people were employed at these times?

Section 5.2 *In Exercises 13–14, use the derivative and other optimization techniques to find the optimal value requested.*

13. **M** **Company Revenue and Cost**   Gatorade/Tropicana North America, a subsidiary of PepsiCo, produces fruit juices and other flavored beverages. Based on data from 1999 to 2001, the net sales (revenue) of Gatorade/Tropicana North America may be modeled by

$$R(t) = -107t^2 + 496t + 3452$$

million dollars and the operating profit (earnings before interest and taxes) may be modeled by

$$P(t) = -18.5t^2 + 85.5t + 433$$

million dollars, where t is the number of years since 1999. (**Source:** Modeled from 2001 PepsiCo Annual Report, p. 44.)

(a) In what year are net sales projected to reach a maximum?

(b) Find the cost function for Gatorade/Tropicana North America.

(c) According to the model, in what year are costs expected to reach a maximum?

(d) Compare the results of (a) and (c). Do the results seem reasonable?

14. **Ice Cream Cone Design** The volume of an ice cream cone (minus the ice cream) is given by

$$V = \frac{1}{3}\pi r^2 h$$

cubic inches, where r is the radius of the top of the cone and h is the height of the cone (both in inches). The surface area of the exterior of the cone is given by

$$A = \pi r \sqrt{r^2 + h^2}$$

square inches.

For a cone with a surface area of $47.12(15\pi)$ square inches, what are the dimensions of the cone with maximum volume? Should ice cream cones be constructed with these dimensions? Explain.

Section 5.3 *In Exercises 15–16, indicate on the graph where the function is concave up and where the function is concave down. Mark the graph where you think inflection points occur.*

15. $f(x) = \dfrac{x^2 + 9}{x^2 + 10}$

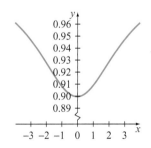

16. $y = x^3 - x^2 - 8x + 12$

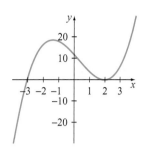

In Exercises 17–18, use the second derivative to find the inflection points of the function algebraically. You may find it helpful to refer to the graphs of the functions in Exercises 15–16.

17. $f(x) = \dfrac{x^2 + 9}{x^2 + 10}$

18. $y = x^3 - x^2 - 8x + 12$

In Exercises 19–20, find the stationary points of the function. Then use the Second Derivative Test to determine where relative extrema occur. If the Second Derivative Test is inconclusive, use the First Derivative Test. You may find it helpful to refer to the graphs of the functions in Exercises 15–16 and your work in Exercises 17–18.

19. $f(x) = \dfrac{x^2 + 9}{x^2 + 10}$

20. $y = x^3 - x^2 - 8x + 12$

In Exercises 21–22, use the second derivative to determine where the maximum and minimum rates of change occur for the specified function. Be sure to evaluate the rate of change function at the endpoints to ensure that you have found the absolute extrema.

21. **Exxon Valdez Oil Spill** Based on data from the fourth to the fifty-sixth day after the accident, the spread of the oil from the site of the *Exxon Valdez* oil spill may be modeled by

$$D(t) = 0.004108t^3 - 0.4432t^2 + 21.07t - 42.07$$

miles from the spill site t days after the accident. (**Source:** Modeled from www.oilspill.state.ak.us/facts/spillmap.html.)

On what day was the oil spreading at the quickest rate? At what rate was it spreading?

22.   **Minivan Speed**   Based on 2.5 minutes of data, the distance of the author's minivan from a stoplight at the beginning of a highway may be modeled by

$$M(t) = -0.0591t^3 + 0.3399t^2 + 0.5019t + 0.005594$$

miles, where t is the amount of time (in minutes) that the minivan has been moving away from the stoplight.

     Rounded to the nearest mile per hour, what were the highest and lowest speeds of the minivan?

Section 5.4   *In Exercises 23–24, differentiate the function with respect to t. Explain the physical meaning of each of the rates. (A is area and s is the length of a side. Lengths are measured in centimeters, areas in square centimeters, and time in seconds.)*

23. $A = 5s^2$ (area of an equilateral cross)

24. $A = \dfrac{\sqrt{3}}{4}s^2$ (area of an equilateral triangle)

In Exercises 25–30, solve the related-rate problem.

25. Swimming Pool Depth   A circular above-ground swimming pool has a 4-foot radius and may be filled to a maximum depth of 3 feet. A woman is filling the pool with a faucet that releases water at a rate of 10 gallons per minute. How quickly is the depth of the water in the pool changing 2 minutes after she begins to fill the pool?

26. Swimming Pool Depth   How rapidly is the depth of the water in the pool in Exercise 25 increasing 5 minutes, 10 minutes, and 15 minutes after the woman begins to fill the pool?

27. Swimming Pool Depth   A rectangular above-ground swimming pool has a 12-foot length and a 6-foot width and may be filled to a maximum depth of 4 feet. A teen is filling the pool with a faucet that releases water at a rate of 9 gallons per minute. How quickly is the depth of the water in the pool changing 2 minutes after he begins to fill the pool?

28. Swimming Pool Depth   How rapidly is the depth of the water in the pool in Exercise 27 increasing 5 minutes, 10 minutes, and 15 minutes after the teen begins to fill the pool?

29. Oil Spill Spread   The surface area of a circular oil spill is increasing at a rate of 250 square feet per hour. How quickly is the radius increasing when it is 2500 feet?

30. Oil Spill Spread   The radius of a circular oil spill is increasing at a rate of 10 feet per hour. How quickly is the surface area changing when the radius is 200 feet?

Make It Real

What to do

1. Recruit a friend, neighbor, or family member to go on a drive with you.
2. Drive to the base of a freeway (or major highway) on-ramp.
3. As you drive up the on-ramp and onto the freeway, have your friend record the odometer reading every 15 seconds for at least 2 minutes. If the odometer is between tenths of a mile, round to the nearest twentieth. (That is, if it is between 2.0 and 2.1, round to 2.05.)
4. Find the equation of the function that best fits the data.
5. Use differentiation to find the velocity and acceleration functions.
6. Determine your maximum velocity and maximum acceleration.

How to get a good data set

1. As you're driving, try to avoid any rapid changes in your acceleration. (This will make a smoother scatter plot.)
2. Collect several minutes of data. You can eliminate extra points at the end of your data set if the model doesn't fit those points well.
3. Use as many data points as possible, yet aim for a model coefficient of determination (r^2) of 0.99 or better.
4. Avoid excessive speeds, erratic driving, and driving under the influence.
5. If you see flashing lights behind you, pull over to the right side of the road and abort the project!.

Chapter
6
The Integral

O n March 24, 1989, the *Exxon Valdez* oil tanker ran aground in Prince William Sound. The ensuing oil spill was one of the most widely publicized environmental disasters in history. If a function modeling the variable rate at which oil is spilling from a tanker is known, the Fundamental Theorem of Calculus may be used to calculate the total amount of oil spilled over a specified time period.

6.1 Indefinite Integrals
- Use integration rules to find the antiderivative of a function

6.2 Integration by Substitution
- Use the method of substitution to integrate functions

6.3 Using Sums to Approximate Area
- Estimate the area between a curve and the *x*-axis by using left- and right-hand sums

6.4 The Definite Integral
- Apply definite integral properties
- Calculate the exact area between a curve and the *x*-axis by using definite integrals

6.5 The Fundamental Theorem of Calculus
- Apply the Fundamental Theorem of Calculus
- Find the function for accumulated change given a rate of change function

6.1 Indefinite Integrals

- Use integration rules to find the antiderivative of a function

A speedometer measures a car's velocity. If we record our car's velocity at 15-second intervals, we will generate a table of data that may be used to find a model for the velocity of the car. Can we determine how far the car traveled from the velocity equation alone? Yes. By using a process called *integration*, we can determine the distance equation if we know the velocity equation.

In this section, we will introduce integration: the process of finding the equation of a function that has a given derivative. We will also define and demonstrate how to use basic rules of integration.

We will begin by looking at three functions: $f(x) = x^3 - x^2 - 8x + 17$, $g(x) = x^3 - x^2 - 8x + 12$, and $h(x) = x^3 - x^2 - 8x + 7$. The equations of these functions look remarkably similar. In fact, their equations differ only by a constant. Similarly, the basic shape of the graphs is the same; they differ only in their vertical placement on the coordinate system (see Figure 6.1).

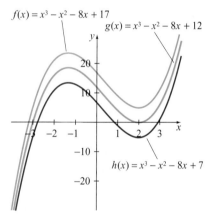

FIGURE 6.1

Let's differentiate each of the functions.

$$f'(x) = 3x^2 - 2x - 8$$
$$g'(x) = 3x^2 - 2x - 8$$
$$h'(x) = 3x^2 - 2x - 8$$

The functions all have the same derivative! In fact, any function of the form

$$f(x) = x^3 - x^2 - 8x + C \ (C \text{ is any constant})$$

will have the derivative

$$f'(x) = 3x^2 - 2x - 8$$

Thus, an infinite number of functions share the same derivative. The graphs of all of these functions will have the same basic shape, but their vertical placement on the coordinate system will differ.

Often we are given the rate of change function (derivative) and asked to find the equation of the function that has the given derivative. As previously illustrated, there is an infinite number of functions that share the same derivative. To represent all such functions, we will write the letter C in the place of the constant term.

EXAMPLE 1

Finding the Antiderivative of a Function

Find the general form of the function that has the derivative $f(x) = 2x$.

SOLUTION We are looking for a function $F(x)$ such that $\dfrac{d}{dx}[F(x)] = f(x)$.

Let's consider the function $F(x) = x^2 + C$.

$$\frac{d}{dx}[F(x)] = \frac{d}{dx}(x^2 + C)$$

$$= 2x + 0$$

$$= 2x$$

Therefore, any function of the form $F(x) = x^2 + C$ has the derivative $f(x) = 2x$. Specific functions include $F(x) = x^2 + 2$, $F(x) = x^2 - 12$, $F(x) = x^2 + 256$, and so on.

DEFINITION OF TERMS

Integration: The process of finding the general form of a function that has a given derivative. Also referred to as *antidifferentiation*.

Indefinite integral: The general form of a function that has a given derivative. Also referred to as the *antiderivative*.

Integral sign: The symbol $\int$, which indicates that integration is to be performed.

Integrand: The function written to the right of the integral sign. This is the function that is to be integrated.

The notation $\int f(x)\, dx$ is read "the integral of $f(x)$ with respect to x" or "the antiderivative of $f(x)$ with respect to x." Just as $\dfrac{d}{dx}[f(x)]$ means "find the derivative of $f(x)$ with respect to x," $\int f(x)\, dx$ means "find the general form of the function that has derivative $f(x)$ with respect to x." The integral sign $\int$ must always be written together with a *differential* (e.g., dx) to indicate the variable of integration. (We discuss differentials in more detail in the next section.) A function f is said to be **integrable** if there exists a function F such that $F' = f$.

EXAMPLE 2

Finding the Antiderivative of a Function

Find $\int 3x^2 \, dx$.

SOLUTION   We are to find the general equation of the function whose derivative is $f(x) = 3x^2$. Recall that $\dfrac{d}{dx}(x^3) = 3x^2$. Therefore, the indefinite integral of $f(x) = 3x^2$ is $F(x) = x^3 + C$.

We can check our work by differentiating the function F with respect to x.

$$\frac{d}{dx}[F(x)] = \frac{d}{dx}(x^3 + C)$$

$$= 3x^2 + 0$$

$$= 3x^2$$

You may have noticed that we have shifted our notation slightly from our earlier treatment of derivatives. Since we are given the rate of change function as the original function, we are using the notation $f(x)$ instead of $f'(x)$ to represent the rate of change function. We use the notation $F(x)$ to represent the function whose derivative is $f(x)$. That is, $F'(x) = f(x)$. Admittedly, this shift in notation is a bit confusing. However, since the notation shift is so widely adopted by people who use calculus, we will also use this commonly accepted notation. Making the transition may be difficult at first, but by the end of the chapter, you will be using the new notation with ease.

Basic Integration Rules

As you might expect, the rules for derivatives and the rules for integrals are intimately related. In fact, many of the rules even share the same names.

POWER RULE FOR INTEGRALS

Let $f(x) = x^n$, where x is a variable and n is a constant with $n \neq -1$. Then

$$\int x^n \, dx = \frac{x^{n+1}}{n+1} + C, \ n \neq -1$$

The rule can be easily verified by differentiating the resultant function.

$$\frac{d}{dx}\left(\frac{x^{n+1}}{n+1} + C\right) = \frac{1}{n+1}\frac{d}{dx}(x^{n+1}) + 0 \qquad \text{Constant Multiple Rule}$$

$$= \frac{n+1}{n+1}x^{n+1-1} \qquad \text{Power Rule}$$

$$= x^n$$

Notice that $n \neq -1$. If $n = -1$, the denominator of $\dfrac{x^{n+1}}{n+1} + C$ is equal to 0, which makes the expression undefined. We will deal with the case of $n = -1$ later.

CONSTANT MULTIPLE RULE FOR INTEGRALS

Let $f(x) = a \cdot g(x)$, where x is a variable and a is a constant. Then

$$\int [a \cdot g(x)]dx = a \cdot \int g(x)\,dx$$
$$= a \cdot G(x) + C$$

where $g(x) = G'(x)$.

EXAMPLE **3**

Integrating a Function

Integrate $f(x) = 6x^2$.

SOLUTION

$$\int 6x^2 dx = 6 \int x^2 dx \qquad \text{Constant Multiple Rule}$$

$$= 6\left(\frac{x^3}{3} + C_1\right) \qquad \text{Power Rule}$$

$$= \frac{6x^3}{3} + 6C_1$$

$$= 2x^3 + C$$

Notice that we rewrote $6C_1$ as C. Since C_1 is a constant, multiplying it by 6 will still result in a constant. The same would be true if we were to add two different constants C_1 and C_2. The result would still be a constant, which we could write as C. Therefore, when doing integration, we will write a single constant C to represent a multiple of a constant or a sum of constants.

Remember that the derivative represents the slope of the tangent line. The derivative of a linear function $f(x) = mx + b$ is $f'(x) = m$. That is, for a linear function, the derivative of the function is the slope of the line. Similarly, the antiderivative of a constant function $f(x) = m$ is the linear function $F(x) = mx + b$. This relationship is summarized in the Constant Rule for Integrals.

CONSTANT RULE FOR INTEGRALS

Let k be a constant. Then

$$\int k\,dx = kx + C.$$

The Sum and Difference Rule is another extremely useful rule to use when working with integrals.

SUM AND DIFFERENCE RULE FOR INTEGRALS

Let $f(x)$ and $g(x)$ be integrable functions of x. Then

$$\int [f(x) \pm g(x)]dx = \int f(x)dx \pm \int g(x)dx$$

EXAMPLE **4**

Antidifferentiating a Function

Find the antiderivative of $h(x) = 2x + 4$.

SOLUTION

$$\int (2x + 4)\,dx = \int 2x\,dx + \int 4\,dx \qquad \text{Sum and Difference Rule}$$

$$= 2\int x\,dx + 4\int 1\,dx \qquad \text{Constant Multiple Rule}$$

$$= 2\left(\frac{x^2}{2}\right) + 4(x) + C \qquad \text{Power and Constant Rules}$$

$$= x^2 + 4x + C$$

EXPONENTIAL RULE FOR INTEGRALS

Let $a > 0$ and $a \neq 1$. Then

$$\int a^x\,dx = \frac{1}{\ln(a)}a^x + C$$

The Exponential Rule may be easily verified by differentiating $y = \frac{1}{\ln(a)}a^x + C$.

$$\frac{d}{dx}\left(\frac{1}{\ln(a)}a^x + C\right) = \frac{d}{dx}\left(\frac{1}{\ln(a)}a^x\right) + \frac{d}{dx}(C) \qquad \text{Sum and Difference Rule}$$

$$= \frac{1}{\ln(a)}\left[\frac{d}{dx}(a^x)\right] + 0 \qquad \begin{array}{l}\text{Constant and Constant}\\ \text{Multiple Rules}\end{array}$$

$$= \frac{1}{\ln(a)}\left[\ln(a)a^x\right] \qquad \text{Exponential Rule}$$

$$= \frac{\ln(a)}{\ln(a)}(a^x)$$

$$= a^x$$

It immediately follows from the Exponential Rule for Integrals that

$$\int e^x\,dx = \frac{1}{\ln(e)}e^x + C$$

$$= \frac{1}{1}e^x + C$$

$$= e^x + C$$

EXAMPLE 5   **Integrating an Exponential Function**

Find $\int 2 \cdot 3^x \, dx$.

SOLUTION

$$
\begin{aligned}
\int 2 \cdot 3^x \, dx &= 2 \int 3^x \, dx && \text{Constant Multiple Rule for Integrals} \\
&= 2 \cdot \frac{1}{\ln(3)} (3^x) + C && \text{Exponential Rule for Integrals} \\
&= \frac{2}{\ln(3)} (3^x) + C
\end{aligned}
$$

Recall that $x^{-1} = \frac{1}{x}$. We saw previously that we cannot use the Power Rule for Integrals to integrate $f(x) = x^{-1}$. However, the function is integrable.

INTEGRAL RULE FOR $\dfrac{1}{x}$

If $x \neq 0$, then

$$
\int \frac{1}{x} \, dx = \ln|x| + C
$$

The absolute value of x is necessary to ensure that the two functions $f(x) = \frac{1}{x}$ and $F(x) = \ln|x| + C$ have the same domain ($\ln x$ is defined only for $x > 0$, but $\ln|x|$ is defined for all $x \neq 0$). We can convince ourselves that the rule holds true by differentiating $\ln|x| + C$.

$$
\begin{aligned}
\frac{d}{dx}(\ln|x| + C) &= \frac{d}{dx}(\ln|x|) + \frac{d}{dx}(C) && \text{Sum and Difference Rule} \\
&= \frac{d}{dx}\left(\begin{cases} \ln(x) & x > 0 \\ \ln(-x) & x < 0 \end{cases} \right) + 0 && \begin{array}{l} \text{Definition of absolute value} \\ \text{Constant Rule} \end{array} \\
&= \begin{cases} \dfrac{1}{x}, & x > 0 \\[2mm] \dfrac{1}{-x}(-1), & x < 0 \end{cases} && \begin{array}{l} \text{Logarithm Rule} \\[2mm] \text{Logarithm and Chain Rules} \end{array} \\
&= \begin{cases} \dfrac{1}{x}, & x > 0 \\[2mm] \dfrac{1}{x}, & x < 0 \end{cases} \\
&= \frac{1}{x}, \quad x \neq 0
\end{aligned}
$$

We see that for all nonzero values of x, $\dfrac{d}{dx}(\ln|x| + C) = \dfrac{1}{x}$.

EXAMPLE 6 **Integrating a Function of the Form $y = \dfrac{c}{x}$**

Integrate $\dfrac{3}{x}$ with respect to x.

SOLUTION

$$\int\left(\frac{3}{x}\right)dx = 3\int\left(\frac{1}{x}\right)dx \qquad \text{Constant Multiple Rule for Integrals}$$

$$= 3\ln|x| + C \qquad \text{Integral Rule for } \tfrac{1}{x}$$

Indefinite Integral Applications

Recall that the units of $\dfrac{dy}{dx}$ are the units of y over the units of x. Since $\int\left(\dfrac{dy}{dx}\right)dx = y + C$, the units of the antiderivative of $\dfrac{dy}{dx}$ are the units of y. In real-world applications, we are often given enough information to figure out the specific value of C, as demonstrated in Example 7.

EXAMPLE 7 **Using Integration to Find a Position Function**

Based on 2.5 minutes of data, the velocity of the author's minivan as he drove from a stoplight onto a highway may be modeled by $V(t) = -0.1773t^2 + 0.6798t + 0.5019$ miles per minute, where t is the number of minutes since he left the stoplight. Find the position function $D(t)$ that shows his distance from the stoplight at time t. Find the total distance traveled during the first minute and during the first two minutes.

SOLUTION The position function $D(t)$ is the antiderivative of the velocity function $V(t)$.

$$\int[V(t)]dt = \int(-0.1773t^2 + 0.6798t + 0.5019)\,dt$$

$$D(t) = -0.1773\frac{t^3}{3} + 0.6798\frac{t^2}{2} + 0.5019t + C \qquad \substack{\text{Constant Multiple, Power, and}\\ \text{Sum and Difference Rules}}$$

$$= -0.0591t^3 + 0.3399t^2 + 0.5019t + C \text{ miles}$$

You may ask, "What is the meaning of C in this context?" Let's see. $D(0)$ represents the distance traveled after 0 minutes. Intuitively, we know that after 0 minutes, the minivan hasn't traveled at all, so $D(0) = 0$.

$$D(0) = -0.0591(0)^3 + 0.3399(0)^2 + 0.5019(0) + C$$

$$= C$$

But $D(0) = 0$, so $C = 0$. $D(t) = -0.0591t^3 + 0.3399t^2 + 0.5019t$ is the distance function for the minivan. According to our distance model, $D(1) = 0.78$ and $D(2) = 1.89$. That is, the total distance traveled during the first minute is 0.78 mile. The total distance traveled during the first two minutes is 1.89 miles.

EXAMPLE **8**

Using Antidifferentiation to Find a Cost Function

Have you ever dreamed of being a published author? Golden Pillar Publishing (www.goldenpillarpublishing.com) publishes books for aspiring authors. In 2003, the marginal cost for printing an 8.25″ × 11″ soft-cover book was

$$c(p) = 0.018 \text{ dollar per page}$$

where p is the number of pages in the book. Including the setup costs, a 1000-page book has a $19.30 production cost. Find the book production cost function.

SOLUTION Recall that marginal cost is the derivative of the cost function. Therefore, the antiderivative of the marginal cost is the cost function.

$$\int c(p)\, dp = \int 0.018\, dp$$
$$C(p) = 0.018p + k \qquad \text{Constant Rule for Integrals}$$

(Note: We used k as a constant, since the function was named C.)
The cost of a 1000-page book is $19.30, so

$$C(1000) = 0.018(1000) + k$$
$$19.30 = 18 + k$$
$$k = 1.30$$

The constant $k = 1.30$ represents the fixed setup cost. The cost to produce a book with p pages is $C(p) = 0.018p + 1.30$ dollars.

6.1 Summary

In this section you learned how to find the equation of some functions that have a given derivative through the process of integration. You also discovered how to use basic integration rules to integrate a variety of functions.

6.1 Exercises

In Exercises 1–20, find the general antiderivative of the function.

1. $f(x) = 2$

2. $g(x) = -2x + 9$

3. $r(t) = t^{-1} - 3t$

4. $f(t) = \dfrac{1.21}{t}$

5. $v(t) = 0.5t + 20$

6. $a(t) = -32$

7. $s(t) = t^3 - 3t^2 + 3t - 1$

8. $s(x) = 2^x$

9. $h(x) = \dfrac{2.3}{x} + 1$

10. $f(x) = 0$

11. $p(x) = 3^x - x^3$

12. $r(x) = 200 - \dfrac{225}{x}$

13. $f(t) = 4t^{-2} + 2t^{-1} + 1$

14. $g(x) = 2^x - x^2$

15. $h(x) = 2(3^x) - 3(2^x)$

16. $q = 2500 - \dfrac{1000}{p^2}$

17. $q = \dfrac{2000}{p} - \dfrac{500}{p^2}$

18. $f(x) = 3(2^x) + 4x^{-1}$

19. $y = 5t^{-2} + 2^t$

20. $h(x) = \dfrac{4x^3 - 15}{x^2}$

In Exercises 21–35, calculate the indefinite integral.

21. $\int (3x - 5)\, dx$

22. $\int (-x^2 + 4x)\, dx$

23. $\int \dfrac{t - 2}{t}\, dt$ (*Hint:* Divide each term in the

numerator by the denominator before integrating.)

24. $\int \dfrac{25u - 1}{u}\, du$

25. $\int (3x^{-2} - 4x^3 + 2)\, dx$

26. $\int (4t^{-1} - 12)\, dt$

27. $\int (5t^{-2} - 16t^2 - 9)\, dt$

28. $\int [5(2^t) - 3t^2]\, dt$

29. $\int \dfrac{5u^2 - 4u + 1}{u^2}\, du$

30. $\int (4t^{-2} - 2^t)\, dt$

31. $\int \left(\dfrac{400}{x^2} + \dfrac{200}{x} + 50 \right) dx$

32. $\int (9^t - 3^t)\, dt$

33. $\int \left(\dfrac{4}{x} - \dfrac{5}{x^2} \right) dx$

34. $\int \left(20 - \dfrac{5u + 2}{u} \right) du$

35. $\int (4t + 4^t)\, dt$

In Exercises 36–39, apply the concept of integration in solving the real-world applications.

36. **Publishing Costs**   Golden Pillar Publishing (www.goldenpillarpublishing .com) publishes books for aspiring authors. In 2003, the marginal printing cost for a 7.5″ × 9.25″ soft-cover book was

$$c(p) = 0.013$$

dollar per page, where p is the number of pages in the book. Including setup costs, a 1000-page book has a $13.90 production cost. Find the book production cost function.

37. **Publishing Costs**   Golden Pillar Publishing (www.goldenpillarpublishing. com) publishes books for aspiring authors. In 2003, the marginal printing cost for a 6″ × 9″ hard-cover book was

$$c(p) = 0.013$$

dollar per page, where p is the number of pages in the book. Including setup costs, a 1000-page book has an $18.00 production cost. Find the book production cost function.

38. **Cost Analysis**   The marginal printing cost of the soft-cover and hard-cover books in Exercises 36 and 37 was the same, yet their cost functions were different. Why?

39. **Minivan Position**   Based on 1.5 minutes of data, the velocity of the author's minivan as he drove from a stoplight into a sparsely populated residential area may be modeled by

$$V(t) = -2.133t^3 + 3.999t^2 - 0.1533t + 0.3167$$

miles per minute, where t is the number of minutes since he left the stoplight. Find the position function $D(t)$ that shows his distance from the stoplight at time t.

Exercises 40–45 are intended to challenge your understanding of the basic integration rules.

40. Let $F_1(x)$ and $F_2(x)$ be specific antiderivatives of $f(x)$. What is the graphical interpretation of $|F_1(x) - F_2(x)|$?

41. Let $F_1(x)$ and $F_2(x)$ be specific antiderivatives of $f(x)$, with $F_1(x) \neq F_2(x)$. Do the graphs of $F_1(x)$ and $F_2(x)$ ever intersect?

42. Let $f(x)$ be a polynomial of degree n with $f(x) = F'(x)$. At most, how many x-intercepts does $F(x)$ have?

43. Let $F'(x) = 3 + \dfrac{1}{x^2}$. Is $F(x)$ an increasing or a decreasing function? Explain.

44. Let $F'(x) = 3x^2$ and $F(0) = 0$. On what interval(s) is $F(x) \geq 0$?

45. Let $F'(x) = 4 - 2x$ and $F(0) = -4$. On what interval(s) is $F(x) \geq 0$?

6.2 Integration by Substitution

- Use the method of substitution to integrate functions

GETTING STARTED   The process of *integration by substitution* is a clever way to integrate a function that initially doesn't appear to be integrable. (Recall that a function f is said to be integrable if there exists a function F such that $F' = f$.) Because of the relatively complex nature of the process, we will defer illustrations of how this process may be used in real-life applications to later sections.

In this section, we will introduce the concept of differentials. We will then demonstrate how to do integration by substitution.

Differentials

Recall that $\dfrac{d}{dx}(y) = \dfrac{dy}{dx}$. The derivative $\dfrac{dy}{dx}$ is a single term, not the quotient of dy and dx. However, it would be useful for us to be able to treat $\dfrac{dy}{dx}$ as if it were a fraction. For this purpose, we introduce the concept of *differentials*. We define the **differential** dx to be an arbitrarily small real number and

$$dy = y'\,dx$$

Observe that dividing both sides of the equation $dy = y'\,dx$ by dx yields

$$\frac{dy}{dx} = y'$$

which is consistent with our earlier notation for the derivative. Thus, by using differentials, we can treat dy and dx in $\dfrac{dy}{dx}$ as separate entities and use the rules for fractions in manipulating the terms.

EXAMPLE **1**   **Calculating the Differential *dy***

Find dy given $y = x^2$.

SOLUTION
$$dy = 2x\,dx$$

EXAMPLE **2**   **Calculating the Differential *dy***

Find dy given $y = 2x^3 - 3x$.

SOLUTION
$$dy = (6x^2 - 3)\,dx$$

We may also use differentials with other variables, such as u and t, as demonstrated in the next two examples.

EXAMPLE 3

Calculating the Differential *du*

Find *du* given $u = e^{3t} + 2$.

SOLUTION

$$du = 3e^{3t}dt$$

EXAMPLE 4

Calculating the Differential *du*

Find *du* given $u = \ln(t)$.

SOLUTION

$$du = \frac{1}{t}dt$$

Integration by Substitution

The technique of **integration by substitution** is used to rewrite a function that doesn't appear to be readily integrable in such a way that it can be integrated using the basic integration rules.

EXAMPLE 5

Integrating by Substitution

Integrate $f(x) = \frac{2\ln(x)}{x}$.

SOLUTION We must find $\int \frac{2\ln(x)}{x} dx$. This function cannot be readily integrated using any of the rules covered in Section 6.1. This is because $f(x) = \frac{2\ln(x)}{x}$ is the product of $2\ln(x)$ and $\frac{1}{x}$, and we haven't yet discussed how to integrate a product. (Contrary to what you might think,

$$\int [f(x)g(x)]\, dx \neq \int f(x)\, dx \cdot \int g(x)\, dx.$$

We will let $u = \ln(x)$. We take the derivative of both sides and use the concept of differentials to find *du*.

$$u = \ln(x)$$

$$\frac{d}{dx}(u) = \frac{d}{dx}[\ln(x)]$$

$$\frac{du}{dx} = \frac{1}{x}$$

$$du = \frac{1}{x}dx$$

We will now rewrite the integral in terms of *u*.

$$\int 2\underbrace{[\ln(x)]}_{u}\underbrace{\left(\frac{1}{x}dx\right)}_{du} = \int 2u\, du$$

But we know that

$$\int 2u\, du = u^2 + C$$

(Although the differential du represents an arbitrarily small number and the du of an integral means "with respect to u," the two uses of the notation du may be interchanged. A more in-depth explanation may be obtained from an advanced calculus text.)

Since the variable u was something that we introduced into the problem, we must rewrite the solution in terms of x. Recall that $u = \ln(x)$.

$$u^2 + C = [\ln(x)]^2 + C$$

Thus, we have

$$\int \frac{2\ln(x)}{x}\, dx = [\ln(x)]^2 + C$$

It is not always possible to find a substitution of variables that makes a difficult function integrable. However, if the integral can be rewritten in any of the following forms, it may be integrated using the basic integration rules.

- $\int u^n du$
- $\int e^u du$
- $\int \frac{1}{u} du$

EXAMPLE 6   **Integrating by Substitution**

Find the antiderivative of $f(x) = 2x(x^2 + 5)^{10}$.

SOLUTION We need to determine $\int 2x(x^2 + 5)^{10} dx$. The function is a product of two functions, so we hope to be able to write it in the form $u^n du$. We'll try $u = x^2 + 5$. Then

$$u = x^2 + 5$$

$$\frac{du}{dx} = 2x$$

$$du = 2x\, dx$$

Let's see if we can rewrite the integral in terms of u and du. Recall that $u = x^2 + 5$ and $du = 2x\, dx$.

$$\int 2x(x^2 + 5)^{10} dx = \int [(x^2 + 5)^{10}](2x\, dx) \qquad \text{Switch the order of the factors and group}$$

$$= \int u^{10} du \qquad \text{Since } u = x^2 + 5 \text{ and } du = 2x\, dx$$

$$= \frac{u^{11}}{11} + C$$

Rewriting the result in terms of x, we have

$$\int 2x(x^2 + 5)^{10} dx = \frac{(x^2 + 5)^{11}}{11} + C$$

We can check our result by differentiating the antiderivative and comparing it to $f(x)$.

$$\frac{d}{dx}\left[\frac{(x^2+5)^{11}}{11}+C\right]=\frac{1}{11}\frac{d}{dx}[(x^2+5)^{11}]+0 \quad \text{Constant and Constant Multiple Rules}$$

$$=11\cdot\frac{1}{11}(x^2+5)^{10}\cdot 2x \quad \text{Power and Chain Rules}$$

$$=2x(x^2+5)^{10}$$

The derivative of the antiderivative is equal to $f(x)$, so we are confident that we integrated the function correctly.

EXAMPLE 7 Integrating by Substitution

Integrate $g(t)=e^{2t-4}$.

SOLUTION We must find $\int e^{2t-4}\,dt$. We hope to be able to rewrite the integral as $\int e^u\,du$. We'll let $u=2t-4$. Then

$$\frac{du}{dt}=2$$

$$du=2\,dt$$

Looking at the integrand (the function to be integrated), we see that we have dt but not $2\,dt$. Can we solve $du=2\,dt$ for dt? Yes!

$$2\,dt=du$$

$$dt=\frac{1}{2}du$$

Substituting, we get

$$\int e^{2t-4}dt=\int e^u\left(\frac{1}{2}du\right) \quad \text{Since } u=2t-4 \text{ and } dt=\frac{1}{2}du$$

$$=\frac{1}{2}\int e^u\,du \quad \text{Constant Multiple Rule for Integrals}$$

$$=\frac{1}{2}e^u+C \quad \text{Exponential Rule}$$

$$=\frac{1}{2}e^{2t-4}+C \quad \text{Rewrite in terms of } t$$

EXAMPLE 8 Integrating by Substitution

Find the general form of a function that has the derivative $f(x)=\frac{2x+3}{x^2+3x+4}$.

SOLUTION We must find $\int \frac{2x+3}{x^2+3x+4}\,dx$. We observe that the numerator is the derivative of the denominator. Consequently, we hope to be able to rewrite the integral in the form $\int \frac{1}{u}\,du$. We'll let $u=x^2+3x+4$. Then

$$\frac{du}{dx} = 2x + 3$$

$$du = (2x + 3)\,dx$$

Rewriting the integral in terms of u and solving, we get

$$\int \frac{2x + 3}{x^2 + 3x + 4}\, dx = \int \frac{(2x + 3)\,dx}{x^2 + 3x + 4}$$

$$= \int \frac{du}{u} \qquad \text{Since } u = x^2 + 3x + 4$$
$$\text{and } du = (2x + 3)\,dx$$

$$= \int \frac{1}{u}\, du$$

$$= \ln|u| + C \qquad \text{Integral Rule for } \frac{1}{u}$$

$$= \ln|x^2 + 3x + 4| + C \qquad \text{Rewrite in terms of } x$$

All functions of the form $F(x) = \ln|x^2 + 3x + 4| + C$ have the derivative $f(x) = \frac{2x + 3}{x^2 + 3x + 4}$.

In each of the previous examples, we selected the "correct" u each time. However, it is very common to attempt integration by substitution with the "wrong" function for u. If you rewrite a function in terms of u and the resulting function isn't integrable, go back and select a different function to be u. This is part of the process of integration by substitution and should not be viewed as an error. As you gain experience with these types of problems, you will more easily recognize what is the best choice for u. In Example 9, we will demonstrate a type of problem that requires an especially clever selection of u.

EXAMPLE 9

Integrating by Substitution a Function Containing a Radical

Find $\int x\sqrt{x - 1}\, dx$.

SOLUTION   We begin by rewriting the integrand with a rational exponent.

$$\int x(x - 1)^{1/2}\, dx$$

Let's pick $u = x - 1$. Then

$$\frac{du}{dx} = 1$$

$$du = dx$$

We make the substitution.

$$\int x u^{1/2}\, du$$

What do we do with the extra x? The integrand must be written in terms of u. (We can't use the Constant Multiple Rule to move the x to the left-hand side of the integral sign because x is a variable, not a constant.) Recall that $u = x - 1$; therefore, $x = u + 1$. Thus we may rewrite the integral as

$$\int xu^{1/2}\,du = \int (u+1)u^{1/2}\,du$$

Since we have completely rewritten the integral in terms of u, we may continue. We will multiply out the integrand and then integrate each term.

$$\int \left(u^{3/2}+u^{1/2}\right)du = \int u^{3/2}\,du + \int u^{1/2}\,du$$

$$= \frac{u^{5/2}}{\frac{5}{2}} + \frac{u^{3/2}}{\frac{3}{2}} + C \qquad \text{Power Rule for Integrals}$$

$$= \frac{2}{5}u^{5/2} + \frac{2}{3}u^{3/2} + C$$

$$= \frac{2}{5}(x-1)^{5/2} + \frac{2}{3}(x-1)^{3/2} + C$$

Thus $\int x\sqrt{x-1}\,dx = \frac{2}{5}(x-1)^{5/2} + \frac{2}{3}(x-1)^{3/2} + C.$

EXAMPLE 10

Determining When Integration by Substitution Won't Work

Integrate $f(x) = xe^x$.

SOLUTION We must find $\int xe^x\,dx$. If we pick $u = x$, then $du = dx$ and the integral becomes

$$\int ue^u\,du$$

Observe that this is identical to the original function written, except that it is written in terms of u. Using the integral techniques we've covered, we don't know how to integrate a function in this form.

Let's try $u = e^x$. Then

$$u = e^x$$
$$\ln(u) = \ln(e^x)$$
$$\ln(u) = x$$
$$x = \ln(u)$$

Since $\frac{du}{dx} = e^x$, $du = e^x dx$. The resultant integral is

$$\int xe^x\,dx = \int [\ln(u)]du$$

We haven't yet learned how to integrate $\ln(u)$, so we are unable to find the solution using this choice of u.

As a final attempt, we'll try $u = xe^x$. Then $du = (1 \cdot e^x + e^x \cdot x)dx$, by the Product Rule. If we attempt to make a substitution, we'll have a mixture of u and x in the integrand, leaving us with a function that we can't integrate $\left(\int xe^x dx = \int u\frac{du}{(1 \cdot e^x + e^x \cdot x)}\right)$.

The function $f(x) = xe^x$ cannot be integrated using integration by substitution. Nevertheless, there is a method that can be used to integrate $f(x) = xe^x$. In Section 7.1, we will show how to integrate this function using integration by parts.

It is important to note that, despite the many advances in calculus, there are some functions that we still don't know how to integrate. Nevertheless, we must attempt to integrate a function using all known methods before we can conclude that it is not integrable.

6.2 Summary

In this section you learned how to work with differentials. You used integration by substitution to find the antiderivative of a function that didn't initially appear to be integrable. You also learned that there are functions that we can't integrate simply by reversing the derivative rules.

6.2 Exercises

In Exercises 1–40, integrate the functions. Some integrals will require integration by substitution, while others may be integrated using basic integration rules. All of the functions may be integrated using the techniques we've covered.

1. $f(x) = 2x(x^2 + 3)^5$ **2.** $g(x) = 6x^2(x^3 + 7)^2$

3. $h(x) = 2x(3x + 5)$

4. $f(t) = (2t + 3)(t^2 + 3t + 1)^3$

5. $s(x) = e^x(e^x + 1)$ **6.** $v(t) = 3^t + t^{-1}$

7. $f(x) = 1.2x^2 - 2.4x + 0.6$

8. $w(t) = 10te^{5t^2 - 4}$ **9.** $p(x) = (x - 2)e^{x^2 - 4x}$

10. $y = 3e^{2x}$ **11.** $y = 3xe^{3x^2 - 6}$

12. $f(x) = (2x - 9)e^{x^2 - 9x + 21}$

13. $g(t) = 4e^{3t}$ **14.** $y = 3(2^x)$

15. $s(x) = \dfrac{\ln(x^2)}{x}$ **16.** $y = 4x(x^2 + 1)^3$

17. $h(t) = \dfrac{2t - 1}{t^2 - t + 2}$ **18.** $g(x) = \dfrac{x - 4}{x^2 - 8x + 9}$

19. $f(x) = \dfrac{6x^2}{x^3 - 9}$ **20.** $h(t) = \dfrac{2t - 1}{t^2 - t + 2}$

21. $p(x) = \dfrac{1}{x \ln x}$ **22.** $s(x) = \dfrac{\ln x}{x}$

23. $y = 3x\sqrt{x - 2}$ **24.** $f(x) = x\sqrt{x + 2}$

25. $g(x) = 2x\sqrt{x^2 + 1}$

26. $h(x) = (x - 1)\sqrt{x^2 - 2x}$

27. $f(t) = t\sqrt{t^2 - 1}$ **28.** $s(x) = \dfrac{x}{\sqrt{x^2 - 1}}$

29. $f(x) = \dfrac{3x^2 - 2x + 1}{x^3 - x^2 + x - 1}$

30. $g(t) = \dfrac{e^{\sqrt{t}}}{2\sqrt{t}}$ **31.** $f(x) = \dfrac{x}{2x - 1}$

32. $g(x) = \dfrac{e^x - \ln 2(2^x)}{e^x - 2^x}$

33. $h(x) = \dfrac{2x}{4x^2 - 5}$ **34.** $v(t) = \ln 2(3^t)$

35. $a(t) = (4t - 1)(t^2 - 2t)$

36. $q = \dfrac{1000}{p} - p\sqrt{p - 2}$

37. $q = \dfrac{4p}{2p^2 + 1} - 4p\sqrt{2p^2 + 1}$

38. $s(t) = 6t(t^2 + 1)^3$

39. $s(t) = (4t - 2)(t^2 - t)^{-1}$

40. $v(x) = x(2^{x^2 - 1})$

Exercises 41–45 are intended to challenge your understanding of integration by substitution.

41. Determine if $f(x) = 2x^3(x^2 - 1)^4$ may be integrated using integration by substitution.

42. Determine if $f(x) = \dfrac{\ln(x) + 1}{x \ln(x)}$ may be integrated using integration by substitution.

43. Determine if $f(x) = \dfrac{x}{\ln(x)}$ may be integrated using integration by substitution.

44. Determine if $f(x) = \dfrac{\ln(x^2)}{4x}$ may be integrated using integration by substitution.

45. Determine if $f(x) = 3x^2(x^2 - 1)^7$ may be integrated using integration by substitution.

6.3 Using Sums to Approximate Area

- Estimate the area between a curve and the x-axis by using left- and right-hand sums

GETTING STARTED River rafting or kayaking is an exciting sport that requires both physical skill and an intellectual understanding of the hydraulics of a river. One factor that enthusiasts must consider before navigating a river is the flow rate of the water. The flow rate indicates how much water (in cubic feet) will pass a fixed point in a second. The flow rate fluctuates constantly, ever changing the intensity of the river. For the inexperienced, flow-rate values mean little. One way to add conceptual meaning to the numbers is to determine how much water has passed a given point over a period of time. As will be explained in the section, we can do this by estimating the area between a graph of the flow rate and the horizontal axis.

In this section, we will demonstrate how left- and right-hand sums may be used to approximate the area between the graph of a function and the x-axis. This will prepare us to link the notion of area with the concept of the definite integral in Section 6.4. The relationship between the two concepts is truly remarkable.

Suppose that the flow rate $f(x)$ of a river is a constant 3000 cubic feet per second. Let's consider the graph of the flow-rate function $f(x) = 3$, shown in Figure 6.2.

The graph is a horizontal line with y-intercept $(0, 3)$. We ask the question, "What is the area of the region between the graph of the function and the x-axis on the interval $[1, 3]$?

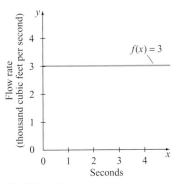

FIGURE 6.2

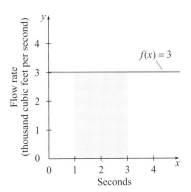

FIGURE 6.3

That is, what is the area of the shaded region in Figure 6.3? We quickly recognize that the region is a rectangle. To calculate its area, we multiply its height by its width.

$$A = 3 \cdot 2$$
$$= 6$$

The area of the region is 6 square units. What does this mean in the context of the problem? Let's recalculate A, this time keeping track of the physical meaning of the length and width of the rectangle.

$$A = \left(3 \, \frac{\text{thousand cubic feet}}{\text{second}} \right)(2 \text{ seconds})$$
$$= 6 \text{ thousand cubic feet}$$

Between the first and third second after we started timing, 6000 cubic feet of water passed our water flow measurement station. In general, if $f(x)$ is a positive rate of change function, then the area between the graph of $f(x)$ and the x-axis on the interval $[a, b]$ represents the total change in $F(x)$ [an antiderivative of $f(x)$] from $x = a$ to $x = b$.

When the shaded region is a familiar geometric shape (triangle, rectangle, and so on), it is easy to calculate the area using known formulas. However, what do we do if the shaded region is not a familiar geometric shape? Can we still calculate the area? In Example 1, we will estimate the area of an irregularly shaped region by drawing a series of rectangles. As you will see, the accuracy of our estimate will increase as we increase the number of rectangles.

EXAMPLE 1

Using Riemann Sums to Estimate an Area

Estimate the area of the region between $f(x) = \sqrt{x} + 2$ and the x-axis on the interval $[0, 4]$.

SOLUTION We begin by drawing the graph of $f(x) = \sqrt{x} + 2$ (see Figure 6.4).

We will first estimate the area by drawing a single rectangle whose upper left corner touches the graph of f (Figure 6.5).

The area of the rectangle is $A = 2 \cdot 4 = 8$ square units. Since the entire region below the graph is not completely shaded, 8 square units is an underestimate of the actual area. To improve our accuracy, we will draw two rectangles that are each two units wide. The upper left-hand corner of each rectangle will touch the graph (see Figure 6.6).

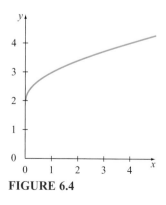

FIGURE 6.4

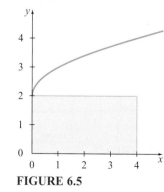

FIGURE 6.5

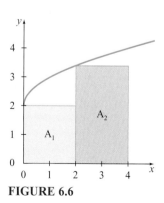

FIGURE 6.6

The area of the first rectangle is easily calculated. $A_1 = 2 \cdot 2 = 4$ square units. We can easily see that the second rectangle has a width of 2 units, but it is difficult to tell from the graph what the height of the rectangle is. However, since the upper left-hand corner of the rectangle is on the graph of f, we can use the equation of $f(x)$ to calculate the height of the rectangle. Since the rectangle touches the graph when $x = 2$, the height of the rectangle is

$$f(2) = \sqrt{2} + 2$$
$$\approx 1.414 + 2$$
$$\approx 3.414 \text{ units}$$

Consequently, the approximate area of the second rectangle is

$$A_2 = 3.414 \cdot 2$$
$$= 6.828$$

We can estimate the area of the region by adding together the areas of the two rectangles.

$$A = A_1 + A_2$$
$$= 4 + 6.828$$
$$= 10.828 \text{ square units}$$

This estimate is substantially better than our first estimate because it has increased the size of the shaded region below the graph. However, it is still an underestimate. How can we increase the accuracy of our estimate? Add more rectangles. This time we'll use four rectangles of width 1 unit, as shown in Figure 6.7.

The width of each rectangle is 1 unit. Since the upper left-hand corner of each rectangle touches the graph of f, the heights of the rectangles are $f(0)$, $f(1)$, $f(2)$, and $f(3)$, respectively. The areas of the rectangles are given by

$$A_1 = f(0) \cdot 1$$
$$= 2 \cdot 1$$
$$= 2 \text{ square units}$$
$$A_2 = f(1) \cdot 1$$
$$= 3 \cdot 1$$
$$= 3 \text{ square units}$$
$$A_3 = f(2) \cdot 1$$
$$= \left(\sqrt{2} + 2\right) \cdot 1$$
$$= 3.414 \text{ square units}$$
$$A_4 = f(3) \cdot 1$$
$$= \left(\sqrt{3} + 2\right) \cdot 1$$
$$= 3.732 \text{ square units}$$

The area of the entire shaded region is the sum of the individual areas.

$$A = A_1 + A_2 + A_3 + A_4$$
$$= 2 + 3 + 3.414 + 3.732$$
$$= 12.146 \text{ square units}$$

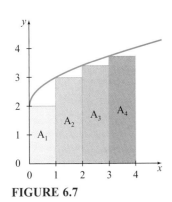

FIGURE 6.7

How can we improve our estimate? Increase the number of rectangles. This time we'll use eight rectangles of width 0.5 unit (see Figure 6.8).

The width of each rectangle is 0.5 unit. The heights of the rectangles are found by evaluating $f(x)$ at $x = 0.0$, $x = 0.5$, $x = 1.0$, $x = 1.5$, $x = 2.0$, $x = 2.5$, $x = 3.0$, and $x = 3.5$ (see Table 6.1).

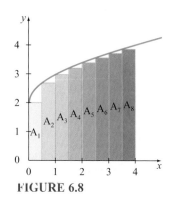

FIGURE 6.8

TABLE 6.1

x	$f(x) = \sqrt{x} + 2$
0.0	2.000
0.5	2.707
1.0	3.000
1.5	3.225
2.0	3.414
2.5	3.581
3.0	3.732
3.5	3.871

The area of the shaded region is given by

$$A = A_1 + A_2 + A_3 + A_4 + A_5 + A_6 + A_7 + A_8$$

$$= 2(0.5) + 2.707(0.5) + 3(0.5) + 3.225(0.5) + 3.414(0.5) + 3.581(0.5)$$
$$+ 3.732(0.5) + 3.871(0.5)$$

$$= (2.000 + 2.707 + 3.000 + 3.225 + 3.414 + 3.581 + 3.732 + 3.871)\,(0.5)$$

$$= (25.530)\,(0.5)$$

$$= 12.765 \text{ square units}$$

We will next estimate the area using 16 rectangles. We know that our area estimate will continue to improve as we increase the number of rectangles; however, as the number of rectangles increases and their width decreases, it becomes increasingly difficult to draw the rectangles. We need a better method.

Since the width of the entire region is 4 units, the width of each of the 16 rectangles will be $\frac{4}{16} = 0.25$ unit. We'll make a table for $f(x)$ with the x-values starting at $x = 0$ and continuing to $x = 4$, with values spaced 0.25 unit apart (see Table 6.2).

TABLE 6.2

x	$f(x) = \sqrt{x} + 2$
0.00	2.000
0.25	2.500
0.50	2.707
0.75	2.866
1.00	3.000
1.25	3.118
1.50	3.225
1.75	3.323
2.00	3.414
2.25	3.500
2.50	3.581
2.75	3.658
3.00	3.732
3.25	3.803
3.50	3.871
3.75	3.936
4.00	4.000

We know that the height of each left-hand rectangle is determined by evaluating $f(x)$ at the x-value on the leftmost edge of the rectangle. The leftmost edge of the first rectangle occurs at $x = 0$. The leftmost edge of the second rectangle occurs at $x = 0.25$. Continuing on down, we see that the leftmost edge of the sixteenth rectangle occurs at $x = 3.75$. The heights of the rectangles are highlighted in the red box in Table 6.3.

TABLE 6.3

Rectangle Number	x	$f(x) = \sqrt{x} + 2$
1	0.00	2.000
2	0.25	2.500
3	0.50	2.707
4	0.75	2.866
5	1.00	3.000
6	1.25	3.118
7	1.50	3.225
8	1.75	3.323
9	2.00	3.414
10	2.25	3.500
11	2.50	3.581
12	2.75	3.658
13	3.00	3.732
14	3.25	3.803
15	3.50	3.871
16	3.75	3.936
	4.00	4.000

Since each rectangle has the same width, we may add up all of the heights of the rectangles and multiply them by the width instead of having to multiply the height of each rectangle individually by its width. (Both methods will yield the same result, but the first method requires fewer computations.)

$$A = (2 + 2.5 + 2.707 + 2.866 + 3 + 3.118 + 3.225 + 3.323 + 3.414$$
$$+ 3.5 + 3.581 + 3.658 + 3.732 + 3.803 + 3.871 + 3.936)(0.25)$$
$$= (52.234)(0.25)$$
$$\approx 13.06 \text{ square units}$$

This method is easier than drawing the rectangles; however, even it can become cumbersome when there are large numbers of rectangles. As it turns out, the exact area of the region is $13\frac{1}{3}$ square units. Our final estimate was fairly close to the exact value.

Each of the estimates in Example 1 is called a **left-hand sum,** since the upper *left* corner of each of the rectangles used in the sum touched the graph of f. The notion of using sums is credited to the famous mathematician Bernhard Riemann. Consequently, these sums are commonly called **Riemann sums.**

EXAMPLE 2

Using Riemann Sums to Estimate an Area

Estimate the area of the region between $f(x) = 3x^2$ and the x-axis on the interval $[1, 3]$ using a left-hand sum with $n = 4$.

SOLUTION We are asked to estimate the area of the shaded region in Figure 6.9 by using a left-hand sum.

Note that the shaded region is constrained to the interval $[1, 3]$.

Since $n = 4$, we are to use 4 rectangles. The width of each rectangle is

$$\frac{3 - 1}{4} = \frac{2}{4}$$

$$= 0.5 \, \text{unit}$$

(In general, if we want to use n rectangles on a region between $x = a$ and $x = b$, the width of each rectangle is $\dfrac{b - a}{n}$.)

We will create a table of data for $f(x)$ starting at $x = 1$ and continuing to $x = 3$ with x-values spaced 0.5 unit apart (see Table 6.4).

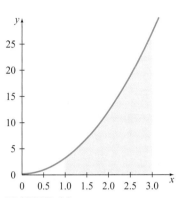

FIGURE 6.9

TABLE 6.4

Rectangle Number	x	$f(x) = 3x^2$
1	1.0	3.00
2	1.5	6.75
3	2.0	12.00
4	2.5	18.75
	3.0	27.00

The left-hand sum estimate of the area is

$$A = (3.00 + 6.75 + 12.00 + 18.75)(0.5)$$

$$= 20.25 \, \text{square units}$$

By drawing the rectangles, as shown in Figure 6.10, we can see that this is an underestimate of the actual area.

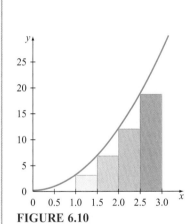

FIGURE 6.10

EXAMPLE 3

Using Riemann Sums to Estimate an Area

Estimate the area of the region between $f(x) = \dfrac{1}{x}$ and the x-axis on the interval $[1, 2]$ using a left-hand sum with $n = 5$.

SOLUTION We are asked to estimate the area of the shaded region in Figure 6.11 by using a left-hand sum with five rectangles.

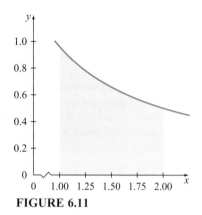

FIGURE 6.11

The width of each rectangle is $\dfrac{2-1}{5} = 0.2$ unit. We will create a table of values for $f(x)$ starting at $x = 1$ and continuing to $x = 2$ with x-values spaced 0.2 unit apart (see Table 6.5).

TABLE 6.5

Rectangle Number	x	$f(x) = \dfrac{1}{x}$
1	1.0	1.000
2	1.2	0.833
3	1.4	0.714
4	1.6	0.625
5	1.8	0.556
	2.0	0.500

The left-hand sum estimate of the area is

$$A = (1.000 + 0.833 + 0.714 + 0.625 + 0.556)\,(0.2)$$
$$= (3.728)\,(0.2)$$
$$= 0.746 \text{ square unit}$$

By drawing the rectangles as shown in Figure 6.12, we can see that this is an overestimate of the actual area.

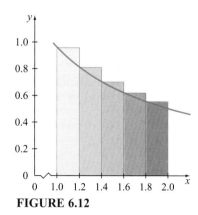

FIGURE 6.12

HOW TO **Finding the Left-Hand Sum for Small Values of *n***
(*n* ≤ 10)

Let f be a function defined on the interval $[a, b]$ whose graph lies above the x-axis. The left-hand sum estimate for the area between the graph of f and the x-axis may be found using the following steps.

1. Calculate the width of each rectangle, $\Delta x = \dfrac{b - a}{n}$. "$\Delta x$" is read "delta x" and is the distance between consecutive x values.

2. Create a table of values for $f(x)$ starting at $x = a$ and ending at $x = b$, with intermediate x-values spaced Δx units apart.

3. Add up the first to the penultimate (second to last) values of $f(x)$ listed in the table. (This is the sum of the heights of the rectangles.)

4. Multiply the result of Step 3 by Δx to get the left-hand sum approximation of the area.

(*Note:* This method may be used for any value of n; however, for values of n larger than 10, the process becomes extremely tedious.)

EXAMPLE 4

Using a Left-Hand Sum to Estimate an Area

Estimate the area between the graph of $f(x) = -x^2 + 4x$ and the x-axis on the interval $[0, 4]$ using a left-hand sum with eight rectangles.

SOLUTION We are asked to estimate the area of the region shown in Figure 6.13.

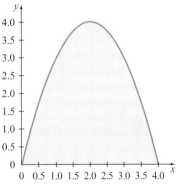

FIGURE 6.13

The width of each rectangle is

$$\Delta x = \frac{4 - 0}{8}$$

$$= \frac{1}{2}$$

$$= 0.5$$

We will create a table of values starting at $x = 0$ and continuing to $x = 4$ with the x-values spaced 0.5 unit apart (see Table 6.6).

TABLE 6.6

Rectangle Number	x	$f(x) = -x^2 + 4x$
1	0.0	0.00
2	0.5	1.75
3	1.0	3.00
4	1.5	3.75
5	2.0	4.00
6	2.5	3.75
7	3.0	3.00
8	3.5	1.75
	4.0	0.00

The sum of the heights of the rectangles is

$$0 + 1.75 + 3 + 3.75 + 4 + 3.75 + 3 + 1.75 = 21$$

We now multiply the sum of the heights by the width of a rectangle.

$$A = 21(0.5)$$
$$= 10.5 \text{ square units}$$

This left-hand sum estimate is close to the exact area ($10\frac{2}{3}$ square units).

As you may have guessed, we can also calculate a right-hand sum. The right-hand sum is the sum of the areas of rectangles whose upper right-hand corner touches the graph of f. The process for finding the right-hand sum is similar to the process for finding the left-hand sum, as will be shown in Example 5.

EXAMPLE 5

Using a Right-Hand Sum to Estimate an Area

Estimate the area between the graph of $f(x) = x^3 - 3x^2 + 3x$ and the x-axis on the interval $[0, 2]$ using a right-hand sum with four rectangles.

SOLUTION We are asked to estimate the area of the shaded region in Figure 6.14.

For the right-hand sum, the upper right-hand corner of each rectangle will touch the graph of $f(x)$, as shown in Figure 6.15. The width of each rectangle will be

$$\Delta x = \frac{2 - 0}{4}$$
$$= 0.5$$

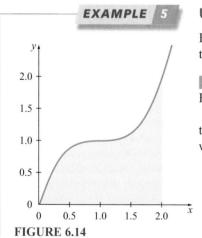

FIGURE 6.14

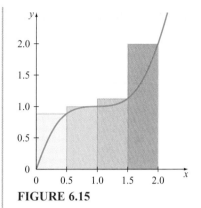

FIGURE 6.15

We will create a table of values for $f(x)$ starting at $x = 0$ and continuing to $x = 2$ with the x-values spaced 0.5 unit apart (see Table 6.7). For the right-hand sum, we select the second through the last term in the table. These are the respective heights of the rectangles.

TABLE 6.7

Rectangle Number	x	$f(x) = x^3 - 3x^2 + 3x$
	0	0.000
1	0.5	0.875
2	1.0	1.000
3	1.5	1.125
4	2.0	2.000

We will sum the heights together and multiply them by the width of a rectangle.

$$A = (0.875 + 1.000 + 1.125 + 2.000)(0.5)$$
$$= (5)(0.5)$$
$$= 2.5 \text{ square units}$$

The right-hand sum estimate for the area is 2.5 square units (using four rectangles).

The process for finding the right-hand sum is identical to the process for finding the left-hand sum, with the exception of Step 3. For the left-hand sum, you add up the first through the penultimate term in the table. For the right-hand sum, you add up the second through the last term.

HOW TO | **Finding the Right-Hand Sum for Small Values of n ($n \leq 10$)**

Let f be a function defined on the interval $[a, b]$ whose graph lies above the x-axis. The right-hand sum estimate for the area between the graph of f and the x-axis may be found using the following steps.

1. Calculate the width of each rectangle, $\Delta x = \dfrac{b - a}{n}$.
2. Create a table of values for $f(x)$ starting at $x = a$ and ending at $x = b$, with intermediate x-values spaced Δx units apart.
3. Add up the second through the last values of $f(x)$ listed in the table. (This is the sum of the heights of the rectangles.)
4. Multiply the result of Step 3 by Δx to get the right-hand sum approximation of the area.

(*Note:* This method may be used for any value of n; however, for values of n larger than 10, the process becomes extremely tedious.)

The table method for finding the left- and right-hand sums works well for small values of n. However, for large values of n (say $n = 100{,}000$), the table method is not practical. In Section 6.4, we will demonstrate how summation notation may be used for left- and right-hand sums with large values of n.

You may wonder, "Is the left-hand or the right-hand sum a better estimate of the area?" Actually, it varies from function to function. Often you can get one of the best area estimates by averaging the left- and right-hand sums.

EXAMPLE 6 Using Riemann Sums to Estimate an Area

Estimate the area between the graph of $f(x) = x^3 - 5x^2 + 6x + 1$ and the x-axis on the interval $[0, 4]$ with $n = 4$ using a left-hand sum, a right-hand sum, and the average of the sums.

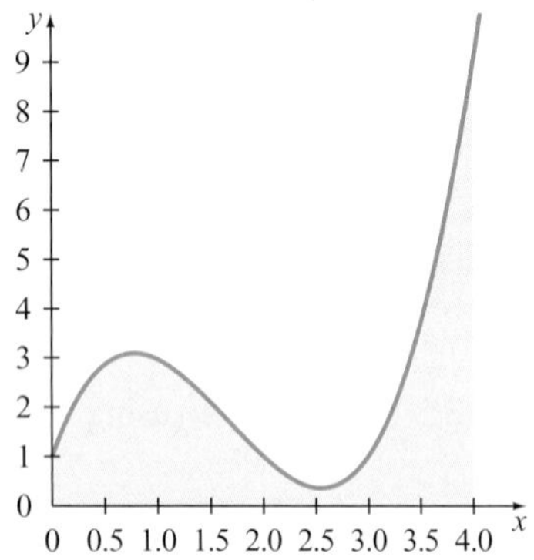

FIGURE 6.16

SOLUTION We are asked to estimate the area of the shaded region in Figure 6.16.

We generate the table of values for $f(x)$ given in Table 6.8.

TABLE 6.8

x	$f(x) = x^3 - 5x^2 + 6x + 1$
0	1
1	3
2	1
3	1
4	9

The left-hand sum is given by

$$\begin{aligned} LHS &= (1 + 3 + 1 + 1)(1) \\ &= 6 \text{ square units} \end{aligned}$$

The right-hand sum is given by

$$\begin{aligned} RHS &= (3 + 1 + 1 + 9)(1) \\ &= 14 \text{ square units} \end{aligned}$$

Averaging the two sums, we get

$$\begin{aligned} \frac{LHS + RHS}{2} &= \frac{6 + 14}{2} \\ &= 10 \text{ square units} \end{aligned}$$

Our estimate is fairly close to the exact area of $9\frac{1}{3}$ square units. (In Section 6.5, we will show you how to calculate the exact area.)

EXAMPLE 7

Using Riemann Sums to Estimate Distance Traveled

The velocity of the author's minivan as he drove from a traffic light onto a highway may be modeled by

$$V(t) = -0.1773t^2 + 0.6798t + 0.5019 \text{ miles per minute}$$

where t is time in minutes. Estimate the area between the velocity graph and the t-axis on the interval $[0.0, 2.5]$ using five rectangles. Then interpret the real-world meaning of the result.

SOLUTION We are asked to estimate the area of the shaded region in Figure 6.17 using five rectangles.

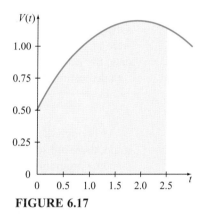

FIGURE 6.17

We construct a table of values for $V(t)$ starting at $t = 0.0$ and ending at $t = 2.5$ with t-values spaced $\Delta t = \dfrac{2.5 - 0}{5} = 0.5$ minute apart (see Table 6.9).

TABLE 6.9

t	$V(t) = -0.1773t^2 + 0.6798t + 0.5019$
0.0	0.502
0.5	0.797
1.0	1.004
1.5	1.123
2.0	1.152
2.5	1.093

Since the height of the rectangles represents velocity in miles per minute and the width of the rectangles represents minutes, the units of the area will be

UNITS

$$\frac{\text{miles}}{\text{minute}} \cdot \text{minute} = \text{miles}$$

We first find the left-hand sum.

$$LHS = (0.502 + 0.797 + 1.004 + 1.123 + 1.152)(0.5)$$
$$= (4.578)(0.5)$$
$$\approx 2.289 \text{ miles}$$

Then we find the right-hand sum.

$$RHS = (0.797 + 1.004 + 1.123 + 1.152 + 1.093)(0.5)$$
$$= (5.169)(0.5)$$
$$\approx 2.585 \text{ miles}$$

Now we average the sums.

$$\frac{LHS + RHS}{2} = \frac{2.289 + 2.585}{2}$$
$$= 2.437 \text{ miles}$$
$$\approx 2.45 \text{ (rounded to the nearest twentieth of a mile)}$$

We estimate that in the first 2.5 minutes of the highway trip, the author traveled a distance of 2.45 miles. (We rounded to twentieths of a mile because the raw data were recorded accurate to the nearest twentieth of a mile.) According to the raw data, the actual distance traveled by the author was 2.45 miles. In this case, our rounded estimate was right on!

Although the TI-83 Plus doesn't come preloaded with a Riemann sum program, there are several excellent programs available at www.ticalc.org. One outstanding program is *Riemann.8xp*, written by Mike Miller of Corban College. We transferred the downloaded program from our PC to our calculator using the TI-CONNECT software. In the following Technology Tip, we will repeat Example 7 using the Riemann.8xp program. (Users of the TI-83 calculator should download the file Rieman83.83P.)

TECHNOLOGY **TIP**

Finding Riemann Sums with Riemann.8xp

1. Press the PRGM key to bring up the list of programs. Select the program RIEMANN. Press ENTER twice.

2. The introduction and credit screen is displayed. Press ENTER.

(Continued)

3. To find a Riemann sum of a function f on an interval $[a, b]$ with n rectangles, enter the following:

$f(x)$: (Enter the function f.)
Lower Bound: a
Upper Bound: b
Partitions: n

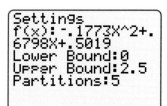

4. The calculator draws the graph of f and draws vertical lines at $x = a$ and $x = b$. Press ENTER.

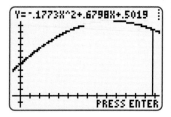

5. The calculator displays an option screen. To calculate the left-hand sum, select `2:LEFT SUM` and press ENTER.

6. The calculator graphs the function, draws the left-hand rectangles, and displays the value of the left-hand sum. Press ENTER to return to the option screen.

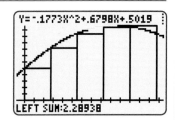

7. Select `3:RIGHT SUM`. The calculator graphs the function, draws the right-hand rectangles, and displays the value of the right-hand sum. Press ENTER to return to the option screen.

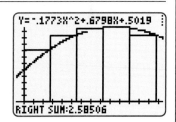

8. Select `5:TRAPEZOID SUM` to calculate the average of the left- and right-hand sums. Press ENTER to return to the option screen.

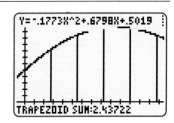

(Continued)

9. Select **1:Settings** to change the function, interval, or number of rectangles. Otherwise, select **7:QUIT** to quit the program.

EXAMPLE 8

Using Riemann Sums to Estimate Water Volume

Based on data from November 22 to 24, 2002, the flow rate of the Rogue River near Agness, Oregon, at noon may be modeled by

$$f(t) = 25t^2 - 75t + 1800 \text{ cubic feet per second}$$

where t is the number of 24-hour periods since noon on November 22, 2002. (**Source:** waterdata.usgs.gov.)

How much water (in gallons) passed by this point near Agness, Oregon, between noon on November 22 and noon on November 24, 2002?

SOLUTION We must initially find the area of the shaded region in Figure 6.18.

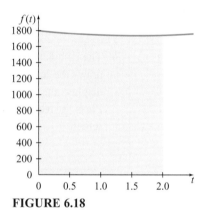

FIGURE 6.18

We decide to use four rectangles and generate the appropriate table of values (Table 6.10).

TABLE 6.10

t	$f(t) = 25t^2 - 75t + 1800$
0.0	1,800
0.5	1,769
1.0	1,750
1.5	1,744
2.0	1,750

This problem is a bit tricky, since the flow rate is given in cubic feet per second and the time intervals are 24-hour periods. Additionally, we're asked to give our answer in gallons, not cubic feet. We'll need to do some unit conversion before reaching our final answer. Since the flow rate is given in cubic feet per second and t is in terms of 24-hour periods, the units of the area of each rectangle will be

$$\frac{\text{cubic feet}}{\text{second}} \cdot \text{24-hour periods}$$

We'll convert these to the appropriate units after finding the left- and right-hand sums.

$$\begin{aligned} LHS &= (1800 + 1769 + 1750 + 1744)(0.5) \\ &= (7063)(0.5) \\ &= 3531 \ (\text{accurate to four significant figures}) \\ RHS &= (1769 + 1750 + 1744 + 1750)(0.5) \\ &= (7013)(0.5) \\ &= 3506 \ (\text{accurate to four significant figures}) \\ \frac{LHS + RHS}{2} &= \frac{3531 + 3506}{2} \end{aligned}$$

UNITS

$$= 3519 \frac{\text{cubic feet}}{\text{second}} \cdot \text{24-hour periods}$$

$$= 3519 \text{ cubic feet} \cdot \frac{\text{24-hour periods}}{\text{second}}$$

How many seconds are there in a 24-hour period?

UNITS

$$\begin{aligned} \text{24-hour periods} &= 24 \cdot 1 \text{ hour} \\ &= 24 \cdot 60 \text{ minutes} \\ &= 1440 \cdot 1 \text{ minute} \\ &= 1440 \cdot 60 \text{ seconds} \\ &= 86{,}400 \text{ seconds} \end{aligned}$$

Thus we have

UNITS

$$= 3519 \text{ cubic feet} \cdot \frac{86{,}400 \text{ seconds}}{\text{second}}$$

$$= 304{,}041{,}600 \text{ cubic feet}$$

Since 1 gallon equals 0.13368 cubic foot, we have

UNITS

$$304{,}041{,}600 \text{ cubic feet} \cdot \frac{1 \text{ gallon}}{0.13368 \text{ cubic feet}} \approx 2{,}274{,}398{,}564 \text{ gallons}$$

Between noon on November 22 and noon on November 24, we estimate that more than 2,274 million gallons of water passed by the measurement station on the Rogue River near Agness, Oregon. This is enough water to fill more than 267 Olympic-size swimming pools!

As noted earlier in the section, if $f(x)$ is a rate of change function, then the area between the graph of $f(x)$ and the x-axis on the interval $[a, b]$ represents the total change in $F(x)$ [an antiderivative of $f(x)$] from $x = a$ to $x = b$. We will discuss this in greater detail in Section 6.5.

6.3 Summary

In this section, you learned how to use left- and right-hand sums to approximate the area between the graph of a function and the horizontal axis. A solid understanding of these concepts will help you grasp the material presented in the next section.

6.3 Exercises

In Exercises 1–5, draw the rectangles used to calculate the left-hand sum estimate of the area between the graph of the function and the horizontal axis on the specified interval. (In each case, use four rectangles.) Then calculate the left-hand sum.

1. $f(x) = 2; [1, 5]$

2. $g(x) = -2x + 9; [0, 4]$

3. $h(x) = 2^x + x; [0, 2]$

4. $f(t) = \dfrac{3}{t}; [1, 3]$

5. $v(t) = 0.5t + 20; [3, 7]$

In Exercises 6–15, use the left-hand sum to estimate the area between the graph of the function and the horizontal axis on the specified interval. For each exercise, calculate the sum with $n = 2$, $n = 4$, and $n = 10$ rectangles.

6. $s(t) = -3t^2 + 3t; [0, 1]$

7. $s(t) = t^3 - 3t^2 + 3t - 1; [1, 2]$

8. $g(x) = -x^2 + 4; [0, 2]$

9. $h(x) = \dfrac{2}{x} + 1; [1, 5]$

10. $f(x) = -2x + 20; [0, 10]$

11. $s(x) = \ln(x); [e, e^2]$

12. $y = x^2 - 2^x; [2, 4]$

13. $y = 4t^2 - 1; [2, 4]$

14. $q = 20 - \dfrac{1000}{p}; [100, 200]$

15. $f(t) = \dfrac{t + 2}{t - 2}; [3, 11]$

In Exercises 16–35, use left- and right-hand sums (with $n = 4$) to estimate the area between the graph of the function and the horizontal axis on the specified interval. In each exercise, calculate the left-hand sum, the right-hand sum, and the average of the two sums. The exact area, A, is given so that you can compare your estimates to the actual area.

16. $f(x) = 6x + 1$ on $[2, 4]$; $A = 38$

17. $g(x) = x^2 - 2x + 2$ on $[3, 5]$; $A = 20\frac{2}{3}$

18. $h(x) = 3x^2 + 3$ on $[0, 2]$; $A = 14$

19. $f(x) = -(x - 2)^3 + 2$ on $[1, 2]$; $A = 2.25$

20. $g(x) = x^3 - 10x^2 + 33x - 36$ on $[5, 9]$;
$A = 250\frac{2}{3}$

21. $h(x) = x^3 - 3x^2 + 2x + 1$ on $[0, 2]$; $A = 2$

22. $f(x) = x^4 - 5x^2 + 8$ on $[0, 3]$; $A = 27.6$

23. $f(x) = x^3 - 9x^2 + 23x$ on $[2, 6]$; $A = 64$

24. $g(x) = x^2 - 7x + 16$ on $[3, 5]$; $A = 8\frac{2}{3}$

25. $h(x) = x^3 - 9x^2 + 26x$ on $[1, 5]$; $A = 96$

26. $y = \ln 4(4^x) - \ln 2(2^x)$ on $[0, 1]$;
$A = 2$

27. $f(x) = 4x - x^4$ on $[0, 1]$; $A = 1.8$

28. $g(x) = x^4 - 4x^3$ on $[5, 6]$; $A = 259.2$

29. $h(x) = e^{2x} - e^x$ on $[0, \ln 2]$; $A = 0.5$

30. $f(t) = e^{3t} - e^t$ on $[0, \ln 4]$; $A = 18$

31. $s(t) = \ln(t)$ on $[1, e]$; $A = 1$

32. $p(x) = \ln 3(3^x)$ on $[0, 1]$; $A = 2$

33. $y = e^x$ on $[\ln 2, \ln 5]$; $A = 3$

34. $y = 3x^2 - \ln 2(2^x)$ on $[3, 8]$; $A = 237$

35. $y = 5x^4 - \ln 5(5^x)$ on $[2, 4]$; $A = 392$

In Exercises 36–40, estimate the area between the graph of the function and the horizontal axis. Then interpret the physical meaning of the result.

36. **M** **Minivan Position** The velocity of the author's minivan as he drove from a traffic light onto a highway may be modeled by

$$V(t) = -0.1773t^2 + 0.6798t + 0.5019$$

miles per minute, where t is time in minutes. Use left- and right-hand sums with $n = 8$ to estimate how far he traveled in the first 2.0 minutes of his trip.

37. **M** **River Flow** Between 5:45 a.m. and 7:45 a.m. on November 22, 2002, the stream flow rate of the Columbia River below the Priest Rapids Dam dropped dramatically. The flow rate during that time period may be modeled by

$$F(t) = 14.50t^4 - 80.15t^3 + 168.3t^2 - 180.8t + 158.9$$

cubic feet per second, where t is the number of hours since 5:45 a.m. (**Source:** waterdata.usgs.gov.)

(a) Convert the units of the flow-rate function from cubic feet per second to cubic feet per hour.

(b) Use left- and right-hand sums with $n = 8$ to approximate the total amount of water that flowed past the flow-rate measurement station between 5:45 a.m. and 7:45 a.m.

(c) Why do you think the flow rate decreased so substantially?

38. **M** **River Flow** Based on data from November 22 to 24, 2002, the flow rate of the Wenatchee River at Peshastin, Washington, at noon may be modeled by

$$f(t) = -100t + 1300$$

cubic feet per second, where t is the number of 24-hour periods since noon on November 22, 2002. (**Source:** waterdata.usgs.gov.)

How much water (in gallons) passed by Peshastin between noon on November 22 and noon on November 24, 2002?

39. **M** **River Flow** Based on data from November 22 to 24, 2002, the flow rate of the Snake River near Irwin, Idaho, at noon may be modeled by

$$f(t) = -50t^2 + 100t + 1050$$

cubic feet per second, where t is the number of 24-hour periods since noon on November 22, 2002. (**Source:** waterdata.usgs.gov.)

How much water (in gallons) passed by this point near Irwin, Idaho, between noon on November 22 and noon on November 24, 2002? (Use $n = 8$.)

40. **M** **River Flow** Based on data from November 22 to 24, 2002, the flow rate of the North Snoqualmie River near Snoqualmie Falls, Washington, at noon may be modeled by

$$f(t) = 25t^2 - 145t + 490$$

cubic feet per second, where t is the number of 24-hour periods since noon on November 22, 2002. (**Source:** waterdata.usgs.gov.)

How much water (in gallons) passed by this point near Snoqualmie Falls between noon on November 22 and noon on November 24, 2002? (Use $n = 8$.)

Exercises 41–45 are intended to challenge your understanding of Riemann sums.

41. Give an example of a positive function f on an interval $[a, b]$ that has the property that the left-hand sum and right-hand sum approximation of the area between the graph of f and the x-axis are equal for all values of Δx.

42. The left-hand sum and right-hand sum of the function $f(x) = -x^2 + 4x$ on the interval $[0, 4]$ are both equal to 10 when four rectangles are used. Does this mean that the area between the graph of $f(x) = -x^2 + 4x$ and the x-axis is equal to 10? Explain.

43. A positive function f is an increasing function. (That is, $f(a) \leq f(b)$ whenever $a \leq b$.) Which sum will best approximate the area between the graph of f and the x-axis on the interval $[a, b]$: the left-hand sum, the right-hand sum, or the average of the left- and right-hand sums? Explain.

44. Calculate the left-hand sum for the function $f(x) = |-2x + 2|$ on the interval $[0, 2]$ using $n = 1, n = 2, n = 3, n = 4, n = 5$, and $n = 6$. For which value(s) of n does the left-hand sum best approximate the actual area between the graph of f and the x-axis on the interval $[a, b]$?

45. For a nonnegative function f on the interval $[a, b]$, does reducing the width of the rectangles used in a Riemann sum ever *worsen* the Riemann sum estimate of the area between the graph of f and the x-axis on the interval $[a, b]$? Explain. (*Hint:* Consider the function $f(x) = |-2x + 2|$ using left-hand sums with $n = 2$ and $n = 3$.)

6.4 The Definite Integral

- Apply definite integral properties
- Calculate the exact area between a curve and the x-axis by using definite integrals

GETTING STARTED   Based on data from 1999 to 2001, the marginal revenue of the Coca-Cola Company may be modeled by

$$M = -0.8058s + 14.44 \text{ dollars per unit case}$$

where s is the number of unit cases (in billions). A unit case is equivalent to 24 eight-ounce servings of finished beverage. In 1999, 16.5 billion unit cases were sold. In 2001, 17.8 billion unit cases were sold. (**Source:** Modeled from Coca-Cola Company 2001 Annual Report, pp. 46, 57.) By how much did the revenue of the Coca-Cola Company grow as the number of unit cases sold increased from 16.5 billion to 17.8 billion? If you were a financial consultant for Coca-Cola, would you encourage the company to reduce its prices so that it could increase unit case sales? We can address questions such as these using the notion of the definite integral.

In this section, we will continue our discussion of area estimates by introducing *summation notation* and the *definite integral*. We will also show you several helpful definite integral properties.

Summation Notation

We use *summation notation* in order to represent the sum of a large number of terms easily. The notation may feel a bit awkward at first; however, as you become skilled in using it, you will come to appreciate its usefulness. Let's return to the table we introduced in Example 5 of Section 6.3 (Table 6.7), reproduced here as Table 6.11.

TABLE 6.11

x	$f(x) = x^3 - 3x^2 + 3x$
0	0.000
0.5	0.875
1.0	1.000
1.5	1.125
2.0	2.000

Recall that for the left-hand sum, the sum of the rectangle heights may be written as $f(0) + f(0.5) + f(1.0) + f(1.5)$. Notice that the x values are $\Delta x = 0.5$ unit apart. An alternative way to write this sum is

$$\sum_{i=0}^{3} f(0 + 0.5i)$$

The Greek letter *sigma* (Σ) tells us that we are to sum the values of the form $f(0 + 0.5i)$. The i is the *index of summation*. It is a variable, and it is always equal to a whole number. The numbers above and below Σ are the *limits of summation*. In this case, the initial value of i is 0 and the ending value is 3.

We will first substitute the initial value of i into $f(0 + 0.5i)$.

$$f[0 + 0.5(0)] = f(0)$$

We will then increase i from 0 to 1 and substitute $i = 1$ into $f(0 + 0.5i)$.

$$f[0 + 0.5(1)] = f(0.5)$$

The summation symbol Σ tells us to sum these two values together. So far we have

$$f(0) + f(0.5)$$

However, the summation tells us that we must continue to increase the index until $i = 3$, each time summing the values.

$$f[0 + 0.5(2)] = f(1)$$

$$f[0 + 0.5(3)] = f(1.5)$$

Therefore, $\displaystyle\sum_{i=0}^{3} f(0 + 0.5i) = f(0) + f(0.5) + f(1.0) + f(1.5)$.

In the next two examples we will temporarily move away from our discussion of heights and areas in order to give you practice using summation notation.

EXAMPLE 1

Using Summation Notation

Calculate $\displaystyle\sum_{i=1}^{10} i^2$.

SOLUTION We are to sum terms of the form i^2 by substituting in the values 1 through 10 for i.

$$\sum_{i=1}^{10} i^2 = (1)^2 + (2)^2 + (3)^2 + (4)^2 + (5)^2 + (6)^2 + (7)^2 + (8)^2 + (9)^2 + (10)^2$$

$$= 1 + 4 + 9 + 16 + 25 + 36 + 49 + 64 + 81 + 100$$

$$= 385$$

EXAMPLE 2

Using Summation Notation

Calculate $\displaystyle\sum_{i=0}^{6} (i - 3)$.

SOLUTION We must sum terms of the form $i - 3$ by substituting in the values 0 through 6 for i.

$$\sum_{i=0}^{6} (i - 3) = (0 - 3) + (1 - 3) + (2 - 3) + (3 - 3) + (4 - 3) + (5 - 3) + (6 - 3)$$

$$= (-3) + (-2) + (-1) + 0 + 1 + 2 + 3$$

$$= 0$$

We can now describe the left- and right-hand sums in terms of summation notation.

SUMMATION NOTATION FOR THE LEFT-HAND SUM

Let f be a function defined on the interval $[a, b]$. If the graph of f is above the x-axis, then the area of the region between the graph of f and the x-axis may be approximated by

$$\sum_{i=0}^{n-1} f(x_i)\,\Delta x = f(x_0)\,\Delta x + f(x_1)\,\Delta x + \cdots + f(x_{n-1})\,\Delta x$$

where $\Delta x = \dfrac{b - a}{n}$ and

$$x_0 = a$$
$$x_1 = a + \Delta x$$
$$\vdots$$
$$x_{n-1} = a + (n - 1)\,\Delta x$$

$\sum_{i=0}^{n-1} f(x_i)\,\Delta x$ is the left-hand sum estimate of the area using n rectangles.

SUMMATION NOTATION FOR THE RIGHT-HAND SUM

Let f be a function defined on the interval $[a, b]$. If the graph of f is above the x-axis, then the area of the region between the graph of f and the x-axis may be approximated by

$$\sum_{i=1}^{n} f(x_i)\,\Delta x = f(x_1)\,\Delta x + f(x_2)\,\Delta x + \cdots + f(x_n)\,\Delta x$$

where $\Delta x = \dfrac{b - a}{n}$ and

$$x_1 = a + \Delta x$$
$$x_2 = a + 2 \cdot \Delta x$$
$$\vdots$$
$$x_n = b$$

$\sum_{i=1}^{n} f(x_i)\,\Delta x$ is the right-hand sum estimate of the area using n rectangles.

In general, a Riemann sum does not require that each rectangle have the same width. However, by making the rectangles have the same width, we reduce the number of calculations needed to find the solution.

Let's work an area example using summation notation.

EXAMPLE 3

Using Summation Notation in Calculating a Riemann Sum

Use a left-hand sum with $n = 6$ to estimate the area between the graph of $f(x) = x^2 + 2$ and the x-axis on the interval $[1, 3]$.

SOLUTION We are asked to estimate the area of the shaded region in Figure 6.19 using a left-hand sum with six rectangles.

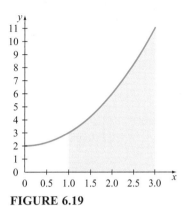

FIGURE 6.19

We will first find the width of each rectangle.

$$\Delta x = \frac{3 - 1}{6}$$

$$= \frac{1}{3} \text{ unit}$$

The left-hand sum is given by

$$\sum_{i=0}^{n-1} f(x_i)\,\Delta x = \sum_{i=0}^{6-1} f(x_i)\left(\frac{1}{3}\right)$$

$$= \sum_{i=0}^{5} [(x_i)^2 + 2]\left(\frac{1}{3}\right)$$

$$= \sum_{i=0}^{5} [(1 + i \cdot \Delta x)^2 + 2]\left(\frac{1}{3}\right) \qquad \text{Since } x_i = a + i\Delta x \text{ and } a = 1$$

$$= \sum_{i=0}^{5} \left\{[1 + i\left(\frac{1}{3}\right)]^2 + 2\right\}\left(\frac{1}{3}\right) \qquad \text{Since } \Delta x = \frac{1}{3}$$

$$= \left\{[1 + 0\left(\frac{1}{3}\right)]^2 + 2\right\}\left(\frac{1}{3}\right) + \left\{[1 + 1\left(\frac{1}{3}\right)]^2 + 2\right\}\left(\frac{1}{3}\right) + \left\{[1 + 2\left(\frac{1}{3}\right)]^2 + 2\right\}\left(\frac{1}{3}\right)$$

$$+ \left\{[1 + 3\left(\frac{1}{3}\right)]^2 + 2\right\}\left(\frac{1}{3}\right) + \left\{[1 + 4\left(\frac{1}{3}\right)]^2 + 2\right\}\left(\frac{1}{3}\right) + \left\{[1 + 5\left(\frac{1}{3}\right)]^2 + 2\right\}\left(\frac{1}{3}\right)$$

$$= \left\{[(1)^2 + 2] + [\left(\frac{4}{3}\right)^2 + 2] + [\left(\frac{5}{3}\right)^2 + 2] + [\left(\frac{6}{3}\right)^2 + 2] + [\left(\frac{7}{3}\right)^2 + 2] + [\left(\frac{8}{3}\right)^2 + 2]\right\}\left(\frac{1}{3}\right)$$

$$= \left[3 + \left(\frac{34}{9}\right) + \left(\frac{43}{9}\right) + (6) + \left(\frac{67}{9}\right) + \left(\frac{82}{9}\right)\right]\left(\frac{1}{3}\right)$$

$$\approx 11.37$$

Admittedly, formulating the sum using summation notation was more complicated than our table method. However, as we use an increasingly large number of rectangles, using summation notation will be the easiest way to represent the left- and right-hand sums symbolically.

The Definite Integral

What would happen if we used infinitely many rectangles to estimate the area between the graph of a function and the horizontal axis? Would the estimate be exact? As we have seen, when we increase the number of rectangles, the width of each rectangle is reduced and our area estimate improves. The definite integral captures the idea of infinitely many rectangles.

THE DEFINITE INTEGRAL

Let f be a continuous function defined on the interval $[a, b]$. The **definite integral of f from a to b** is given by

$$\int_a^b f(x)\,dx = \lim_{n \to \infty} \{[f(x_1) + f(x_2) + \cdots + f(x_n)]\Delta x\}$$

where $x_1, x_2, \ldots, x_n$ are the right-hand endpoints of subintervals of length

$$\Delta x = \frac{b-a}{n}.$$

The use of infinitely many rectangles may also be represented using summation notation in conjunction with limit notation. We have $\int_a^b f(x)\,dx = \lim_{n \to \infty} \sum_{i=1}^{n} f(x_i)\,\Delta x$ and $\int_a^b f(x)\,dx = \lim_{n \to \infty} \sum_{i=0}^{n-1} f(x_i)\,\Delta x$. That is, when we use infinitely many rectangles, the left- and right-hand sums become equal. When we write the symbol $\lim_{n \to \infty}$, we mean that n is made to be infinitely large.

For the definite integral $\int_a^b f(x)\,dx$, a is called the **lower limit of integration** and b is called the **upper limit of integration.** As with indefinite integrals, $f(x)$ is called the **integrand.** If $\int_a^b f(x)\,dx$ is defined, then f is said to be **integrable** on $[a, b]$.

The connection between the indefinite integral $\int f(x)\,dx$ and the definite integral $\int_a^b f(x)\,dx$ is not immediately obvious. For a nonnegative function $f(x)$, $\int f(x)\,dx$ is a family of functions with the same derivative, while $\int_a^b f(x)\,dx$ is the area between the graph of f and the x-axis. However, despite the dramatically different meanings of indefinite and definite integrals, the two are closely related, as will be discussed in Section 6.5.

In our interpretation of the definite integral, we required f to be a nonnegative function. In general, the expression $\int_a^b f(x)\,dx$ is the sum of the shaded regions above the x-axis minus the sum of the shaded regions below the x-axis. In other words, $\int_a^b f(x)\,dx$ is the **sum of the signed areas.**

Consider the graph of $f(x) = x^3 - 4x^2 + 3x$ on the interval $[0, 3]$ (Figure 6.20).

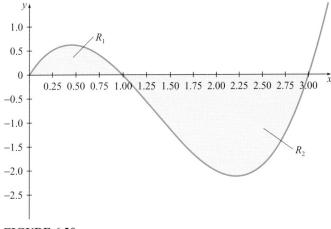

FIGURE 6.20

The first region, R_1, has area $\dfrac{5}{12}$ square unit. The second region, R_2, has area $2\dfrac{2}{3}$ square units. Therefore, $\displaystyle\int_0^3 (x^3 - 4x^2 + 3x)\,dx = \dfrac{5}{12} - 2\dfrac{2}{3} = -2\dfrac{1}{4}$. The fact that the number is negative indicates that more area lies below the x-axis than above it.

In considering the relationship between $\displaystyle\int_a^b f(x)\,dx$ and the left- and right-hand sums, it is helpful to think of $f(x)$ as the height of an infinitely narrow rectangle and dx as the width of that rectangle. The integral sign $\displaystyle\int$ is a somewhat distorted S, meaning "sum," and tells us to sum the areas of infinitely many adjacent rectangles of width dx between $x = a$ and $x = b$. (From our earlier discussion of differentials, we know that we can choose the value of dx to be as small as we like.) The height of each rectangle varies as x moves from a to b in steps of length dx.

THE MEANING OF $\displaystyle\int_a^b f(x)\,dx$ FOR A NONNEGATIVE FUNCTION f

Let f be a nonnegative function on $[a, b]$. The definite integral

$$\int_a^b f(x)\,dx$$

is the area of the region between the graph of f and the x-axis.

EXAMPLE 4

Interpreting the Graphical Meaning of a Definite Integral

Interpret the graphical meaning of $\displaystyle\int_0^3 (-x^2 + 3x)\,dx = 4.5$.

SOLUTION We know that $\displaystyle\int_0^3 (-x^2 + 3x)\,dx$ represents the sum of the signed areas between the graph of $y = -x^2 + 3x$ and the x-axis. However, since $y \geq 0$

for all values of x in the interval $[0, 3]$, $\int_0^3 (-x^2 + 3x)\,dx$ gives the area of the shaded region in Figure 6.21.

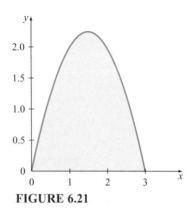

FIGURE 6.21

The area of the shaded region is 4.5 square units.

EXAMPLE 5

Interpreting the Meaning of a Definite Integral in an Applied Setting

Based on data from 1985 to 1999, the rate of change in the per capita consumption of breakfast cereal may be modeled by

$$R(t) = -0.014154t^2 + 0.233t - 0.3585 \text{ pounds per person per year}$$

where t is the number of years since the end of 1980. The graph of R is depicted in Figure 6.22.

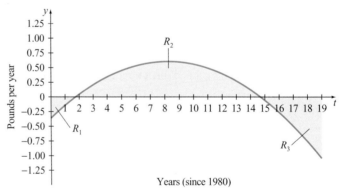

FIGURE 6.22

The areas of the three shaded regions are 0.296 square unit, 5.214 square units, and 2.034 square units, respectively. Calculate and interpret the meaning of

$$\int_0^{19} (-0.014154t^2 + 0.233t - 0.3585)\,dt.$$

SOLUTION $\int_0^{19} (-0.014154t^2 + 0.233t - 0.3585)\,dt$ is the sum of the signed areas. Since regions R_1 and R_3 are below the t-axis, their signed areas are negative.

Therefore,

$$\int_{0}^{19}(-0.014154t^2 + 0.233t - 0.3585)\,dt = -0.296 + 5.214 - 2.034 = 2.884$$

Recall that the units of the integral are found by multiplying the units of the independent variable by the units of the dependent variable. That is,

$$\text{years since 1980} \cdot \frac{\text{pounds}}{\text{year since 1980}} = \text{pounds}$$

So from the end of 1980 through 1999, the annual per capita breakfast cereal consumption increased by a net of 2.884 pounds. That is, taking into account the increases and decreases in cereal consumption rates, annual breakfast cereal consumption increased by 2.884 pounds per person from the end of 1980 through 1999.

Definite Integral Properties

Some of the definite integral properties are directly related to the indefinite integral rules. You will readily note the similarities.

DEFINITE INTEGRAL PROPERTIES

For integrable functions f and g, the following properties hold:

- $$\int_{a}^{a} f(x)\,dx = 0$$

- $$\int_{a}^{b} f(x)\,dx = -\int_{b}^{a} f(x)\,dx$$

- $$\int_{a}^{b} k \cdot f(x)\,dx = k \int_{a}^{b} f(x)\,dx \text{ for constant } k$$

- $$\int_{a}^{b} [f(x) \pm g(x)]\,dx = \int_{a}^{b} f(x)\,dx \pm \int_{a}^{b} g(x)\,dx$$

- $$\int_{a}^{b} f(x)\,dx = \int_{a}^{c} f(x)\,dx + \int_{c}^{b} f(x)\,dx \text{ for } a \le c \le b$$

EXAMPLE 6 **Applying Integration Properties**

Given $\displaystyle\int_{-2}^{0}(2x + 3)\,dx = 2$, determine $\displaystyle\int_{0}^{-2}(2x + 3)\,dx$.

SOLUTION Switching the limits of integration changes the sign of the definite integral. Therefore, $\displaystyle\int_{0}^{-2}(2x + 3)\,dx = -2$.

EXAMPLE 7

Applying Integration Properties

Given $\int_2^4 2x\,dx = 12$ and $\int_2^4 (2x + 5)\,dx = 22$, find $\int_2^4 5\,dx$.

SOLUTION We know that $\int_2^4 2x\,dx + \int_2^4 5\,dx = \int_2^4 (2x + 5)\,dx$. Therefore,

$$12 + \int_2^4 5\,dx = 22$$

$$\int_2^4 5\,dx = 10$$

EXAMPLE 8

Applying Integration Properties

Given $\int_{-1}^1 3x^2\,dx = 2$ and $\int_1^3 3x^2\,dx = 26$, find $\int_{-1}^3 3x^2\,dx$.

SOLUTION The only difference between the three integrals is the limits of integration. Since the upper limit of integration of $\int_{-1}^1 3x^2\,dx$ is the same as the lower limit of integration of $\int_1^3 3x^2\,dx$, the two integrals may be written together as a single integral $\int_{-1}^3 3x^2\,dx$. The value of this integral is found by summing the values of the two integrals from which it was created. That is,

$$\int_{-1}^3 3x^2\,dx = \int_{-1}^1 3x^2\,dx + \int_1^3 3x^2\,dx$$

$$= 2 + 26$$

$$= 28$$

EXAMPLE 9

Interpreting the Meaning of a Definite Integral in an Applied Setting

The rate of change in the annual per capita consumption of chicken from the end of 1985 through 1999 may be modeled by

$R(t) = 0.010492t^3 - 0.22635t^2 + 1.3212t - 0.4688$ pounds per person per year

where t is the number of years since the end of 1985. (**Source:** Modeled from *Statistical Abstract of the United States, 2001*, Table 202, p. 129.)

The sum of the signed areas of the shaded regions of the graph in Figure 6.23 is $\int_0^{15} R(t)\,dt$ and equals 19.75.

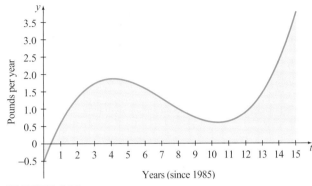

FIGURE 6.23

Given $\int_{10}^{15} R(t)\,dt = 7.60$, calculate and interpret the meaning of $\int_0^{10} R(t)\,dt$.

SOLUTION $\int_0^{10} R(t)\,dt$ is the net change in the annual per capita consumption of chicken from the end of 1985 through the end of 1995 in pounds per person. We know that $\int_0^{15} R(t)\,dt = 19.75$ and $\int_{10}^{15} R(t)\,dt = 7.60$, so

$$\int_0^{10} R(t)\,dt + \int_{10}^{15} R(t)\,dt = \int_0^{15} R(t)\,dt$$
$$\int_0^{10} R(t)\,dt = \int_0^{15} R(t)\,dt - \int_{10}^{15} R(t)\,dt$$
$$= 19.75 - 7.60$$
$$= 12.15$$

From the end of 1985 through 1995, the annual per capita consumption of chicken increased by 12.15 pounds. That is, at the end of 1995, on average, a person was eating 12.15 pounds more chicken per year than he or she was at the end of 1985.

For poultry producers, this information is great news. For beef producers, it may be a cause for concern. Over the same time period, annual per capita beef consumption dropped by 10.2 pounds. (**Source:** *Statistical Abstract of the United States, 2001*, Table 202, p. 129.)

EXAMPLE 10

Using a Definite Integral to Calculate an Accumulated Change

Based on data from 1999 to 2001, the marginal revenue of the Coca-Cola Company may be modeled by

$$M = -0.8058s + 14.44 \text{ dollars per unit case}$$

where s is the number of unit cases (in billions). A unit case is equivalent to 24 eight-ounce servings of finished beverage. In 1999, 16.5 billion unit cases were sold. In 2001, 17.8 billion unit cases were sold. (**Source:** Modeled from Coca-Cola Company 2001 Annual Report, pp. 46, 57.) By how much did the revenue of Coca-Cola Company grow as the number of unit cases sold increased from 16.5 billion to 17.8 billion?

SOLUTION We begin by drawing a graph of the marginal revenue function on the interval $[16.5, 17.8]$ and shading the region between the graph and the horizontal axis (see Figure 6.24).

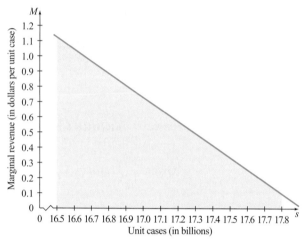

FIGURE 6.24

We are asked to find $\displaystyle\int_{16.5}^{17.8} (-0.8058s + 14.44)\,ds$. Since the marginal revenue function is positive on the interval $[16.5, 17.8]$, $\displaystyle\int_{16.5}^{17.8} (-0.8058s + 14.44)\,ds$ is the area of the shaded region. We recognize that the shaded region is a trapezoid.

The area of a trapezoid is given by $A = \frac{1}{2}(b_1 + b_2)h$, where h is the height of the trapezoid and b_1 and b_2 are the respective lengths of the bases. Our graph is a trapezoid turned on its side. We can calculate the length of each base by evaluating M at $s = 16.5$ and $s = 17.8$. The height of the trapezoid is the length of the interval. We have $M(16.5) = 1.144$ and $M(17.8) = 0.097$.

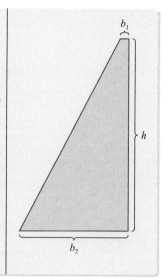

The length of the interval is $17.8 - 16.5 = 1.3$. Therefore, the area of the trapezoid is

$$A = \frac{1}{2}(b_1 + b_2)h$$

$$= \frac{1}{2}(1.144 + 0.097)(1.3)$$

$$= 0.8067$$

Thus $\displaystyle\int_{16.5}^{17.8} (-0.8058s + 14.44)\,ds \approx 0.8$ billion dollars. (The raw data were accurate to the nearest tenth of a billion dollars.) Increasing sales from 16.5 billion to 17.8 billion unit cases increased revenue by \$0.8 billion.

When working definite integral problems, it is often helpful to use a graphing calculator to verify the accuracy of our results. The following Technology Tip describes how to find a definite integral graphically.

TECHNOLOGY TIP

Finding a Definite Integral Graphically

1. Enter the function $f(x)$ as Y1 by pressing the [Y=] button and typing the equation.

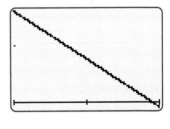

2. Graph the function over the specified domain. (You may need to press [WINDOW] and adjust the Xmin and Xmax settings. Press [ZOOM] then [0] to automatically adjust the y-values so that the entire graph will appear on the screen.)

3. Press [2nd] then [TRACE] to bring up the CALCULATE menu. Select item

   $7:\displaystyle\int f(x)\,dx$ and press [ENTER].

4. The calculator asks, Lower Limit? Enter the value of the lower limit of integration. Then press [ENTER]. (*Warning:* The lower limit must be within the domain of the viewing rectangle.)

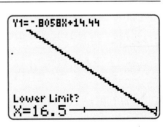

(Continued)

5. The calculator asks, Upper Limit? Enter the value of the upper limit of integration. Then press ENTER. (*Warning:* The upper limit must be within the domain of the viewing rectangle.)

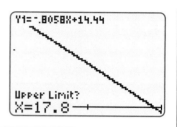

6. The calculator shades the region between the graph of the function and the *x*-axis and then displays the value of the definite integral. In this case, we found

$$\int_{16.5}^{17.8} (-0.8058s + 14.44)\, ds = 0.8067.$$

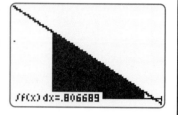

6.4 Summary

In this section, you learned how to use summation notation. You discovered that the definite integral is the sum of the signed areas. You also learned how to manipulate definite integrals using definite integral properties.

6.4 Exercises

In Exercises 1–10, calculate the sum.

1. $\displaystyle\sum_{i=1}^{6} i$

2. $\displaystyle\sum_{i=3}^{7} (i^2 - i)$

3. $\displaystyle\sum_{i=0}^{5} (-1)^i i$

4. $\displaystyle\sum_{i=0}^{4} [(3i - 1) \cdot 4]$

5. $\displaystyle\sum_{i=2}^{3} (4i^2 - 2i - 3)$

6. $\displaystyle\sum_{i=1}^{4} [4(i^2 - 2i)]$

7. $\displaystyle\sum_{i=1}^{4} f(1 + i \cdot \Delta x)(\Delta x)$ given that

$f(x) = x^2$ and $\Delta x = 0.25$

8. $\displaystyle\sum_{i=0}^{3} f(1 + i \cdot \Delta x)(\Delta x)$ given that

$f(x) = x^2$ and $\Delta x = 0.25$

9. $\displaystyle\sum_{i=1}^{4} f(2 + i \cdot \Delta x)(\Delta x)$ given that

$f(x) = 3x - 1$ and $\Delta x = 0.25$

10. $\displaystyle\sum_{i=0}^{3} f(2 + i \cdot \Delta x)(\Delta x)$ given that

$f(x) = 3x - 1$ and $\Delta x = 0.25$

In Exercises 11–20, draw the graph of the integrand function and shade the region corresponding to the definite integral. Then calculate the definite integral using geometric formulas for area.

11. $\displaystyle\int_{5}^{7} 1\, dx$

12. $\displaystyle\int_{2}^{5} 3\, dx$

13. $\displaystyle\int_{0}^{4} x\, dx$

14. $\displaystyle\int_{0}^{3} (-x + 3)\, dx$

15. $\displaystyle\int_{1}^{2} (-2x + 5)\, dx$

16. $\displaystyle\int_{1}^{3} (4x - 4)\, dx$

17. $\displaystyle\int_{-1}^{1} (2x + 2)\, dx$

18. $\displaystyle\int_{-1}^{1} (3x + 4)\, dx$

19. $\displaystyle\int_{-2}^{-1} (-2x + 1)\, dx$

20. $\displaystyle\int_{-2}^{3} (-x + 5)\, dx$

In Exercises 21–30, use the definite integral properties to find the numeric value of each definite integral given that $\int_2^4 f(x)\,dx = 5,$

$\int_2^6 f(x)\,dx = 9,$ $\int_2^4 g(x)\,dx = 2,$ *and* $\int_2^6 g(x)\,dx = 1.$

21. $\int_2^4 [2f(x)]\,dx$ **22.** $\int_2^6 [3f(x)]\,dx$

23. $\int_2^6 [f(x) + g(x)]\,dx$ **24.** $\int_2^6 [f(x) - g(x)]\,dx$

25. $\int_4^6 f(x)\,dx$ **26.** $\int_4^6 g(x)\,dx$

27. $\int_4^6 [f(x) + g(x)]\,dx$ **28.** $\int_4^6 [2f(x) - g(x)]\,dx$

29. $\int_6^2 [3f(x) - 4g(x)]\,dx$

30. $\int_6^2 [f(x) + g(x)]\,dx$

In Exercises 31–35, write the definite integral that represents the sum of the signed areas of the shaded regions of the graph.

31. $f(x) = -x^2 + 5x - 6$

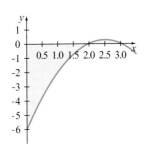

32. $f(x) = x^3 - 4x^2$

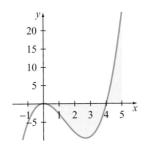

33. $f(x) = -x^3 + 4x + 20$

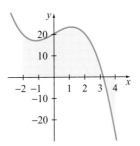

34. $f(x) = \dfrac{x + 1}{x^2 + 1}$

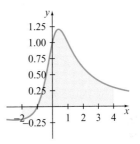

35. $f(x) = -|x - 3| + 3$

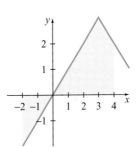

In Exercises 36–40, shade the regions of the graph that will be used in calculating the definite integral.

36. $\int_2^4 (x - 3)\,dx$

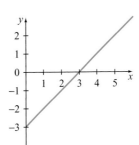

37. $\int_1^3 (-x^2 + 4x - 3)\, dx$

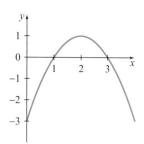

38. $\int_0^2 \frac{x^2 - 1}{x^2 + 1}\, dx$

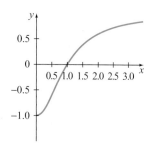

39. $\int_{-1}^1 e^{-x^2 + 1}\, dx$

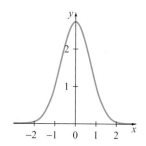

40. $\int_0^2 \ln\left(\frac{1}{x^2 + 1}\right) dx$

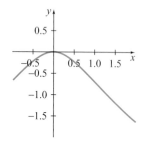

In Exercises 41–45, use the properties of definite integrals to find the solution.

41. **M** **Bottled Water Consumption** Based on data from 1980 to 1999, the rate of change in the annual per capita consumption of bottled water may be modeled by

$$R(t) = 0.2612(1.106)^t$$

gallons per person per year, where t is the number of years since the end of 1980. (**Source:** Modeled from *Statistical Abstract of the United States, 2001*, Table 204, p. 130.) The graph of $R(t)$ is shown.

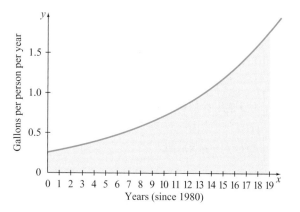

The area of the shaded region of the graph is $\int_0^{19} [0.2612(1.106)^t]\, dt$ and is equal to 15.0.

Additionally, $\int_0^{10} [0.2612(1.106)^t]\, dt = 4.5.$

Calculate and interpret the meaning of $\int_{10}^{19} [0.2612(1.106)^t]\, dt.$

42. **M** **Margarine Consumption** Based on data from 1980 to 1999, the rate of change in the annual per capita consumption of margarine may be modeled by

$$R(t) = 0.001147t^3 - 0.03445t^2 + 0.2658t - 0.5098$$

pounds per person per year, where t is the number of years since the end of 1980. (**Source:** Modeled from *Statistical Abstract of the United States, 2001*, Table 202, p. 129.) The graph of $R(t)$ is shown.

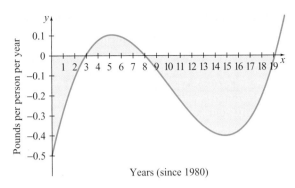

Years (since 1980)

The sum of the signed areas of the shaded regions is $\int_0^{19}\left(\begin{array}{c}0.001147t^3 - 0.03445t^2 \\ + 0.2658t - 0.5098\end{array}\right)dt$ and is equal to -3.10. Additionally,

$$\int_0^8\left(\begin{array}{c}0.001147t^3 - 0.03445t^2 \\ + 0.2658t - 0.5098\end{array}\right)dt = -0.28.$$

Calculate and interpret the meaning of

$$\int_8^{19}\left(\begin{array}{c}0.001147t^3 - 0.03445t^2 \\ + 0.2658t - 0.5098\end{array}\right)dt.$$

43. **M** **Mozzarella Cheese Consumption**
Based on data from 1980 to 1999, the rate of change in the annual per capita consumption of mozzarella cheese may be modeled by

$$R(t) = 0.0009778t^3 - 0.02958t^2 \\ + 0.2392t - 0.06311$$

pounds per person per year, where t is the number of years since the end of 1980. (**Source:** Modeled from *Statistical Abstract of the United States, 2001*, Table 202, p. 129.) The graph of $R(t)$ is shown.

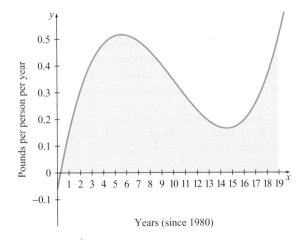

Years (since 1980)

The sum of the signed areas of the shaded regions is $\int_0^{19}\left(\begin{array}{c}0.0009778t^3 - 0.02958t^2 \\ + 0.2392t - 0.06311\end{array}\right)dt$ and is equal to 6.20. Additionally,

$$\int_{15}^{19}\left(\begin{array}{c}0.0009778t^3 - 0.02958t^2 \\ + 0.2392t - 0.06311\end{array}\right)dt = 1.14.$$

Calculate and interpret the meaning of

$$\int_0^{15}\left(\begin{array}{c}0.0009778t^3 - 0.02958t^2 \\ + 0.2392t - 0.06311\end{array}\right)dt.$$

44. **M** **Flounder Production** Based on data from 1990 to 1999, the rate of change in flounder steak and filet production in the continental United States may be modeled by

$$R(t) = 0.2842t - 4.755$$

million pounds per year, where t is the number of years since the end of 1990. (**Source:** Modeled from *Statistical Abstract of the United States, 2001*, Table 863, p. 553.)

Graph R and shade the region whose area is given by $\int_0^9 (0.2842t - 4.755)\,dt$. Then calculate

$$\int_0^9 (0.2842t - 4.755)\,dt$$

and interpret its real-world meaning. (*Hint:* Use the area of a trapezoid formula to calculate the area between the graph of the function and the horizontal axis.)

45. **M** **AP Calculus Participants** Based on data from 1969 to 2002, the rate of change in the number of AP Calculus AB exam participants may be modeled by

$$R(t) = 315.70t - 770.64$$

people per year, where t is the number of years since the end of 1969. (**Source:** Modeled from The College Board data.)

Graph R and shade the region whose area is given by $\int_{23}^{33} (315.70t - 770.64)\,dt$. Then calculate

$$\int_{23}^{33} (315.70t - 770.64)\,dt$$

and interpret its real-world meaning. (*Hint:* Use the area of a trapezoid formula to calculate the value of the definite integral.)

Exercises 46 to 50 are intended to challenge your understanding of Riemann sums and definite integrals.

46. A function f on $[a, b]$ has the property that
$$\sum_{i=0}^{3} f(x_i)\Delta x = \sum_{i=1}^{4} f(x_i)\Delta x. \text{ (That is, the left-hand}$$
sum is equal to the right-hand sum.) Does this mean that $\int_a^b f(x)\,dx = \sum_{i=0}^{3} f(x_i)\Delta x$? Explain.

47. Suppose that $f(x)$ is *increasing* on the interval $[a, b]$. Is the following statement true for all values of n?
$$\sum_{i=0}^{n-1} f(x_i)\Delta x \le \int_a^b f(x)\,dx \le \sum_{i=1}^{n} f(x_i)\Delta x$$
Justify your answer.

48. Suppose that $f(x)$ is *decreasing* on the interval $[a, b]$. Is the following statement true for all values of n?
$$\sum_{i=0}^{n-1} f(x_i)\Delta x \le \int_a^b f(x)\,dx \le \sum_{i=1}^{n} f(x_i)\Delta x$$
Justify your answer.

49. Find three different functions f that have the property that $\int_2^4 f(x)\,dx = 0$.

50. Give an example of an integrable function f that has the property that $\int_a^b f(x)\,dx = f(b) - f(a)$.

6.5 The Fundamental Theorem of Calculus

- Apply the Fundamental Theorem of Calculus
- Find the function for accumulated change given a rate of change function

GETTING STARTED Based on data from 1993 to 2002, the rate of change in the sales income of the Starbucks Corporation may be modeled by

$$R(t) = 58.46t + 79.33 \text{ million dollars per year}$$

where t is the number of years since 1993.

According to the model, by how much did the sales income increase between the end of 1996 and the end of 2002?

Although we may use the methods previously covered to answer this question, in this section we will demonstrate how the Fundamental Theorem of Calculus can greatly simplify our computations. This theorem is one of the most powerful tools in calculus and gives us a remarkably easy way to calculate the sum of the signed areas between the graph of a function and the x-axis.

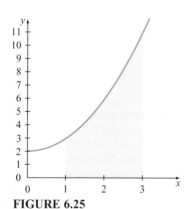

FIGURE 6.25

In Example 3 of Section 6.4, we used a left-hand sum with six rectangles to estimate the area between the graph of $f(x) = x^2 + 2$ and the x-axis on the interval $[1, 3]$ (see Figure 6.25).

The left-hand sum approximation of the area was approximately 11.37 square units, which we noted was an underestimate of the actual area.

The right-hand sum approximation of the area between $f(x) = x^2 + 2$ and the x-axis on $[1, 3]$ using six rectangles is 14.04 square units. This is an overestimate of the actual area. Averaging the two sums, we estimate that the shaded region has an approximate area of 12.70 square units.

Let's find the antiderivative of the function f.

$$\int f(x)\,dx = \int (x^2 + 2)\,dx$$

$$F(x) = \frac{x^3}{3} + 2x + C$$

We'll evaluate the function F at the endpoints of the interval.

$$F(3) = \frac{(3)^3}{3} + 2(3) + C$$

$$= 15 + C$$

$$F(1) = \frac{(1)^3}{3} + 2(1) + C$$

$$= 2\frac{1}{3} + C$$

We next find the difference between the two values.

$$F(3) - F(1) = 15 + C - \left(2\frac{1}{3} + C\right)$$

$$= 12\frac{2}{3}$$

$$\approx 12.67$$

Notice that the constant C was eliminated in the computation. This will happen anytime we calculate the difference between two different values of the anti-derivative. This value, 12.67, is extremely close to our estimate of 12.70. In fact, the exact area of the region is $12\frac{2}{3}$ square units. Thus, the definite integral $\int_{1}^{3} f(x)\,dx$ not only represents the area of the region between $f(x)$ and the x-axis on the interval $[1, 3]$, it also represents the total change in $F(x)$ from $x = 1$ to $x = 3$, where $F'(x) = f(x)$. This remarkable result is a key component of the Fundamental Theorem of Calculus.

FUNDAMENTAL THEOREM OF CALCULUS

Let f be a continuous function on $[a, b]$. Then

$$\int_{a}^{b} f(x)\,dx = F(b) - F(a)$$

where F is any antiderivative of f.

Since F may be *any* antiderivative of f, we will choose F to be the anti-derivative of f that has the constant $C = 0$. This choice will make all of our com-putations easier. We'll work several examples to illustrate the power of the Fundamental Theorem of Calculus.

EXAMPLE   1

Using the Fundamental Theorem of Calculus to Calculate a Definite Integral

Calculate $\displaystyle\int_{3}^{7} (2x - 4)\, dx$.

SOLUTION   We must first find an antiderivative of $f(x) = 2x - 4$ and then evaluate the antiderivative at $x = 7$ and $x = 3$.

$$\int_{3}^{7} (2x - 4)\, dx = (x^2 - 4x) \Big|_{3}^{7}$$

An antiderivative of $f(x) = 2x - 4$ is $F(x) = x^2 - 4x$. The vertical bar tells us that we must evaluate this antiderivative at $x = 7$ and $x = 3$ and then find the difference of the two values.

$$(x^2 - 4x) \Big|_{3}^{7} = [(7)^2 - 4(7)] - [(3)^2 - 4(3)]$$
$$= (49 - 28) - (9 - 12)$$
$$= 21 - (-3)$$
$$= 24$$

Therefore, $\displaystyle\int_{3}^{7} (2x - 4)\, dx = 24$. Since $f(x) = 2x - 4$ is positive on the interval $[3, 7]$, the total area between the graph of f and the x-axis on $[3, 7]$ is 24 square units.

EXAMPLE   2

Using the Fundamental Theorem of Calculus to Calculate a Definite Integral

Calculate $\displaystyle\int_{-1}^{1} (x^3 - x)\, dx$.

SOLUTION

$$\int_{-1}^{1} (x^3 - x)\, dx = \left(\frac{x^4}{4} - \frac{x^2}{2}\right) \Big|_{-1}^{1}$$
$$= \left[\frac{(1)^4}{4} - \frac{(1)^2}{2}\right] - \left[\frac{(-1)^4}{4} - \frac{(-1)^2}{2}\right]$$
$$= \left(\frac{1}{4} - \frac{1}{2}\right) - \left(\frac{1}{4} - \frac{1}{2}\right)$$
$$= 0$$

This result is due to the fact that the area of the shaded region above the x-axis and the area of the shaded region below the x-axis are equal.

EXAMPLE   3

Using the Fundamental Theorem of Calculus to Find the Area of a Region

Find the area of the region between the graph of the function $f(x) = x^3 - x$ and the x-axis over the interval $[-1, 1]$.

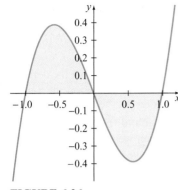

FIGURE 6.26

We are asked to find the combined area of the regions enclosed by the graph of f and the x-axis. We begin by graphing f and shading the enclosed regions (see Figure 6.26).

The graph of f bordering the first region intersects the x-axis at $x = -1$ and $x = 0$. The area of the region is given by

$$\int_{-1}^{0} (x^3 - x)\, dx = \frac{x^4}{4} - \frac{x^2}{2}\bigg|_{-1}^{0}$$

$$= \left[\frac{(0)^4}{4} - \frac{(0)^2}{2}\right] - \left[\frac{(-1)^4}{4} - \frac{(-1)^2}{2}\right]$$

$$= 0 - \left(\frac{1}{4} - \frac{1}{2}\right)$$

$$= \frac{1}{4}$$

The graph of f bordering the second region intersects the x-axis at $x = 0$ and $x = 1$. The region lies below the x-axis, so $\int_{0}^{1} (x^3 - x)\, dx$ will be a negative number equal to -1 times the area.

$$\int_{0}^{1} (x^3 - x)\, dx = \frac{x^4}{4} - \frac{x^2}{2}\bigg|_{0}^{1}$$

$$= \left[\frac{(1)^4}{4} - \frac{(1)^2}{2}\right] - \left[\frac{(0)^4}{4} - \frac{(0)^2}{2}\right]$$

$$= \left(\frac{1}{4} - \frac{1}{2}\right) - 0$$

$$= -\frac{1}{4}$$

$$= -1 \cdot \frac{1}{4}$$

The area of the second region is also $\frac{1}{4}$ square unit. Therefore, the area bounded by the graph of f and the x-axis is

$$A = \frac{1}{4} + \frac{1}{4}$$

$$= \frac{1}{2} \text{ square unit}$$

Notice in Example 3 that we used separate definite integrals to calculate the area above the x-axis and the area below the x-axis. Using a single definite integral would have led to an erroneous result. If we calculate $\int_{-1}^{1} (x^3 - x)\, dx$, we get the sum of the *signed* areas, not the actual area.

$$\int_{-1}^{1} (x^3 - x)\, dx = \frac{1}{4}x^4 - \frac{1}{2}x^2 \bigg|_{-1}^{1}$$

$$= \left[\frac{1}{4}(1)^4 - \frac{1}{2}(1)^2\right] - \left[\frac{1}{4}(-1)^4 - \frac{1}{2}(-1)^2\right]$$

$$= \left(\frac{1}{4} - \frac{1}{2}\right) - \left(\frac{1}{4} - \frac{1}{2}\right)$$

$$= 0$$

A common error among beginning calculus students is to assume that the definite integral *always* yields the area between the graph of the function and the *x*-axis. As was just illustrated, this is not the case.

Although we are often given an interval over which to find the area, we are sometimes required to find the interval ourselves. When we see the phrase "Find the area of the region bounded by $f(x)$ and the *x*-axis," we are being asked to calculate the area of the region(s) enclosed by the graph of f and the *x*-axis. As long as f is a continuous function, we can determine the area by setting up definite integrals with the limits of integration representing each consecutive pair of *x*-intercepts, as demonstrated in Example 4.

EXAMPLE **4**

Using the Fundamental Theorem of Calculus to Find the Area of a Region

Find the area of the region bounded by $f(x) = 4x^3 - 2x$ and the *x*-axis.

SOLUTION We must first determine where f crosses the *x*-axis. To do this, we set $f(x)$ equal to zero and solve.

$$0 = 4x^3 - 2x$$

$$0 = 4x\left(x^2 - \frac{1}{2}\right)$$

$$0 = 4x\left(x - \sqrt{\frac{1}{2}}\right)\left(x + \sqrt{\frac{1}{2}}\right)$$

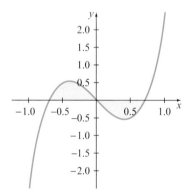

FIGURE 6.27

The function has *x*-intercepts at $x = -\sqrt{\frac{1}{2}}$, $x = 0$, and $x = \sqrt{\frac{1}{2}}$. Let's graph the function so that we can see the location of the bounded regions (Figure 6.27).

On the interval $\left(-\infty, -\sqrt{\frac{1}{2}}\right)$, the graph of f and the *x*-axis do not enclose a region. The first enclosed region occurs on the interval $\left(-\sqrt{\frac{1}{2}}, 0\right)$. The second enclosed region occurs on the interval $\left(0, \sqrt{\frac{1}{2}}\right)$. On the interval $\left(\sqrt{\frac{1}{2}}, \infty\right)$, the graph of f and the *x*-axis do not enclose a region. The area we're looking for is

$$A = \left| \int_{-\sqrt{1/2}}^{0} (4x^3 - 2x)\, dx \right| + \left| \int_{0}^{\sqrt{1/2}} (4x^3 - 2x)\, dx \right|$$

Notice that we placed absolute values on each of the definite integrals. This will guarantee that the value returned for each region is that region's area instead of its *signed* area.

$$A = \left| \int_{-\sqrt{1/2}}^{0} (4x^3 - 2x)\, dx \right| + \left| \int_{0}^{\sqrt{1/2}} (4x^3 - 2x)\, dx \right|$$

$$= \left| (x^4 - x^2)\, \right|_{-\sqrt{1/2}}^{0} \left| + \right| (x^4 - x^2)\, \right|_{0}^{\sqrt{1/2}} \left|$$

$$= \left| \left\{ [(0)^4 - (0)^2] - \left[\left(-\sqrt{\frac{1}{2}}\right)^4 - \left(-\sqrt{\frac{1}{2}}\right)^2 \right] \right\} \right| + \left| \left\{ \left[\left(\sqrt{\frac{1}{2}}\right)^4 - \left(\sqrt{\frac{1}{2}}\right)^2 \right] - [(0)^4 - (0)^2] \right\} \right|$$

$$= \left| \left[(0 - 0) - \left(\frac{1}{4} - \frac{1}{2} \right) \right] \right| + \left| \left[\left(\frac{1}{4} - \frac{1}{2} \right) - (0 - 0) \right] \right|$$

$$= \left| 0 - \left(-\frac{1}{4} \right) \right| + \left| \left(-\frac{1}{4} - 0 \right) \right|$$

$$= \frac{1}{2} \text{ square unit}$$

The area of the region bounded by $f(x) = 4x^3 - 2x$ and the x-axis is $\frac{1}{2}$ square unit.

The relationship between a function and its antiderivative becomes especially meaningful if the function represents a rate of change.

ACCUMULATED CHANGE OF A FUNCTION

Let f be the rate of change function (derivative) of F on $[a, b]$. Then $\int_a^b f(x) \, dx = F(b) - F(a)$ is the accumulated change in F over the interval $[a, b]$.

EXAMPLE 5

Using the Fundamental Theorem of Calculus to Calculate an Accumulated Change In Revenue

Based on data from 1999 to 2001, the marginal revenue of the Coca-Cola Company may be modeled by

$$m(s) = -0.8058s + 14.44 \text{ dollars per unit case}$$

where s is the number of unit cases (in billions). A unit case is equivalent to 24 eight-ounce servings of finished beverage. In 1999, 16.5 billion unit cases were sold. In 2001, 17.8 billion unit cases were sold. (**Source:** Modeled from Coca-Cola Company 2001 Annual Report, pp. 46, 57.) By how much did the revenue of Coca-Cola Company grow as the number of unit cases sold increased from 16.5 billion to 17.8 billion?

SOLUTION Marginal revenue is the rate of change in the revenue function. The accumulated change in revenue is given by

$$\int_{16.5}^{17.8} (-0.8058s + 14.44) \, ds = (-0.4029s^2 + 14.44s) \Big|_{16.5}^{17.8}$$

$$= [-0.4029(17.8)^2 + 14.44(17.8)] - [-0.4029(16.5)^2 + 14.44(16.5)]$$

$$= (-127.65 + 257.03) - (-109.69 + 238.26)$$

$$= (129.38) - (128.57)$$

$$= 0.81$$

Increasing the number of cases sold from 16.5 billion to 17.8 billion increased revenue by 0.8 billion dollars. (The raw data were accurate to the nearest tenth of a billion dollars.)

Let's look at the functions m and M graphically. The graph of m is a line, as shown in Figure 6.28. As previously shown in Section 6.4, the area of the shaded region between the graph of m and the horizontal axis is 0.8.

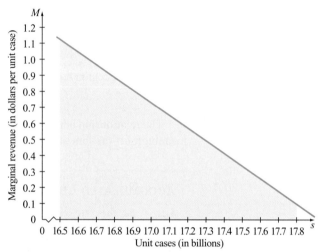

FIGURE 6.28

The graph of M is a parabola (Figure 6.29).

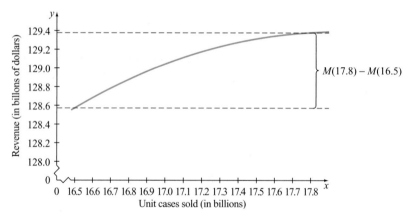

FIGURE 6.29

The vertical distance between the point $[17.8, M(17.8)]$ and the point $[16.5, M(16.5)]$ is given by $M(17.8) - M(16.5)$.

$$M(17.8) - M(16.5) = 129.38 - 128.57$$
$$= 0.81$$

Amazingly, the area between the graph of m and the horizontal axis on the interval $[16.5, 17.8]$ is easily calculated by finding the difference $M(17.8) - M(16.5)$, where M is an antiderivative of m.

A common error of beginning calculus students is to assume that the revenue earned by the company was 0.8 billion dollars. The value 0.8 billion dollars is the accumulated *change* in revenue, not the revenue value itself.

EXAMPLE 6

Using the Fundamental Theorem of Calculus to Calculate an Accumulated Change in Sales

Based on data from 1993 to 2002, the rate of change in the sales income of the Starbucks Corporation may be modeled by

$$r(t) = 58.46t + 79.33 \text{ million dollars per year}$$

where t is the number of years since 1993.

According to the model, by how much did the sales income increase between the end of 1996 and the end of 2002?

SOLUTION   We are asked to find $\int_3^9 (58.46t + 79.33) \, dt$.

$$\int_3^9 (58.46t + 79.33) \, dt = (29.23t^2 + 79.33t) \Big|_3^9$$

$$= [29.23(9)^2 + 79.33(9)] - [29.23(3)^2 + 79.33(3)]$$

$$= 3081.6 - 501.06$$

$$\approx 2580 \text{ million dollars}$$

Between the end of 1996 and the end of 2002, the sales income increased by 2580 million dollars. (Based on the raw data, the actual change was 2592 million. Our model estimate was off by about $12 million. Although that may seem like a lot, our model estimate was within 0.5 percent of the actual value.)

Changing Limits of Integration

When the integrand function in a definite integral requires integration by substitution, the limits of integration must be changed along with the variable. This technique will be demonstrated in Examples 7 and 8.

EXAMPLE 7

Changing the Limits of Integration of a Definite Integral

Calculate $\int_0^1 x^2(x^3 + 1)^3 \, dx$.

SOLUTION   Let $u = x^3 + 1$. Then

$$\frac{d}{dx}(u) = \frac{d}{dx}(x^3 + 1)$$

$$\frac{du}{dx} = 3x^2$$

$$du = 3x^2 dx$$

$$\frac{du}{3} = x^2 dx$$

Therefore,

$$\int_0^1 x^2(x^3 + 1)^3 \, dx = \int_a^b u^3 \left(\frac{du}{3}\right)$$

$$= \frac{1}{3} \int_a^b u^3 du$$

What are the values of a and b? Recall that the limits of integration of the original function were $x = 0$ and $x = 1$. We will substitute these values of x into the equation relating u and x to find the new limits of integration. Since $u = x^3 + 1$, the new limits of integration are $a = 0^3 + 1$ and $b = 1^3 + 1$. Therefore,

$$\int_0^1 x^2(x^3 + 1)^3 \, dx = \frac{1}{3} \int_1^2 u^3 du$$

$$= \frac{1}{3} \frac{u^4}{4} \Big|_1^2$$

$$= \frac{2^4}{12} - \frac{1^4}{12}$$

$$= 1.25$$

EXAMPLE **8**   **Changing the Limits of Integration of a Definite Integral**

Calculate $\displaystyle\int_1^3 \frac{\ln(x)}{x} \, dx$.

SOLUTION   At first, the function does not appear to be integrable; however, if we write $f(x) = \frac{\ln(x)}{x}$ as $f(x) = \ln(x) \cdot \frac{1}{x}$, we recognize that f is of the form $u \cdot u'$. Let $u = \ln x$. Then

$$\frac{d}{dx}(u) = \frac{d}{dx}(\ln x)$$

$$\frac{du}{dx} = \frac{1}{x}$$

$$du = \frac{1}{x} dx$$

Therefore, $\displaystyle\int_1^3 \left(\ln x \cdot \frac{1}{x}\right) dx = \int_a^b u \, du$. What are the values of a and b? Recall that the limits of integration of the original function were $x = 1$ and $x = 3$. We will substitute these values of x into the equation relating u and x to find the new limits of integration. Since $u = \ln(x)$, the new limits of integration are $a = \ln(1)$ and $b = \ln(3)$. Since $\ln(1) = 0$, $a = 0$. Therefore,

$$\int_1^3 \frac{\ln(x)}{x}\,dx = \int_0^{\ln(3)} u\,du$$

$$= \frac{u^2}{2}\bigg|_0^{\ln(3)}$$

$$= \frac{[\ln(3)]^2}{2} - \frac{(0)^2}{2}$$

$$= \frac{[\ln(3)]^2}{2}$$

$$\approx 0.6035$$

In Section 6.4, we showed how to use the TI-83 Plus to determine the value of a definite integral graphically. In the following Technology Tip, we will demonstrate an alternative way to calculate the definite integral.

TECHNOLOGY **TIP**

Calculating a Definite Integral with *fnInt*

1. Press (MATH) to bring up the MATH menu. Select 9:fnInt. Press (ENTER).

```
MATH NUM CPX PRB
4↑³√(
5: ˣ√
6:fMin(
7:fMax(
8:nDeriv(
9█fnInt(
0:Solver...
```

2. The calculator program *fnInt* requires four values: the function, the variable of integration, the lower limit, and the upper limit. Enter each of the values and press (ENTER). To calculate $\int_1^3 \frac{\ln(x)}{x}\,dx$, we enter *fnInt* $(\ln(x)/x, x, 1, 3)$.

```
fnInt(ln(X)/X,X,
1,3)
```

3. The calculator displays the value of the definite integral.

```
fnInt(ln(X)/X,X,
1,3)
        .6034744804
```

6.5 Summary

In this section, you learned how to use one of the most powerful tools in calculus: the Fundamental Theorem of Calculus. You used this theorem to calculate a bounded area quickly and to determine the accumulated change of a rate of change function. In future sections, you will continue to use this powerful tool.

6.5 Exercises

In Exercises 1–20, use the Fundamental Theorem of Calculus to calculate the definite integral.

1. $\int_5^7 1 \, dx$

2. $\int_2^5 3 \, dx$

3. $\int_0^4 x \, dx$

4. $\int_0^3 (-x + 3) \, dx$

5. $\int_1^2 (-3x^2 + 5) \, dx$

6. $\int_1^3 (4t^2 - 4t) \, dt$

7. $\int_{-1}^1 (8x^3 + 2x) \, dx$

8. $\int_{-1}^1 (5t^4 - 4t^3) \, dt$

9. $\int_1^3 \left(\frac{1}{x}\right) dx$

10. $\int_1^3 \left(\frac{4}{x}\right) dx$

11. $\int_2^5 (2^t) \, dt$

12. $\int_{-3}^3 (3^t) \, dt$

13. $\int_{-2}^4 (5 \cdot 2^x) \, dx$

14. $\int_{-4}^{-2} (7 \cdot 3^x) \, dx$

15. $\int_0^1 \left(\frac{2x}{x^2 + 1}\right) dx$

16. $\int_1^4 \left(\frac{e^x}{e^x + 1}\right) dx$

17. $\int_0^2 (x^2 - 1)(x^3 - 3x)^3 \, dx$

18. $\int_{-1}^1 (3x + 4) \, dx$

19. $\int_{-2}^{-1} (-2x + 1) \, dx$

20. $\int_{-2}^3 (-x + 5) \, dx$

In Exercises 21–30, calculate the area of the region bounded by the graph of the function and the x-axis. You may find it helpful to graph the function before attempting each exercise.

21. $h(x) = x^3 + x^2 - 12x$

22. $g(x) = -x^2 + 4$

23. $f(x) = x^3 - x^2$

24. $g(x) = x^4 - 4x^2$

25. $f(x) = x^3 - 1$ between $x = -1$ and $x = 1$

26. $h(x) = x^2 - 4$ between $x = -1$ and $x = 1$

27. $g(x) = 1 - 3^x$ between $x = 0$ and $x = 2$

28. $f(x) = e^x - x^2$ between $x = 0$ and $x = 2$

29. $h(x) = x - x^2$ between $x = -1$ and $x = 2$

30. $g(x) = x^3 - \frac{1}{x}$ between $x = 1$ and $x = 2$

In Exercises 31–35, use the Fundamental Theorem of Calculus to find the answer to the question.

31. **M** **AP Calculus Exam** Based on data from 1969 to 2002, the rate of change in the number of AP Calculus AB exam participants may be modeled by

$$R(t) = 315.70t - 770.64$$

people per year, where t is the number of years since the end of 1969. (**Source:** Modeled from The College Board data.)

If the model is an accurate predictor of the future, by how much will the number of people taking the exam each year increase between the end of 2002 and the end of 2012?

32. **M** **Bottled Water Consumption** Based on data from 1980 to 1999, the rate of change in the annual per capita consumption of bottled water may be modeled by

$$R(t) = 0.2612(1.106)^t$$

gallons per person per year, where t is the number of years since the end of 1980. (**Source:** Modeled from *Statistical Abstract of the United States, 2001,* Table 204, p. 130.)

Between the end of 1995 and the end of 1999, by how much did per capita bottled water consumption increase?

33. **M** **Minivan Position** The velocity of the author's minivan as he drove from a traffic light onto a highway may be modeled by

$$V(t) = -0.1773t^2 + 0.6798t + 0.5019$$

miles per minute, where t is time in minutes. How far did he travel in the first 1.5 minutes of his trip?

34. **M** **McDonald's Sales** Based on data from 1990 to 2001, the rate of change of franchised sales of McDonald's Corporation may be modeled by

$$R(t) = -82.586t + 1729.0$$

million dollars per year, where t is the number of years since the end of 1990. (**Source:** Modeled from www.mcdonalds.com data.)

Between the end of 1990 and the end of 2001, by how much did McDonald's franchised sales increase?

35. **M** **Mozzarella Cheese Consumption** Based on data from 1980 to 1999, the rate of change in the annual per capita consumption of mozzarella cheese may be modeled by

$$R(t) = 0.0009778t^3 - 0.02958t^2 + 0.2392t - 0.06311$$

pounds per person per year, where t is the number of years since the end of 1980. (**Source:** Modeled from *Statistical Abstract of the United States, 2001*, Table 202, p. 129.)

Between the end of 1990 and the end of 1999, by how much did per capita mozzarella cheese consumption change?

Exercises 36–40 are intended to challenge your understanding of the Fundamental Theorem of Calculus.

36. A function $F(x)$ has the properties that $F'(x) = 3x^2 + 2x + 1$ and $F(1) = 5$. Find $F(2)$.

37. A function $F(x)$ has the properties that $F'(x) = \dfrac{1}{x}$ and $F(1) = 1 + \ln(4)$. Find $F(4)$.

38. The velocity of a car is modeled by the function $v(t) = 66$ miles per hour, where t is the number of hours that the car has been traveling. Calculate $\displaystyle\int_{0.5}^{1} v(t)\,dt$ and interpret the meaning of the result.

39. Calculate the area of the region bounded by the graph of $g(x) = 2^x$ and the graph of $h(x) = x^2$. (*Hint:* The graphs of g and h intersect in exactly three places.)

40. Find a function $f(x) \neq 0$ such that $\displaystyle\int_{-a}^{a} f(x)\,dx = 0$ on any interval $[-a, a]$. Then discuss the relationship between $F(-a)$ and $F(a)$ given that $F'(x) = f(x)$.

Chapter 6 Review Exercises

Section 6.1 *In Exercises 1–4, find the general antiderivative of the function.*

1. $g(x) = -2x + 5$

2. $f(x) = 3^x + x - 1$

3. $h(x) = 4x - \dfrac{3}{x}$

4. $g(x) = x^4 - 4x^3$

In Exercises 5–8, calculate the indefinite integral.

5. $\displaystyle\int (4x^3 - 5x - 1)\,dx$

6. $\displaystyle\int \frac{u + 2}{u}\,du$

7. $\displaystyle\int \frac{2t + 1}{2t}\,dt$

8. $\displaystyle\int (x^2 - 2x + 3)\,dx$

In Exercises 9–10, apply the concept of integration in solving the real-world applications.

9. **Publishing Costs** Golden Pillar Publishing (www.goldenpillarpublishing.com) publishes books for aspiring authors. In 2003, the marginal printing cost for a 8.25″ × 11″ soft-cover book was

$$c(p) = 0.018$$

dollars per page, where p is the number of pages in the book. Including setup costs, a 100-page book had a $3.10 production cost. Find the book production cost function.

10.   **Free Fall**   The Giant Drop is one of the most popular amusement park rides manufactured by Intamin. Riders are hoisted to a height of 227 feet and then dropped into a free fall. (**Source:** www.sixflags.com.)

The velocity of an object that is dropped into a free fall may be modeled by

$$v(t) = -32t$$

feet per second, where t is the number of seconds since the object was dropped. A negative velocity indicates that the object is moving toward the ground. Given that position is the antiderivative of velocity, find the position function for the free-fall portion of the Giant Drop thrill ride.

Section 6.2   *In Exercises 11–15, integrate the functions. Some integrals will require integration by substitution, while others may be integrated using basic integration rules.*

11. $f(x) = x(x^2 - 9)^5$ **12.** $f(t) = 2t\sqrt{t - 4}$

13. $h(x) = \dfrac{2\ln x}{x}$ **14.** $f(x) = \dfrac{2x}{x^2 + 1}$

15. $g(t) = \dfrac{6 + t}{2t}$

Section 6.3   *In Exercises 16–20, use the left-hand sum to estimate the area between the graph of the function and the horizontal axis on the specified interval. For each exercise, calculate the sum with $n = 2$, $n = 4$, and $n = 10$ rectangles.*

16. $f(x) = x^2 - 1; [1, 5]$

17. $g(x) = -4x + 16; [0, 4]$

18. $h(x) = \ln x; [1, 2]$ **19.** $f(t) = t - \dfrac{1}{t}; [3, 5]$

20. $g(t) = \sqrt{t + 2}; [0, 1]$

In Exercises 21–25, use left- and right-hand sums (with $n = 4$) to estimate the area between the graph of the function and the horizontal axis on the specified interval. In each exercise, calculate the left-hand sum, the right-hand sum, and the average of the two sums. The exact area, A, is given so that you can compare your estimates to the actual area.

21. $f(t) = 3t^2 - 2t + 1$ on $[0, 1]; A = 1$

22. $g(x) = x - \sqrt{x}$ on $[1, 4]; A = 2\dfrac{5}{6}$

23. $h(x) = 2x + \dfrac{1}{x}$ on $[1, 2]; A = 3 + \ln 2 \approx 3.693$

24. $g(x) = \sqrt{x + 4}$ on $[0, 5]; A = 12\dfrac{2}{3}$

25. $g(x) = -3x^2 + 3$ on $[0, 1]; A = 2$

In Exercises 26–27, estimate the area between the graph of the function and the horizontal axis. Then interpret the physical meaning of the result.

26.   **Free Fall**   The Giant Drop is one of the most popular amusement park rides manufactured by Intamin. Riders are hoisted to a height of 227 feet and then dropped into a free fall. (**Source:** www.sixflags.com.)

The velocity of an object that is dropped into a free fall may be modeled by

$$v(t) = -32t$$

feet per second, where t is the number of seconds since the object was dropped. A negative velocity indicates that the object is moving toward the ground. Estimate the area of the region between the graph of v and the horizontal axis on the interval $[0, 2]$ and interpret the meaning of the result. (*Hint:* The function v is entirely negative on the interval. Consequently, the left- and right-hand sums will yield a negative value. Since the entire region is below the horizontal axis, you can simply drop the negative sign to get the area estimate.)

27.   **River Flow**   Between 5:00 p.m. and 9:45 p.m. on November 22, 2002, the stream flow rate of the Columbia River below the Priest Rapids Dam increased dramatically. The flow rate during that time period may be modeled by

$$f(t) = 714.05t^3 - 5282.6t^2 + 21{,}738t + 61{,}656$$

cubic feet per second, where t is the number of hours since 5:00 p.m. (**Source:** waterdata.usgs.gov.)

(a) Convert the units of the flow-rate function from cubic feet per second to cubic feet per hour.

(b) Estimate the area of the region between the graph of f and the horizontal axis on the interval $[0, 4.75]$ and interpret its meaning.

(c) Depending upon the pool dimensions, an Olympic-size swimming pool can hold between 94,000 and 114,000 cubic feet of water. How many 100,000-cubic-foot swimming pools could be filled with the volume of water that flowed past the checkpoint below the Priest Rapids Dam on the Columbia River between 5:00 p.m. and 9:45 p.m.?

Section 6.4   *In Exercises 28–30, calculate the sum.*

28. $\displaystyle\sum_{i=1}^{6} i^2 - 2i + 1$ **29.** $\displaystyle\sum_{i=3}^{7} (i-4)^2$

30. $\displaystyle\sum_{i=3}^{7} 4(2i-9)^2$

In Exercises 31–35, draw the graph of the integrand function. Then calculate the definite integral using geometric formulas for area.

31. $\displaystyle\int_{2}^{5} 2x\,dx$ **32.** $\displaystyle\int_{0}^{2} (-3x+6)\,dx$

33. $\displaystyle\int_{0}^{6} (-|x-3|+3)\,dx$

34. $\displaystyle\int_{1}^{2} (-2x+4)\,dx$ **35.** $\displaystyle\int_{4}^{6} 2\,dx$

In Exercises 36–40, use the definite integral properties to find the numeric value of each definite integral, given that $\displaystyle\int_{1}^{3} f(x)\,dx = 2$, $\displaystyle\int_{1}^{4} f(x)\,dx = 5$,

$\displaystyle\int_{1}^{3} g(x)\,dx = 1$, *and* $\displaystyle\int_{3}^{4} g(x)\,dx = 3$.

36. $\displaystyle\int_{3}^{4} [2f(x)]\,dx$ **37.** $\displaystyle\int_{1}^{4} [3g(x)]\,dx$

38. $\displaystyle\int_{3}^{4} [2f(x)+4g(x)]\,dx$

39. $\displaystyle\int_{1}^{4} [f(x)-g(x)]\,dx$

40. $\displaystyle\int_{1}^{4} [4f(x)-5g(x)]\,dx$

In Exercises 41–42, shade the regions of the graph that will be used in calculating the definite integral.

41. $\displaystyle\int_{1}^{2} \left(\frac{x}{x^2+1}\right) dx$

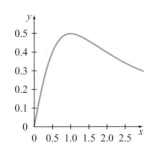

42. $\displaystyle\int_{1}^{2} \left(\frac{x^2+2}{x^2+1}\right) dx$

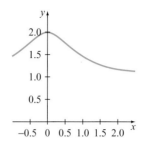

In Exercises 43–44, use the properties of definite integrals to find the solution.

43. **Milk Consumption**   Based on data from 1985 to 1999, the rate of change in the annual per capita consumption of milk as a beverage may be modeled by

$$R(t) = 0.001326t^2 - 0.02192t - 0.1541$$

gallons per person per year, where t is the number of years since the end of 1985. (**Source:** Modeled from *Statistical Abstract of the United States, 2001, Table 204*, p. 130.) The graph of $R(t)$ is shown.

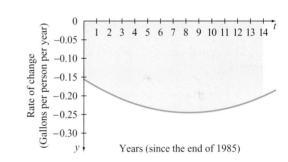

The signed area of the shaded region of the graph is

$$\int_{0}^{14} (0.001326t^2 - 0.02192t - 0.1541)\,dt$$

and is equal to -3.093. Additionally,

$$\int_{0}^{8} (0.001326t^2 - 0.02192t - 0.1541)\,dt$$
$$= -1.708$$

(a) Calculate and interpret the meaning of

$$\int_{8}^{14} (0.001326t^2 - 0.02192t - 0.1541)\,dt$$

(b) In 1993, California milk processors launched the "Got milk?" advertising campaign in an effort to turn around the 20-year trend of declining milk consumption. (**Source:** www.gotmilk.com.) Considering the results of part (a) together with the graph of R, do you think that the "Got milk?" advertising campaign has been successful?

44. **M** **Personal Income**   Based on data from 1993 to 2000, the rate of change in the annual per capita personal income in New Mexico may be modeled by

$$R(t) = 8.4546t^2 + 45.944t + 723.00$$

dollars per person per year, where t is the number of years since the end of 1993. (**Source:** Modeled from Bureau of Economic Analysis data.) The graph of $R(t)$ is shown.

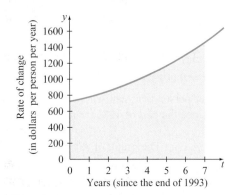

The signed area of the shaded region of the graph is

$$\int_0^7 (8.4546t^2 + 45.944t + 723.00)\,dt$$

and is equal to 7153. Additionally,

$$\int_4^7 (8.4546t^2 + 45.944t + 723.00)\,dt = 3713$$

Calculate and interpret the meaning of

$$\int_0^4 (8.4546t^2 + 45.944t + 723.00)\,dt.$$

Section 6.5   *In Exercises 45–50, use the Fundamental Theorem of Calculus to calculate the definite integral.*

45. $\displaystyle\int_2^3 (3x + 1)\,dx$

46. $\displaystyle\int_{-2}^2 (x^2)\,dx$

47. $\displaystyle\int_{-1}^1 (-x^3)\,dx$

48. $\displaystyle\int_0^1 (3e^x + 3x^2)\,dx$

49. $\displaystyle\int_2^3 \frac{2x}{x^2 + 1}\,dx$

50. $\displaystyle\int_1^2 \frac{\ln x}{x}\,dx$

In Exercises 51–55, calculate the area of the region bounded by the graph of the function and the x-axis. You may find it helpful to graph the function before attempting the exercise.

51. $h(x) = x^3 - 8$ between $x = 1$ and $x = 5$

52. $g(x) = 2.5 - \dfrac{1}{x} - x$ between $x = 0.5$ and $x = 4$

53. $f(x) = \dfrac{2x}{x^2 + 1}$ between $x = -3$ and $x = 3$

54. $f(t) = 2^t - t^2$ between $t = 0$ and $t = 4$

55. $g(x) = (x^2 - 1)(x^3 - 3x)^3$ between $x = -\sqrt{3}$ and $x = \sqrt{3}$.

In Exercises 56–57, use the Fundamental Theorem of Calculus to determine the answer to the question.

56. **M** **Personal Income**   Based on data from 1993 to 2000, the rate of change in the annual per capita personal income in Louisiana may be modeled by

$$R(t) = -28.940t + 883.18$$

dollars per person per year, where t is the number of years since the end of 1993. (**Source:** Modeled from Bureau of Economic Analysis data.)
Between the end of 1995 and the end of 2000, by how much did the annual per capita personal income increase?

57. **M** **Personal Income**   Based on data from 1993 to 2000, the rate of change in the annual per capita personal income in Florida may be modeled by

$$R(t) = -18.914t^2 + 209.16t + 645.4$$

dollars per person per year, where t is the number of years since the end of 1993. (**Source:** Modeled from Bureau of Economic Analysis data.)
Between 1995 and 2000, by how much did the annual per capita personal income increase?

Make It Real

What to do

1. Collect at least five years' worth of data that are reported annually (for example, annual income or annual tuition cost).

2. Find a model for the data. This model represents the rate of change in the cumulative sum of the data (for example, the rate of change in one's total earned income in dollars per year).

3. Find the antiderivative of the model equation.

4. Use the Fundamental Theorem of Calculus to determine the projected accumulated change over a period of at least four years, including at most one future year.

5. Interpret the meaning of the results from Step 4 and explain why the information is useful.

Example

Data Points

Johnson and Johnson Pharmaceutical Operating Profit

Years (Since 1997) (t)	Operating Profit (millions of dollars per year) (P)
0	2,332
1	3,114
2	3,735
3	4,394
4	4,928

Meaning

From the end of 1998 through 2002, the cumulative operating profit earned by Johnson and Johnson is projected to increase by 17,350 million dollars. When making investment choices, long-term investors often look at cumulative profit over time instead of focusing on fluctuations in annual profits.

Fundamental Theorem

$$A(5) - A(1) = 20{,}070 - 2719.2$$
$$\approx 17{,}350$$

Model

$$P = -32.71t^2 + 778.1t + 2341$$

Antiderivative

$$A = -10.90t^3 + 389.1t^2 + 2341t$$

405

Chapter 7

Advanced Integration Techniques and Applications

Fashion trends come and go. When a new product is first introduced to the market, sales are often slow. However, as the popularity of the product increases, sales increase rapidly. After a period of time, the market becomes saturated with the product, and sales taper off. The cumulative sales of a product that demonstrates the type of sales growth just described may often be modeled by a logistic function.

7.1 Integration by Parts
- Use the integration by parts method to find the antiderivative of a function

7.2 Area Between Two Curves
- Calculate the area of the region between two continuous curves

7.3 Differential Equations and Applications
- Solve introductory-level differential equations
- Apply Newton's Law of Heating and Cooling
- Analyze and interpret the real-life meaning of differential equations

7.4 Differential Equations: Limited Growth and Logistic Models
- Find the equation of a limited growth model
- Find the equation of a logistic growth model
- Analyze and interpret the real-life meaning of differential equations

7.1 Integration by Parts

■ Use the integration by parts method to find the antiderivative of a function

GETTING ***STARTED***   As we saw in Chapter 6, many derivative rules have a corresponding integral rule. Although there is no product rule for integrals, the *integration by parts method* is directly related to the product rule for derivatives.

In this section, we will demonstrate how to use the method of integration by parts to integrate a function. This technique is among the more challenging methods to use, so we will use it only after basic integration methods and the substitution method have failed.

Recall that the Product Rule for derivatives states that

$$\frac{d}{dx}[u(x) \cdot v(x)] = u'(x) \cdot v(x) + v'(x) \cdot u(x)$$

where u and v are differentiable functions. If we integrate both sides of the equation with respect to x, we get

$$\int \left\{ \frac{d}{dx}[u(x) \cdot v(x)] \right\} dx = \int [u'(x) \cdot v(x) + v'(x) \cdot u(x)] dx$$

$$u(x) \cdot v(x) = \int [u'(x) \cdot v(x)] dx + \int [v'(x) \cdot u(x)] dx$$

$$= \int \left[\frac{du}{dx} \cdot v(x) \right] dx + \int \left[\frac{dv}{dx} \cdot u(x) \right] dx$$

$$u(x) \cdot v(x) = \int v(x) \, du + \int u(x) \, dv$$

For notational ease, we will write $u(x)$ as u and $v(x)$ as v, keeping in mind that both u and v are functions of x.

$$uv = \int v \, du + \int u \, dv$$

We'll solve this equation for $\int u \, dv$.

$$\int u \, dv = uv - \int v \, du$$

This equation is the basis of the integration by parts method.

INTEGRATION BY PARTS

Let u and v be differentiable functions of x. Then

$$\int u \, dv = uv - \int v \, du$$

Consider the integral $\int (xe^x) \, dx$. We attempted to integrate this function by substitution in Section 6.2, but we were unable to do so. We will now attempt to integrate the function using integration by parts. In order to use this method, we must first identify a function u and a function v' such that $uv' = xe^x$. We typically will pick the function with the simplest derivative to be u or the most complex function that may be easily integrated to be v'. In this case, we pick

$$u(x) = x \qquad \text{and} \qquad v'(x) = e^x$$

Next we'll differentiate u with respect to x and, using the concept of differentials, solve for du.

$$\frac{d}{dx}(u) = \frac{d}{dx}(x)$$

$$\frac{du}{dx} = 1$$

$$du = 1 \ dx$$

Next we'll integrate both sides of the equation $v'(x) = e^x$ with respect to x.

$$v'(x) = e^x$$

$$\int v'(x) \, dx = \int e^x dx$$

$$v(x) = e^x \qquad \text{We will add in the constant } C \text{ at the end of the problem}$$

We'll now return to the formula $\int u \, dv = uv - \int v \, du$ and plug in the corresponding pieces. (Recall that, by definition, $dv = v'(x) \, dx$. In this case, $dv = e^x dx$.)

$$\int u \, dv = uv - \int v \, du$$

$$\int (x)(e^x dx) = (x) \cdot (e^x) - \int (e^x)(dx)$$

$$= xe^x - \int e^x \, dx$$

$$= xe^x - e^x + C$$

(In this example, as in previous examples, the constant C represents the sum of all constants generated throughout the integration process.)

Let's check our work by differentiating $F(x) = xe^x - e^x + C$.

$$\frac{d}{dx}[F(x)] = \frac{d}{dx}(xe^x - e^x + C)$$

$$f(x) = [1 \cdot e^x + (e^x) \cdot (x)] - e^x + 0$$

$$= e^x + xe^x - e^x$$

$$= xe^x$$

The result checks out, so we did the problem correctly.

Now that we've demonstrated the basic theory that surrounds the integration by parts method, we will refine the process to increase our efficiency. We'll demonstrate the streamlined process in Example 1 and then detail the steps of the process.

EXAMPLE 1   **Integrating by Parts**

Integrate $f(x) = 3x(2^x)$ with respect to x.

SOLUTION We are asked to find $\int 3x(2^x) \, dx$. Since $\frac{d}{dx}(3x)$ is a constant, we will select $u = 3x$. Consequently, $dv = 2^x dx$. Observe that $\int 3x(2^x)\,dx$ may be rewritten as $\int u\,dv$. We are now ready to do integration by parts.

$u = 3x$	$dv = 2^x\,dx$	Choose u and dv
$du = 3\,dx$	$v = \dfrac{2^x}{\ln(2)}$	Differentiate u and integrate dv with respect to x

$$\int 3x(2^x)\,dx = \int u\,dv$$

$$= uv - \int v\,du$$

$$= (3x)\left[\frac{2^x}{\ln(2)}\right] - \int \frac{2^x}{\ln(2)}(3\,dx)$$

$$= \frac{3x(2^x)}{\ln(2)} - \frac{3}{\ln(2)}\int 2^x dx \qquad \text{Constant Multiple Rule}$$

$$= \frac{3x(2^x)}{\ln(2)} - \frac{3}{\ln(2)}\left[\frac{2^x}{\ln(2)}\right] + C \qquad \text{Exponential Rule}$$

$$= \left[\frac{2^x}{\ln(2)}\right]\left[3x - \frac{3}{\ln(2)}\right] + C \qquad \text{Factor out } \frac{2^x}{\ln(2)}$$

Therefore, $\int 3x(2^x)\,dx = \left[\dfrac{2^x}{\ln(2)}\right]\left[3x - \dfrac{3}{\ln(2)}\right] + C$.

HOW TO | **Detailed Steps for the Integration by Parts Method**

Let u and v be differentiable functions of x. To integrate $\int u\,dv$, do the following:

1. Identify u and dv. The product of these two factors must equal the entire integrand coupled with the associated dx.
2. Differentiate u and integrate dv with respect to x.
3. Write the expression $uv - \int v\,du$ in terms of x.
4. Integrate $\int v\,du$ to eliminate the integral sign.
5. Simplify.

We'll demonstrate these steps in the next several examples.

EXAMPLE 2

Integrating by Parts

Integrate $f(x) = x\ln(x)$.

SOLUTION We must find $\int x \ln(x)\, dx$. We'll pick $u = \ln(x)$ and $dv = x\, dx$.

$u = \ln(x)$	$dv = x\, dx$	Choose u and dv
$du = \dfrac{1}{x}\, dx$	$v = \dfrac{x^2}{2}$	Differentiate u and integrate dv with respect to x

$$uv - \int v\, du = \ln(x) \cdot \frac{x^2}{2} - \int \left(\frac{x^2}{2} \cdot \frac{1}{x} \right) dx$$

$$= \frac{1}{2} x^2 \ln(x) - \frac{1}{2} \int x\, dx \qquad \text{Constant Multiple Rule}$$

$$= \frac{1}{2} x^2 \ln(x) - \frac{1}{2} \left(\frac{1}{2} x^2 \right) + C \qquad \text{Power Rule}$$

$$= \frac{1}{2} x^2 \left[\ln(x) - \frac{1}{2} \right] + C \qquad \text{Factor out } \frac{1}{2} x^2$$

Therefore, $\int x \ln(x)\, dx = \frac{1}{2} x^2 \left[\ln(x) - \frac{1}{2} \right] + C.$

EXAMPLE 3

Integrating by Parts

Simplify $\int (2x + 1)(x - 3)^3 dx$.

SOLUTION We'll pick $u = 2x + 1$ and $dv = (x - 3)^3 dx$.

$u = 2x + 1$	$dv = (x - 3)^3 dx$	
$du = 2\, dx$	$v = \dfrac{(x - 3)^4}{4}$	By the substitution method

$$uv - \int v\, du = (2x + 1)\left[\frac{(x - 3)^4}{4} \right] - \int \left[\frac{(x - 3)^4}{4} \right](2\, dx)$$

$$= \frac{(2x + 1)(x - 3)^4}{4} - \frac{2}{4} \int (x - 3)^4 dx \qquad \text{Constant Multiple Rule}$$

$$= \frac{1}{4}(2x + 1)(x - 3)^4 - \frac{1}{2} \int (x - 3)^4 dx \qquad \text{Simplify fractions}$$

$$= \frac{1}{4}(2x + 1)(x - 3)^4 - \frac{1}{2}\left[\frac{(x - 3)^5}{5} \right] + C \qquad \text{Power Rule}$$

$$= \frac{1}{4}(2x + 1)(x - 3)^4 - \frac{1}{10}(x - 3)^5 + C \qquad \text{Simplify fractions}$$

$$= \frac{5}{20}(2x + 1)(x - 3)^4 - \frac{2}{20}(x - 3)^5 + C \qquad \text{Rewrite fractions}$$

$$= \frac{1}{20}(x - 3)^4[5(2x + 1) - 2(x - 3)] + C \qquad \text{Factor out } \frac{1}{20}(x - 3)^4$$

$$= \frac{1}{20}(x-3)^4(10x + 5 - 2x + 6) + C \qquad \text{Simplify}$$

$$= \frac{1}{20}(x-3)^4(8x + 11) + C \qquad \text{Group like terms}$$

Therefore, $\displaystyle\int (2x+1)(x-3)^3 dx = \frac{1}{20}(x-3)^4(8x+11) + C$.

EXAMPLE 4

Integrating by Parts

Simplify $\displaystyle\int (x^2)(x+1)^3 dx$.

SOLUTION Let $u = x^2$ and $dv = (x+1)^3\, dx$.

$u = x^2$	$dv = (x+1)^3\,dx$
$du = 2x\,dx$	$v = \dfrac{(x+1)^4}{4}$

By the substitution method

$$uv - \int v\,du = (x^2)\left[\frac{(x+1)^4}{4}\right] - \int\left[\frac{(x+1)^4}{4}\right](2x\,dx)$$

$$= \frac{1}{4}x^2(x+1)^4 - \frac{1}{2}\int x(x+1)^4\,dx \qquad \text{Constant Multiple Rule}$$

Observe that $\displaystyle\int x(x+1)^4 dx$ is not readily integrable. We'll integrate this piece separately using integration by substitution and then substitute the result back into the equation.

Let $w = x + 1$. Then $dw = dx$ and $x = w - 1$. Thus

$$\int x(x+1)^4 dx = \int (w-1)(w^4)\,dw$$

$$= \int (w^5 - w^4)\,dw$$

$$= \frac{w^6}{6} - \frac{w^5}{5} + C \qquad \text{Power Rule}$$

$$= \frac{5w^6}{30} - \frac{6w^5}{30} + C \qquad \text{Get a common denominator}$$

$$= \frac{1}{30}w^5(5w - 6) + C \qquad \text{Factor out } \tfrac{1}{30}w^5$$

$$= \frac{1}{30}(x+1)^5[5(x+1) - 6] + C \qquad \text{Replace } w \text{ with } x+1$$

$$= \frac{1}{30}(x+1)^5(5x + 5 - 6) + C \qquad \text{Simplify}$$

$$= \frac{1}{30}(x+1)^5(5x - 1) + C \qquad \text{Group like terms}$$

We'll substitute this result back into the integration by parts equation.

$$\frac{1}{4}x^2(x+1)^4 - \frac{1}{2}\int x(x+1)^4\,dx = \frac{1}{4}x^2(x+1)^4 - \frac{1}{2}\left[\frac{1}{30}(x+1)^5(5x-1)\right] + C$$

$$= \frac{15}{60}x^2(x+1)^4 - \frac{1}{60}(x+1)^5(5x-1) + C$$

$$= \frac{1}{60}(x+1)^4[15x^2 - (x+1)(5x-1)] + C$$

$$= \frac{1}{60}(x+1)^4[15x^2 - (5x^2 + 4x - 1)] + C$$

$$= \frac{1}{60}(x+1)^4(10x^2 - 4x + 1) + C$$

Therefore, $\displaystyle\int (x^2)(x+1)^3\,dx = \frac{1}{60}(x+1)^4(10x^2 - 4x + 1) + C$. Phew!

Although we used the substitution method to integrate the second integral in Example 4, we could have used integration by parts to integrate the second integral. In Example 5, we will apply the integration by parts method multiple times.

EXAMPLE 5

Integrating by Parts

Integrate $f(x) = x^2 e^x$.

SOLUTION   We're asked to find $\displaystyle\int x^2 e^x\,dx$. We'll pick $u = x^2$ and $dv = e^x\,dx$.

$u = x^2$	$dv = e^x\,dx$
$du = 2x\,dx$	$v = e^x$

$$uv - \int v\,du = (x^2)(e^x) - \int 2xe^x\,dx$$

$$= x^2 e^x - 2\int xe^x\,dx$$

The remaining integral, $\displaystyle\int xe^x\,dx$, is not readily integrable. We will integrate this function using integration by parts. We will select new values for u and dv; however, it is important to note that although the variables are the same, the functions are not equal to the u and v identified previously.

$u = x$	$dv = e^x\,dx$
$du = dx$	$v = e^x$

$$x^2 e^x - 2\int xe^x\,dx = x^2 e^x - 2\left(xe^x - \int e^x\,dx\right)$$

$$= x^2 e^x - 2(xe^x - e^x) + C$$

$$= x^2 e^x - 2xe^x + 2e^x + C$$

$$= e^x(x^2 - 2x + 2) + C$$

Integration by parts can also be used to integrate functions that formerly did not appear integrable. In Example 6, we will integrate the natural log function.

EXAMPLE 6

Integrating the Natural Logarithm Function

Simplify $\int \ln(x)\,dx$.

SOLUTION Let $u = \ln(x)$ and $dv = dx$.

$u = \ln(x)$	$dv = dx$
$du = \dfrac{1}{x}dx$	$v = x$

$$uv - \int v\,du = [\ln(x)](x) - \int x\left(\frac{1}{x}dx\right)$$
$$= x\ln(x) - \int 1\,dx$$
$$= x\ln(x) - x + C$$
$$= x[\ln(x) - 1] + C$$

Therefore, $\int \ln(x)\,dx = x[\ln(x) - 1] + C$.

EXAMPLE 7

Using Integration to Forecast Motor Vehicle Deaths

Based on data from 1980, 1985, 1990, and 1995, the number of motor vehicle deaths in South Carolina may be modeled by

$$m(t) = 843.3 + 65.92\ln(t)$$

where t is the number of years since 1979. According to the model, how many motor vehicle deaths occurred in South Carolina between 1980 and 1995?

SOLUTION Since $m(t)$ is the annual death rate, we must calculate $\int_1^{16}[843.3 + 65.92\ln(t)]\,dt$ to determine the cumulative number of deaths between 1980 and 1995.

$$\text{Deaths} = \int_1^{16}[843.3 + 65.92\ln(t)]\,dt$$

$$= 843.3t\,\Big|_1^{16} + 65.92\int_1^{16}\ln(t)\,dt \qquad \text{Power, Constant Multiple Rules}$$

$$= 843.3t\,\Big|_1^{16} + 65.92\{t[\ln(t) - 1]\}\Big|_1^{16} \qquad \text{Integral of natural log function}$$

$$= [843.3(16) - 843.3(1)]$$
$$\quad + 65.92\{16[\ln(16) - 1]\} - 65.92\{1[\ln(1) - 1]\} \qquad \text{Fundamental Theorem of Calculus}$$

$$= 12{,}649.5 + 1935.5$$

$$= 14{,}585$$

According to the model, the cumulative number of motor vehicle deaths in South Carolina between 1980 and 1995 was 14,585 deaths.

7.1 Summary

In this section, you learned how to use the method of integration by parts to integrate a function. You discovered that this technique is among the more challenging integration methods and should be used only after basic integration methods and the substitution method have failed.

7.1 Exercises

In Exercises 1–10, integrate the function using integration by parts.

1. $f(x) = xe^{-x}$

2. $g(x) = 5x \ln(x)$

3. $h(t) = (3t - 5) \ln(t)$

4. $f(t) = \dfrac{6}{t} + \ln(t)$

5. $h(t) = t^3 \ln(t)$

6. $h(t) = \dfrac{\ln(t)}{t} - 3$

7. $f(x) = 4xe^{2x+1}$

8. $f(x) = \ln(x^2)$

9. $h(x) = e^{2x}(2x - 1)$

10. $f(t) = (3t + 4)e^{2t}$

In Exercises 11–25, integrate the function using the simplest possible method.

11. $g(t) = \dfrac{3^t}{3^t - 1}$

12. $h(t) = (2t^2 - 9)(e^t)$

13. $f(x) = 3x(e^x - x)$

14. $g(x) = (4x - 1)[\ln(x)]$

15. $h(x) = 2x(x^2 - 9)$

16. $f(t) = (t - 1)^5(2t - 7)$

17. $g(t) = (3t^2)(t^3 - 1)$

18. $h(x) = x^6(3x - 4)$

19. $g(x) = 3^x(x^2 - 3)$

20. $g(x) = \dfrac{[\ln(x)]^2}{x}$

21. $h(x) = 3xe^{3x^2}$

22. $g(t) = t - t\ln(t)$

23. $f(t) = e^2 t^2$

24. $h(t) = (2t - 3) \ln(t)$

25. $f(x) = \dfrac{2^{\ln x}}{x}$

In Exercises 26–30, use the integration by parts method to determine the answer to the question.

26. Textbook Sales A new college textbook edition typically generates most of its sales in the year of its publication. Sales drop off in subsequent years as a result of competition from the used book market. Suppose that the annual sales of a particular textbook may be modeled by

$$S(t) = 30,000 \, te^{-1.5t}$$

textbooks, where t is the number of years since the edition was published. According to the model, how many textbooks will be sold in the first three years of the edition?

27. **M** **Real Networks Licensing Revenue** Based on data from 1999 to 2003, the annual net revenue from *software licensing fees* for Real Networks, Inc., may be modeled by

$$R(t) = 1208(t - 0.88)e^{-1.34t} + 56.7$$

million dollars, where t is the number of years since the end of 1998. (**Source:** Modeled from Real Networks, Inc., 2003 Report, p. 14.)

According to the model, what was the accumulated net revenue from software licensing fees between the end of 2000 and the end of 2003?

28. **M** **Real Networks Advertising Revenue** Based on data from 1999 to 2003, the annual net revenue from *advertising* for Real Networks, Inc., may be modeled by

$$A(t) = 1430(t - 0.95)e^{-1.89t} + 6$$

million dollars, where t is the number of years since the end of 1998. (**Source:** Modeled from Real Networks, Inc., 2003 Report, p. 14.)

According to the model, what was the accumulated net revenue from advertising between the end of 2000 and the end of 2003?

29. **M** **Electronic Arts Profit** Based on data from 2000 to 2004, the annual gross profit for Electronic Arts, Inc., may be modeled by

$$P(t) = 274(t - 0.6)e^{-0.232t} + 596$$

million dollars, where t is the number of years since the end of 1998. (**Source:** Modeled from Electronic Arts 2004 Annual Report, p. 19.)

According to the model, what was the accumulated profit between the end of 2000 and the end of 2004?

30. **M** **Real Networks Operating Expenses** Based on data from 1999 to 2003, the annual total operating expenses for Real Networks, Inc., may be modeled by

$$C(t) = 5970(t - 1.04)e^{-1.73t} + 150$$

million dollars, where t is the number of years since the end of 1998. (**Source:** Modeled from Real Networks, Inc., 2003 Report, p. 14.)

According to the model, what was the accumulated operating expense between the end of 1999 and the end of 2003?

Exercises 31–35 are intended to challenge your understanding of integration by parts.

31. Integrate $h(x) = 4x^2 e^x$.

32. Integrate $f(t) = 3^t \sqrt{t}$.

33. Integrate $g(t) = 6t^{0.5}\ln(t)$.

34. Explain the relationship between the integration by parts method for integrals and the Product Rule for derivatives.

35. Explain an effective strategy for determining when to use the integration by parts method.

7.2 Area Between Two Curves

- Calculate the area of the region between two continuous curves

GETTING STARTED Electronic Arts, Inc., is arguably the most dominant force in the electronic gaming software market. Sixteen of the games the company shipped in 2002 sold more than one million copies each. Its 2002 title "Harry Potter and the Sorcerer's Stone"™ generated more than $206 million in net revenues!

The rate of change in the net revenues at Electronic Arts, Inc., may be modeled by

$$r(t) = 162.840t^3 - 823.299t^2 + 961.788t + 65.839 \text{ million dollars per year}$$

and the rate of change in cost of goods sold may be modeled by

$$c(t) = 66.288t^3 - 328.464t^2 + 355.674t + 61.435 \text{ million dollars per year}$$

where t is the number of years since the end of fiscal year 1998. (**Source:** Modeled from Electronic Arts, Inc., Annual Report; March 31, 2002.) By how much did the company's gross profits increase between the end of fiscal year 1998 and the end of fiscal year 2002? This problem may be interpreted graphically by calculating the area of the region between the graph of r and the graph of c.

In this section, we will discuss how to calculate the area of the bounded region between two graphs. We will also explain the meaning of the definite integral of the difference between two rate of change functions.

AREA BETWEEN TWO CURVES

If the graph of f lies above the graph of g on an interval $[a, b]$, then the area of the region between the two graphs from $x = a$ to $x = b$ is given by

$$\int_a^b [f(x) - g(x)]\,dx$$

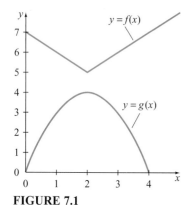

FIGURE 7.1

This result is relatively easy to verify graphically. Consider the graphs of the functions f and g shown in Figure 7.1. We see that on the interval $[0, 4]$, $f(x) > g(x)$. That is, the graph of f lies above the graph of g.

Consider $A = \int_1^3 [f(x) - g(x)]\,dx$. By the Sum and Difference Rule for integrals, we know that this equation is equivalent to

$$A = \int_1^3 f(x)\,dx - \int_1^3 g(x)\,dx$$
$$= (\text{area between } f \text{ and } x\text{-axis on } [1, 3])$$
$$- (\text{area between } g \text{ and } x\text{-axis on } [1, 3])$$

The shaded region in Figure 7.2 is the area between the graph of f and the x-axis on the interval $[1, 3]$.

The shaded region in Figure 7.3 is the area between the graph of g and the x-axis on the interval $[1, 3]$.

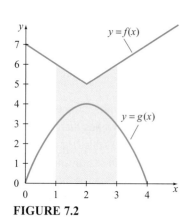

FIGURE 7.2

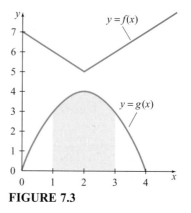

FIGURE 7.3

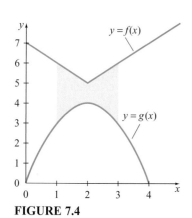

FIGURE 7.4

Subtracting the second area from the first area results in the area between the graphs of f and g on the interval $[1, 3]$ (Figure 7.4).

We'll do a few examples before returning to the Electronic Arts profit question.

EXAMPLE 1

Finding the Area of an Enclosed Region

Calculate the area of the region between the graphs of $f(x) = x^2 - 2x$ and $g(x) = x + 2$ on the interval $[0, 3]$.

SOLUTION   We begin by graphing the functions together and shading the appropriate region (see Figure 7.5).

We see that the graph of $g(x) = x + 2$ lies above the graph of $f(x) = x^2 - 2x$ on the interval $[0, 3]$. We will calculate the area of the region bounded by the two graphs.

$$\int_0^3 [g(x) - f(x)]\,dx = \int_0^3 [(x + 2) - (x^2 - 2x)]\,dx$$

$$= \int_0^3 (-x^2 + 3x + 2)\,dx$$

$$= \left(-\frac{1}{3}x^3 + \frac{3}{2}x^2 + 2x\right)\Big|_0^3$$

$$= \left[-\frac{1}{3}(3)^3 + \frac{3}{2}(3)^2 + 2(3)\right] - (0)$$

$$= -9 + \frac{27}{2} + 6$$

$$= \frac{21}{2}$$

$$= 10.5$$

The area of the region between the two graphs is 10.5 square units.

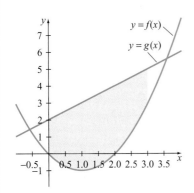

FIGURE 7.5

EXAMPLE 2

Finding the Area of an Enclosed Region

Calculate the area of the bounded region between the graphs of $f(x) = -x^2 + 2x + 1$ and $g(x) = x^2 - 2x + 1$.

SOLUTION   We begin by graphing the functions together and shading the bounded region (Figure 7.6).

It appears that the graphs intersect at $x = 0$ and $x = 2$; however, we must confirm this algebraically.

$$f(x) = g(x)$$
$$-x^2 + 2x + 1 = x^2 - 2x + 1$$
$$0 = 2x^2 - 4x$$
$$2x(x - 2) = 0$$
$$x = 0, x = 2$$

Our algebraic solution confirms our graphical observation.

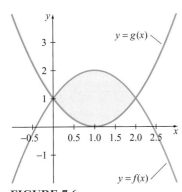

FIGURE 7.6

We see that the graph of f lies above the graph of g on the interval $[0, 2]$. We will calculate the area of the "eye-shaped" region between the two graphs.

$$A = \int_0^2 [f(x) - g(x)]\, dx$$

$$= \int_0^2 [(-x^2 + 2x + 1) - (x^2 - 2x + 1)]\, dx$$

$$= \int_0^2 (-x^2 + 2x + 1 - x^2 + 2x - 1)\, dx$$

$$= \int_0^2 (-2x^2 + 4x)\, dx$$

$$= \left(-\frac{2}{3}x^3 + 2x^2 \right)\Big|_0^2$$

$$= \left[-\frac{2}{3}(2)^3 + 2(2)^2 \right] - 0$$

$$= \frac{8}{3}$$

The area of the region between the two graphs is $2\frac{2}{3}$ square units.

EXAMPLE 3

Finding the Area of an Enclosed Region

Find the area of the bounded region between $f(x) = x^3 - 3x^2 + 3x$ and $g(x) = -x^2 + 4x - 2$.

SOLUTION We begin by graphing the functions together and shading the bounded regions (see Figure 7.7).

The graphs intersect in three different places: $x = -1$, $x = 1$, and $x = 2$. We can confirm our graphical observation by evaluating both f and g at these points and ensuring that the function values are equal at these x values.

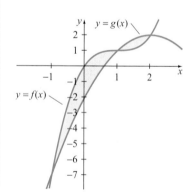

y = g(x)

y = f(x)

FIGURE 7.7

TABLE 7.1

x	$f(x) = x^3 - 3x^2 + 3x$	$g(x) = -x^2 + 4x - 2$
-1	-7	-7
1	1	1
2	2	2

Our computational results, shown in Table 7.1, confirm our graphical observation.

There are two bounded regions. The first bounded region occurs on the interval $[-1, 1]$. On this interval, the graph of f lies above the graph of g. The second bounded region occurs on the interval $[1, 2]$. On this interval, the graph of g lies above the graph of f. The combined area of the regions is given by

$$A = \int_{-1}^{1} [f(x) - g(x)]\, dx + \int_{1}^{2} [g(x) - f(x)]\, dx$$

$$= \int_{-1}^{1} [f(x) - g(x)]\, dx - \int_{1}^{2} [f(x) - g(x)]\, dx$$

Factor out a -1 to make the integrands equal

Since both integrands are equal, we will simplify the expression $f(x) - g(x)$ and substitute the result into each integrand.

$$f(x) - g(x) = x^3 - 3x^2 + 3x - (-x^2 + 4x - 2)$$
$$= x^3 - 3x^2 + 3x + x^2 - 4x + 2$$
$$= x^3 - 2x^2 - x + 2$$

$$A = \int_{-1}^{1} [f(x) - g(x)]\, dx - \int_{1}^{2} [f(x) - g(x)]\, dx$$

$$= \int_{-1}^{1} (x^3 - 2x^2 - x + 2)\, dx - \int_{1}^{2} (x^3 - 2x^2 - x + 2)\, dx$$

$$A = \left(\frac{1}{4}x^4 - \frac{2}{3}x^3 - \frac{1}{2}x^2 + 2x\right)\Big|_{-1}^{1} - \left(\frac{1}{4}x^4 - \frac{2}{3}x^3 - \frac{1}{2}x^2 + 2x\right)\Big|_{1}^{2}$$

$$= \left\{\left[\frac{1}{4}(1)^4 - \frac{2}{3}(1)^3 - \frac{1}{2}(1)^2 + 2(1)\right] - \left[\frac{1}{4}(-1)^4 - \frac{2}{3}(-1)^3 - \frac{1}{2}(-1)^2 + 2(-1)\right]\right\}$$

$$- \left\{\left[\frac{1}{4}(2)^4 - \frac{2}{3}(2)^3 - \frac{1}{2}(2)^2 + 2(2)\right] - \left[\frac{1}{4}(1)^4 - \frac{2}{3}(1)^3 - \frac{1}{2}(1)^2 + 2(1)\right]\right\}$$

$$= \left[\left(\frac{13}{12}\right) - \left(-\frac{19}{12}\right)\right] - \left[\left(\frac{8}{12}\right) - \left(\frac{13}{12}\right)\right]$$

$$= \frac{37}{12} = 3\frac{1}{12}$$

The combined area of the bounded regions between the two graphs is $3\frac{1}{12}$ square units.

Difference of Accumulated Changes

In Chapter 6, we saw that the Fundamental Theorem of Calculus had special meaning when working with a rate of change function f. If the units of f were miles per hour and t was in terms of hours, then $\int_{a}^{b} f(t)\, dt$ indicated that the total change in distance (in miles) between $t = a$ hours and $t = b$ hours was $F(b) - F(a)$. This notion may be applied to the difference of two rate of change functions as detailed in the following box.

DIFFERENCE OF TWO ACCUMULATED CHANGES

If f and g are continuous rate of change functions defined on the interval $[a, b]$, then the difference of the accumulated change of f from $x = a$ to $x = b$ and the accumulated change of g from $x = a$ to $x = b$ is given by

$$\int_{a}^{b} [f(x) - g(x)]\, dx$$

EXAMPLE 4

Calculating an Accumulated Change

Based on data from 1998 to 2002, the rate of change in the annual net revenues of Electronic Arts, Inc., may be modeled by

$$r(t) = 162.840t^3 - 823.299t^2 + 961.788t + 65.839 \text{ million dollars per year}$$

and the rate of change in annual cost of goods sold may be modeled by

$$c(t) = 66.288t^3 - 328.464t^2 + 355.674t + 61.435 \text{ million dollars per year}$$

where t is the number of years since the end of fiscal year 1998. (**Source:** Modeled from Electronic Arts, Inc., Annual Report, March 31, 2002.) By how much did the company's annual net revenues, cost of goods sold, and gross profits change between the end of fiscal year 1998 and the end of fiscal year 2002?

SOLUTION The total change in the annual net revenues of Electronic Arts, Inc., between the end of 1998 and the end of 2002 may be determined by calculating the sum of the signed areas of the shaded regions between the graph of r and the horizontal axis on the interval $[0, 4]$ (see Figure 7.8).

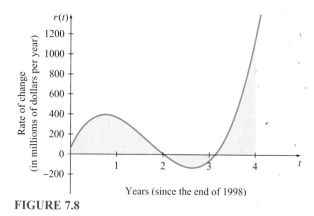

FIGURE 7.8

In other words, the change in the annual net revenues is given by

$$\int_0^4 r(t)\, dt = \int_0^4 (162.840t^3 - 823.299t^2 + 961.788t + 65.839)\, dt$$

$$= (40.710t^4 - 274.433t^3 + 480.894t^2 + 65.839t)\Big|_0^4$$

$$= [40.710(4)^4 - 274.433(4)^3 + 480.894(4)^2 + 65.839(4)] - (0)$$

$$= 10{,}421.76 - 17{,}563.712 + 7694.304 + 263.356$$

$$= 815.708 \text{ million dollars}$$

Between the end of fiscal year 1998 and the end of fiscal year 2002, annual net revenues increased by about $815,708,000.

The total change in the annual cost of goods sold between the end of fiscal year 1998 and the end of fiscal year 2002 may be determined by calculating the sum of the signed areas of the shaded regions between the graph of c and the horizontal axis on the interval $[0, 4]$ (see Figure 7.9).

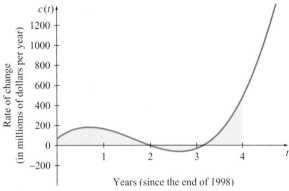

FIGURE 7.9

In other words, the total change in the annual cost of goods sold is given by

$$\int_0^4 c(t)\,dt = \int_0^4 (66.288t^3 - 328.464t^2 + 355.674t + 61.435)\,dt$$

$$= (16.572t^4 - 109.488t^3 + 177.837t^2 + 61.435t)\Big|_0^4$$

$$= \left[16.572(4)^4 - 109.488(4)^3 + 177.837(4)^2 + 61.435(4)\right] - (0)$$

$$= 4242.432 - 7007.232 + 2845.392 + 245.740$$

$$= 326.332 \text{ million dollars}$$

Between the end of fiscal year 1998 and the end of fiscal year 2002, the annual cost of goods sold increased by about $326,332,000.

The change in annual gross profit between the end of fiscal year 1998 and the end of fiscal year 2002 is given by

$$\int_0^4 [r(t) - c(t)]\,dt$$

since the rate of change in profit is the difference of the rate of change in revenue and the rate of change in cost. However,

$$\int_0^4 [r(t) - c(t)]\,dt = \int_0^4 r(t)\,dt - \int_0^4 c(t)\,dt$$

Therefore,

$$\int_0^4 [r(t) - c(t)]\,dt = 815.708 - 326.332$$

$$= 489.376 \text{ million dollars}$$

Between the end of fiscal year 1998 and the end of fiscal year 2002, the annual profit increased by about $489,376,000.

We have drawn the graphs of r and c together and shaded the bounded regions between the two graphs, as shown in Figure 7.10. These regions represent the accumulated change in annual profit.

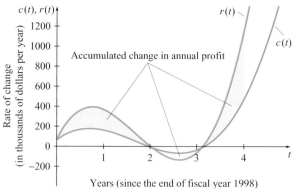

FIGURE 7.10

Note that although annual profits increased from the start of fiscal year 1998 through the end of fiscal year 2000 and from early in fiscal year 2002 through the end of fiscal year 2002, they decreased between the end of fiscal year 2000 and early in fiscal year 2002.

EXAMPLE 5

Calculating an Accumulated Change

Based on data from 1959 to 1989, the rate of change in the annual per capita income in Washington state may be modeled by

$$w(t) = -5.96t + 327.4 \text{ dollars per year}$$

and the rate of change in the per capita income in the United States may be modeled by

$$u(t) = 239.0 \text{ dollars per year}$$

where t is the number of years since the end of 1959. (**Source:** Modeled from www.census.gov.) In 1959, the per capita income in Washington state exceeded the national per capita income.

From the end of 1959 until the end of 1989, by how much did the pay gap between the Washington state and the national per capita income change?

SOLUTION We can calculate $\int_0^{30} w(t)\,dt$ and $\int_0^{30} u(t)\,dt$ independently and then calculate their difference, as was demonstrated in Example 4. However, it is often easier to calculate $\int_0^{30} [w(t) - u(t)]\,dt$.

$$w(t) - u(t) = (-5.96t + 327.4) - (239.0)$$
$$= -5.96t + 88.4$$

$$\int_0^{30} [w(t) - u(t)]\,dt = \int_0^{30} (-5.96t + 88.4)\,dt$$

$$= -2.98t^2 + 88.4t \Big|_0^{30}$$

$$= [-2.98(30)^2 + 88.4(30)] - [-2.98(0)^2 + 88.4(0)]$$
$$= -30$$

Between 1959 and 1989, the pay gap between the annual per capita income in Washington state and the national average annual per capita income changed by $30. Since the Washington state per capita income exceeded the national per capita income in 1959, the pay gap between the two was *reduced* by $30 between 1959 and 1989. (If the Washington state per capita income had been less than the national per capita income in 1959, the pay gap would have *increased* by $30 between 1959 and 1989.)

Let's look at the problem graphically. We begin by graphing both u and w (see Figure 7.11). The sum of the signed areas between the graph of u and the graph of w represents the change in the pay gap.

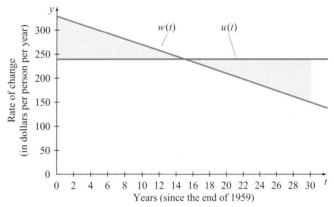

FIGURE 7.11

The sum of the signed areas of the shaded regions shown in the graph is -30.

Another way to look at the problem graphically is to graph the function $g(t) = w(t) - u(t)$, which represents the rate of change in the pay gap between the Washington state and the national per capita income (see Figure 7.12).

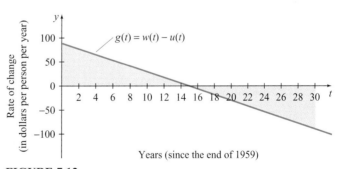

FIGURE 7.12

When $g(t) > 0$, the Washington state per capita income is growing faster than the national per capita income. When $g(t) < 0$, the Washington state per capita income is growing more slowly than the national per capita income. The sum of the signed areas of the shaded regions shown in the graph is -30. This means that, over the 30-year period from 1959 to 1989, the gap between the Washington state per capita income and the national per capita income was reduced by $30.

7.2 Summary

In this section, you learned how to calculate the area of the bounded region between two graphs. You also learned that the definite integral of the difference between two rate of change functions is equal to the difference between the accumulated changes in each of the individual functions.

7.2 Exercises

In Exercises 1–20, calculate the combined area of the region(s) bounded by the graphs of the two functions. (On some of the exercises, you may find it helpful to use your graphing calculator to locate the intersection points of the graphs.)

1. $f(x) = x^2; g(x) = x$

2. $f(x) = x^2 - 3x + 2; g(x) = 2$

3. $f(x) = x^2 - 4; g(x) = -x^2 + 4$

4. $f(x) = 2x^2 - 1; g(x) = 1$

5. $f(x) = x^2 - 4x - 5; g(x) = -3x - 5$

6. $f(x) = 2x^2 - 1; g(x) = x^3 - 1$

7. $f(x) = -x; g(x) = -x^2 + 3x$

8. $f(x) = x; g(x) = x^3 - x$

9. $f(x) = x^3 - 6x^2 + 5x; g(x) = 6x^2 + 50x$

10. $f(x) = 2x^2 + 2x + 4; g(x) = x^3 + 2x + 4$

11. $f(x) = x^2 - 2x; g(x) = x^3 - 3x^2$

12. $f(x) = x - 1; g(x) = x^2 - 2x$

13. $f(x) = x^3 - x; g(x) = -x^3 + x$

14. $f(x) = 2^x + x; g(x) = 3^x - x$

15. $f(x) = 3^x; g(x) = x^3$

16. $f(x) = 2^x; g(x) = x^3 + 1$

17. $f(x) = 2^x; g(x) = x^2$

18. $f(x) = 3\ln(x); g(x) = 2x - 4$

19. $f(x) = 2 + \ln(x); g(x) = 5x^2$

20. $f(x) = x^2 - 4; g(x) = \dfrac{\ln(x)}{x}$

In Exercises 21–25, graph the functions together on the same axes. Then determine the solution by using the integral techniques demonstrated in this section. (You may integrate each function separately and then calculate the difference, or, if you prefer, you may calculate the difference between the functions and then integrate the result.)

21. **M**   **Company Cost and Profit**   Based on data from 1999 to 2001, the annual gross revenue of the Johnson & Johnson company and its subsidiaries may be modeled by
$$R(t) = 665t^2 + 1150t + 27{,}357$$
million dollars, and the annual cost of goods sold may be modeled by
$$C(t) = 103t^2 + 315t + 8539$$
million dollars, where t is the number of years since the end of 1999. (**Source:** Modeled from Johnson & Johnson 2001 Annual Report, p. 6.)
   According to the models, what was the total profit earned between 1999 and 2001?

22. **M**   **Company Sales and Profit**   Based on data from 1999 to 2001, the annual net sales of Gatorade/Tropicana North America may be modeled by
$$R(t) = -107t^2 + 496t + 3452$$
million dollars, and the annual operating profit may be modeled by
$$P(t) = -18.5t^2 + 85.5t + 433$$
million dollars, where t is the number of years since the end of 1999. (**Source:** Modeled from PepsiCo 2001 Annual Report, p. 44.)
   According to the models, what were the total costs of the company between 1999 and 2001? (Assume that *annual net sales* is the same as *annual revenue.*)

23. **M** **Prison vs. College Population**   Based on data from 1980 to 1998, the rate of change in the number of adults in prison, in jail, on probation, or on parole may be modeled by

$$r(t) = 238.1$$

thousand people per year, where t is the number of years since 1980. (**Source:** Modeled from *Statistical Abstract of the United States, 2001*, Table 335, p. 202.)

Based on data from 1980 to 1998, the rate of change in the number of students enrolled in private colleges may be modeled by

$$s(t) = 1.839t + 23.25$$

thousand students per year, where t is the number of years since 1980. (**Source:** Modeled from *Statistical Abstract of the United States, 2001*, Table 205, p. 133.)

Determine the accumulated change in the difference of the two populations from the end of 1980 through 1998. Interpret the meaning of the result.

24. **M** **Company Operating Costs**   Based on data from 1999 to 2001, the rate of change in the annual net sales of Pepsi-Cola North America may be modeled by

$$R(t) = -131t - 749.5$$

million dollars per year, and the rate of change in the annual operating profit may be modeled by

$$P(t) = 12t + 76$$

million dollars per year, where t is the number of years since the end of 1999. (**Source:** Modeled from PepsiCo 2001 Annual Report, pp. 23 and 24.)

Determine the accumulated change in annual operating costs from the end of 1999 through 2001.

25. **M** **Company Profit**   Based on data from 1999 to 2001, the rate of change in the annual net revenue of the Coca-Cola Company may be modeled by

$$r(t) = -402t + 806$$

million dollars, and the rate of change in the annual cost of goods sold may be modeled by

$$c(t) = -355t + 372.5$$

million dollars, where t is the number of years since the end of 1999. (**Source:** Modeled from Coca-Cola Company 2001 Annual Report.)

Determine the accumulated change in annual profit from the end of 1999 through 2001.

Exercises 26–30 are intended to challenge your understanding of calculating the area between curves. In each exercise, we are interested in the region that is bordered by all of the given functions.

26. What is the combined area of the regions bordered by all three functions $f(x) = x^2$, $g(x) = -x^2 + 4$, and $h(x) = 2x$?

27. What is the combined area of the regions bordered by the three functions $f(x) = 2^x$, $g(x) = 2^{-x}$, and $h(x) = -x + 4$?

28. What is the combined area of the regions bordered by the three functions $f(x) = x^2$, $g(x) = x$, and $h(x) = -x^2 + 1$?

29. What is the area of the region bordered by the three functions $f(x) = x^2$, $g(x) = 5x - 6$, and $h(x) = -x^2$?

30. What is the area of the region bordered by the four functions $f(x) = x$, $g(x) = -x + 4$, $h(x) = (x - 2)^2$, and $j(x) = 1.5$?

7.3 Differential Equations and Applications

- Solve introductory-level differential equations
- Apply Newton's Law of Heating and Cooling
- Analyze and interpret the real-life meaning of differential equations

GETTING STARTED   In February 1992, a 79-year-old woman sustained third-degree burns on over 6 percent of her body when she spilled a cup of McDonald's coffee on herself. She requested that McDonald's pay the associated medical expenses, and when the company denied her request, she filed suit. During the trial, a McDonald's quality assurance manager testified that the company required that coffee in the pot be held at between 180 and 190 degrees Fahrenheit (°F). Liquids at 180 degrees will cause third-degree burns in two to seven seconds. If the temperature of the coffee had been 25 degrees cooler, the woman might have avoided serious injury. The jury initially awarded the woman $2.7 million in damages; however, a judge later reduced this amount to $480,000. Ultimately, the woman and Mc-Donald's settled the case for an undisclosed amount. (**Sources:** www.lectlaw.com; www.consumerrights.net.)

How long does it take a 186-degree cup of coffee to cool to 155 degrees? Questions such as these may be answered using differential equations.

In this section, we will introduce differential equations and show how integral calculus may be used to solve them. Additionally, we'll discuss Newton's Law of Heating and Cooling and other applications of differential equations.

A **first-order differential equation** is an equation that relates an unknown function and its first derivative. In the first few examples, we will look at some simple first-order differential equations and their solutions.

EXAMPLE 1   **Solving a Separable Differential Equation**

Find a function $F(t)$ given that $\dfrac{dF}{dt} = F$.

SOLUTION   We are looking for a function that is its own derivative.

$$\frac{dF}{dt} = F$$

$$\frac{dF}{F} = dt \qquad\qquad \text{Multiply by } dt \text{ and divide by } F$$

$$\frac{1}{F}\,dF = dt \qquad\qquad \text{Rewrite } \tfrac{dF}{F} \text{ as } \tfrac{1}{F}\,dF$$

$$\int \frac{1}{F}\,dF = \int dt \qquad\qquad \text{Take the integral of both sides}$$

$$\ln|F| = t + C \qquad\qquad \text{Integrate}$$

$$e^{t+C} = |F| \qquad\qquad \text{Recall } \ln|y| = x \text{ is equivalent to } e^x = |y|$$

$$F = \pm e^t \cdot e^C \qquad \text{Rules of exponents and definition of absolute value}$$

$$= \pm e^C e^t$$

Letting $k = e^C$, we get the *general solution* of the differential equation:

$$F(t) = \pm k e^t$$

Any function of this form is a solution to the differential equation $\dfrac{dF}{dt} = F$. For example, $F(t) = 2e^t$, $F(t) = -3e^t$, and $F(t) = e^t$ are all *particular solutions* to the differential equation.

EXAMPLE 2

Finding a Particular Solution for a Separable Differential Equation

Find the particular solution of the differential equation $\dfrac{dF}{dt} = F^2$ given $F(0) = 4$.

SOLUTION

$$\frac{dF}{dt} = F^2$$

$$\frac{dF}{F^2} = dt \qquad \text{Divide by } F^2 \text{ and multiply by } dt$$

$$\frac{1}{F^2}dF = dt \qquad \text{Rewrite } \frac{dF}{F^2} \text{ as } \frac{1}{F^2}dF$$

$$F^{-2}dF = dt \qquad \text{Rewrite } \frac{1}{F^2}dF \text{ as } F^{-2}dF$$

$$\int F^{-2}dF = \int dt \qquad \text{Take the integral of both sides}$$

$$\frac{F^{-1}}{-1} = t + C \qquad \text{Integrate with Power Rule}$$

$$F^{-1} = -t - C \qquad \text{Multiply both sides by } -1$$

$$\frac{1}{F} = -t - C \qquad \text{Rewrite } F^{-1} \text{ as } \frac{1}{F}$$

$$F = \frac{1}{-t - C} \qquad \text{Divide by } -t - C \text{ and multiply by } F$$

This is the *general solution* to the differential equation. However, since we know that $F(0) = 4$, we will be able to find the *particular solution*.

$$F = \frac{1}{-t - C}$$

$$4 = \frac{1}{-0 - C} \qquad \text{Substitute } F = 4 \text{ and } t = 0$$

$$4 = -\frac{1}{C}$$

$$C = -\frac{1}{4}$$

Therefore, the particular solution is $F(t) = \dfrac{1}{-t + \frac{1}{4}}$. We can verify the accuracy of our work by differentiating this function.

$$\frac{d}{dt}[F(t)] = \frac{d}{dt}\left(\frac{1}{-t + \frac{1}{4}}\right)$$

$$\frac{dF}{dt} = \frac{d}{dt}\left(-t + \frac{1}{4}\right)^{-1}$$

$$= -1\left(-t + \frac{1}{4}\right)^{-2}(-1)$$

$$= \left(-t + \frac{1}{4}\right)^{-2}$$

$$= \frac{1}{\left(-t + \frac{1}{4}\right)^2}$$

$$= \left(\frac{1}{-t + \frac{1}{4}}\right)^2$$

$$= F^2$$

Since the result is equivalent to the differential equation we were given at the start of the problem, we believe that our calculated particular solution, $F(t) = \frac{1}{-t + \frac{1}{4}}$, is correct. To verify, we check to see that $F(0) = 4$.

$$F(0) = \frac{1}{-(0) + \frac{1}{4}}$$

$$= 4$$

We used the **separation of variables** process to solve the differential equations in Examples 1 and 2. We will now formally describe the process.

SOLVING SEPARABLE FIRST-ORDER DIFFERENTIAL EQUATIONS

A first-order differential equation is said to be **separable** if it may be written as

$$\frac{dy}{dx} = \frac{f(x)}{g(y)}$$

for some functions $f(x)$ and $g(y)$. The solution to the differential equation is obtained by moving the x and y variables to opposite sides of the equal sign and integrating. That is,

$$\int g(y)\,dy = \int f(x)\,dx$$

It is essential that all x terms be grouped with the dx and all y terms be grouped with the dy.

EXAMPLE 3

Finding a General Solution to a Separable Differential Equation

The rate of change of an investment account earning compound interest may be given by the first-order differential equation $\frac{dA}{dt} = kA$, where k is a positive constant. (If $k = 0$, the value of the investment remains constant. If k is negative, the investment is losing value.) Find the general solution to the differential equation.

SOLUTION We will solve the differential equation using the separation of variables process.

$$\frac{dA}{dt} = kA$$

$$\frac{dA}{A} = k\,dt \qquad\qquad \text{Separate the variables}$$

$$\int \frac{1}{A}\,dA = \int k\,dt \qquad\qquad \text{Take the integral of both sides}$$

$$\ln|A| = kt + C \qquad\qquad \text{Integrate}$$

$$\ln A = kt + C \qquad\qquad A,\text{ the account value, is assumed to be positive}$$

$$A = e^{kt+C} \qquad\qquad \text{Since, in general, } \ln A = b \text{ is equivalent to } A = e^b$$

$$= e^{kt} \cdot e^{C} \qquad\qquad \text{Rules of exponents}$$

$$= Pe^{kt} \text{ where } P = e^{C}$$

The general solution to the differential equation is $A = Pe^{kt}$.

We immediately recognize the result from Example 3 as the continuous compound interest formula. The standard compound interest formula $A = P\left(1 + \frac{r}{n}\right)^{nt}$ may be converted to the form $A = Pe^{kt}$ by letting

$$\left(1 + \frac{r}{n}\right)^{n} = e^{k}$$

$$k = \ln\left[\left(1 + \frac{r}{n}\right)^{n}\right]$$

A DIFFERENTIAL EQUATION FOR CONTINUOUS COMPOUND INTEREST

The value of an investment account earning continuous compound interest is

$$A = Pe^{kt} \text{ dollars}$$

where A is the value of the investment after t years, P is the initial value of the investment (in dollars), and k is the continuous interest rate.

The rate of change in the value of the investment is given by

$$\frac{dA}{dt} = kA \frac{\text{dollars}}{\text{year}}$$

EXAMPLE 4

Finding a Particular Solution to a Differential Equation

The rate of change of an investment account earning compound interest is given by $\frac{dA}{dt} = kA$, where k is a positive constant. The initial account value was $1000. At the end of the third year, the account value is $1120. Find the particular solution to the differential equation.

SOLUTION The general solution to the differential equation $\dfrac{dA}{dt} = kA$ is $A = Pe^{kt}$. We also know that $A(0) = 1000$ and $A(3) = 1120$.

$$A = Pe^{kt}$$
$$1000 = Pe^{k(0)}$$
$$1000 = P \cdot 1$$
$$P = 1000$$

Thus $A = 1000e^{kt}$. We will find the value of k by substituting in the second point, $(3, 1120)$.

$$A = 1000e^{kt}$$
$$1120 = 1000e^{k(3)}$$
$$1.12 = e^{3k}$$
$$\ln(1.12) = \ln(e^{3k})$$
$$0.1133 = 3k$$
$$k = 0.03778$$

Therefore, $A = 1000e^{0.03778t}$ is the particular solution to the differential equation. This equation may also be written as $A = 1000(1.0385)^t$, since $e^{0.03778} = 1.0385$. The account is earning interest at a rate of 3.778 percent compounded continuously or a rate of 3.85 percent compounded annually.

Newton's Law of Heating and Cooling

The world-renowned scientist and scholar Sir Isaac Newton determined that the rate of change in an object's temperature is proportional to the difference between the constant temperature of the environment surrounding the object and the object's temperature. This observation is summarized in **Newton's Law of Heating and Cooling.**

NEWTON'S LAW OF HEATING AND COOLING

Let T be the temperature of an object at time t and A be the temperature of the environment surrounding the object (ambient temperature). Then

$$\frac{dT}{dt} = k(T - A)$$

where k is a constant that varies depending upon the physical properties of the object.

According to Newton's Law of Heating and Cooling,

$$\frac{dT}{dt} = k(T - A)$$

We will use substitution to rewrite the equation.

We define the function $y = T - A$ and differentiate this function with respect to t.

$$\frac{d}{dt}(y) = \frac{d}{dt}(T - A)$$

$$\frac{dy}{dt} = \frac{d}{dt}(T) - \frac{d}{dt}(A)$$

$$= \frac{dT}{dt} - \frac{dA}{dt}$$

$$= \frac{dT}{dt} - 0 \qquad\qquad \frac{dA}{dt} = 0 \text{ since } A \text{ is a constant}$$

$$= \frac{dT}{dt}$$

So $\dfrac{dy}{dt} = \dfrac{dT}{dt}$. Rewriting the equation, $\dfrac{dT}{dt} = k(T - A)$, in terms of y, we get

$$\frac{dy}{dt} = ky$$

since $\dfrac{dT}{dt} = \dfrac{dy}{dt}$ and $T - A = y$. Separating the variables, we rewrite the equation as

$$\frac{1}{y}dy = k\,dt$$

Integrating both sides of the equation, we get

$$\int \frac{1}{y}dy = \int (k\,dt)$$

$$\ln|y| = kt + C$$

Recall that if $x = \ln|y|$, then $e^x = |y|$. Consequently,

$$\ln|y| = kt + C$$

$$|y| = e^{kt+C}$$

$$y = \pm(e^{kt} \cdot e^C)$$

Now we'll back substitute $y = T - A$.

$$T - A = (e^{kt} \cdot e^C)$$

$$T = A + (e^{kt} \cdot e^C)$$

$$T = A + Se^{kt} \text{ where } S = e^C$$

Thus the general solution to the differential equation $\dfrac{dT}{dt} = k(T - A)$ is $T = A + Se^{kt}$.

EXAMPLE 5   ## Using a Differential Equation to Model Water Temperature

To test Newton's Law of Heating and Cooling, the author heated a cup of water to 186°F and placed it in a 76°F room. The temperature of the water was recorded at two-minute time intervals as shown in Table 7.2.

TABLE 7.2 Hot Water Cooling in 76°F Room

Minutes (t)	Temperature (degrees Fahrenheit) (T)
0	186
2	175
4	167
6	159
8	153
10	147
12	142
14	138

Use Newton's Law of Heating and Cooling to find a model for the water temperature.

SOLUTION We know that the general solution to the differential equation given in Newton's Law of Heating and Cooling is $T = A + Se^{kt}$. We'll substitute in values from the table to determine the numeric values of k and C. We know that $A = 76$, since the ambient temperature is 76 degrees Fahrenheit. Additionally, we know that at time $t = 0$, $T = 186$.

$$T = 76 + Se^{kt}$$
$$186 = 76 + Se^{k(0)}$$
$$110 = Se^0$$
$$110 = S$$

Our equation $T = A + Se^{kt}$ may now be rewritten as

$$T(t) = 76 + 110e^{kt}$$

We'll use the table values $t = 14$ and $T = 138$ to find the value of k. (We can select any pair of table values. Since we're developing a model, selecting a different pair of table values would slightly alter the model equation.)

$$138 = 76 + 110e^{k(14)}$$
$$62 = 110e^{14k}$$
$$0.5636 = e^{14k}$$
$$\ln(0.5636) = 14k \qquad \text{Log both sides}$$
$$k = -0.04095$$

Since k is negative, the temperature of the water is dropping at a continuous rate of 4.095 percent per minute. The temperature of the water may be modeled by

$$T(t) = 76 + 110e^{-0.04095t} \text{ degrees Fahrenheit}$$

Let's see how well this function fits the table of data.

TABLE 7.3 Hot Water Cooling in 76°F Room

Minutes (t)	Actual Temperature (degrees Fahrenheit) (T)	Model Temperature (degrees Fahrenheit) (M)
0	186	186
2	175	177
4	167	169
6	159	162
8	153	155
10	147	149
12	142	143
14	138	138

Newton's Law of Heating and Cooling gave a remarkably good estimate for the measured temperature. The discrepancies between the two may be due to measurement device error or other physical phenomena. As with any mathematical model, some error is to be expected.

EXAMPLE 6

Using a Differential Equation to Calculate Cooling Rates

According to the model developed in Example 5, at what rate was the water cooling 4, 8, and 12 minutes into the cooling period?

SOLUTION The model for the water temperature was given by

$$T(t) = 76 + 110e^{-0.04095t} \text{ degrees}$$

t seconds after the temperature of the water was measured to be 186 degrees. The rate of change in the temperature is given by the differential equation

$$\frac{dT}{dt} = k(T - A)$$
$$= -0.04095(T - 76)$$

We're asked to evaluate this function at $t = 4$, $t = 8$, and $t = 12$.

$$\left.\frac{dT}{dt}\right|_{t=4} = -0.04095[T(4) - 76]$$
$$= -0.04095(169.4 - 76)$$
$$= -3.82$$

Four minutes into the cooling period, the water was cooling at a rate of 3.82 degrees per minute.

$$\left. \frac{dT}{dt} \right|_{t=8} = -0.04095[T(8) - 76]$$
$$= -0.04095(155.3 - 76)$$
$$= -3.25$$

Eight minutes into the cooling period, the water was cooling at a rate of 3.25 degrees per minute.

$$\left. \frac{dT}{dt} \right|_{t=12} = -0.04095[T(12) - 76]$$
$$= -0.04095(143.3 - 76)$$
$$= -2.76$$

Twelve minutes into the cooling period, the water was cooling at a rate of 2.76 degrees per minute.

Assuming that coffee cools at the same rate as water, we estimate that it would take a cup of McDonald's coffee eight minutes to cool from 186 degrees to 155 degrees Fahrenheit. However, the size and shape of the coffee cup, the addition of cream and sugar, and other environmental variables could dramatically affect the accuracy of our estimate.

7.3 Summary

In this section, you learned how to use integral calculus to solve first-order differential equations. You also learned how to apply differential equations in Newton's Law of Heating and Cooling and in the continuous compound interest formula.

7.3 Exercises

In Exercises 1–10, find the general solution of the separable differential equation.

1. $\dfrac{dy}{dx} = \dfrac{x}{y}$

2. $\dfrac{dA}{dt} = 3A^2$

3. $\dfrac{dy}{dx} = \sqrt{y}$

4. $\dfrac{dA}{dt} = \dfrac{3A}{t}$

5. $\dfrac{dy}{dx} = \dfrac{y^2}{x^2}$

6. $\dfrac{dA}{dt} = 4(20 - A)$

7. $\dfrac{dy}{dx} = 0.25(y - 4)$

8. $\dfrac{dy}{dx} = 4y$

9. $\dfrac{dy}{dx} = 4y - y^2$

(*Hint:* $\displaystyle\int \dfrac{dy}{y(M - y)} = \dfrac{1}{M} \ln \left| \dfrac{y}{M - y} \right|$.)

10. $\dfrac{dA}{dt} = 6 - 3A$

In Exercises 11–20, find the particular solution for the differential equations.

(Hint: $\displaystyle\int \frac{dy}{y(M-y)} = \frac{1}{M}\ln\left|\frac{y}{M-y}\right|.$)

11. $\dfrac{dy}{dx} = \dfrac{2x}{y}$; $y(1) = 4$

12. $\dfrac{dA}{dt} = 4A^2$; $A(0) = -1$

13. $\dfrac{dy}{dx} = 3y$; $y(0) = -1$ **14.** $\dfrac{dA}{dt} = \dfrac{2A}{t}$; $A(1) = 1$

15. $\dfrac{dy}{dx} = 25(y-5)$; $y(0) = 0$

16. $\dfrac{dA}{dt} = 4 - 2A$; $A(0) = 0$

17. $\dfrac{dy}{dx} = y(4-y)$; $y(0) = \dfrac{1}{2}$

18. $\dfrac{dA}{dt} = A^2 - 4A$; $A(0) = 8$

19. $\dfrac{dy}{dx} = 2y - y^2$; $y(0) = 100$

20. $\dfrac{dA}{dt} = A(10-A)$; $A(0) = 5$

In Exercises 21–28, use Newton's Law of Heating and Cooling to find the equation of the model that best fits the data. Then answer the given questions.

21. **Water Temperature** The author placed a cup of ice water in a 76°F room and recorded the temperature of the water at 10-minute intervals as shown in the table.

Cold Water Warming in a 76°F Room

Minutes (*t*)	Temperature (degrees Fahrenheit) (*T*)
0	34.3
10	40.3
20	43.8
30	48.2
40	51.8
50	54.7

(a) Find a model for the temperature of the water at time *t* minutes.

(b) Determine the rate of change in the water's temperature at 10 minutes and 50 minutes into the warming period.

22. **Water Temperature** How long will it take to freeze a 60°F cup of water in a 10°F freezer? (At 32°F, water freezes.) (Since the result will vary depending upon the quantity of water and the shape of the container, assume that for the container in question, the water temperature will be 46°F at the end of the eighth minute.)

23. **Cake Temperature** Cakes are typically baked at 350°F. If a pan filled with cake batter is moved from a 75°F kitchen into a 350°F oven, how long will it take for the cake batter to heat up to 200°F? (Since the result will vary depending upon the quantity of the batter and the shape of the pan, assume that for the container in question, the batter temperature will be 100°F at the end of the third minute.)

24. **Cake Temperature** A cake is removed from a 350°F oven and placed on a cooling rack in a 75°F room. If the temperature of the cake was 350°F when it was removed from the oven and its temperature is 250°F two minutes later, how long will it take for the cake to cool to 100°F?

25. **Time of Death** When the forensic investigator arrives upon the scene of a homicide, one of the most important things to determine is the time of death of the victim. Suppose that the victim's internal body temperature was 84.5°F when the body was discovered in a 70°F room. Thirty minutes later, the person's body temperature had dropped to 83.6°F. If the person's internal body temperature was 98°F at the time of death, how long had the person been dead when the forensic investigator arrived on the scene?

26. According to the model developed in Exercise 25, at what rate was the corpse's temperature dropping 1 hour, 2 hours, and 4 hours into the cooling period?

27. **Time of Death** Suppose that a homicide victim's internal body temperature was 59.1°F when the body was discovered in a 30°F room. Thirty minutes later, the person's body temperature had dropped to 56.9°F. If the person's internal body temperature was 98°F at the time of death, how long had the person been dead when the forensic investigator arrived on the scene?

28. According to the model developed in Exercise 27, at what rate was the corpse cooling 4 hours, 6 hours, and 12 hours into the cooling period?

Exercises 29–32 deal with the notion of continuous growth.

29. Investment Value A $1000 investment has a continuous interest rate of 10 percent. Write an equation for the value of the investment after t years.

30. Investment Value The value of an investment is modeled by $A = 2000e^{0.08t}$ dollars, where t is the number of years since 2006. What will be the annual rate of change in the value of the investment when the account value is $3000? $4000? $5000?

31. Investment Value The rate of change of an investment account earning continuous compound interest is given by

$$\frac{dA}{dt} = kA$$

where k is a positive constant. The initial account value was $2500. At the end of the third year, the account value was $4200. Find the particular solution to the differential equation.

32. Investment Value The rate of change of an investment account earning continuous compound interest is given by

$$\frac{dA}{dt} = kA$$

where k is a positive constant. The initial account value was $420. At the end of the fifth year, the account value was $630. Find the particular solution to the differential equation.

Exercises 33–35 are intended to challenge your understanding of differential equations.

33. Show that the derivative of $A = Pe^{kt}$ is equivalent to kA.

34. Find the general solution of the differential equation.

$$2y + y'x^2 = 5$$

35. Find the general solution of the differential equation.

$$3x^2y^2 + 2yy'x^3 = 0$$

7.4 Differential Equations: Limited Growth and Logistic Models

- Find the equation of a limited growth model
- Find the equation of a logistic growth model
- Analyze and interpret the real-life meaning of differential equations

GETTING STARTED On September 11, 2001, many of us woke up to the horrific news of the terrorist attack on the World Trade Center. News bulletins announcing the attack flooded the radio and television channels. Initially, the news of the attack spread very rapidly; however, as the day progressed, the news spread more slowly, since most people had already received word of the attack. The spread of information, such as the news of the September 11 attack, may often be modeled mathematically. In this section, we will continue our discussion of differential equations and show how they can be used to find limited growth and logistic models.

Limited Growth Model

Sociologists often assume that the rate at which news spreads via mass media is proportional to the number of people who have not yet heard the news. This type of growth is called **limited growth.** For a limited growth function y with a maximum value of M, the rate of change, $\frac{dy}{dt}$, is proportional to the difference between the present value of y and M. That is, $\frac{dy}{dt} = k(M - y)$.

EXAMPLE 1

Using Differential Equations to Forecast the Spread of Information

At 8:46 a.m. on September 11, 2001, the hijacked American Airlines Flight 11 slammed into the North Tower of the World Trade Center in New York City. Eighteen minutes later, United Airlines Flight 175 crashed into the South Tower of the World Trade Center. By the end of the day, thousands of people were dead as a result of the deadliest terrorist attack in American history.

Suppose that in a community of 10,000 people, 3000 people were watching television or listening to the radio when the news broke on September 11, 2001. For the purpose of the model, we'll assume that all 3000 people heard the news simultaneously *five* minutes after the first attack. (It takes time for news crews to reach the site of any newsworthy event.) Find the equation for the limited growth model that models the spread of the news of the attack. Then determine how long it took for 90 percent of the community to hear the news and at what rate people were hearing the news at that time.

SOLUTION The maximum number of people that can be informed is 10,000, so $M = 10,000$. Additionally, we know that when $t = 5$, $y = 3000$. We have

$$\frac{dy}{dt} = k(M - y)$$

$$\frac{dy}{dt} = k(10,000 - y)$$

We can use the separation of variables process to find the equation of the general solution and then use the point $(5, 3000)$ to find the particular solution.

$$\frac{dy}{dt} = k(10,000 - y)$$

$$\frac{dy}{(10,000 - y)} = k\,dt$$

Let $u = 10,000 - y$; then $du = -dy$. Rewriting the left-hand side of the equation in terms of u, we get

$$\frac{-du}{u} = k\,dt$$

$$-\int \frac{1}{u}du = \int k\,dt$$

$$-\ln|u| = kt + C$$

$$-\ln u = kt + C \qquad \text{Since } u > 0$$

$$\ln u = -kt - C$$

$$u = e^{-kt - C}$$

$$10,000 - y = e^{-kt}e^{-C} \qquad \text{Since } u = 10,000 - y$$

$$y = 10,000 - Se^{-kt} \qquad \text{Where } S = e^{-C}$$

At $t = 0$, nobody had heard the news of the attack because the information hadn't yet been broadcasted to the community. (It wasn't until five minutes after the crash that the first 3000 people in the community heard the news.) We can use this information to find the value of S.

$$y(0) = 10{,}000 - Se^{-k(0)}$$
$$0 = 10{,}000 - Se^0$$
$$S = 10{,}000$$

A general solution to the differential equation is

$$y = 10{,}000 - 10{,}000e^{-kt}$$
$$= 10{,}000(1 - e^{-kt})$$

We'll now use the point (5, 3000) to find the particular solution.

$$3000 = 10{,}000(1 - e^{-k(5)})$$
$$0.3 = 1 - e^{-5k}$$
$$-0.7 = -e^{-5k}$$
$$0.7 = e^{-5k}$$
$$\ln(0.7) = -5k$$
$$k = \frac{\ln(0.7)}{-5}$$
$$= 0.07133$$

The limited growth model that fits the data is given by

$$y = 10{,}000(1 - e^{-0.07133t})$$

Since 90 percent of 10,000 is 9000, we need to determine when 9000 people had heard the news.

$$9000 = 10{,}000(1 - e^{-0.07133t})$$
$$0.9 = 1 - e^{-0.07133t}$$
$$-0.1 = -e^{-0.07133t}$$
$$0.1 = e^{-0.07133t}$$
$$\ln(0.1) = -0.07133t$$
$$t = \frac{\ln(0.1)}{-0.07133}$$
$$= 32.28$$

We estimate that 32 minutes after the attack, 90 percent of the people in the community had heard the news. We'll now determine at what rate the word was spreading.

$$\frac{dy}{dt} = k(10{,}000 - y)$$
$$= 0.07133(10{,}000 - y)$$
$$\left.\frac{dy}{dt}\right|_{t=32.28} = 0.07133(10{,}000 - 9000) \qquad \text{[Note: } y(32.28) = 9000\text{]}$$
$$= 71.33$$

Approximately 32 minutes after the attack, the news was spreading at a rate of roughly 71 people per minute.

In Example 1, we solved the given differential equation by using the separation of variables method. Using this same approach, we can derive the general solution to the differential equation $\frac{dy}{dt} = k(M - y)$.

LIMITED GROWTH

Assume that the rate of growth of a function y with maximum value M is proportional to the difference between the present value of y and M. That is,

$$\frac{dy}{dt} = k(M - y)$$

The solution to the differential equation with initial condition

$$y(0) = 0$$

is given by

$$y = M(1 - e^{-kt})$$

and has the graph shown in Figure 7.13.

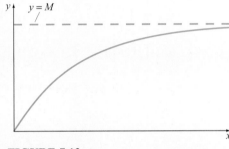

FIGURE 7.13

A limited growth model may be used to model the cumulative number of people infected by a disease during a bioterrorist attack once quarantine efforts are underway. In the journal article *Modeling Potential Responses to Smallpox as a Bioterrorist Weapon*, scientists at the Centers for Disease Control and Prevention and at Don Millar and Associates demonstrated the catastrophic effect of a smallpox attack. Table 7.4 shows how rapidly the cumulative number of smallpox cases would increase should there be no intervention.

TABLE 7.4   Estimates of Cumulative Total Smallpox Cases After 365 Days with No Intervention

Number Initially Infected[a]	Number Infected per Infectious Person[b]	Cumulative Total Number of Smallpox Cases, Days Postrelease[c]			
		30 Days	90 Days	180 Days	365 Days
10	1.5	31	214	2,190	224 thousand
10	3.0	64	4,478	2.2 million	774 billion
1,000	1.5	3,094	21,372	219,006	22 million
1,000	3.0	6,397	447,794	222 million	77 trillion

[a]Number initially infected refers to those who are exposed during a release so that they subsequently become infectious to others. This scenario excludes those who are exposed but either do not become ill (i.e., are immune or are not exposed to an infectious dose) or do not become infectious (residual immunity from prior vaccination may be sufficient to prevent onward transmission).
[b]The number of persons infected per infectious person is the transmission rate.
[c]Assumes an unlimited supply of smallpox-susceptible persons.

Source: *Emerging Infectious Diseases,* vol. 7, no. 6, November–December 2001, p. 959.

Figure 7.14 shows the effectiveness of adequate intervention once smallpox is detected. If 25 percent of the people infected are quarantined daily, the cumulative number of people infected will continue to increase at an increasing rate; however, if 50 percent of the people infected are quarantined daily, the cumulative number of people infected will increase at a decreasing rate. The graphs in the bottom right-hand corner of the figure show that the cumulative number of people infected may be modeled by a limited growth function when 50 percent of the infected individuals are quarantined daily.

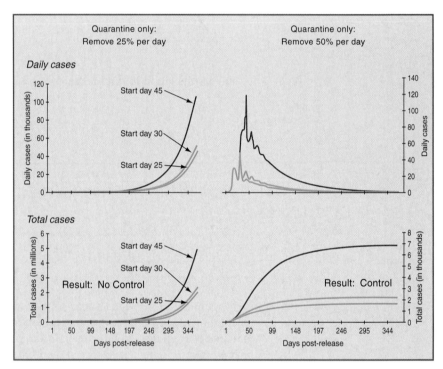

Figure 4. Daily and total cases of smallpox after quarantining infectious persons at two daily rates and three post-release start dates. The graphs demonstrate that if quarantine is the only intervention used, a daily removal rate of ≥ 50% is needed to stop transmission within 365 days post-release. At a 25% daily removal rate of infectious persons by quarantine, a cohort of all those entering the first day of overt symptoms (i.e., rash) is entirely removed within 17 days (18 to 20 days post-incubation) after the first day of overt symptoms, with 90% removed within 9 days. At a 50% daily removal of infectious persons by quarantine, a cohort of all those entering their first day of overt symptoms (i.e., rash) is entirely removed within 7 days (8 to 10 days post-incubation) after the first day of overt symptoms, with 90% removed within 4 days. The daily rate of removal (quarantine) relates only to the removal of those who are infectious (i.e., overtly symptomatic). The rate does not include any persons who may be quarantined along with overtly symptomatic patients, such as unvaccinated household contacts. Data generated by assuming 100 persons initially infected and a transmission rate of 3 persons infected per infectious person. For clarity, the graphs of daily cases do not include the assumed 100 initially infected persons. The graphs of total cases include the 100 initially infected.

FIGURE 7.14

Source: *Emerging Infectious Diseases,* vol. 7, no. 6, November–December 2001, p. 959.

EXAMPLE 2

Using Differential Equations to Forecast the Effect of a Bioterrorist Attack

Reading from the graph in the bottom right corner of Figure 7.14, we can construct a limited growth model. It appears that a maximum of 7000 people will be infected with smallpox as a result of a bioterrorist attack. It also appears that 99 days after the attack, 4750 people are infected. Do the following:

(a) Determine the equation of the limited growth model. (This models the cumulative number of people infected if 50 percent of the infected people are quarantined daily beginning on the 45th day after the attack.)
(b) Determine the rate at which the disease is spreading on the 50th and 100th days after the attack.

SOLUTION

(a) The equation of the limited growth model is given by

$$y = M(1 - e^{-kt})$$

Since a maximum of 7000 people will be infected, $M = 7000$.

$$y = 7000(1 - e^{-kt})$$

Additionally, we know that $(99, 4750)$ lies on the graph of the function.

$$4750 = 7000(1 - e^{-k(99)})$$
$$0.6786 = 1 - e^{-k(99)}$$
$$-0.3214 = -e^{-99k}$$
$$0.3214 = e^{-99k}$$
$$\ln(0.3214) = -99k$$
$$k = \frac{\ln(0.3214)}{-99}$$
$$= 0.01147$$

Thus $y = 7000(1 - e^{-0.01147t})$ models the number of people infected t days after an attack. (*Note:* Since we are modeling the equation from the graph, our model lacks precision. In addition, the smallpox graph doesn't appear to pass through $(0, 0)$, and our model does. However, despite the limitations of our model, it does give us a rough idea of what would happen in a bioterrorist smallpox attack.)

(b) We can use the differential equation to estimate the rate at which the disease is spreading.

$$\frac{dy}{dt} = k(M - y)$$
$$= 0.01147(7000 - y)$$

$$\frac{dy}{dt}\bigg|_{t=50} = 0.01147[7000 - y(50)]$$
$$= 0.01147(7000 - 3055)$$
$$= 45$$

On the 50th day, we estimate that the disease is spreading at a rate of 45 people per day. That is, approximately 45 additional people are infected between the 50th and 51st days.

$$\frac{dy}{dt}\bigg|_{t=100} = 0.01147[7000 - y(100)]$$
$$= 0.01147(7000 - 4777)$$
$$= 25$$

On the 100th day, we estimate that the disease is spreading at a rate of 25 people per day. That is, approximately 25 additional people are infected between the 100th and 101st days.

Commercial fish growers seek to create a fish habitat that will maximize the size of each fish in the least amount of time. The Von Bertalanffy Limited Growth model is widely used in the fish industry. The model assumes that there is a maximal length, L_∞, that a fish will attain under optimal conditions and that the rate of change in the length of the fish, $\dfrac{dL}{dt}$, is proportional to the difference between the maximal length and the current length. That is, $\dfrac{dL}{dt} = k(L_\infty - L)$. Solving this differential equation, we can find the general form of the Von Bertalanffy Limited Growth model.

$$\frac{dL}{dt} = k(L_\infty - L)$$

$$\frac{dL}{L_\infty - L} = k\,dt \qquad \text{Separation of variables}$$

$$\int \frac{1}{L_\infty - L}\,dL = \int k\,dt$$

$$-\ln|L_\infty - L| = kt + C \qquad \text{Since } \tfrac{d}{dL}[-\ln(L_\infty - L)] = \tfrac{1}{L_\infty - L}$$

$$\ln(L_\infty - L) = -kt - C \qquad \text{Since } L_\infty - L > 0$$

$$L_\infty - L = e^{-k(t+C)}$$

$$L = L_\infty - e^{-k(t+C)}$$

The solution to the differential equation is commonly written as $L = L_\infty(1 - e^{-k(t-t_0)})$, where t_0 is a negative constant that varies from species to species. The constant t_0 has the property that $L(t_0) = 0$. Our solution may be converted from $L = L_\infty - e^{-k(t+C)}$ to $L = L_\infty(1 - e^{-k(t-t_0)})$ by letting $C = \dfrac{-\ln(L_\infty)}{k} - t_0$. (No matter what the value of C is, we can always find a t_0 that makes this equality true.)

EXAMPLE　3

Forecasting Fish Size with a Von Bertalanffy Limited Growth Model

Turkish scientists conducted a study on the Eastern Black Sea between 1991 and 1996 and determined the values of k, L_∞, and t_0 for economically important fish species in the region. (**Source:** www.tagem.gov.tr.) The whiting fish species has the Von Bertalanffy growing constants given in Table 7.5.

TABLE 7.5

Whiting	k	L_∞	t_0
Female	0.11	43.3 cm	-1.91 years
Male	0.136	34.2 cm	-2.02 years

Source: www.fishbase.org.

(a) According to the model, how long is a two-year-old female whiting?
(b) According to the model, how long is a two-year-old male whiting?
(c) Does the two-year-old female whiting grow faster than the two-year-old male?

SOLUTION

(a) We know that $L(t) = L_\infty(1 - e^{-k(t-t_0)})$. We have $L_\infty = 43.3$, $k = 0.11$, and $t_0 = -1.91$. Substituting these values into the Von Bertalanffy Limited Growth model yields $L(t) = 43.3(1 - e^{-0.11[t-(-1.91)]})$. We determine the length of a two-year-old female whiting by evaluating this function at $t = 2$.

$$L(2) = 43.3(1 - e^{-0.11[2-(-1.91)]})$$
$$= 43.3(1 - e^{-0.11(3.91)})$$
$$= 43.3(0.3496)$$
$$\approx 15.1$$

We estimate the length of a two-year-old female whiting to be about 15.1 cm.

(b) We know that $L(t) = L_\infty(1 - e^{-k(t-t_0)})$. We have $L_\infty = 34.2$, $k = 0.136$, and $t_0 = -2.02$. Substituting these values into the Von Bertalanffy Limited Growth model yields $L(t) = 34.2(1 - e^{-0.136[t-(-2.02)]})$. We determine the length of a two-year-old male whiting by evaluating this function at $t = 2$.

$$L(2) = 34.2(1 - e^{-0.136[2-(-2.02)]})$$
$$= 34.2(1 - e^{-0.136(4.02)})$$
$$= 34.2(0.4212)$$
$$\approx 14.4$$

We estimate the length of a two-year-old male whiting to be about 14.4 cm.

(c) Finally, we need to evaluate each of the differential equations at $t = 2$. Recall that the general form of the differential equation is $\dfrac{dL}{dt} = k(L_\infty - L)$.

Female:

$$\left.\frac{dL}{dt}\right|_{t=2} = 0.11[43.3 - L(2)]$$
$$= 0.11(43.3 - 15.1)$$
$$= 3.1$$

A two-year-old female whiting grows at a rate of 3.1 cm per year.

Male:

$$\left.\frac{dL}{dt}\right|_{t=2} = 0.136[34.2 - L(2)]$$
$$= 0.136(34.2 - 14.4)$$
$$= 2.7$$

A two-year-old male whiting grows at a rate of 2.7 cm per year.

From our calculations we conclude that the two-year-old female whiting grows faster than the two-year-old male whiting.

Logistic Growth

When a successful product is introduced to the market, cumulative product sales often increase very slowly initially; however, as the popularity of the product increases, sales increase rapidly. Then, as the market becomes saturated with the product, cumulative sales growth again slows. This type of growth is called **logistic growth.**

LOGISTIC GROWTH

Assume that the rate of growth of a function y with maximum value M is proportional to the product of the present value of y and the difference between the present value of y and M. That is,

$$\frac{dy}{dt} = ky(M - y)$$

The solution to this differential equation with initial condition

$$y(0) = \frac{M}{1 + S}$$

is given by

$$y = \frac{M}{1 + Se^{-kMt}}$$

The graph of the solution is an s-shaped (sigmoidal) graph (see Figure 7.15).

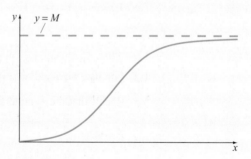

FIGURE 7.15

For example, consider the digital video disc (DVD) player. The DVD player was introduced into the market in the first quarter of 1997. By the end of 2002, nearly 50 percent of U.S. homes had DVD capability. (**Source:** DVD Entertainment Group.) Looking at a scatter plot of the cumulative number of DVD players sold (Figure 7.16), it initially looks as if the cumulative number of DVD players sold is increasing exponentially.

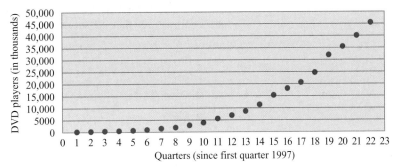

FIGURE 7.16

However, when we attempt to model the data with an exponential function (Figure 7.17), we see that the exponential function ends up growing much more rapidly than the data.

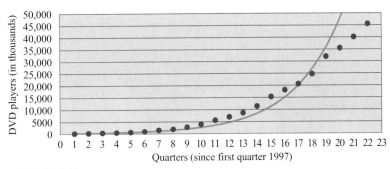

FIGURE 7.17

However, if we model the data with a logistic function (Figure 7.18), we see that DVD player sales appear to exhibit logistic growth behavior.

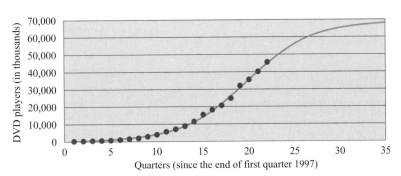

FIGURE 7.18

Using logistic regression on the TI-83 Plus, we determine that the logistic model for the given DVD player data is given by

$$D(t) = \frac{68{,}571}{1 + 288.8e^{-0.28708t}} \text{ thousand DVD players}$$

where t is the number of quarters since the end of the first quarter of 1997.

Plotting the graph of D together with the scatter plot of the data (Figure 7.19), we see that the solution fits the data remarkably well.

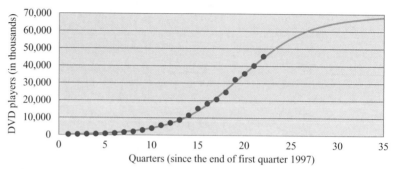

FIGURE 7.19

How rapidly were DVD player sales increasing at the end of the first quarter of 2000? We could differentiate $D(t) = \dfrac{68,571}{1 + 288.8e^{-0.28708t}}$ and evaluate the derivative at $t = 12$. However, it may be easier to calculate $D'(12)$ using the differential equation.

$$\frac{dD}{dt} = kD(M - D)$$

Since $-0.28708t = -Mkt$, $k = \dfrac{0.28708}{M}$. Since $M = 68,571$, $k = \dfrac{0.28708}{68,571}$. The differential equation may now be written as follows:

$$\frac{dD}{dt} = \left(\frac{0.28708}{68,571}\right)D(68,571 - D)$$

$$= 0.28708D - 0.000004187D^2$$

$$\left.\frac{dD}{dt}\right|_{t=12} = 0.28708D(12) - 0.000004187[D(12)]^2$$

$$D'(12) = 0.28708(6713.1) - 0.000004187(6713.1)^2$$

$$= 1927.20 - 188.69$$

$$\approx 1739$$

At the end of the first quarter of 2000, the cumulative number of DVD players sold was increasing at a rate of 1739 thousand DVD players per quarter. That is, from the end of the first quarter of 2000 to the end of the second quarter of 2000, we estimate that 1739 thousand DVD players were sold.

How rapidly will DVD player sales be increasing at the end of the first quarter of 2006? That is, what is $D'(36)$?

$$\left.\frac{dD}{dt}\right|_{t=36} = 0.28708D(36) - 0.000004187[D(36)]^2$$

$$D'(36) = 0.28708(67,934) - 0.000004187(67,934)^2$$

$$= 19,502 - 19,323$$

$$= 179$$

At the end of the first quarter of 2006, we forecast that the cumulative number of DVD players sold will be increasing at a rate of 179 thousand DVD players per quarter. That is, from the end of the first quarter of 2006 to the end of the second quarter of 2006, about 179 thousand DVD players will be sold. We are more skeptical of this estimate than of the 2000 estimate, since the first quarter of 2006 ($t = 36$) is so far outside the raw data set ($1 \leq t \leq 22$).

In practice, it is easiest to create a logistic model by using logistic regression. The following Technology Tip demonstrates how to find a logistic model for a set of data.

TECHNOLOGY TIP

Logistic Regression

1. Enter the data using the Statistics Menu List Editor. (Refer to Section 1.3 if you've forgotten how to do this.)

L1	L2	L3	3
0	1		
1	2		
2	4		
3	8		
4	10		
5	11		

L3(1) =

2. Bring up the Statistics CALC menu, select item B:Logistic, and press ENTER.

```
EDIT CALC TESTS
7↑QuartReg
8:LinReg(a+bx)
9:LnReg
0:ExpReg
A:PwrReg
B▮Logistic
C:SinReg
```

3. If you want to automatically paste the regression equation into the Y= editor, press the key sequence VARS Y-VARS; 1:Function; 1:Y1 and press ENTER. Otherwise press ENTER.

```
Logistic
y=c/(1+ae^(-bx))
a=16.65350584
b=1.161751976
c=11.62174904
```

EXAMPLE 4

Using a Differential Equation to Forecast the Price of a VCR

Based on data from 1997 to 2002, the average price of a video cassette recorder (VCR) may be modeled by

$$P(t) = \frac{110.0}{1 + 0.1955e^{1.150t}} + 65 \text{ dollars}$$

where t is the number of years since the end of 1997. The dramatic drop in price may be attributed in part to the introduction of the DVD player to the market in the first quarter of 1997. (**Source:** Modeled from Consumer Electronics Association data.) How quickly was the average price of a VCR changing at the end of 1999 and the end of 2001?

SOLUTION The function $P(t)$ is a logistic function shifted vertically 65 units. We'll define a new function $N(t) = P(t) - 65$. We have

$$N(t) = \frac{110.0}{1 + 0.1955e^{1.150t}}$$

This function is a solution to the differential equation

$$\frac{dN}{dt} = kN(110.0 - N)$$

$$= \left(\frac{-1.150}{110.0}\right)N(110.0 - N)$$

$$= -1.15N + 0.01045N^2$$

Since $\dfrac{dN}{dt} = \dfrac{dP}{dt}$,

$$\frac{dP}{dt} = -1.15N + 0.01045N^2$$

$$\left.\frac{dP}{dt}\right|_{t=2} = -1.15N(2) + 0.01045[N(2)]^2$$

$$= -1.15(37.29) + 0.01045(37.29)^2$$

$$= -28.35$$

At the end of 1999, the price of a VCR was dropping at a rate of about $28.35 per year.

$$\left.\frac{dP}{dt}\right|_{t=4} = -1.15N(4) + 0.01045[N(4)]^2$$

$$= -1.15(5.379) + 0.01045(5.379)^2$$

$$= -5.88$$

At the end of 2001, the price of a VCR was dropping at a rate of about $5.88 per year. The introduction of the DVD player had a dramatic impact on the price of a VCR.

In the past several examples, we have demonstrated the use of both limited and logistic growth models. Both types of models approach a constant value as the domain values grow large; however, near the origin the models behave quite differently. A limited growth model grows very rapidly at first, while a logistic model initially grows very slowly. Consequently, when choosing a model to use, it is important to consider the expected growth behavior of the function at the start of the time period in addition to looking at the shape of the scatter plot.

7.4 Summary

In this section, you learned how to use differential equations to find limited growth and logistic growth models. You also practiced interpreting the results of differential equations.

7.4 Exercises

In Exercises 1–6, determine the equation of the limited growth model for each species of fish and then answer the associated questions. You may find it helpful to refer to Example 3 in this section.

1.   **Red Mullet**

Red Mullet	k	L_∞	t_0
Female	0.23	25.6 cm	−1.38 years
Male	0.21	22.2 cm	−2.08 years

Source: www.fishbase.org.

  (a) How long is a male, three-year-old red mullet?
  (b) How long is a female, three-year-old red mullet?
  (c) Is the three-year-old male or the three-year-old female red mullet growing faster?

2.   **Turbot**

Turbot	k	L_∞	t_0
Female	0.115	103.4 cm	−0.93 year
Male	0.143	77.3 cm	−1.22 years

  (a) At what rate does a 50.0-cm-long male turbot grow?
  (b) At what rate does a 50.0-cm-long female turbot grow?
  (c) Which is older, a 50.0-cm-long male turbot or a 50.0-cm-long female turbot? Explain.

3.   **Flounder**

Flounder	k	L_∞	t_0
Female	0.197	44.98 cm	−1.43 years
Male	0.22	33.0 cm	−2.24 years

Source: www.fishbase.org.

  (a) At what rate does a 20.0-cm-long male flounder grow?
  (b) At what rate does a 20.0-cm-long female flounder grow?
  (c) Which is older, a 20-cm-long male flounder or a 20-cm-long female flounder? Explain.

4.   **Picarel Fish**

Pickerel	k	L_∞	t_0
Female	0.232	17.6 cm	−1.15 years
Male	0.793	20.6 cm	0.08 years

  (a) How long is a male, four-year-old picarel?
  (b) How long is a female, four-year-old picarel?
  (c) Is the four-year-old male or the four-year-old female picarel growing faster?

5.   **Copper Shark**

Copper Shark	k	L_∞	t_0
	0.038	385.0 cm	-3.477

Source: www.fishbase.org.

(a) How long is a three-year-old copper shark?

(b) At what rate is the length of a three-year-old copper shark changing?

6.  **Cowcod**

Cowcod	k	L_∞	t_0
	0.052	86.9 cm	-1.94

(a) How long is a two-year-old cowcod?

(b) At what rate is the length of a two-year-old cowcod changing?

In Exercises 7–12, use the logistic growth model and associated differential equation to answer the questions.

7. **M** **Pediatric AIDS** Based on data from 1992 to 2001, the estimated number of new pediatric AIDS cases in the United States in year t may be modeled by

$$P(t) = \frac{919.9}{1 + 0.0533e^{0.07758t}} + 90$$

cases, where t is the number of years since the end of 1992. (**Source:** Modeled from Centers for Disease Control and Prevention data.)

The rate of change of the pediatric AIDS function P is the same as the rate of change of the function $N(t) = \dfrac{919.9}{1 + 0.0533e^{0.07758t}}$.

(a) How many new pediatric AIDS cases are estimated to have occurred in 2000?

(b) What is the differential equation whose solution is $N(t)$?

(c) At what rate was the estimated number of new pediatric AIDS cases changing at the end of 2000?

(d) Based on the results of part (c) and the graph of $P(t)$, does it look like efforts to reduce pediatric AIDS in the United States are generating positive results? Explain.

8. **M** **Adult and Adolescent AIDS** Based on data from 1981 to 1995, the number of adult and adolescent deaths due to AIDS in a given year in the United States may be modeled by

$$P = \frac{53{,}955}{1 + 38.834e^{-0.45127t}}$$

deaths, where t is the number of years since the end of 1981. (**Source:** Modeled from Centers for Disease Control and Prevention data.)

(a) According to the model, how many adult and adolescent AIDS deaths occurred in the United States in 1995?

(b) According to the model, at what rate were adult and adolescent AIDS deaths increasing at the end of 1995?

(c) The number of adult and adolescent AIDS deaths in the United States has decreased every year since 1995, in part because of the increased availability of drug treatments and AIDS prevention efforts. In 2001, there were 8063 deaths. If the number of AIDS deaths had followed the logistic model, how many deaths would have occurred in 2002?

9. **M** **School Internet Access** Based on data from 1994 to 2000, the rate of change in the percentage of public-school classrooms with Internet access may be modeled by

$$\frac{dy}{dt} = 0.01077y(85.88 - y)$$

percentage points per year, where t is the number of years since the end of 1994. In 2000, 77 percent of public-school classrooms had Internet access. (**Source:** Modeled from *Statistical Abstract of the United States, 2001*, Table 243, p. 155.)

(a) Find the particular solution to the differential equation and interpret its real-world meaning.

(b) At what rate was the percentage of public schools with Internet access changing at the end of 2000?

10. **M** **National Internet Usage** Based on data from 1995 to 1999 and U.S. Census Bureau projections for 2000 to 2004, the rate of change in annual per capita Internet usage may be modeled by

$$\frac{dy}{dt} = 0.002882y(232.2 - y)$$

hours per year, where t is the number of years since the end of 1995. (**Source:** *Statistical Abstract of the United States, 2001*, Table 1125, p. 704.) In 1999, the annual per capita Internet usage was 99 hours.

(a) Find the particular solution to the differential equation and interpret its real-world meaning.

(b) At what rate was the annual per capita Internet usage changing at the end of 2000?

11.   **Hotel/Motel Occupancy** Based on data from 1990 to 1999, the average hotel/motel room rate may be modeled by

$$R(t) = \frac{27.80}{(1 + 47.23e^{-0.6425t})} + 57$$

dollars per day, where t is the number of years since the end of 1990. (**Source:** *Statistical Abstract of the United States, 2001*, Table 1266, p. 774.)

The rate of change of the room rate function R is the same as the rate of change of the function

$$N(t) = \frac{27.80}{(1 + 47.23e^{-0.6425t})}$$

(a) Find the differential equation whose solution is N.

(b) Determine at what rate the hotel/motel room rate was changing at the end of 1997 and at the end of 1999.

12.   **Information Technology** Based on data from 1990 to 2000, the percentage of the economy attributed to the information technology sector may be modeled by

$$P(t) = \frac{2.956}{1 + 30.15e^{-0.5219t}} + 7.9$$

percent, where t is the number of years since the end of 1990. (**Source:** Modeled from *Statistical Abstract of the United States, 2001*, Table 1122, p. 703.)

The rate of change in the information technology percentage of the economy function P is the same as the rate of change of the function

$$N(t) = \frac{2.956}{1 + 30.15e^{-0.5219t}}$$

(a) Find the differential equation whose solution is N.

(b) Determine at what rate the information technology percentage of the economy was changing at the end of 1995 and at the end of 2000.

In Exercises 13–14, use a limited growth function to model the data. Then answer the given questions.

13. **Spread of Information** Suppose that in a city of 350,000 people, 40,000 people were watching television or listening to the radio when the news of a bioterrorist attack was first broadcast. For the purpose of the model, assume that all 40,000 people heard the news simultaneously five minutes after the attack. Model the spread of the news of the attack. How long did it take for 75 percent of the city to hear the news, and at what rate was the news spreading at that time?

14.   **Video Game Sales** Blizzard Entertainment released the engaging real-time strategy game "Warcraft III: Reign of Chaos" on July 3, 2002. By July 22, 2002, more than a million copies of the game had been sold. (**Source:** www.pcgameworld.com.)

Model the "Warcraft III" game sales, assuming that at the end of the twentieth day after the game was released, 1 million copies of the game had been sold. Furthermore, assume that a total of 8 million copies of the game will be sold over the life of the game. According to the model, how long did it take for 4 million copies to be sold, and at what rate were sales increasing at that time?

In Exercises 15–20, use logistic regression to find the logistic model for the data. Then answer the given questions.

15.   **Deadly Fights over Money**

Homicides Due to Arguments over Money or Property

Years since 1990 (t)	Homicides (H)
1	520
2	483
3	445
4	387
5	338
6	328
7	287
8	241
9	213
10	206

Source: *Crime in the United States 1995, 2000*, Uniform Crime Report, FBI.

According to the model, at what rate were money-related homicides decreasing in 2001?

16. **Cassette Tape Market Share**

Years Since 1993 (t)	Percent of Music Market (percentage points) (P)
0	38.0
1	32.1
2	25.1
3	19.3
4	18.2
5	14.8
6	8.0
7	4.9
8	3.4
9	2.4

Source: Recording Industry of America.

According to the model, in what year will cassette tape market share drop below 1 percent? At what rate will the cassette tape market share be decreasing at that time?

   17. **Movie Box Office Sales**

My Big Fat Greek Wedding

Weekend	Week #	Cumulative Gross Box Office Sales (dollars)
Apr. 19–21 (2002)	1	597,362
June 28–30	11	19,340,988
Sept. 6–8	21	95,824,732
Nov. 15–17	31	199,574,370
Jan. 24–26 (2003)	41	236,448,697
Apr. 4–6	51	241,437,427

Source: www.boxofficeguru.com.

(a) According to the model, were cumulative box office sales for *My Big Fat Greek Wedding* increasing at a higher rate in Week 1 or Week 51?

(b) IFC Films was the distributor for *My Big Fat Greek Wedding*. (**Source:** www.boxofficeguru.com.) If you were a marketing consultant to IFC Films, what would you tell the company about forecasted box office sales beyond Week 51?

  18. **Resource Value**

Value of Fabricated Metals Shipments

Years Since 1992 (t)	Shipment Value (millions of dollars) (V)
0	170,403
1	177,967
2	194,113
3	212,444
4	222,995
5	242,812
6	253,720
7	256,900
8	258,960

Source: *Statistical Abstract of the United States, 2001*, Table 982, p. 624.

(*Hint:* Before creating the model, align the data by subtracting 170,000 from each value of V. After doing logistic regression, add back the 170,000 to the resultant model equation.)

According to the model, was the value of fabricated metals shipments increasing more rapidly in 1995 or in 1997?

19.   **TV Homes with Cable**

Years Since 1970 (t)	TV Homes with Cable (percent) (C)
0	6.7
5	12.6
10	19.9
15	42.8
16	45.6
17	47.7
18	49.4
19	52.8
20	56.4
21	58.9
22	60.2
23	61.4
24	62.4
25	63.4
26	65.3
27	66.5
28	67.2
29	67.5

Source: *Statistical Abstract of the United States, 2001*, Table 1126, p. 705.

According to the model, what percentage of TV homes will have cable in 2006, and at what rate will that percentage be increasing?

 20.   **U.S. Army Personnel**

Years Since 1980 (t)	Personnel (thousands) (P)
0	777
2	780
4	780
6	781
8	772
10	732
12	610
14	541
16	491
18	484
20	482

Source: *Statistical Abstract of the United States, 2001*, Table 500, p. 329.

(*Hint:* Before creating the model, align the data by subtracting 480 from each value of P. After doing logistic regression, add back the 480 to the resultant model equation.)

According to the model, how many people were in the U.S. Army in 2005, and at what rate was that number changing?

Exercises 21–23 are intended to challenge your understanding of differential equations.

21. Prove that if $\dfrac{dy}{dt} = k(M - y)$ and $y(0) = 0$, then $y = M(1 - e^{-kt})$.

22. Prove that if $y = \dfrac{M}{1 + Se^{-kMt}}$, then $\dfrac{dy}{dt} = ky(M - y)$.

23. Show that if $D(t) = \dfrac{100}{1 + 200e^{-0.1t}}$, then
$$D'(t) = 0.001[D(t)]\{100 - [D(t)]\}.$$

Chapter 7 Review Exercises

Section 7.1   *In Exercises 1–10, integrate the function using the simplest possible method.*

1. $f(x) = 2xe^{-x}$

2. $g(x) = -4x\ln(x)$

3. $h(x) = 3x^2 e^x$

4. $g(t) = 2t\ln(t)$

5. $f(x) = \dfrac{[\ln(x)]^3}{x}$

6. $h(x) = (2x - 1)e^{x^2 - x + 1}$

7. $g(x) = (5x - 3)[\ln(x)]$

8. $f(t) = (t - 1)^4 (2t - 5)$ **9.** $h(x) = 4xe^{4x}$

10. $g(t) = 2t - t\ln(t)$

Section 7.2   *In Exercises 11-15, calculate the combined area of the region(s) bounded by the graphs of the two functions. (On some of the exercises, you may find it helpful to use your graphing calculator to locate the intersection points of the graphs.)*

11. $f(x) = x^2 - 1; g(x) = -x^2 + 17$

12. $f(x) = x^2 - 1; g(x) = x^3 - 3x^2 - x + 3$

13. $f(x) = 2^x; g(x) = x^2$

14. $f(x) = 3^x; g(x) = 3x^3$

15. $f(x) = \ln(x); g(x) = 0.1x$

In Exercises 16–20, graph each of the functions. Then determine the solution by using the integral techniques demonstrated in Section 7.2.

16.   **Adult and Adolescent AIDS**   Based on data from 1981 to 1992, the rate of change in the annual number of adult and adolescent AIDS cases diagnosed in the United States may be modeled by

$$R(t) = 75.48t^3 - 1237t^2 + 6652t - 4882$$

cases per year, and the rate of change in the annual number of adult and adolescent deaths in the United States due to AIDS may be modeled by

$$D(t) = -107.3t^2 + 1641t - 975.3$$

deaths per year, where t is the number of years since the end of 1981. (**Source:** Modeled from Centers for Disease Control and Prevention data.)

Determine the accumulated change in the difference of the two populations from the end of 1981 through 1992. Interpret the meaning of the result.

17.   **Company Profit**   Based on data from 1999 to 2001, the rate of change in the revenue of the Johnson & Johnson company and its subsidiaries may be modeled by

$$r(t) = 1330t + 1150$$

million dollars per year, and the rate of change in the cost of goods sold may be modeled by

$$c(t) = 206t + 315$$

million dollars, where t is the number of years since the end of 1999. (**Source:** Modeled from Johnson & Johnson 2001 Annual Report.)

Determine the accumulated change in annual profit from the end of 1999 through 2001.

18.   **Public- and Private-School Enrollment**   Based on U.S. National Center for Education Statistics enrollment projections for 2000 to 2010, the rate of change in the annual number of students enrolled in public schools may be modeled by

$$g(t) = 0.4971t^2 - 3.022t + 51.17$$

thousand students per year, and the rate of change in the number of students enrolled in private schools may be modeled by

$$p(t) = -19.99t + 280.5$$

thousand students per year, where t is the number of years since the end of 2000. (**Source:** Modeled from *Statistical Abstract of the United States, 2001,* Table 205, p. 133.) In 2000, 58,821,000 students were enrolled in public schools and 9,303,000 students were enrolled in private schools.

Determine the accumulated change in the annual enrollment gap between private and public schools over the period 2000 to 2010. Interpret the meaning of the result.

19.   **Public- and Private-School Teachers**   Based on U.S. National Center for Educational Statistics projections for 2000 to

2010, the rate of change in the number of public-school teachers may be modeled by

$$g(t) = -1.4242t + 16.11$$

thousand teachers per year, and the rate of change in the number of private-school teachers may be modeled by

$$p(t) = -0.8158t + 1.453$$

thousand teachers per year, where t is the number of years since the end of 2000. (**Source:** Modeled from *Statistical Abstract of the United States, 2001,* Table 207, p. 134.) In 2000, there were 2,850,000 public-school teachers and 402,000 private-school teachers.

Determine the predicted accumulated change in the annual gap between private- and public-school teachers over the period 2000 to 2010. Interpret the meaning of the result.

20. **M** **Wages of Clothing Producers** Based on data from 1980 to 2000, the rate of change in the average hourly wage of a men's and boys' furnishings production worker may be modeled by

$$m(t) = 0.007032t + 0.1446$$

dollars per year, and the rate of change in the average hourly wage of a women's and misses' outerwear production worker may be modeled by

$$w(t) = 0.006394t + 0.1333$$

dollars per year, where t is the number of years since 1980. (**Source:** Modeled from *Statistical Abstract of the United States, 2001,* Table 609, p. 394.) In 1980, the hourly wage of a men's and boys' furnishings production worker was \$4.23, and the average hourly wage of a women's and misses outerwear production worker was \$4.61.

Determine the accumulated change in the hourly wage gap between men's and boys' furnishings and women's and misses outerwear production workers over the period 1980 to 2000.

Section 7.3 *In Exercises 21–25, find the general solution of the separable differential equation. Assume that k is a constant. (Hint:*

$$\int \frac{dy}{y(M-y)} = \frac{1}{M} \ln\left(\frac{y}{M-y}\right).)$$

21. $\dfrac{dy}{dx} = \dfrac{y}{x}$

22. $\dfrac{dA}{dt} = -5A^2$

23. $\dfrac{dA}{dt} = 2(16 - A)$

24. $\dfrac{dy}{dx} = 0.1(y - 10)$

25. $\dfrac{dA}{dt} = 4A - A^2$

In Exercises 26–30, find the particular solution for the differential equations.

26. $\dfrac{dy}{dx} = \dfrac{2x}{y}; \; y(0) = -1$

27. $\dfrac{dA}{dt} = -5A^2; \; A(0) = -\dfrac{1}{2}$

28. $\dfrac{dA}{dt} = 2(16 - A); \; A(0) = 15$

29. $\dfrac{dy}{dx} = 0.1(y - 10); \; y(0) = 11$

30. $\dfrac{dA}{dt} = 4A - A^2; \; A(0) = 1$

In Exercise 31, use Newton's Law of Heating and Cooling.

31. **Time of Death** A crime scene investigator finds that the temperature of a corpse in a 70°F room is 80°F. Thirty minutes later, the temperature of the corpse is 77°F. Assuming that the woman's temperature was 98.6°F when she died, find a function for the temperature of the body t minutes after the woman died. Then use the model to estimate how long the woman had been dead when the crime scene investigator first took her temperature.

Section 7.4 *In Exercises 32–33, find the model indicated and answer the associated questions.*

32. **Textbook Sales** Cumulative textbook sales may often be modeled by a limited growth model. A publisher's marketing analyst predicted that a new edition of a textbook would sell 12,900 copies in its first year and 20,500 copies over the life of the edition.

(a) Find the limited growth model for the cumulative number of sales of the edition.

(b) Calculate the cumulative number of books sold by the end of the second year.

(c) Determine the rate of change in the cumulative number of textbook sales at the end of the second year.

33. **Electronic Game Sales** Based on data from 1996 to 2000, the number of copies of computer and video games sold worldwide may be modeled by

$$S(t) = \frac{121.8}{1 + 17.23e^{-1.794t}} + 100$$

million units, where t is the number of years since the end of 1996. (**Source:** Modeled from Interactive Digital Software Association State of the Industry Report, 2000-2001.)

The rate of change in the number of game units sold, S, is the same as the rate of change of the function

$$N(t) = \frac{121.8}{1 + 17.23e^{-1.794t}}$$

(a) Find the differential equation whose solution is N.
(b) Determine at what rate computer and video game sales were changing in 1997 and in 1999.

In Exercises 34–35, determine the equation of the limited growth model for each species of fish and then answer the associated questions.

34. **Rainbow Trout in Australia**

Rainbow Trout	k	L_∞	t_0
	0.340	52.0 cm	−0.08 year

(a) How long is a three-year-old rainbow trout?
(b) At what rate is the length of a three-year-old rainbow trout changing?

35. **Albacore Tuna off California**

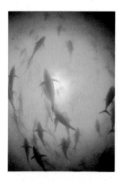

Albacore Tuna	k	L_∞	t_0
	0.233	109.0 cm	−2.31 years

Source: www.fishbase.org.

(a) How long is a two-year-old albacore tuna?
(b) At what rate is the length of a two-year-old albacore tuna changing?

In Exercises 36–37, use logistic regression to find the logistic model for the data. Then answer the given questions.

36. **Number of Different Banks**

Years Since 1984 (t)	Banks (B)
0	17,900
1	18,033
2	17,876
3	17,325
4	16,562
5	15,829
6	15,192
7	14,517
8	13,891
9	13,261
10	12,641
11	12,002
12	11,478
13	10,923
14	10,463
15	10,221
16	9,908

Source: *Statistical Abstract of the United States, 2001,* Table 1173, p. 728.

(*Hint:* Align the data by subtracting 9900 from B. After finding the model using logistic regression, add back the 9900.)

According to the model, at what rate was the number of banks declining in 2000?

37.  **U.S. Homicide Rate**

Years Since 1990 (t)	Homicides per 100,000 People (H)
0	9.4
1	9.8
2	9.3
3	9.5
4	9
5	8.2
6	7.4
7	6.8
8	6.3
9	5.7
10	5.5

Source: FBI Uniform Crime Reports, 2000.

(*Hint:* Align the data by subtracting 5.4 from *H*. After finding the model using logistic regression, add back the 5.4.)

According to the model, at what rate was the number of homicides per hundred thousand people declining in 2000?

Make It Real

What to do

1. Identify a public company that you would like to work for in the future.
2. Obtain a copy of the company's annual financial report.
3. Using five or more years of data, model the revenue of the company and the costs of the company as a function of time.
4. Graph the revenue and cost models on the same axes.
5. Calculate the accumulated change in revenue and the accumulated change in costs over a five-year period using the models.
6. Shade the region that represents the accumulated change in profit.
7. Using the results from Steps 5 and 6, calculate the accumulated change in profit over the five-year period

Where to look for data

Publicly held companies are required to release financial data to the public through annual reports. Many of these annual reports are available online and may be accessed through the investor information page on the company's website. Here are links to the homepages of some prominent companies.

Safeway
www.safeway.com

McDonald's
www.mcdonalds.com

Kellogg's
www.kelloggs.com

Revlon
www.revlon.com

draw the surface. Additionally, purchased graphing programs such as Maple, Mathematica, and Autograph can create beautiful 3D surfaces that can be easily rotated.

To help you better visualize the surface, the graph of the cylinder volume function is shown from two different angles in Figure 8.4.

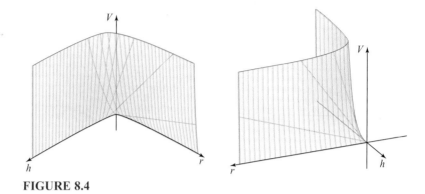

FIGURE 8.4

EXAMPLE 1

Graphing a Multivariable Function

Generate a table of data for the multivariable function $f(x, y) = x^2 + y^2$ and use a graphing utility to draw the surface.

SOLUTION To calculate the values in Table 8.3, we substitute each ordered pair (x, y) into the function f. For example,

$$f(x, y) = x^2 + y^2$$
$$f(-3, -2) = (-3)^2 + (-2)^2$$
$$= 9 + 4$$
$$= 13$$

TABLE 8.3

x ↓	$y \rightarrow$						
	−3	−2	−1	0	1	2	3
−3	18	13	10	9	10	13	18
−2	13	8	5	4	5	8	13
−1	10	5	2	1	2	5	10
0	9	4	1	0	1	4	9
1	10	5	2	1	2	5	10
2	13	8	5	4	5	8	13
3	18	13	10	9	10	13	18

Notice that the low point of this surface occurs at the origin, $(0, 0, 0)$. As we move away from the origin in any direction, the value of z increases. Two different views of the graph of the surface are shown in Figure 8.5.

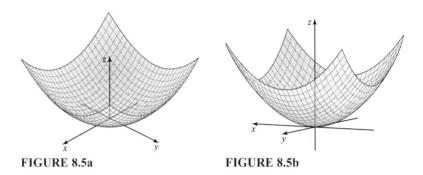

FIGURE 8.5a   **FIGURE 8.5b**

Figure 8.5a is a graph in standard position. The x-axis points to the left, the y-axis points to the right, and the z-axis points upward. In Figure 8.5b, we have rotated the axes so that we can see the surface from a different point of view.

EXAMPLE 2

Graphing a Multivariable Function

Generate a table of data for the multivariable function $f(x, y) = y^3 - yx^2$ and use a graphing utility to draw the surface.

SOLUTION

TABLE 8.4

x ↓ \ y →	−3	−2	−1	0	1	2	3
−3	0	10	8	0	−8	−10	0
−2	−15	0	3	0	−3	0	15
−1	−24	−6	0	0	0	6	24
0	−27	−8	−1	0	1	8	27
1	−24	−6	0	0	0	6	24
2	−15	0	3	0	−3	0	15
3	0	10	8	0	−8	−10	0

The graph of the function on the specified domain is shown in Figure 8.6 in standard position and with the axes rotated.

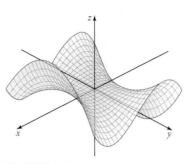
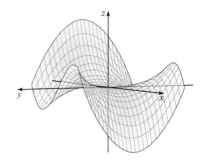

FIGURE 8.6

EXAMPLE 3

Using a Multivariable Function to Forecast the Value of an Investment

A $1000 investment is to be made into an account paying 6 percent compound interest. The value of the investment compounded n times per year after t years is given by

$$A(n, t) = 1000\left(1 + \frac{0.06}{n} \right)^{nt}$$

Generate a table of data for the multivariable function using $n = 1, 2, 4, 12, 365$ and $t = 0, 1, 2, 3$. Graph the function and interpret the meaning of $(12, 2, 1127.16)$.

SOLUTION

TABLE 8.5

$n \downarrow$	$t \rightarrow$			
	0	1	2	3
1	1000.00	1060.00	1123.60	1191.02
2	1000.00	1060.90	1125.51	1194.05
4	1000.00	1061.36	1126.49	1195.62
12	1000.00	1061.68	1127.16	1196.68
365	1000.00	1061.83	1127.49	1197.20

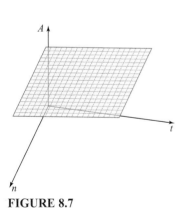

FIGURE 8.7

When we graph the function using a window with $0 \le A \le 1200$, the graph looks like a plane parallel to the nt plane (Figure 8.7).

However, when we use the window $1000 \le A \le 1200$, we get a better look at what is happening (Figure 8.8).

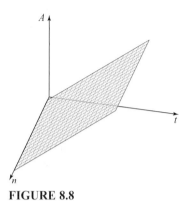

FIGURE 8.8

The compounding frequency, n, has a greater effect on the value of the investment as t increases. The meaning of $(12, 2, 1127.16)$ is that a $1000 investment earning 6 percent interest compounded monthly ($t = 12$) at the end of two years ($n = 2$) will have a value of $1127.16.

EXAMPLE 4 **Using a Multivariable Function to Forecast Business Sales**

Based on data from 1998 to 2001, the net sales of Kellogg Company and subsidiaries may be modeled by

$$S(a, e) = -0.36661a + 0.16556e + 4671.3 \text{ million dollars}$$

where a is the amount of money spent on advertising (in millions of dollars) and e is the number of employees working for the company. (**Source:** Modeled from Kellogg Company 2001 Annual Report, p. 26.) In 2001, the company had 26,424 employees, spent $519.2 million on advertising, and had $8853.3 million in net sales. If the company were to increase advertising spending to $700 million and decrease the number of employees to 25,000, what would be the predicted net sales?

SOLUTION We are asked to calculate $S(700, 25{,}000)$.

$$S(700, 25{,}000) = -0.36661(700) + 0.16556(25{,}000) + 4671.3$$
$$= 8553.7$$

Net sales are predicted to be $8553.7 million when advertising spending is $700 million and the number of employees is 25,000.

Let's look a little more closely at the coefficients of the model equation. The units of the output of the function are *millions of sales dollars*. Consequently, the units of each term in the sum must also be *millions of sales dollars*. The units of a are *millions of advertising dollars*. What are the units of the coefficient -0.36661? The units are *sales dollars per advertising dollar*, since

UNITS

$$\frac{\text{sales dollars}}{\text{advertising dollar}} \cdot \text{millions of } \cancel{\text{advertising dollars}} = \text{millions of sales dollars}$$

According to the model, for every advertising dollar, sales decrease by 0.36661 million dollars ($366,610). What are the units of the coefficient 0.16556? The units are *millions of sales dollars per employee*, since

UNITS

$$\frac{\text{millions of sales dollars}}{\cancel{\text{employee}}} \cdot \cancel{\text{employees}} = \text{millions of sales dollars}$$

According to the model, adding an employee will increase sales by 0.16556 million dollars ($165,560).

Although the coefficients of the model indicate that increasing advertising spending will decrease net sales, we know that if Kellogg stops advertising altogether ($a = 0$), its net sales will eventually decrease, not increase. However, since the coefficient of a in the model is negative, the model does seem to suggest that the company might want to look more closely at its advertising expenditures to ensure that it is getting the desired return on its investment.

The **Cobb-Douglas production function,**

$$f(L, C) = kL^m C^n$$

is widely used in economics. The model assumes that the output of a company, industry, or country is a function of its labor L and its capital C. (Capital may be interpreted as the dollar value of the money invested in the company, including equipment, material, and building costs.) The model requires that m, n, and k are positive constants and $m + n = 1$.

EXAMPLE 5   **Using a Cobb-Douglas Production Model**

Based on data from 2000 and 2001, a Cobb-Douglas production model for the Coca-Cola Company is given by

$$P(L, C) = 0.02506L^{0.0481} C^{0.9519} \text{ billion unit cases of liquid beverage}$$

where L is the labor cost (including sales, payroll, and other taxes) in millions of dollars and C is the cost of capital expenditures (property, plant, and equipment) in millions of dollars. One unit case of beverage equals 24 eight-ounce servings. (**Source:** Modeled from Coca-Cola Company 2000 and 2001 Annual Reports.) According to the model, would it be better to spend $800 million on capital expenditures and $200 million on labor costs or $700 million on capital expenditures and $300 million on labor costs? Explain.

SOLUTION   We are asked to calculate $P(200, 800)$ and $P(300, 700)$.

$$P(200, 800) = 0.02506(200)^{0.0481}(800)^{0.9519}$$
$$= 18.8$$
$$P(300, 700) = 0.02506(300)^{0.0481}(700)^{0.9519}$$
$$= 16.8$$

Spending $200 million on labor and $800 million on capital expenditures is expected to produce 18.8 billion unit cases of beverage, while spending $300 million on labor and $700 million on capital expenditures is expected to produce 16.8 billion unit cases of beverage. The model suggests that it may be better to spend $800 million on capital expenditures and $200 million on labor; however, there may be additional constraints to take into consideration. For example, if the Coca-Cola Company were to purchase additional equipment with its capital, it might need to hire additional people to operate the new equipment. Hiring additional people would result in a corresponding increase in the labor cost.

Using a Multivariable Function to Find the Surface Area of a Box

A box manufacturer designed a closeable box with a square base, as shown in Figure 8.9. Find the surface area equation for the unassembled box. How many square inches of material are required to construct a box 6 inches wide, 6 inches long, and 18 inches high?

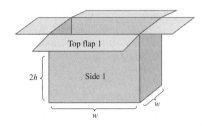

	h	Top flap 1	Top flap 2	Top flap 3	Top flap 4	
$2h$		Side 1	Side 2	Side 3	Side 4	
h		Bottom flap 1	Bottom flap 2	Bottom flap 3	Bottom flap 4	$\frac{1}{4}w$
		w	w	w	w	

FIGURE 8.9

SOLUTION   The surface area of the unassembled box is given by

$$A(h, w) = (h + 2h + h)(w + w + w + w) + \frac{1}{4}w(2h)$$

$$= (4h)(4w) + \frac{1}{2}hw$$

$$= 16hw + \frac{1}{2}hw$$

$$= 16.5hw \text{ square inches}$$

The height of the box is $2h$. So

$$2h = 18$$
$$h = 9$$

The width and length of the box are both 6 inches, so $w = 6$.

$$A(h, w) = 16.5(9)(6)$$
$$= 891$$

The surface area of the unassembled box is 891 square inches.

8.1 Summary

In this section, you learned how to evaluate multivariable functions and saw how functions of two variables may be represented graphically. You also discovered some applications of multivariable functions.

8.1 Exercises

In Exercises 1–10, generate a table of data for the multivariable function using $x = -2, 0, 2$ and $y = -1, 0, 1$. Then use a graphing utility to graph the function. Compare the coordinates of the points in the table to the coordinates of the points on the graph of the function.

1. $z = 3x - 2y$ **2.** $z = xy - y^2$

3. $z = 4xy + 4$

4. $f(x, y) = x^2 + 2xy + y^2$

5. $f(x, y) = e^{-xy}$

6. $f(x, y) = x^2 - 2xy + y^2$

7. $f(x, y) = \dfrac{x^2 + 1}{y^2 + 1}$ **8.** $f(x, y) = \dfrac{xy}{y^2 + 1}$

9. $z = (x + y)^3$ **10.** $z = \ln(x^2 y^2 + 1)$

In Exercises 11–20, use the multivariable function to answer the given question.

11. **Cobb-Douglas Function** Based on data from 1999 and 2000, a Cobb-Douglas model for Ford Motor Company is given by

$$P(L, C) = 0.5319L^{0.4558} C^{0.5442}$$

thousand vehicles, where L is the total labor cost and C is the capital expenditure cost, both in millions of dollars. In 2000, Ford Motor Company spent $25,783 million on labor and produced 7424 thousand vehicles. (**Source:** Modeled from Ford Motor Company 2001 Annual Report.)

According to the model, if Ford spends $20,000 million on labor and $9000 million on capital expenditures, how many vehicles will it produce?

12. **Housing Costs** The U.S. Department of Housing and Urban Development (HUD) published a research study in January 2002 comparing the cost of building a steel-frame home to the cost of building a wood-frame home. Two homes with identical floor plans were built side by side in Beaufort, South Carolina.

The material and labor costs of constructing each home are given in the following table.

	Material Cost	Labor Cost
Steel Frame	$97,447.52	$32,228.94
Wood Frame	$91,589.38	$25,236.83

Source: HUD.

Find an equation for the total cost of constructing w identical wood-frame homes and s steel-frame homes in Beaufort, South Carolina. How much will it cost to construct ten wood-frame and five steel-frame homes?

13. **Personal Data Assistant Software** SplashData is a leading developer of virtual wallet applications for Palm OS handhelds. As of December 2002, SplashData offered consumers six products: SplashWallet Suite ($39.95), SplashClock ($5.00), SplashID ($19.95), SplashMoney ($14.95), SplashPhoto ($19.95), and SplashShopper ($19.95). (**Source:** www.splashdata.com.)

Find the formula for the SplashData revenue function. How much revenue will SplashData generate if it sells 20 SplashWallet Suites, 50 SplashClocks, 10 SplashIDs, 100 SplashMoneys, 40 SplashPhotos, and 70 SplashShoppers?

14. **Future Value of Investment** Find the equation for the future value of an investment of P dollars into an account paying 6 percent interest compounded monthly t years after the investment is made. What is the value of a $2500 investment after 10 years?

15. **Surface Area** Referring to Example 6 in the section, find a formula for the exterior surface area of the assembled box with a square base. (Assume that the top and bottom of the box are closed.) What is the exterior surface area of the box with a 6-inch width, 6-inch length, and 18-inch height?

16. **Surface Area Analysis**   In Example 6 we calculated the surface area of an unassembled box, and in Exercise 15 you calculated the exterior surface area of the assembled box. Of what practical value are each of these results?

17.   **Body Mass Index**   The Body Mass Index (BMI) is a helpful indicator of obesity or underweight in adults. A high BMI is predictive of death from cardiovascular disease. The Centers for Disease Control and Prevention indicate that a healthy BMI for adults is between 18.5 and 24.9. They use the following guidelines.

Underweight	BMI less than 18.5
Overweight	BMI of 25.0 to 29.9
Obese	BMI of 30.0 or more

The following formula is used to calculate body mass index.

$$BMI = \frac{703 \cdot W}{H^2}$$

where W is your weight in pounds and H is your height in inches. What is your BMI?

18.   **Relative Humidity**   The relative humidity of the air alters our perception of the temperature. As the humidity increases, the air feels hotter to us. The heat index (apparent temperature) is a function of air temperature and relative humidity, as shown in the table below.

When exposed to apparent temperatures exceeding 105°F for prolonged periods of time, people are at high risk of sunstroke, a potentially fatal condition. The heat index values were devised for shady, light wind conditions. Exposure to full sunshine can increase apparent temperature by up to 15°F. What is the apparent temperature when the air temperature is 90°F and the relative humidity is 80 percent?

	Relative Humidity (%)												
Air Temperature (°F)	40	45	50	55	60	65	70	75	80	85	90	95	100
110	136												
108	130	137											
106	124	130	137										
104	119	124	131	137					**Heat Index**				
102	114	119	124	130	137				**(Apparent Temperature)**				
100	109	114	118	124	129	136							
98	105	109	113	117	123	128	134						
96	101	104	108	112	116	121	126	132					
94	97	100	103	106	110	114	119	124	129	135			
92	94	96	99	101	105	108	112	116	121	126	131		
90	91	93	95	97	100	103	106	109	113	117	122	127	132
88	88	89	91	93	95	98	100	103	106	110	113	117	121
86	85	87	88	89	91	93	95	97	100	102	105	108	112
84	83	84	85	86	88	89	90	92	94	96	98	100	103
82	81	82	83	84	84	85	86	88	89	90	91	93	95
80	80	80	81	81	82	82	83	84	84	85	86	86	87

Source: National Weather Service.

19. *M* **Wind Chill** Wind alters our perception of the temperature. As the wind strength increases, the air feels colder to us. This phenomenon is referred to as *wind chill*. Wind chill temperature is a function of wind speed and air temperature and is modeled by

$$W(T, V) = 35.74 + 0.6215T$$
$$- 35.75V^{0.16} + 0.4275TV^{0.16}$$

degrees Fahrenheit, where T is the actual air temperature (in degrees Fahrenheit) and V is the velocity of the wind (in miles per hour). (**Source:** National Climatic Data Center.) When the air is 10°F and the wind is blowing at 40 mph, what is the wind chill temperature?

20. **Ice Cream Cones** The GE-500 Semi-Automatic is a machine that is designed to create a variety of sizes of cones, ranging from a diameter of 20 millimeters (mm) to a diameter of 56 mm. The number of cones that can be made in an hour depends upon the diameter of the cone. The production rate of 20-mm cones is 4400 per hour, while the production rate of 42-mm cones is 2000 per hour. (**Source:** www.maneklalexports.com.)

Let s be the number of hours spent producing 20-mm cones and l be the number of hours spent producing 42-mm cones. Find a formula for the cumulative number of cones produced as a function of s and l. How many cones are produced if 1.5 hours are spent making 20-mm cones and 3 hours are spent making 42-mm cones?

Exercises 21–23 are intended to challenge your understanding of multivariable functions.

21. What is the domain of the multivariable function?

$$f(x, y) = \frac{\sqrt{x^2 - 1}}{y}$$

22. What are the domain and range of the multivariable function?

$$f(x, y) = 3 + \sqrt{x - 2} + \sqrt{y^2 - 4}$$

23. The amount of a person's take-home pay is a multivariable function. Identify *four* different variables that contribute to the amount of a person's take-home pay.

8.2 Partial Derivatives

- Analyze and interpret multivariable mathematical models
- Calculate first- and second-order partial derivatives of multivariable functions
- Find the equation of a cross section of a graph of a surface
- Determine the meaning of partial rates of change

GETTING STARTED The wind chill temperature is a function of the air temperature and the wind velocity. If the wind speed is increasing at a rate of 2 mph per hour, how quickly is the wind chill temperature changing? Questions such as these may be answered by using *partial derivatives*.

In this section, we will demonstrate how to calculate partial derivatives. These partial rates of change allow us to determine the effect of a change in one of the input variables on the output.

Recall that the operator $\dfrac{d}{dx}$ means "take the derivative with respect to x." The operator $\dfrac{\partial}{\partial x}$ means "take the partial derivative with respect to x." When calculating a partial derivative, all variables except the variable of differentiation are treated as constants. Consider the function $z = x^2 - xy$.

$$\frac{\partial}{\partial x}(z) = \frac{\partial}{\partial x}(x^2 - xy)$$

$$\frac{\partial z}{\partial x} = 2x - y$$

Observe that we treated the y in xy as if it were a constant. That is, when differentiating $x^2 - xy$, we treated the expression as if it were $x^2 - cx$ (where c is a constant). Since $\dfrac{d}{dx}(x^2 - cx) = 2x - c$, $\dfrac{\partial z}{\partial x} = 2x - y$. Let's now take the partial derivative of the function with respect to y.

$$\frac{\partial}{\partial y}(z) = \frac{\partial}{\partial y}(x^2 - xy)$$

$$\frac{\partial z}{\partial y} = -x$$

Since the variable of differentiation is y, we treated x as a constant. That is, when differentiating $x^2 - xy$, we treated the expression as if it were $c^2 - cy$ (where c is a constant). Since $\dfrac{d}{dy}(c^2 - cy) = -c$, $\dfrac{\partial}{\partial y}(x^2 - xy) = -x$. Observe that $\dfrac{\partial z}{\partial x} \neq \dfrac{\partial z}{\partial y}$. The partial derivative of a function varies based upon the variable of differentiation.

PARTIAL DERIVATIVES

The **partial derivatives** of a function $f(x, y)$ are given by

$\dfrac{\partial f}{\partial x}$ (read "the partial of f with respect to x")

$\dfrac{\partial f}{\partial y}$ (read "the partial of f with respect to y")

To calculate $\dfrac{\partial f}{\partial x}$, differentiate the function treating the y variable as a constant.

To calculate $\dfrac{\partial f}{\partial y}$, differentiate the function treating the x variable as a constant.

EXAMPLE 1

Finding Partial Derivatives

Find the partial derivatives of $f(x, y) = 3xy + y^2 + 5$.

SOLUTION

$$\begin{aligned}
\frac{\partial f}{\partial x} &= \frac{\partial}{\partial x}(3xy + y^2 + 5) \\[2mm]
&= \frac{d}{dx}(3cx + c^2 + 5) \qquad \text{\small c is a constant representing y} \\[2mm]
&= 3c \\[2mm]
&= 3y \qquad\qquad\qquad \text{\small Replace c with y}
\end{aligned}$$

Replacing the variable y with the constant c is an optional step used to help you apply the partial differentiation rules correctly. As you become skilled at applying the rules, the step will become unnecessary.

$$\frac{\partial f}{\partial y} = \frac{\partial}{\partial y}(3xy + y^2 + 5)$$

$$= \frac{d}{dy}(3cy + y^2 + 5) \qquad c \text{ is a constant representing } x$$

$$= 3c + 2y$$

$$= 3x + 2y \qquad\qquad \text{Replace } c \text{ with } x$$

PARTIAL DERIVATIVES (ALTERNATIVE NOTATION)

The **partial derivatives** of a function $f(x, y)$ may alternatively be written as

f_x (read "the partial of f with respect to x")

f_y (read "the partial of f with respect to y")

EXAMPLE 2

Finding Partial Derivatives

Find the partial derivatives of $f(x, y) = x^2 y^2$.

SOLUTION

$$f_x = \frac{\partial}{\partial x}(x^2 y^2)$$

$$= \frac{d}{dx}(x^2 c^2) \qquad c \text{ is a constant representing } y$$

$$= 2xc^2$$

$$= 2xy^2 \qquad\qquad \text{Replace } c \text{ with } y$$

$$f_y = \frac{\partial}{\partial y}(x^2 y^2)$$

$$= \frac{d}{dy}(c^2 y^2) \qquad c \text{ is a constant representing } x$$

$$= 2c^2 y$$

$$= 2x^2 y \qquad\qquad \text{Replace } c \text{ with } x$$

The units of $\dfrac{\partial f}{\partial x}$ are the units of f divided by the units of x. Similarly, the units of $\dfrac{\partial f}{\partial y}$ are the units of f divided by the units of y.

EXAMPLE 3

Interpreting the Meaning of the Partial Derivatives of the Compound Interest Function

Find the partial derivatives of the compound interest function $A(r, t) = 1000\left(1 + \dfrac{r}{4}\right)^{4t}$. Then evaluate the partial derivatives at the point $(0.12, 3, 1425.76)$ and interpret the meaning of the result.

SOLUTION We will first find the partial derivatives.

$$\frac{\partial}{\partial t}(A) = \frac{\partial}{\partial t}\left[1000\left(1 + \frac{r}{4}\right)^{4t}\right]$$

$$\frac{\partial A}{\partial t} = 1000 \cdot \ln\left(1 + \frac{r}{4}\right) \cdot \left(1 + \frac{r}{4}\right)^{4t} \cdot 4 \qquad \text{Exponential, Constant Multiple, and Chain Rules}$$

$$A_t = 4000 \ln\left(1 + \frac{r}{4}\right) \cdot \left(1 + \frac{r}{4}\right)^{4t}$$

$$\frac{\partial A}{\partial r} = 1000 \cdot 4t\left(1 + \frac{r}{4}\right)^{4t-1} \cdot \left(\frac{1}{4}\right) \qquad \text{Power, Constant Multiple, and Chain Rules}$$

$$A_r = 1000t\left(1 + \frac{r}{4}\right)^{4t-1}$$

Next, we will evaluate each of the partial derivatives at the point (0.12, 3, 1425.76).

$$A_t(0.12, 3) = 4000 \ln\left(1 + \frac{0.12}{4}\right) \cdot \left(1 + \frac{0.12}{4}\right)^{4(3)}$$

$$= 4000 \cdot \ln(1.03)(1.03)^{12}$$

$$= 168.58$$

The units of the partial derivative are the units of the output (dollars) divided by the units of t (years). An investment account earning 12 percent interest (compounded quarterly) with an initial investment of $1000 is increasing in value at a rate of $168.58 per year at the end of the third year. That is, from the end of the third year to the end of the fourth year, the account value will increase by about $168.58.

$$A_r = 1000t\left(1 + \frac{r}{4}\right)^{4t-1}$$

$$A_r(0.12, 3) = 1000(3)\left(1 + \frac{0.12}{4}\right)^{4(3)-1}$$

$$= 3000(1.03)^{11}$$

$$= 4152.70$$

The units of the partial derivative are the units of the output (dollars) divided by the units of r (100 percentage points). (Observe that $0.12 \cdot 100$ percentage points $= 12\%$.)

$$4152.70\frac{\text{dollars}}{100 \text{ percentage points}} = 41.527 \text{ dollars per percentage point}$$

An investment account earning 12 percent interest per year (compounded quarterly) with an initial investment of $1000 is increasing in value at a rate of $41.53 per percentage point increase in the interest rate at the end of the third year. That is, increasing the interest rate from 12 percent per year to 13 percent per year at the end of the third year will increase the account value at the end of the fourth year by approximately $41.53 over what it would have been without the interest rate increase.

EXAMPLE 4

Interpreting the Meaning of Partial Derivatives

The distance of a cue ball from the eight ball in the game of billiards may be expressed in terms of two perpendicular components parallel to the side or end of the pool table, respectively, as shown in Figure 8.10.

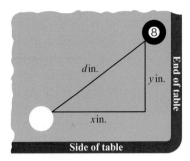

FIGURE 8.10

By the Pythagorean Theorem, $d = \sqrt{x^2 + y^2}$. Find the partial derivative of the distance function. Then evaluate each partial derivative at the point $(3, 4, 5)$ and interpret the meaning of the result.

SOLUTION We will first determine the equations of the partial derivatives.

$$d = \sqrt{x^2 + y^2}$$
$$= (x^2 + y^2)^{1/2}$$

$$d_x = \frac{1}{2}(x^2 + y^2)^{-1/2}(2x) \qquad \text{Chain Rule treating } y \text{ as a constant}$$
$$= x(x^2 + y^2)^{-1/2}$$

$$d_y = \frac{1}{2}(x^2 + y^2)^{-1/2}(2y) \qquad \text{Chain Rule treating } x \text{ as a constant}$$
$$= y(x^2 + y^2)^{-1/2}$$

Evaluating these functions at $(3, 4, 5)$, we determine

$$d_x(3, 4) = (3)(3^2 + 4^2)^{-1/2}$$
$$= \frac{3}{5}$$
$$= 0.6$$

$$d_y(3, 4) = (4)(3^2 + 4^2)^{-1/2}$$
$$= \frac{4}{5}$$
$$= 0.8$$

When the cue ball is five inches away from the eight ball, rolling toward the eight ball one inch (along the path depicted in Figure 8.10) will change the cue ball's distance from the end of the pool table by 0.6 inch and its distance from the side of the pool table by 0.8 inch.

EXAMPLE 5

Using Partial Derivatives to Forecast Changes in the Wind Chill Temperature

Wind chill temperature is a function of wind speed and air temperature and is modeled by

$$W(T, V) = 35.74 + 0.6215T - 35.75V^{0.16} + 0.4275TV^{0.16}$$

degrees Fahrenheit, where T is the actual air temperature (in degrees Fahrenheit) and V is the velocity of the wind (in miles per hour). (**Source:** National Climatic Data Center.) If the wind speed is 20 mph, how much will a 1-mph change in wind speed alter the wind chill temperature if the current temperature remains at 30 degrees? If the current temperature is 30 degrees, how much will a 1-degree increase in temperature alter the wind chill temperature if the current wind speed remains at 20 mph?

SOLUTION   We begin by calculating the partial derivatives of W.

$$W(T, V) = 35.74 + 0.6215T - 35.75V^{0.16} + 0.4275TV^{0.16}$$

UNITS

$$\frac{\partial W}{\partial T} = 0.6215 + 0.4275V^{0.16} \quad \frac{\text{degree of wind chill temperature}}{\text{degree of actual temperature}}$$

$$\frac{\partial W}{\partial V} = 0.16(-35.75)V^{-0.84} + 0.16(0.4275TV^{-0.84})$$

$$= -5.72V^{-0.84} + 0.0684TV^{-0.84}$$

UNITS

$$= 0.0684V^{-0.84}(-83.63 + T) \quad \frac{\text{degree of wind chill temperature}}{\text{miles per hour}}$$

We will now evaluate each of the partial derivatives at $(30, 20)$.

$$\frac{\partial W}{\partial T}(30, 20) = 0.6215 + 0.4275(20)^{0.16}$$

UNITS

$$= 1.312 \quad \frac{\text{degree of wind chill temperature}}{\text{degree of actual temperature}}$$

Increasing the temperature from 30 to 31 degrees will increase the wind chill temperature by about 1.3 degrees if the wind remains at 20 mph. (Since the partial derivative does not contain the variable T, the wind chill temperature will increase by about 1.3 degrees for every 1 degree increase in temperature regardless of the initial temperature.)

$$\frac{\partial W}{\partial V}(30, 20) = 0.0684V^{-0.84}(-83.63 + T)$$

$$= 0.0684(20)^{-0.84}(-83.63 + 30)$$

UNITS

$$= -0.2962 \quad \frac{\text{degree of wind chill temperature}}{\text{miles per hour}}$$

Increasing the wind speed from 20 to 21 miles per hour will decrease the wind chill temperature by about 0.3 degree when the temperature is 30 degrees. The colder the temperature is, the more substantial the impact a change in wind speed will have on the wind chill temperature.

Cross Sections of a Surface

When an apple is sliced in half, the exposed surface is called a **cross section** of the apple. If the cross section is dipped in paint and pressed on a sheet of paper, the resultant image is a two-dimensional figure (see Figure 8.11).

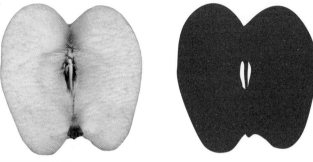

FIGURE 8.11

The shape of the cross section varies depending upon the point at which the apple is sliced. Figure 8.12 shows three different cross sections of the same apple.

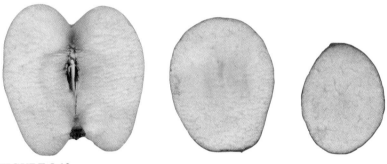

FIGURE 8.12

The equation for the border of the cross section can be derived from the original multivariable equation. (The shape of an apple may be modeled by a multivariable piecewise equation. However, because of the complexity of the equation, we will focus on simpler models as we analyze cross sections.)

Consider the graph of the function $f(x, y) = -x^3 + 4y^2$, shown in Figure 8.13.

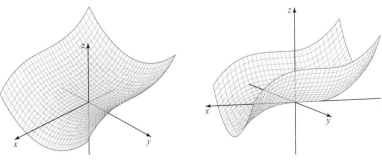

FIGURE 8.13

If we want to know what the cross section of the surface looks like at $y = 0$, we calculate $f(x, 0)$. The equation of the cross-section graph is given by

$$f(x, 0) = -x^3 + 4(0)^2$$
$$= -x^3$$

Slicing the surface at $y = 0$ yields the graph shown in Figure 8.14.

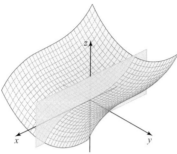 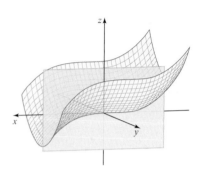

FIGURE 8.14

Note: Looking at the graph from this angle, the x-values go from positive to negative as we read the graph from left to right.

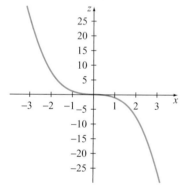

FIGURE 8.15

Graphing the cross section in two dimensions yields the curve $z = -x^3$ (Figure 8.15).

The slope of the cross-section graph is $\dfrac{dz}{dx} = -3x^2$. What is the relationship between the slope of the cross-section graph and the partial derivative? Let's find f_x and evaluate it at $(x, 0)$.

$$f_x = -3x^2$$

Observe that $f_x(x, 0) = -3x^2$. The slope equation for the cross-section graph is the same as the equation of the partial derivative of f with respect to x evaluated at $(x, 0)$. Let's look at another cross section and see if the relationship holds true. This time we'll slice the surface at $y = 2$ (see Figure 8.16).

 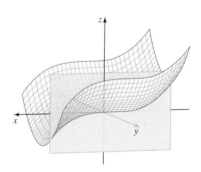

FIGURE 8.16

Note: Looking at the graph from this angle, the x-values go from positive to negative as we read the graph from left to right.

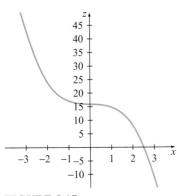

FIGURE 8.17

The equation of the cross-section graph when $y = 2$ is given by $f(x, 2)$.

$$f(x, 2) = -x^3 + 4(2)^2$$
$$= -x^3 + 16$$

Graphing the cross-section graph in two dimensions yields the graph shown in Figure 8.17.

The slope of the cross section graph is $\dfrac{dz}{dx} = -3x^2$. We find f_x and evaluate it at $(x, 2)$.

$$f_x = -3x^2$$

Observe that $f_x(x, 2) = -3x^2$. The slope equation for the cross-section graph is the same as the equation of the partial derivative of f with respect to x evaluated at $(x, 2)$.

GEOMETRIC INTERPRETATION OF PARTIAL DERIVATIVES

The partial derivatives of a function $f(x, y)$ represent rates of change of the graph of the surface.

- $f_x(x, y)$ is the rate of change of the graph of the function $f(x, y)$ when y is held constant. It is referred to as the *slope in the x direction.*
- $f_y(x, y)$ is the rate of change of the graph of the function $f(x, y)$ when x is held constant. It is referred to as the *slope in the y direction.*

Consider the graph of $f(x, y) = -x^2 + 4x - y^2 + y + xy + 100$ (see Figure 8.18). The point $(3, 1, 106)$ lies on the graph.

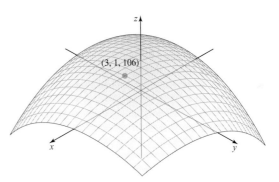

FIGURE 8.18

Let's calculate the partial derivatives.

$$f_x(x, y) = -2x + 4 + y$$
$$f_y(x, y) = -2y + 1 + x$$

Recall that $f_x(x, y)$ is the slope in the x-direction. Thus, $f_x(3, 1)$ is the slope of the graph in the x-direction at the point $(3, 1, 106)$.

$$f_x(x, y) = -2x + 4 + y$$
$$f_x(3, 1) = -2(3) + 4 + 1$$
$$= -1$$

There is a line tangent to the graph at $(3, 1, 106)$ that has slope -1. The line is of the form $z = -1x + b$, since the line is in the x-direction. Substituting in $x = 3$ and $z = 106$, we can find the exact equation of the line.

$$z = -1x + b$$
$$106 = -3 + b$$
$$b = 109$$
$$\text{So } z = -x + 109$$

The equation of the tangent line to the graph of f in the x-direction at the point $(3, 1, 106)$ is $z = -x + 109$. We draw the line on the graph of f (see Figure 8.19).

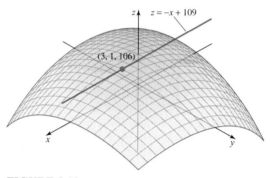

FIGURE 8.19

Recall that $f_y(x, y)$ is the slope in the y-direction. Thus, $f_y(3, 1)$ is the slope of the graph in the y-direction at the point $(3, 1, 106)$.

$$f_y(x, y) = -2y + 1 + x$$
$$f_y(3, 1) = -2(1) + 1 + 3$$
$$= 2$$

There is a line tangent to the graph at $(3, 1, 106)$ that has slope 2. The line is of the form $z = 2y + b$, since the line is in the y-direction. Substituting in $y = 1$ and $z = 106$, we can find the exact equation of the line.

$$z = 2y + b$$
$$106 = 2(1) + b$$
$$106 = 2 + b$$
$$b = 104$$
$$\text{So } z = 2y + 104$$

So the equation of the tangent line to the graph of f in the y-direction at the point $(3, 1, 106)$ is $z = 2y + 104$. We draw the line on the graph of f (see Figure 8.20).

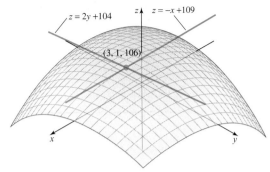

FIGURE 8.20

It is important to note that in three dimensions, the graph of the equation $z = 2y + 104$ is really the *plane* containing the green line that we have drawn, not the line itself. However, by requiring $x = 3$, we constrain the plane to the green line. Similarly, the graph of $z = -x + 109$ is the graph of the *plane* containing the blue line we have drawn, not the line itself. However, by requiring $y = 1$, we constrain the plane to the blue line. Without these additional constraints on x and y, respectively, the graph would look like Figure 8.21.

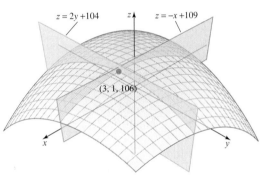

FIGURE 8.21

EXAMPLE 6

Finding Tangent Lines of a Multivariable Function

Given the function $f(x, y) = x^2 - xy^2$, determine the equations of the tangent lines in the x-direction and in the y-direction at the point $(3, 2, -3)$.

SOLUTION The rate of change of f in the x-direction is given by

$$f_x(x, y) = 2x - y^2$$

Thus the slope of the tangent line in the x-direction at $(3, 2, -3)$ is

$$f_x(3, 2) = 2(3) - (2)^2$$
$$= 2$$

Since y is being held constant, the equation will have the form $z = 2x + b$. We find the value of b by substituting in the point $(3, 2, -3)$.

$$z = 2x + b$$
$$-3 = 2(3) + b$$
$$b = -9$$
$$\text{So } z = 2x - 9$$

Thus the equation of the tangent line in the x-direction (when y is constrained to $y = 2$) is $z = 2x - 9$.

The rate of change of f in the y direction is given by

$$f_y(x, y) = -2xy$$

Thus the slope of the tangent line in the y-direction at $(3, 2, -3)$ is

$$f_y(3, 2) = -2(3)(2)$$
$$= -12$$

Since x is being held constant, the equation will have the form $z = -12y + b$. We find the value of b by substituting in the point $(3, 2, -3)$.

$$z = -12y + b$$
$$-3 = -12(2) + b$$
$$b = 21$$
$$\text{So } z = -12y + 21$$

Thus the equation of the tangent line in the y-direction (when x is constrained to $x = 3$) is $z = -12y + 21$.

We can confirm our results graphically by using a graphing utility (see Figure 8.22).

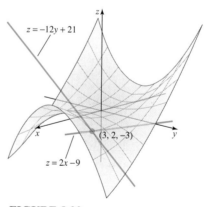

FIGURE 8.22

Second-Order Partial Derivatives

Recall that the derivative of a derivative of a function is called the second derivative of the function. Likewise, a partial derivative of a partial derivative of a function is called a **second-order partial derivative.**

SECOND-ORDER PARTIAL DERIVATIVES

The **second-order partial derivatives** of a function $f(x, y)$ may be written as

$$f_{xx} = \frac{\partial}{\partial x}\left(\frac{\partial f}{\partial x}\right) \text{ (read "the second partial of } f \text{ with respect to } x\text{")}$$

$$f_{yy} = \frac{\partial}{\partial y}\left(\frac{\partial f}{\partial y}\right) \text{ (read "the second partial of } f \text{ with respect to } y\text{")}$$

$$f_{xy} = \frac{\partial}{\partial y}\left(\frac{\partial f}{\partial x}\right) \text{ (read "the mixed second partial derivative of } f \text{ with respect to } x \text{ then } y\text{")}$$

$$f_{yx} = \frac{\partial}{\partial x}\left(\frac{\partial f}{\partial y}\right) \text{ (read "the mixed second partial derivative of } f \text{ with respect to } y \text{ then } x\text{")}$$

For all functions with continuous first and second partial derivatives, $f_{xy} = f_{yx}$. For all multivariable functions in this text, $f_{xy} = f_{yx}$.

EXAMPLE **7**

Finding First and Second Partial Derivatives

The graph of the function $f(x, y) = x^2 + y^2 + 1$ is shown in Figure 8.23. Calculate the value of the first and second partial derivatives of the function at the point $(0, 0, 1)$.

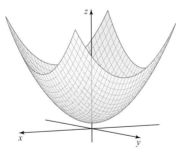

FIGURE 8.23

SOLUTION   We have

$$f_x = 2x$$
$$f_y = 2y$$
$$f_{xx} = 2$$
$$f_{yy} = 2$$
$$f_{xy} = 0$$
$$f_{yx} = 0$$

Evaluating each of the partial derivatives at $(0, 0, 1)$, we get

$$f_x(0, 0) = 2(0)$$
$$= 0$$
$$f_y(0, 0) = 2(0)$$
$$= 0$$

The second partial derivatives remain constant for all values of x and y. Thus

$$f_{xx}(0, 0) = 2$$
$$f_{yy}(0, 0) = 2$$
$$f_{xy}(0, 0) = 0$$
$$f_{yx}(0, 0) = 0$$

EXAMPLE 8

Finding First and Second Partial Derivatives

The graph of the function $f(x, y) = x^3 - xy^2$ is shown in Figure 8.24. Calculate the first and second partial derivatives of the function.

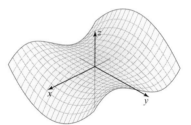

FIGURE 8.24

SOLUTION

$$f_x = 3x^2 - y^2$$
$$f_y = -2xy$$
$$f_{xx} = 6x$$
$$f_{yy} = -2x$$
$$f_{xy} = -2y$$
$$f_{yx} = -2y$$

In the next section, we will demonstrate the key role that second-order partial derivatives play in identifying the location of the maxima and minima of a surface.

8.2 Summary

In this section, you learned how to calculate partial derivatives. You saw that these partial rates of change allow you to determine the effect on the output of a change in one of the input variables.

8.2 Exercises

In Exercises 1–10, calculate the first and second partial derivatives of each of the functions.

1. $f(x, y) = 3xy$ **2.** $f(x, y) = x^2y - xy^2$

3. $f(x, y) = 4xy - y^2$ **4.** $f(x, y) = x^3 - y^3$

5. $W = t\sqrt{v}$ **6.** $W = v^2t^{1.2} + v$

7. $V = \pi r^2 h$ **8.** $V = lw^2$

9. $z = x \ln(xy)$ **10.** $z = xy \ln(y)$

In Exercises 11–17, calculate and interpret the practical meaning of each of the partial derivatives.

11. Volume of a Cylinder   The volume of a cylinder is given by $V(h, r) = \pi r^2 h$, with h and r measured in inches. What is the practical meaning of $V_r(3, 2)$ and $V_h(3, 2)$?

12. Volume of a Box   The volume of a rectangular box with a square base is given by $V(l, w) = lw^2$, with l and w measured in inches. What is the practical meaning of $V_l(4, 5)$ and $V_w(4, 5)$?

13. **M** **Wind Chill Temperature**   Wind chill temperature is a function of wind speed and air temperature and is modeled by
$$W(T, V) = 35.74 + 0.6215T$$
$$- 35.75V^{0.16} + 0.4275TV^{0.16}$$
degrees Fahrenheit, where T is the actual air temperature (in degrees Fahrenheit) and V is the velocity of the wind (in miles per hour). (Source: National Climatic Data Center.) What is the practical meaning of $W_T(27, 5)$ and $W_V(27, 5)$?

14. **M** **Wind Chill Temperature**   Wind chill temperature is a function of wind speed and air temperature and is modeled by
$$W(T, V) = 35.74 + 0.6215T$$
$$- 35.75V^{0.16} + 0.4275TV^{0.16}$$
degrees Fahrenheit, where T is the actual air temperature (in degrees Fahrenheit) and V is the velocity of the wind (in miles per hour). (Source: National Climatic Data Center.) What is the practical meaning of $W_T(-30, 15)$ and $W_V(-30, 15)$?

15. **M** **Body Mass Index**   The following formula is used to calculate the body mass index:
$$B(H, W) = \frac{703 \cdot W}{H^2}$$
where W is weight in pounds and H is height in inches. What is the practical meaning of $B_H(67, 155)$ and $B_W(67, 155)$?

16. **M** **Body Mass Index**   The following formula is used to calculate body mass index:
$$B(H, W) = \frac{703 \cdot W}{H^2}$$
where W is weight in pounds and H is height in inches. What is the practical meaning of $B_H(67, 100)$ and $B_W(67, 100)$?

17. **M** **Price-to-Earnings Ratio**   The price-to-earnings ratio (P/E) is the most common measure of how expensive a stock is, and is given by
$$R(E, P) = \frac{P}{E}$$
where P is the price of a share of the stock and E is the earnings per share. What is the practical meaning of $R_E(4, 32)$ and $R_P(4, 32)$?

In Exercises 18–20, use partial derivatives to determine the answer to each question.

18. **M** **Price-to-Earnings Ratio**   The price-to-earnings ratio (P/E) is the most common measure of how expensive a stock is, and is given by
$$R(E, P) = \frac{P}{E}$$

Recall that relative extrema of single-variable functions could not occur at endpoints. Similarly, relative extrema of two-variable functions may not occur on the border of the domain region.

To help you visualize what relative extrema in three-dimensions look like, we will label the relative extrema on the graphs of several multivariable functions (see Figure 8.25).

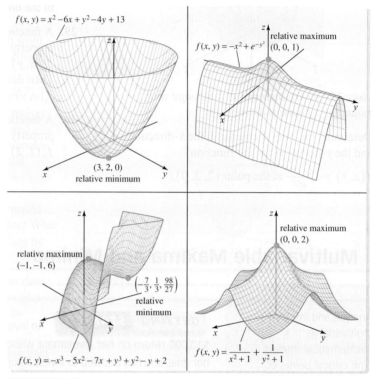

FIGURE 8.25

As was the case in two dimensions, relative extrema have horizontal tangent lines provided that the extrema don't occur at a sharp point on the graph.

In addition to relative extrema, a three-dimensional surface may have a *saddle point*. A **saddle point** is a point (a, b, c) that is a relative maximum of one cross-section graph and a relative minimum of another cross-section graph. The portion of the graph surrounding the saddle point looks like a "saddle." Consider the graph of $f(x, y) = -x^2 + y^2 + 2$, shown in Figure 8.26.

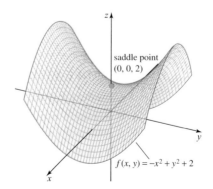

FIGURE 8.26

If we slice the graph with the plane $x = 0$, we see a cross section that looks like a concave up parabola with a vertex at $(0, 0, 2)$ (see Figure 8.27).

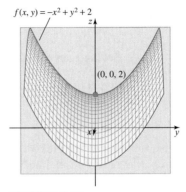

$f(x, y) = -x^2 + y^2 + 2$

$(0, 0, 2)$

FIGURE 8.27

The smallest possible value of z (in the cross section) is reached at the vertex. Similarly, if we slice the graph with the plane $y = 0$, we see a cross section that looks like a concave down parabola with a vertex at $(0, 0, 2)$ (see Figure 8.28).

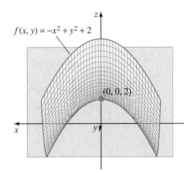

$f(x, y) = -x^2 + y^2 + 2$

$(0, 0, 2)$

FIGURE 8.28

The largest possible value of z (in the cross section) is reached at the vertex. Since the same point is a maximum value of one cross section and a minimum value of a different cross section, the point $(0, 0, 2)$ is a saddle point. Later in the section, we will discuss how to find saddle points using partial derivatives.

Recall from our discussion of single-variable functions that a value c in the domain of f was called a critical value of f if $f'(c) = 0$ or $f'(c)$ was undefined. All relative extrema occurred at critical values.

A similar definition is used for two-variable functions.

CRITICAL POINT OF A TWO-VARIABLE FUNCTION

An ordered pair (c, d) in the domain of f is a **critical point** of f if

$$f_x(c, d) = 0 \text{ and } f_y(c, d) = 0$$

(Both statements must be true in order for (c, d) to be a critical point.)

When we refer to the critical point (c, d), we mean that (c, d) is the pair of input values whose output is $f(c, d)$.

Let's find the critical point of $f(x, y) = x^2 + y^2 + 1$.

$$f_x = 2x$$
$$f_y = 2y$$

Setting each partial derivative equal to zero and solving, we find that

$$f_x = 2x \qquad f_y = 2y$$
$$0 = 2x \qquad 0 = 2y$$
$$x = 0 \qquad y = 0$$

Therefore, the ordered pair $(0, 0)$ is a critical point of f. The domain region of the function is defined by $-\infty < x < \infty$ and $-\infty < y < \infty$. Since $(0, 0)$ is in the interior of the region, it is a relative extremum candidate.

To determine whether a maximum or minimum occurs at a critical point, we will apply the Second Derivative Test for Two-Variable Functions.

SECOND DERIVATIVE TEST FOR TWO-VARIABLE FUNCTIONS (D-TEST)

To determine if a relative extremum of a function f occurs at a critical point (c, d), first calculate

$$D(c, d) = f_{xx}(c, d) \cdot f_{yy}(c, d) - [f_{xy}(c, d)]^2$$

Then

- f has a **relative maximum** at (c, d) if $D(c, d) > 0$ and $f_{xx}(c, d) < 0$.
- f has a **relative minimum** at (c, d) if $D(c, d) > 0$ and $f_{xx}(c, d) > 0$.
- f has a **saddle point** at (c, d) if $D(c, d) < 0$.
- The test is inconclusive if $D(c, d) = 0$.

To apply the D-Test, we must first find the second partial derivatives. For the function $f(x, y) = x^2 + y^2 + 1$, we have the first partial derivatives $f_x = 2x$ and $f_y = 2y$. Therefore,

$$f_{xx} = 2$$
$$f_{yy} = 2$$
$$f_{xy} = 0$$

We evaluate each of the second partial derivatives at the critical point $(0, 0)$. In this case, the second partials are constant for all values of x and y. Thus

$$f_{xx}(0, 0) = 2$$
$$f_{yy}(0, 0) = 2$$
$$f_{xy}(0, 0) = 0$$

We now calculate D.

$$D(0, 0) = 2 \cdot 2 - 0^2$$
$$= 4$$

Since $D > 0$, either a maximum or a minimum occurs at $(0, 0)$. Since $f_{xx}(0, 0) > 0$, a relative minimum occurs at $(0, 0)$. Evaluating f at $(0, 0)$, we determine that

$$f(0, 0) = 0^2 + 0^2 + 1$$
$$= 1$$

The relative minimum occurs at $(0, 0, 1)$.

EXAMPLE **1**

Locating Relative Extrema and Saddle Points

Find the location of the relative extrema and saddle points of $f(x, y) = x^2 + y^2 + xy - 2$. The graph of the surface is shown in Figure 8.29.

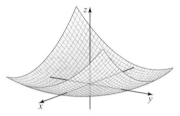

FIGURE 8.29

SOLUTION The partial derivatives of f are

$$f_x = 2x + y$$
$$f_y = 2y + x$$

We set each partial derivative equal to zero and solve.

$$0 = 2x + y \qquad 0 = 2y + x$$
$$y = -2x \qquad y = -0.5x$$

Since both equalities must be true at a critical point, we set the functions equal to each other and solve.

$$-2x = -0.5x$$
$$1.5x = 0$$
$$x = 0$$

Back substituting into the equation $y = -2x$, we determine that

$$y = -2(0)$$
$$= 0$$

The critical point of the function is $(0, 0)$. We now find the second partial derivatives.

$$f_{xx} = 2$$
$$f_{yy} = 2$$
$$f_{xy} = 1$$

Since the second partials are constant, evaluating each second partial at the critical point will result in the same values.

$$D(0, 0) = 2 \cdot 2 - 1 \cdot 1$$
$$= 3$$

Since $D(0, 0) > 0$ and $f_{xx} > 0$, a relative minimum occurs at $(0, 0)$. The value of the function at $(0, 0)$ is $f(0, 0)$.

$$f(0, 0) = (0)^2 + (0)^2 + (0)(0) - 2$$
$$= -2$$

The relative minimum occurs at $(0, 0, -2)$. There are no relative maxima or saddle points.

EXAMPLE **2**   ## Locating Relative Extrema and Saddle Points

Find the location of the relative extrema and saddle points of $f(x, y) = e^{-xy}$. The graph of the surface is shown from two different viewpoints in Figure 8.30.

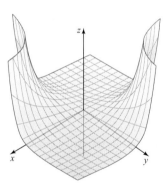

FIGURE 8.30

SOLUTION   We'll first find the partial derivatives of f.

$$f_x = -ye^{-xy}$$
$$f_y = -xe^{-xy}$$

To find the critical points, we'll set each partial equal to zero and solve.

$$0 = -ye^{-xy}$$

We solve the equation by setting each factor to zero.

$$-y = 0 \qquad e^{-xy} = 0$$
$$y = 0$$

Since $e^{-xy} \neq 0$ for all values of x and y, the only value that makes the equation zero is $y = 0$.

We'll now set f_y equal to zero.

$$0 = -xe^{-xy}$$

We solve the equation by setting each factor to zero.

$$x = 0 \qquad e^{-xy} = 0$$

Since $e^{-xy} \neq 0$ for all values of x and y, the only value that makes the equation zero is $x = 0$. Thus, the critical point of the function is $(0, 0)$.

We'll now find the second partials and calculate the value of D at $(0, 0)$.

$$f_{xx} = y^2 e^{-xy}$$
$$f_{yy} = x^2 e^{-xy}$$
$$f_{xy} = -1(e^{-xy}) + (-xe^{-xy})(-y)$$
$$= -e^{-xy}(1 - xy)$$
$$f_{xx}(0, 0) = 0$$
$$f_{yy}(0, 0) = 0$$
$$f_{xy}(0, 0) = -e^{(0)(0)}[1 - (0)(0)]$$
$$= -1$$
$$D(0, 0) = 0 \cdot 0 - (-1)^2$$
$$= -1$$

Since $D < 0$, a saddle point occurs at the critical point $(0, 0)$. The coordinates of the saddle point are $(0, 0, 1)$. There are no relative extrema.

EXAMPLE 3

Finding Relative Extrema and Saddle Points

Find the location of the relative extrema and saddle points of $f(x, y) = 0.1x^2y^2$. Two different views of the graph of the surface are shown in Figure 8.31.

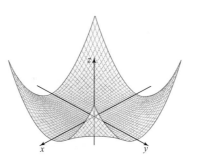

FIGURE 8.31

SOLUTION The partial derivatives of f are given by

$$f_x = 0.2xy^2$$
$$f_y = 0.2x^2y$$

Setting the partial derivatives equal to zero and solving yields

$$0 = 0.2xy^2 \qquad\qquad 0 = 0.2x^2y$$
$$0 = xy^2 \qquad\qquad 0 = x^2y$$
$$x = 0 \quad \text{or} \quad y = 0 \qquad x = 0 \quad \text{or} \quad y = 0$$

Any point of the form $(a, 0)$ or $(0, b)$ is a critical point of the function (a and b are real numbers). The function f has an infinite number of critical points.

The second partial derivatives are given by

$$f_{xx} = 0.2y^2$$
$$f_{yy} = 0.2x^2$$
$$f_{xy} = 0.4xy$$
$$f_{yx} = 0.4xy$$
$$D = (0.2y^2)(0.2x^2) - (0.4xy)^2$$
$$= 0.04x^2y^2 - 0.16x^2y^2$$
$$= -0.12x^2y^2$$

Observe that $D(a, 0) = 0$ and $D(0, b) = 0$ for all values of a and b.

The D-Test is inconclusive for all critical points. However, by evaluating the function at points near the critical points, we can determine what is happening there. We'll generate a data table (Table 8.6) by increasing and decreasing each domain variable on either side of the critical points.

TABLE 8.6

x $\downarrow$	$y \rightarrow$				
	-2	-1	0	1	2
-2	1.6	0.4	0	0.4	1.6
-1	0.4	0.1	0	0.1	0.4
0	0	0	0	0	0
1	0.4	0.1	0	0.1	0.4
2	1.6	0.4	0	0.4	1.6

The values highlighted in Table 8.6 are the function values of the critical points of f. Observe that as we move away from the highlighted values vertically or horizontally, the value of f increases or remains constant. Since $f(a, 0) \leq f(x, y)$ and $f(0, b) \leq f(x, y)$ for all values of a, b, x, and y, relative minima occur at all of the critical points. There are no relative maxima or saddle points.

EXAMPLE 4

Using Multivariable Functions to Forecast Investment Growth

An investor has $3000 to invest in two separate accounts. The first account is a certificate of deposit guaranteed to earn 6 percent annually. The second account is an aggressive growth fund expected to earn between -4 percent and 16 percent annually. The combined dollar value of the return on the investment accounts after 20 years is given by

$$V = x(1.06)^{20} + (3000 - x)(1 + y)^{20} - 3000$$

where x is the amount of money invested in the certificate of deposit and y is the growth rate (as a decimal) of the aggressive growth fund. (If the growth rate is 8 percent, $y = 0.08$.) A graph of the function is shown in Figure 8.32. The green plane is a graph of the function $z = 0$ and is drawn to provide a point of reference. When the graph of V is above the green plane, the investor makes money. When the graph of V is below the green plane, the investor loses money.

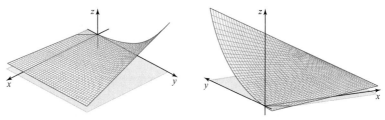

FIGURE 8.32

Find all critical points of V and determine the location of relative extrema and saddle points.

SOLUTION

$$V_x = (1.06)^{20} - (1 + y)^{20}$$
$$V_y = 20(3000 - x)(1 + y)^{19}$$

We set each partial derivative equal to zero.

$$0 = (1.06)^{20} - (1 + y)^{20}$$
$$(1 + y)^{20} = (1.06)^{20}$$
$$1 + y = 1.06$$
$$y = 0.06$$
$$0 = 20(3000 - x)(1 + y)^{19}$$

$$3000 - x = 0 \quad \text{or} \quad (1 + y)^{19} = 0$$
$$x = 3000 \qquad\qquad y = -1$$

Recall that in order to be a critical point, the ordered pair must satisfy both equations. Since $-1 \neq 0.06$, we throw out the value $y = -1$. Thus the only critical point is $(3000, 0.06)$.

We'll now find the second partials.

$$V_{xx} = 0$$
$$V_{yy} = 380(3000 - x)(1 + y)^{18}$$
$$V_{xy} = -20(1 + y)^{19}$$

$$D = (0)[380(3000 - x)(1 + y)^{18}] - [-20(1 + y)^{19}]^2$$
$$= -400(1 + y)^{38}$$

$$D(3000, 0.06) = -400(1 + 0.06)^{38}$$

$$D(3000, 0.06) < 0 \qquad\qquad \text{We only need to know the sign of } D$$

Since $D < 0$, a saddle point occurs at $(3000, 0.06, 6621.41)$. However, from the context of the problem, we know that $0 \le x \le 3000$. Therefore, this point sits on the border of the domain region. As with relative extrema, a saddle point may not sit on the border of the domain region. We conclude that the function does not have any relative extrema or saddle points on the domain region defined by $0 \le x \le 3000$ and $-0.04 \le y \le 0.16$.

Let's look at the table of data for the function in the region surrounding the critical point given in Example 4 (Table 8.7).

TABLE 8.7

x $\downarrow$	$y \rightarrow$						
	−0.04	0.00	0.04	0.06	0.08	0.12	0.16
2,600	5,515.35	5,738.55	6,215.00	6,621.41	7,202.94	9,197.07	13,122.86
2,700	5,791.87	5,959.27	6,316.60	6,621.41	7,057.55	8,553.15	11,497.49
2,800	6,068.38	6,179.98	6,418.20	6,621.41	6,912.17	7,909.24	9,872.13
2,900	6,344.89	6,400.69	6,519.81	6,621.41	6,766.79	7,265.32	8,246.77
3,000	6,621.41	6,621.41	6,621.41	6,621.41	6,621.41	6,621.41	6,621.41

From the table, we see that investing all of the money in the certificate of deposit or earning exactly 6 percent on the amount of money invested in the aggressive growth account will yield a combined return of $6621.41 on the $3000 investment. As the amount of money invested in the certificate of deposit (x) decreases, the amount of the return becomes more volatile. If $2600 is invested in the certificate of deposit and $400 is invested in the aggressive growth account, the combined return could be as low as $5515.35 or as high as $13,122.86. Many investors are willing to risk losing $1106.06 ($6621.41 − $5515.35) for the potential to earn an additional $6501.45 ($13,122.86 − $6621.41). Investors who are willing to risk all $3000 on the aggressive growth account could lose as much as $1673.99 of their initial investment or earn as much as $55,382.28 on top of their initial investment. Most financial advisers encourage investors to diversify their investments among a variety of investment accounts with varying levels of risk in order to achieve a return that meets the investor's goals.

EXAMPLE 5   **Using Multivariable Functions to Forecast Product Sales**

A software company produces two versions of its lead product: one for the Windows operating system and the other for the Macintosh operating system. Through market research, the company determines that at a price of p dollars, approximately q copies of the Windows version will be sold, and at a price of s dollars, approximately v copies of the Macintosh version will be sold. The relationship between the price of each item and sales may be modeled by

$$q = \frac{100,000}{p} + 100 - 0.2p \quad \text{and} \quad v = \frac{20,000}{s} + 50 - 0.1s$$

Determine the price the company should charge for each version in order to maximize the combined revenue from sales.

SOLUTION The revenue from the sale of q copies of the Windows version is given by

$$pq = p\left(\frac{100{,}000}{p} + 100 - 0.2p\right)$$
$$= 100{,}000 + 100p - 0.2p^2$$

Revenue from the sale of v copies of the Macintosh version is given by

$$sv = s\left(\frac{20{,}000}{s} + 50 - 0.1s\right)$$
$$= 20{,}000 + 50s - 0.1s^2$$

The combined revenue from the sales of the two products is given by

$$R(p, s) = 100{,}000 + 100p - 0.2p^2 + 20{,}000 + 50s - 0.1s^2$$
$$= 100p - 0.2p^2 + 50s - 0.1s^2 + 120{,}000$$

From the graph of R on the region defined by $0 \le p \le 500$ and $0 \le s \le 500$ (see Figure 8.33), we can see that the revenue function has a relative maximum.

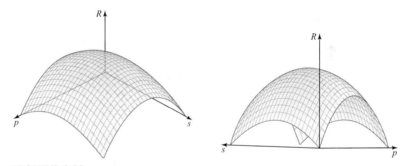

FIGURE 8.33

To find the location of the relative maximum, we first find the partial derivatives.

$$R_p = 100 - 0.4p$$
$$R_s = 50 - 0.2s$$

We then set each partial derivative equal to zero and solve.

$$0 = 100 - 0.4p \qquad 0 = 50 - 0.2s$$
$$0.4p = 100 \qquad\quad 0.2s = 50$$
$$p = 250 \qquad\qquad s = 250$$

The critical point is $(250, 250)$. The second partial derivatives are

$$R_{pp} = -0.4$$
$$R_{ss} = -0.2$$
$$R_{ps} = 0$$

$$D(250, 250) = (-0.4)(-0.2) - 0$$
$$= 0.08$$

Since $D > 0$ and $R_{pp} < 0$, a relative maximum occurs at $(250, 250)$. To maximize revenue, the company should charge $250 for each version of the software. The maximum combined revenue is $138,750, with $112,500 coming from the Windows version and $26,250 coming from the Macintosh version.

EXAMPLE 6

Finding Relative Extrema and Saddle Points

Find the relative extrema and saddle points of $f(x, y) = x^4 - 4x^3 + 4x^2 - y^2$. Two different views of the graph are shown in Figure 8.34.

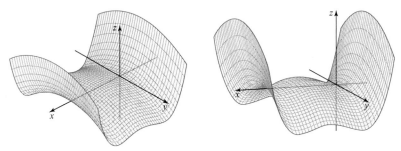

FIGURE 8.34

SOLUTION We will first find the critical points of f. We begin by finding the partial derivatives of f.

$$f_x = 4x^3 - 12x^2 + 8x$$
$$f_y = -2y$$

We then set each partial derivative equal to 0 and solve.

$$4x^3 - 12x^2 + 8x = 0 \qquad -2y = 0$$
$$4x(x^2 - 3x + 2) = 0 \qquad y = 0$$
$$4x(x - 1)(x - 2) = 0$$
$$x = 0, 1, 2$$

We must pair up each x-value with each y-value to form the critical points. The critical points are $(0, 0)$, $(1, 0)$, and $(2, 0)$.

Next we will find the second partial derivatives and perform the D-Test on each critical point.

$$f_{xx} = 12x^2 - 24x + 8$$
$$= 4(3x^2 - 6x + 2)$$
$$f_{yy} = -2$$
$$f_{xy} = 0$$
$$D = (12x^2 - 24x + 8)(-2) - (0)(0)$$
$$= -8(3x^2 - 6x + 2)$$

When there are many critical points, it is often helpful to construct a table of data (see Table 8.8).

TABLE 8.8

Critical Point	$f_{xx} = 4(3x^2 - 6x + 2)$	$D = -8(3x^2 - 6x + 2)$	Graphical Interpretation
$(0, 0)$	8	-16	Saddle point
$(1, 0)$	-4	8	Relative maximum
$(2, 0)$	8	-16	Saddle point

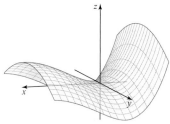

FIGURE 8.35

In order to understand better what a saddle point looks like, we'll zoom in on the region surrounding the critical point at $(0, 0)$ (see Figure 8.35).

Looking at the graph from the xz plane (Figure 8.36), it looks as if a relative minimum occurs at the critical point.

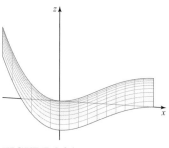

FIGURE 8.36

However, looking at the graph from the yz plane (Figure 8.37), it looks as if a relative maximum occurs at the critical point.

When a graph has a saddle point, the point will look like a minimum from one plane and a maximum from the other plane. (The planes will not necessarily be the xz plane and yz plane.)

To determine the actual coordinates of each of the extrema and saddle points, we evaluate the function at the critical points (see Table 8.9).

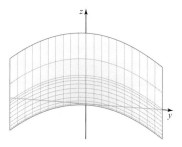

FIGURE 8.37

TABLE 8.9

x $\downarrow$	y
0	0
1	1
2	0

From the table, we see the relative maximum occurs at $(1, 0, 1)$. Saddle points occur at $(0, 0, 0)$ and $(2, 0, 0)$. These results are consistent with our earlier conclusions.

8.3 Summary

In this section, you learned the techniques used to find relative extrema and saddle points of two-variable functions. In the next section, we will look at finding absolute extrema subject to function constraints.

8.3 Exercises

In Exercises 1–25, find the relative extrema and saddle points of the function.

1. $f(x, y) = x^2 + 4y^2 - 10$

2. $f(x, y) = x^2 - 4xy$

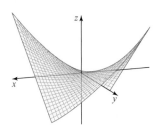

3. $f(x, y) = -x^2 - y^3 + 3y - x$

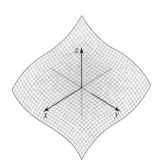

4. $f(x, y) = x^4 - 4x^2 + y^4 - 9y^2 + 1$

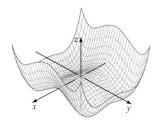

5. $f(x, y) = x^2 y - y^2 + x$

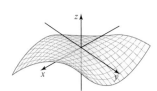

6. $f(x, y) = x^4 - 4x + y^4 - 9y^2 + 1$

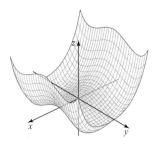

7. $f(x, y) = x^4 - 4x + y^3 - y^2$

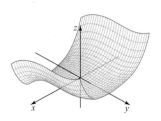

Maximizing Volume with the Lagrange Multiplier Method

A box designer wants to construct a closeable box as shown in Figure 8.41. The designer requires that the sum of the height and width of the piece of cardboard used to make the box be 120 inches. What are the dimensions of the box with the maximum volume?

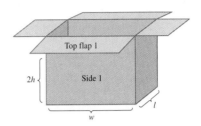

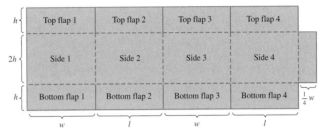

FIGURE 8.41

SOLUTION We are to maximize $V(h, l, w) = 2hlw$ subject to the constraint $4h + 2l + 2.25w = 120$.

We have $g(h, l, w) = 4h + 2l + 2.25w - 120$. Note that $g(h, l, w) = 0$.
The Lagrange function is given by

$$F(h, l, w, \lambda) = V(h, l, w) + \lambda g(h, l, w)$$
$$= 2hlw + \lambda(4h + 2l + 2.25w - 120)$$
$$= 2hlw + 4h\lambda + 2l\lambda + 2.25w\lambda - 120\lambda$$

The partial derivatives are

$$F_h = 2lw + 4\lambda$$
$$F_l = 2hw + 2\lambda$$
$$F_w = 2hl + 2.25\lambda$$
$$F_\lambda = 4h + 2l + 2.25w - 120$$

We will set the partial derivatives F_h, F_l, and F_w equal to 0 and solve each one for λ.

$$0 = 2lw + 4\lambda \qquad 0 = 2hw + 2\lambda$$
$$4\lambda = -2lw \qquad 2\lambda = -2hw$$
$$\lambda = -\frac{1}{2}lw \qquad \lambda = -hw$$

$$0 = 2hl + 2.25\lambda$$
$$2.25\lambda = -2hl$$
$$\lambda = -\frac{8}{9}hl$$

We set the first two equations equal to each other to get h in terms of l.

$$-\frac{1}{2}lw = -hw$$

$$-\frac{1}{2}lw + hw = 0$$

$$w\left(-\frac{1}{2}l + h\right) = 0$$

So $w = 0$ or $h = \frac{1}{2}l$. Since we are looking for the maximum volume, we throw out $w = 0$.

Next we set the second two equations equal to each other to get w in terms of l.

$$-hw = -\frac{8}{9}hl$$

$$hw - \frac{8}{9}hl = 0$$

$$h\left(w - \frac{8}{9}l\right) = 0$$

So $h = 0$ or $w = \frac{8}{9}l$. Since we are looking for the maximum volume, we throw out $h = 0$.

We substitute the values $w = \frac{8}{9}l$ and $h = \frac{1}{2}l$ into $g(l, w, h) = 0$ to solve for l.

$$4h + 2l + 2.25w - 120 = 0$$

$$4\left(\frac{1}{2}l\right) + 2l + 2.25\left(\frac{8}{9}l\right) - 120 = 0$$

$$2l + 2l + \frac{18}{9}l - 120 = 0$$

$$6l = 120$$

$$l = 20$$

Then we back substitute to solve for w and h.

$$w = \frac{8}{9}l$$

$$= \frac{8}{9}(20)$$

$$= \frac{160}{9}$$

$$\approx 17.78$$

$$h = \frac{1}{2}(20)$$

$$= 10$$

13. Ⓜ **Cobb-Douglas Model** Based on data from 1999 and 2000, a Cobb-Douglas model for Ford Motor Company is given by

$$P(L, C) = 0.5319L^{0.4558}C^{0.5442}$$

thousand vehicles, where L is the total labor cost and C is the capital expenditure cost, both in millions of dollars. In 2000, Ford Motor Company spent $25,783 million on labor and produced 7424 thousand vehicles. (**Source:** Modeled from Ford Motor Company 2001 Annual Report.)

Assuming that the model is valid for future years, how much should Ford spend on labor and how much should it spend on capital in order to maximize vehicle production if its annual budget for labor and capital expenditures is $40,000 million?

14. Landscape Design A landscape architect is designing a garden for a client as shown in the following figure.

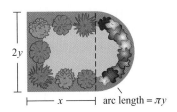

$2y$ | x | arc length $= \pi y$

She intends to line the flat edges of the garden with shrubs and the curved edge of the garden with flowers. The shrubs cost $25 a linear foot, and the flowers cost $20 a linear foot, including labor. The client is willing to pay up to $1000 and wants as large a garden as possible. What are the dimensions of the garden? (*Hint:* The area of the garden is given by

$$A(x, y) = 2xy + \frac{1}{2}\pi y^2.)$$

15. Ⓜ **Cobb-Douglas Model** Based on data from 2000 to 2001, a Cobb-Douglas model for the Coca-Cola Company is given by

$$P(L, C) = 0.02506L^{0.0481}C^{0.9519}$$

billion unit cases of beverage, where L is the amount of money spent on labor (including sales, payroll, and other taxes) and C is the amount of money spent on capital expenditures, both in millions of dollars. (One unit case of beverage equals 24 eight-ounce servings.) In 2001, Coca-Cola spent a combined total of $917 million on labor and capital expenditures and produced 17.8 billion unit cases of beverage. (**Source:** Coca-Cola 2001 Annual Report.) According to the model, what is the maximum number of unit cases that can be produced with a labor and capital budget of $917 million?

Chapter 8 Review Exercises

Section 8.1 *In Exercises 1–3, generate a table of data for the multivariable function using $x = -2, 0, 2$ and $y = -1, 0, 1$. Then use a graphing utility to graph the function. Compare the coordinates of the points in the table to the coordinates of the points on the graph of the function.*

 1. $z = 4x - 5y$

 2. $z = 2xy - 3y^2$

 3. $z = xy + 4y$

In Exercises 4–5, use the multivariable function to answer the given question.

4. Ⓜ **Price-to-Earnings Ratio** The price-to-earnings ratio (P/E) is the most common measure of how expensive a stock is and is given by

$$R = \frac{P}{E}$$

where P is the price of a share of the stock and E is the earnings per share. Growth stocks typically have high P/Es compared to the overall market. Investors are willing to pay more for these stocks because they expect the company's earnings and stock price to rise. Stocks with low P/Es are often considered to be overlooked value stocks.

As of December 28, 2002, the share prices of four major companies were

Apple Computer:	$14.06
General Electric:	$24.70
Microsoft:	$52.97
Electronic Arts:	$51.50

Their earnings per share were

Apple Computer: $0.18
General Electric: $1.41
Microsoft: $1.41
Electronic Arts: $0.71

(**Source:** www.quicken.com.)

Calculate the P/E ratio for each of the companies. According to your calculations, which of the companies would you classify as "growth" stocks?

5. 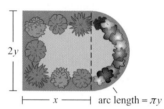   **Debt/Asset Ratio**   The debt/asset ratio shows the proportion of a company's assets that are financed through debt and is calculated by

$$R = \frac{L}{A}$$

where L is the total value of the company's liabilities and A is the total value of its assets. If most of the company's assets are financed through equity (stock ownership interest in a company), the ratio is less than 1. If most of the company's assets are financed through debt, the ratio is greater than 1. Companies, and individuals, with high debt/asset ratios may be in danger if creditors start to demand repayment of debt.

As of December 28, 2002, the total debt of each of three major companies was

Apple Computer: $316.0 million
General Electric: $254,522.0 million
Microsoft: $0.0 million

Their reported debt/equity ratios were

Apple Computer: 0.08
General Electric: 4.25
Microsoft: 0.00

(**Source:** www.quicken.com.)

Determine the value of the assets of Apple Computer and General Electric. Then interpret the meaning of the debt/equity ratio for Microsoft.

Section 8.2   *In Exercises 6–10, calculate the first and second partial derivatives of each of the functions.*

6. $f(x, y) = 6xy$

7. $f(x, y) = 2x^2y - 4xy^2$ 8. $f(x, y) = 5xy - 5y^2$

9. $z = xy \ln(y)$ 10. $z = e^x \ln(y)$

In Exercises 11–12, interpret the practical meaning of each of the partial derivatives.

11. **Cylindrical Can Volume**   The volume of a cylindrical can is given by $V(h, r) = \pi r^2 h$, where r is the radius and h is the height of the can, both measured in inches. Find V_h and V_r. Then evaluate the partial derivatives at $(4, 5)$ and interpret the meaning of the results.

12. **Landscape Design**   The area of the garden shown in the figure is given by

$$A(x, y) = 2xy + \frac{1}{2}\pi y^2.$$

Find A_x and A_y. Then evaluate the partial derivatives at $(6, 4)$ and interpret the meaning of the results.

In Exercises 13–14, a function graph with a cross section drawn at $x = 0$ and $y = -1$ is shown. Determine the equations of the tangent lines of the function in the x-direction and y-direction at the point $(0, -1, f(0, -1))$.

13. $f(x, y) = x - y + 2$

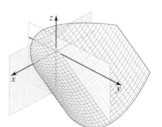

14. $f(x, y) = x^2 - 4y + 2$

Section 8.3 *In Exercises 15–16, find the relative extrema and saddle points of the function.*

15. $f(x, y) = x^2 - xy + 4y^2 + 10$

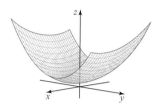

16. $f(x, y) = -x^3 + 3x + 4y - y^2$

Section 8.4 *In Exercises 17–18, solve the constrained optimization problem using the Lagrange Multiplier Method.*

17. Maximize $f(x, y) = -x^3 + 3x + 4y - y^2$ subject to the constraint $-2x + y = 0$.

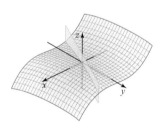

18. Maximize $f(x, y) = -x^3 + 3x + 4y^3$ subject to $-2x + y = -6$.

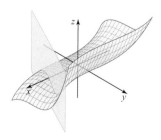

In Exercises 19–20, use the Lagrange Multiplier Method to find the answer to the question.

19. Landscape Design A landscape architect is designing a garden for a client as shown in the following figure.

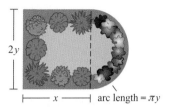

She intends to line the flat edges of the garden with shrubs and the curved edge of the garden with flowers. The shrubs cost \$30 a linear foot, and the flowers cost \$40 a linear foot, including labor. The client is willing to pay up to \$1200 and wants as large a garden as possible. What are the dimensions of the garden?

20. Cobb-Douglas Model Based on data from 2000 to 2001, a Cobb-Douglas model for the Coca-Cola Company is given by

$$P(L, C) = 0.02506L^{0.0481}C^{0.9519}$$

billion unit cases of beverage, where L is the amount of money spent on labor (including sales, payroll, and other taxes) and C is the amount of money spent on capital expenditures, both in millions of dollars. (One unit case of beverage equals 24 eight-ounce servings.) In 2001, Coca-Cola spent a combined total of \$917 million on labor and capital expenditures and produced 17.8 billion unit cases of beverage. (**Source:** Coca-Cola 2001 Annual Report.)

Assuming that the model works for future years, what is the maximum number of unit cases that can be produced with a labor and capital budget of \$998 million?

Make It Real

 in header area of page is the banner

PROJECT 8

What to do

1. Go to www.weather.com and enter your zip code to access your local forecast.

2. Use the wind chill formula to calculate the current wind chill temperature. The wind chill formula is given by

$$W(T, V) = 35.74 + 0.6215T - 35.75V^{0.16} + 0.4275TV^{0.16}$$

where T is the temperature (in degrees Fahrenheit) and V is the wind speed (in miles per hour).

3. Compare the calculated wind chill temperature to the "Feels Like" temperature displayed on the forecast.

4. Estimate how much the "Feels Like" temperature will change if the wind speed increases by 1 mph.

Local weather forecast

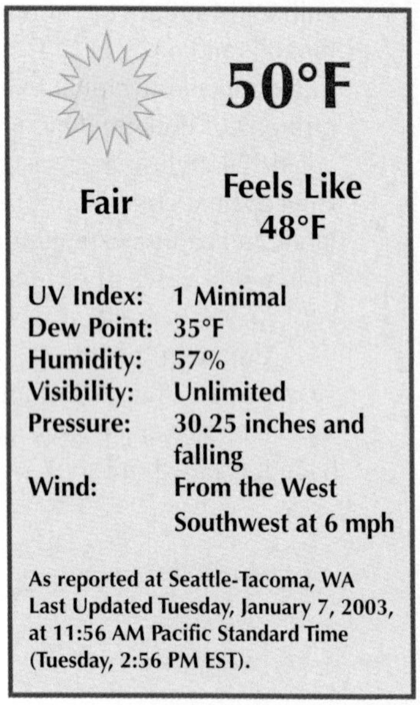

50°F

Fair

Feels Like
48°F

UV Index:	1 Minimal
Dew Point:	35°F
Humidity:	57%
Visibility:	Unlimited
Pressure:	30.25 inches and falling
Wind:	From the West Southwest at 6 mph

As reported at Seattle-Tacoma, WA
Last Updated Tuesday, January 7, 2003,
at 11:56 AM Pacific Standard Time
(Tuesday, 2:56 PM EST).

Source: www.weather.com.
Reprinted by permission.

Answers to Odd-Numbered Exercises

CHAPTER 1

Section 1.1 *(page 12)*

1. Your weight is a function of your age since at any instant in time you have only one weight.

3. Temperature is a function of the time of day.

5. The number of salmon in a catch is not a function of the number of fish caught.

7. $C(4) = 159.80$
   The total cost of four pairs of shoes is $159.80.

9. $H(2) = 56$
   Two seconds after he jumped, the cliff diver is 56 feet above the water.

11. $E(4) = 0.06$
    In the fourth quarter since December 1999, shares in the tortilla company earned $0.06 per share.

13. $P(4) \approx 18.72$
    On November 8, 2001, the closing stock price of the computer company was approximately $18.72.

15. It appears that near $x = 1$, the graph goes vertical. If it does so, a vertical line drawn at that point would touch the graph in multiple locations. However, if the graph doesn't actually go vertical near $x = 1$, then it is a function. One drawback of reading a graph is that it is sometimes difficult to tell if the graph goes vertical or not.

17. The graph is a function.

19. The graph is a function.

21. 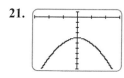   23.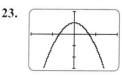

25. $y \approx 4; y = 4$

27. $y \approx 2; y$ is undefined.

29. The domain is all real numbers.

31. The domain is all real numbers.

33. The domain is all real numbers except $t = 1$. That is, $\{t \mid t \neq 1\}$.

35. The domain is all real numbers greater than or equal to -1. That is $\{a \mid a \geq -1\}$.

37. The domain is the set of real numbers greater than or equal to -3. That is, $\{x \mid x \geq -3\}$.

39. The domain is the set of whole numbers. That is, $\{n \mid n$ is a whole number$\}$.

41. Using our current calendar system, the domain of the function is the set of whole numbers between 1 and the current year.

43. The domain is all real numbers except $x = -1$ and $x = 1$.

45. Yes. Even though the domain value $x = 1$ is listed twice in the table, it is linked with the same range value both times, $y = 6$.

Section 1.2 *(page 27)*

1. $m = -1$ 3. $m = -0.2$ 5. $m = 0$

7. y-intercept: $(0, 10)$
   x-intercept: $(-2, 0)$

9. y-intercept: $(0, 11)$
   x-intercept: $(-5.5, 0)$

11. y-intercept: $(0, -4)$
    x-intercept: $\left(\frac{4}{3}, 0\right)$

13. The slope-intercept form of the line is $y = -x + 7$. The standard form of the line is $x + y = 7$. A point-slope form of the line is $y - 5 = -1(x - 2)$.

15. The slope-intercept form of the line is $y = -0.2x + 3.64$. The standard form of the line is $5x + 25y = 91$. A point-slope form of the line is $y - 3.4 = -0.2(x - 1.2)$.

17. The slope-intercept form of the line is
$y = 0x + 2$ and is commonly written as $y = 2$.
The standard form of the line is $0x + y = 2$ and
is also often written as $y = 2$. A point-slope form
of the line is $y - 2 = 0(x + 2)$.

19. $y = 4x - 2$

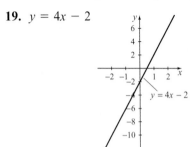

21. $y - 4 = 0.5(x - 2)$

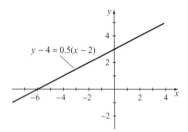

23. $2x - 3y = 5$

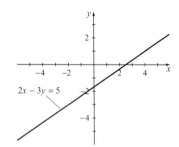

25. $y = -\frac{2}{3}x + \frac{4}{3}$

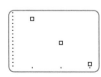

27. The slope-intercept form of the line is

$y = \frac{2}{3}x + 3$.

29. The equation of the line is $x = 3$.

31. The data represents a linear function.

$m = 993.2$ year 2000 dollars per year

Between 1989 and 1999, the U.S. average
personal income (in year 2000 dollars) increased
by an average of $993.20 per year.

33. The data represents a linear function.

$m = 1183.01$ dollars per month

Between September 2001 and October 2001, the
employee's take-home pay increased at a rate of
$1183.01 per month.

35. The data represents a linear function.

$m = \$0.0375$ per pound

It costs an average of $0.0375 per pound to
dispose of clean wood.

37. In order to consume 8 grams of fiber, you would
need to eat 3 servings $\left(2\frac{1}{4}\text{ cups}\right)$ of Wheaties
along with the large banana.

39. The equation of the vertical line is $x = 4$.

41. The slope-intercept form is $y = mx + b$, and
the point-slope form is $y - y_1 = m(x - x_1)$.
For a vertical line, the slope m is undefined and
thus may not be substituted into either of the
two forms.

43. Vertical lines are not functions, since they fail the
Vertical Line Test. However, all nonvertical lines
are functions.

45. $x = 3$

47. The x-intercept is $\left(\frac{c}{a}, 0\right)$.

The y-intercept is $\left(0, \frac{c}{b}\right)$.

The slope of the line is $m = -\frac{a}{b}$.

49. b must equal 23.   **51.** (a, b)

Section 1.3 *(page 45)*

1. (a) The scatter plot shows that the data are near-
linear.

(b) The equation of the line of best fit is
$y = -6.290x + 110.9$.

(c) $m = -6.290$ means that the Harbor Capital Appreciation Fund share price is dropping at a rate of $6.29 per month.

The y-intercept means that in month 0 of 2000, the fund price was $110.94. Since the months of 2000 begin with 1, not 0, the y-intercept does not represent the price at the end of January. It could, however, be interpreted as being the price at the end of December 1999.

(d) This model is a useful tool to show the trend in the stock price between October and December 2000. Since stock prices tend to be volatile, we are somewhat skeptical of the accuracy of data values outside of that domain. (Answers may vary.)

3. (a) The scatter plot shows that the data are near-linear.

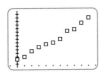

(b) The equation of the line of best fit is
$y = 1165.9x + 80,887$.

(c) $m = 1165.9$ means that Washington State public university enrollment is increasing by about 1166 students per year.

The y-intercept means that, according to the model, Washington State public university enrollment was 80,887 in 1990.

(d) The model fits the data extremely well, as shown by the graph of the line of best fit.

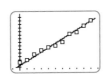

This model could be used by Washington State legislators and university administrators in budgeting and strategic planning. (Answers may vary).

5. (a) The scatter plot shows that although the data are not linear, they are nonincreasing.

(b) The equation of the line of best fit is
$y = -3.4x + 39.6$.

(c) $m = -3.4$ means that the per capita income ranking of North Carolina is changing at a rate of 3.4 places per year. That is, the state is moving up in the rankings by about 3 places per year.

The y-intercept means that in 1995 (year 0), North Carolina was ranked 40th out of the 50 states. (Only positive whole number rankings make sense.)

(d) This model is not a highly accurate representation of the data, as shown by the graph of the line of best fit, so it should be used with caution.

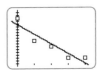

However, an incumbent government official could use the model in a 2000 reelection campaign as evidence that the state's economy had improved during his or her tenure in office. The official might also use the model to claim that the trend of improvement will continue if he or she is reelected. (Answers may vary.)

7. (a)

$$C(x) = \begin{cases} 15.25 & 0 \le x \le 320 \\ 0.04274x + 1.036 & 340 \le x \le 760 \\ 0.04409x & 780 \le x \end{cases}$$

For values of $x \ge 780$, $C(x)$ is directly proportional to x.

(b) $15.25

(c) $23.26

(d) The cost per pound is lowest when 780 or more pounds of trash are disposed of. As a construction company, we would try to keep the weight of our trash deliveries at or above 780 pounds. (Answers may vary.)

9. (a) $C(t) = \begin{cases} 29.99 & 0 \le t \le 200 \\ 0.4t - 50.01 & t > 200 \end{cases}$

(b) For the Qwest plan, we have
$C(300) = \$82.49$

For the Sprint plan, we have
$C(300) = \$69.99$

The Sprint plan is the best deal for a customer who uses 300 anytime minutes.

11. (a) Let t be the production year of a Toyota Land Cruiser 4-Wheel Drive and V be the value of the vehicle in 2001.
$$V = 4505t - 8{,}966{,}350$$

(b) $\$7610$

(c) The model substantially underestimated the value of a 1992 Land Cruiser.

13. (a) Let F be the number of grams of fat in x Chef Salads. We have $F = 8x$.

The number of fat grams is directly proportional to the number of Chef Salads.

(b) less than 10 salads

(c) Up to 7.75 salads

(d) Up to 5.875 salads

15. Let s be the number of segments and C be the maximum total cost of the ticket (in dollars) from Seattle to Phoenix. We have
$$C = 10s + 222$$
Note that $s \ge 1$.

17. a is the price of a cup of coffee, b is the price of a bagel, and c is the price of a muffin.

19. The correlation coefficient $r = 0$ indicates that the model doesn't fit the data well.

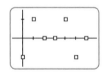

In addition, we can see from the scatter plot that the model doesn't fit well.

21. The y-intercept form of a line passing through the origin is
$$y = mx$$
This implies that y is directly proportional to x.

If the linear model $y = mx$ fits the original data well, it is likely that the dependent variable of the original data is directly proportional to the independent variable. However, if the linear model $y = mx$ does not fit the data well, the dependent

and independent variables of the original data are not directly proportional.

23. The line of best fit is $y = 1$. This model fits the table of data perfectly.

```
LinReg
y=ax+b
a=0
b=1
r²=
r=
```

However, from the calculator display, we see that r is undefined.

The correlation coefficient is formally defined as

$$r = \frac{n(\Sigma xy) - (\Sigma x)(\Sigma y)}{\sqrt{n(\Sigma x^2) - (\Sigma x)^2}\,\sqrt{n(\Sigma y^2) - (\Sigma y)^2}}$$

$$r = \frac{0}{0}$$

Because we can't divide by zero and there is a zero in the denominator of the expression, the correlation coefficient r is undefined.

25. $y = \begin{cases} -2x + 10 & -2 \le x \le 4 \\ 0.5x & 4 < x \le 10 \end{cases}$

CHAPTER 1 REVIEW EXERCISES

Section 1.1 *(page 49)*

1. $C(2) = 99.90$
The cost to buy two pairs of shoes is $\$99.90$.

3. $H(2) = 36$
The cliff diver is 36 feet above the water 2 seconds after he jumps from a 100-foot cliff.

5. The domain is all real numbers.

7. The domain is all real numbers except $r = 1$ and $r = -1$.

9. At $t = 2$, $P \approx 21.03$. The stock price of the computer company at the end of the day two days after December 16, 2001, was about $\$21.03$.

11. The graph represents a function.

Section 1.2 *(page 50)*

13. $m = \dfrac{2}{3}$

15. $m = -11$

17. The y-intercept is $(0, 18)$.
The x-intercept is $(6, 0)$.

19. $y = -1x + 7$
In standard form, $x + y = 7$.

Section 1.3 *(page 50)*

21. (a) Let x be the number of large orders of French fries consumed. The amount of fat consumed is given by $F(x) = 29x$ grams.
    (b) Let y be the number of Big N' Tasty sandwiches consumed. The amount of fat consumed is given by $G(y) = 32y$ grams.
    (c) Only one combination meal may be eaten.

23. (a) Let t be the production year of the Mercedes-Benz Roadster two-door SL500 and $v(t)$ be the value of the car in 2001.
$$v(t) = 7312.5t - 14,558,975$$
    (b) $58,712.50
    (c) The linear model was extremely effective at accurately predicting the value of the 1999 Mercedes-Benz Roadster two-door SL500. The $212.50 difference between the predicted value of the model and the NADA guide average value was negligible.

CHAPTER 2

Section 2.1 *(page 71)*

1. Concave up, y-intercept $(0, 1)$, vertex $(1, 0)$

3. Concave up, y-intercept $(0, 0)$, vertex $(-0.5, -0.75)$

5. Concave up, y-intercept $(0, 2.1)$, vertex $(0.25, 1.925)$

7. $y = -x^2 + 4x - 1$ 9. $y = x^2 - 10x + 5$

11. (a) $W(t) = 36.92t^2 - 43.60t + 2072$
    (b)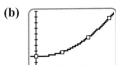
    (c) As shown in the scatter plot, the model appears to fit the data well. It seems reasonable that women's basketball game attendance will continue to increase; consequently, we believe this model is good.

13. (a) $D(t) = 896.67t^2 + 723.33t - 540$
    (b)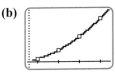

(c) As shown in the scatter plot, the model appears to fit the data well. Since the scatter plot fits the data well and we expect that DVD shipments will continue to increase, we believe this model is a good model.

15. (a) $C(v) = -0.03550v^2 + 85.451v - 47,197$
    (b)

(c) As shown in the scatter plot, the model fits the data perfectly. Although the scatter plot fits the data perfectly, we believe that the model has some definite limitations. According to the model, as the number of blank VHS tapes drops below 1195, the number of CD-R disks will also begin to decrease. This doesn't seem reasonable. We anticipate that CD-R disk sales will continue to increase even as VHS tape sales decrease.

17. (a) (See part d.)
    (b) Since the data set consists of three nonlinear points, a quadratic model will fit the data perfectly.
    (c) The quadratic model is
    $I(t) = -10.31t^2 + 972.4t + 22,530$
    (d)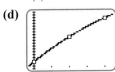

(e) Since the model fits the data and since we expect personal incomes to continue to increase, we anticipate that the model will be relatively accurate in forecasting per capita personal incomes between 1980 and 2010.

19. (a)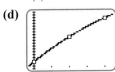

(b) The scatter plot does not seem to resemble a parabola. Consequently, we do not think that a quadratic function will fit the data well.

21. (a) (See part d.)
    (b) The scatter plot roughly resembles a parabola, so we will model the data with a quadratic function.
    (c) The quadratic model is
    $M(t) = 0.02628t^2 + 0.3340t + 47.46$

(d)

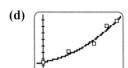

(e) Since the model fits the data and since it seems reasonable to assume that the upward trend in conventional mortgages will continue, we anticipate that the model will be relatively accurate in forecasting the percentage of new privately owned one-family houses financed with a conventional mortgage between 1977 and 2005. (Note: 2005 is arbitrarily selected as a year *near* the last data point.)

(f) When $t \approx 5.4$, $M(t) = 50$. Since t is the number of years since the end of 1970, we estimate that in mid-1976 ($t = 5.4$), 50 percent of mortgages were conventional mortgages.

23. (a)

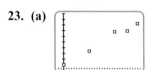

(b) The scatter plot does not seem to resemble a parabola. Consequently, we do not think that a quadratic function will fit the data well.

25. (a) (See part d.)

(b) The scatter plot seems to resemble a portion of a parabola. We think that a quadratic model will fit the data.

(c) The quadratic model is
$$H(t) = -0.01293t^2 + 1.398t + 57.60$$

(d)

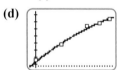

(e) We believe the model is a good model because it seems reasonable to assume that the percentage of homes with a garage will continue to increase.

(f) Since $H(33.6) \approx 90$, we estimate that in mid-2004 ($t = 33.6$), 90 percent of new homes had garages.

27. (a) (See part d.)

(b) Although the scatter plot appears to be near-linear, we can still construct a quadratic model. When quadratic models are used to model near-linear functions, the value of a is relatively close to zero. (If we had our choice

of which type of model to construct, we would construct a linear model because it is simpler.)

(c) The quadratic model is
$$P(t) = -0.02143t^2 + 8.279t + 68.27$$

(d)

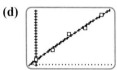

(e) Since the model fits the scatter plot fairly well, and since home sales prices typically increase over time, we believe that our model will relatively accurately predict future home prices in the northeastern United States.

29. (a) (See part d.)

(b) The scatter plot looks concave up. Consequently, we expect that a parabola will fit the data.

(c) The quadratic model is
$$P(t) = 0.0417t^2 + 0.06833t + 2.455$$

(d)

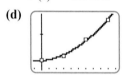

(e) Since the model fits the scatter plot fairly well, and since auto leasing appears to be increasing in popularity, we believe that our model will relatively accurately predict the percentage of households leasing vehicles.

(f) Since $P(7) \approx 5$, we estimate that at the end of 1996, 5 percent of households were leasing vehicles.

31. (a) (See part d.)

(b) The scatter plot looks like a concave up parabola, so we believe that a quadratic model will fit the data well.

(c) The quadratic model is
$$S(t) = 29.23t^2 + 79.33t + 177.4$$

(d)

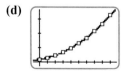

(e) Since the model fits the data extremely well, and since we expect Starbucks sales to continue to increase, we believe we have a good model.

(f) Since $S(11.6) \approx 5000$, we expect that midway through fiscal year 2005 ($t = 11.6$), Starbucks sales will reach $5000 million.

33. (a) (See part d.)
   (b) The scatter plot looks more or less concave up and increasing. A quadratic model will probably be a good choice for the data set.
   (c) The quadratic model is
   $$P(t) = 0.0007346t^2 + 0.009701t + 0.7803$$
   (d)
   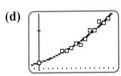
   (e) Since the model fits the data relatively well, and since we expect the price of chicken to continue to increase (because of inflation), we believe we have a good model.

35. (a) (See part d.)
   (b) The scatter plot is generally concave up and increasing. We believe that a quadratic model will fit the data well.
   (c) The quadratic model is
   $$S(t) = 0.9195t^2 + 23.25t + 2652$$
   (d)
   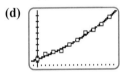
   (e) Since the model fits the data relatively well, and since it seems reasonable to anticipate that private college enrollment will continue to increase, we believe that the model is a good model.

37. (a) (See part d.)
   (b) The scatter plot appears concave up and increasing. A quadratic model may fit the data well.
   (c) The quadratic model is
   $$P(t) = 123.73t^2 + 445.58t + 22,847$$
   (d)
   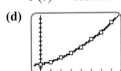
   (e) Since the model fits the data relatively well, and since it seems reasonable to anticipate that personal incomes will continue to increase, we believe the model is a good model.

39. (a) (See part d.)
   (b) The scatter plot appears near-linear. Nevertheless, we can come up with a decent quadratic model for the data.

(c) The quadratic model is
   $$P(t) = -14.470t^2 + 883.18t + 17,620$$
   (d)
   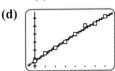
   (e) Since the model fits the data relatively well, and since it seems reasonable to anticipate that personal incomes will continue to increase, we believe that the model is a good model.

41. (a) (See part d.)
   (b) The scatter plot appears slightly concave down and increasing. A quadratic model should fit the data well.
   (c) The quadratic model is
   $$P(t) = -32.71t^2 + 778.1t + 2341$$
   (d)
   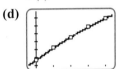
   (e) Since the model fits the data well, and since it seems reasonable that the company's profits will continue to increase, we believe that the model is a good model.
   (f) Since $P(6.5) \approx 6000$, we estimate that midway through 2004 ($t = 6.5$), Johnson & Johnson profits reached $6000 million.

43. (a) (See part d.)
   (b) The scatter plot is concave up and increasing. A quadratic function may fit the data well.
   (c) The quadratic model is
   $$F(t) = 12.66t^2 + 48.52t + 2102$$
   (d)
   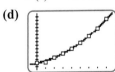
   (e) Since the model fits the data well, and since it seems reasonable that the payroll will continue to grow, we believe that the model is a good model.

45. (a) (See part d.)
   (b) The scatter plot appears more or less concave down and increasing. A quadratic model may fit the data well.
   (c) The quadratic model is
   $$F(t) = -0.04079t^2 + 1.453t + 401.6$$

(d)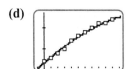

(e) Since the model fits the data well, and since it seems reasonable that the number of private-school teachers will continue to increase, we believe that the model is a good model.

47. (a) (See part d.)
  (b) The scatter plot appears more or less concave up and increasing. A quadratic model may fit the data well.
  (c) The quadratic model is
  $W(t) = 0.003197t^2 + 0.1333t + 4.606$

(d)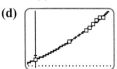

(e) Since the model fits the data well, and since it seems reasonable that wages will continue to increase, we believe that the model is a good model.

49. (a) (See part d.)
  (b) The scatter plot appears concave up. A quadratic model may fit the data well.
  (c) The quadratic model is
  $A(t) = 9.396t^2 - 11.58t + 1063$

(d)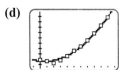

(e) Since the model fits the data well, and since it seems reasonable that advertising expenditures will continue to increase, we believe that the model is a good model.
  (f) Since $A(10.6) \approx 2000$, we estimate that in mid-2001 ($t = 10.6$), billboard advertising expenditures reached $2000 million ($2 billion).

51. (Answers may vary.)
  $y = ax(x - 4)$
  $y = x^2 - 4x$ when $a = 1$
  $y = -x^2 + 4x$ when $a = -1$

53. No. Although any set of three *nonlinear* points defines a unique quadratic function, the three points given all lie on the same line, $y = 2x + 1$.

55. Since the function is nonnegative for all values of x, the graph of the parabola must be entirely above the x axis. Consequently, the graph must be concave up. (If it were concave down, it would cross the x axis.) Since the graph is concave up, $a > 0$.

Section 2.2 *(page 95)*

1. The quartic function that best fits the data is given by $S(t) = -0.53891t^4 + 15.050t^3 - 103.86t^2 + 48.261t + 12{,}393$. $S(16) = 12{,}904$ means that there were 12,904 thousand students enrolled in high school in 2000.

3. The quadratic function that best fits the data is $R(t) = 1.874t^2 - 1.425t + 2165$.
  $R(21.5) \approx 3000$ means that the number of public elementary and secondary teachers reached 3 million (3000 thousand) in mid-2002 ($t = 21.5$).

5. The cubic function that best fits the data is $C(t) = -0.5278t^3 + 5.905t^2 - 4.234t + 4.048$. $C(7) \approx 83$ means that 83 percent of public school classrooms had Internet access in 2001. $C(8) \approx 78$ means that 78 percent of public school classrooms had Internet access in 2002. The 2001 estimate seems reasonable, since we expect Internet access in schools to become increasingly common. The 2002 estimate does not seem reasonable, since it forecasts a decline in Internet access.

7. The cubic model that best fits the data is $C(t) = 0.1563t^3 - 2.268t^2 + 14.84t + 164.0$. $C(15) \approx 404$ means that there will be 404 million debt cards by the end of 2005. Our model estimate is substantially higher than the consulting firm's 270 million card projection.

9. The quartic function that best fits the data is given by $S(t) = 4.807t^4 - 98.56t^3 + 541.3t^2 + 2936t + 10{,}610$. $S(25) \approx 760{,}000$ means that 760 billion shares were traded in 2005. Because of the rapid growth in the market in recent years, this estimate seems reasonable. However, the recession in the early 2000s may have affected the number of shares sold.

11. The quartic function that best fits the data is $W(t) = 100t^4 - 1518.5t^3 + 8244.4t^2 - 17{,}294t + 42{,}667$. $W(6.5) \approx 40{,}000$, meaning that in mid-1999 ($t = 6.5$), the average annual wage per worker for computer and equipment retailers first reached $40,000.

13. The quadratic function of best fit is $W(t) = 1421.4t^2 - 2321.4t + 56,429$. $W(7.6) \approx 121,000$, meaning that in mid-2000 ($t = 7.6$), the average wage per worker in the prepackaged software development industry reached $121,000.

15. The cubic model that best fits the software wholesaler's wages is $W(t) = 113.89t^3 - 251.19t^2 + 277.78t + 52,571$. We estimate that at the end of 2002, the average wage of both software wholesalers and software retailers was $144,000.

17. The quartic model that best fits the data is $W(t) = 51.515t^4 - 701.52t^3 + 3119.7t^2 - 2450.4t + 41,563$. $W(7) \approx 60,000$, meaning that the television broadcasting average wage will first exceed $60,000 at the end of 1999. This estimate may be a bit high, but it is not unreasonable. Since $t = 7$ is close to the end of the data set ($t = 6$) and $60,000 is somewhat close to $54,600, we are comfortable with the estimate.

19. The cubic model that best fits the data is $W(t) = -0.005860t^3 + 0.2430t^2 - 0.06623t + 6.541$. $C(32) \approx 61.2$, meaning that 61.2 percent of TV homes had cable at the end of 2002. We would generally expect the percentage of homes with cable to continue to increase, as it has over the 29 years of the data set. For this reason, the estimate seems unreasonable. However, if people are replacing cable with satellite dishes or some other technology, it is possible that the percentage of homes with cable could decline.

21. A polynomial of degree n may have at most $n - 1$ bends.

23. $f(x) = f(-x)$

25. $y = x^4 - 2x^3 - x^2 + 2x$

Section 2.3 *(page 115)*

1. The graph is decreasing and concave up. The y-intercept is $(0, 4)$.

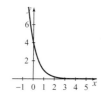

3. The graph is increasing and concave up. The y-intercept is $(0, 0.5)$.

5. The graph is increasing and concave up. The y-intercept is $(0, 0.4)$.

7. The graph is decreasing and concave down. The y-intercept is $(0, -1.2)$.

9. The graph is decreasing and concave up. The y-intercept is $(0, 3)$.

11. $y = 2(3)^x$

13. $y = 5(2)^x$

15. $y = 64\left(\dfrac{1}{2}\right)^x$

17. $y = 0.25(4)^x$

19. $y = 256\left(\dfrac{1}{2}\right)^x$

21. $I(t) = 82.67(1.061)^t$ (rounded model)
$I(25) = 364.5$ (using unrounded model)
At the end of 2005, dental prices are expected to be 364.5 percent of the 1984 price.

23. $I(t) = 106.1(0.9657)^t$ (rounded model)
$I(25) = 44.4$ (using unrounded model)
At the end of 2005, the price of television set is expected to be 44.4 percent of the 1984 price.

25. $I(t) = 86.75(1.051)^t$ (rounded model)
$I(25) = 302.7$ (using unrounded model)
At the end of 2005, the price of admission to an entertainment venue is expected to be 302.7 percent of its 1984 price.

27. $B(t) = 235(1.0232)^t$
where B is the balance (in dollars) and t is the number of years from now. In just over 2 years and 8 months ($t \approx 2.7$), the balance is projected to reach $250.

29. $p(t) = 59.90(1.28)^t$
where p is the Dave Matthews Band concert tickets price and t is the number of years since September 2004. A concert ticket that costs $59.90 in September 2004 is expected to cost $98.14 in September 2006.

31. $x = 8.827$ **33.** $x = 14.21$ **35.** $x = 2$

37. (a) The predicted value of the investment is given by $V(t) = 2000(1.0867)^t + 1000(1.0693)^t$ where V is the investment value (in dollars) and t is the number of years from now.
(b) $V(20) = \$14{,}368.66$
Before changing the amount of money invested in each account, the investor should consider her risk tolerance. (Answers may vary.)

39. (a) Growth account: $699.40
Inflation-linked Bond account: $1163.20
(b) The combined value of the accounts is projected to double by June 2017 ($t = 14.43$).

Section 2.4 *(page 129)*

1. $y = 2$ **3.** $y = 6$ **5.** $y = -2$

7. $y = \log_4(x)$ is concave down and increasing.

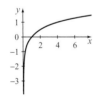

9. $y = \log_{0.7} x$ is concave up and decreasing.

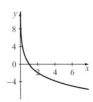

11. $x = 6$ **13.** $x = -1$ **15.** $x = 3$

17. $x = \dfrac{1}{16}$ **19.** $x = 243$ **21.** $4\log(2)$

23. $\log(2x^4)$ **25.** $\log(jam)$

27. $\log\left(\dfrac{1}{324}x^{-16}\right)$ **29.** $-\log(x)$

31. $2\ln(x)$ **33.** $2\ln(x)$ **35.** $\ln(27)$

37. $\ln\left(\dfrac{1}{64}\right)$ **39.** $\ln(729x^5)$

41. $T(i) = -73.69 + 16.71\ln(i)$
$T(125) = 7.0$
According to the model, the Consumer Price Index reached 125 seven years after 1980. That is, at the end of 1987, the price of dental services were 125 percent of their 1984 price.

43. $T(i) = 129.8 - 27.76\ln(i)$
$T(125) = -4.2$
According to the model, the Consumer Price Index was at 125 4.2 years before the end of 1980. That is, in late 1976, the price of a television set was 125 percent of its 1984 price.

45. $T(i) = -88.48 + 19.84\ln(i)$
$T(125) = 7.3$
According to the model, the Consumer Price Index was at 125 7.3 years after the end of 1980. That is, in early 1988, the price of admission to an entertainment venue was 125 percent of its 1984 price.

47. We know that $y = \log_b(x)$ is equivalent to $x = b^y$. From the definition of $y = \log_b(x)$, we also know that b is positive.
We know that raising a positive number to any power yields a positive number. So b^y will always be positive. But $x = b^y$, so x must always be positive. Hence, the domain of $y = \log_b(x)$ is all positive real numbers.

49. No.

Section 2.5 *(page 142)*

1. $S(t) = 8.451t^3 - 111.9t^2 + 470.0t + 965.0$
Since $S(12) \approx 5099$, we forecast that gaming hardware sales will be $5099 million in 2002.

3. $C(t) = 0.3214t^2 + 0.2500t + 37.36$
Since $C(8) \approx 60$, we forecast that the 2005–06 tuition cost per credit will be $60.

5. $S(t) = 1165.9t + 80,887$

Since $S(20) \approx 104,200$, we forecast that the 2010 public university enrollment will be 104,200.

7. $C(t) = \dfrac{1106}{1 + 0.1259e^{0.5491t}}$

Since $S(13) \approx 7$, we forecast that there will be 7 pediatric AIDS cases in the United States in 2005.

9. $A(t) = \dfrac{38.56}{1 + 14.52e^{-0.8032t}} + 16$

Since $A(10) \approx 54$, we forecast that there were 54 air carrier accidents in 2002.

11. $C(t) = 0.60(1.03)^t$ dollars, where t is the number of years from now.

13. $P(t) = 3t + 87$ dollars, where t is the number of years from now.

15. The logistic model for monthly product sales is

$$S(t) = \frac{200.67}{1 + 25.85e^{-0.4973t}}$$

17. The logistic model for the data is

$$P(t) = \frac{6930}{1 + 166.2e^{-0.4019t}} + 2600$$

The maximum projected population of the city is $6930 + 2600 = 9530$. According to the logistic model, the 2003 population was 6258, since $P(13) \approx 6258$.

The exponential model that best fits the aligned data is

$$P(t) = 67.59(1.377)^t + 2600$$

According to the exponential model, the 2003 population was 6925, since $P(13) \approx 6925$. Although this estimate is still below the actual 2003 population (7480), it is substantially better than the logistic model estimate. (Answers may vary.)

19.
$$T(x) = \begin{cases} 0.1x & \text{if } 0 \le x \le 7000 \\ 0.15x - 350 & \text{if } 7000 < x \le 28,400 \\ 0.25x - 3190 & \text{if } 28,400 < x \le 68,800 \end{cases}$$

21. In addition to looking at how well the model fits the data, we must consider the expected future behavior of the thing being modeled. Since the model is being used to forecast future events, we must make sure that the model's prediction seems reasonable based on what we know about the thing being modeled. (Answers may vary.)

23. Businesses can benefit financially by accurately predicting the future. Although mathematical models are imperfect at forecasting the future, they offer an educated guess as to what might happen. In short, they increase the likelihood of an accurate prediction. (Answers may vary.)

25. Because of the dramatic variability in the data set, we don't recognize any trends or patterns in the scatter plot. None of the standard mathematical models will fit this data set well. Our best guess might be to identify the range of y values and state that we would expect future data values to fall within that range. (Answers may vary.)

CHAPTER 2 REVIEW EXERCISES

Section 2.1 *(page 147)*

1. The quadratic model that best fits the data is
$$A(t) = 42.293t^2 + 158.32t + 6574.5$$
Since $A(12) \approx 14,564$, we estimate that $14,564 million was spent on magazine advertising in 2002.

3. Early in the second quarter of 2001 ($t \approx 10.3$).

5. $A(t) = 34.992t^2 + 94.909t + 8963.2$

Since $A(11.8) \approx 15,000$, we estimate that $15 billion was spent on Yellow Pages advertising in the one-year period that ended in late 2002 ($t = 11.8$).

Section 2.2 *(page 148)*

7. $A(t) = -32.085t^3 + 656.16t^2 - 1640.7t + 31,914$

Since $A(11) \approx 50,557$, we estimate that $50,557 million was spent on newspaper advertising in 2001 ($t = 11$).

9. $A(t) = 10.222t^4 - 221.51t^3 + 1616.4t^2 - 2487.8t + 26,671$

Since $A(11.0) \approx 50,000$, we estimate that $50 billion was spent on broadcast TV advertising in 2001 ($t = 11$). Broadcast TV annual advertising expenditures beyond 2001 are expected to exceed $50 billion.

11. $E(t) = 0.001228t^3 - 0.03338t^2 + 0.6439t + 9.844$

Since $E(23) \approx 21.94$, we estimate that average hourly earnings of a manufacturing industry employee in Michigan in 2003 ($t = 23$) was $21.94.

13. The manufacturing wage in Florida is roughly 65 percent of the Michigan manufacturing wage. Thus labor costs in Florida will be substantially less than labor costs in Michigan. For this reason, a new manufacturing business may prefer to start up in Florida as opposed to Michigan. However, additional factors such as business tax rate, infrastructure, availability and cost of raw materials, etc., should also be considered. (Answers may vary.)

Section 2.3 *(page 151)*

15. The function $y = -0.3(2.8)^x$ is concave down and decreasing. The y-intercept is $(0, -0.3)$.

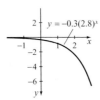

17. $y = 9(2)^x$

19. $N(t) = 22{,}925(1.0727)^t$
$N(10) \approx 46{,}248$ means that there will be 46,248 McDonalds restaurants in 2007.

21. $E(t) = 4500(1.023)^t$
$E(5) \approx 5042$ means that five years from now, my monthly household expenses are expected to be $5042.

23. $x \approx 1.723$

Section 2.4 *(page 151)*

25. $y = -1$

27. The graph of $y = \log_{0.4}(x)$ is decreasing and concave up.

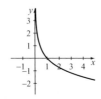

29. $x = 4$ **31.** $x = 0.04$

33. $\log(2x^2)$ **35.** $-\log(3)$

37. $F(t) = 49.96 + 5.112 \ln t$. Since $F(10) \approx 61.7$, we estimate that in 1999, 61.7 percent of auto accident fatalities were not alcohol-related.

Section 2.5 *(page 152)*

39. $P(t) = \dfrac{57.81}{1 + 0.5443e^{0.3900t}}$

CHAPTER 3 EXERCISES

Section 3.1 *(page 162)*

1. 2   **3.** -1 **5.** 0

7. $\dfrac{\ln(5)}{20} \approx 0.0805$ **9.** $\dfrac{\sqrt{2}}{6} \approx 0.2357$

11. $\dfrac{7}{4}$ degrees per hour

13. Between 1996 and 1998, 1366.55 points per year; between 1997 and 1999, 1794.40 points per year

15. 30 words per minute per quarter

17. -0.378 newspaper subscription per cable TV subscriber

19. 1.375 people below poverty level per unemployed person

21. $f(x) = 2^x$

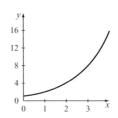

$\dfrac{f(3) - f(1)}{3 - 1} = 3$

23. $f(x) = 5$

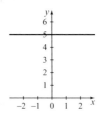

$\dfrac{f(3) - f(1)}{3 - 1} = 0$

25. $f(x) = (x - 2)^2$

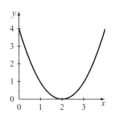

$\dfrac{f(3) - f(1)}{3 - 1} = 0$

27. $\dfrac{f(3) - f(1)}{3 - 1} = 2$ **29.** $\dfrac{f(3) - f(1)}{3 - 1} = -1$

31. ≈ 147 people per month

33. ≈ 2013 people per month

35. 2300 people per month

37. ≈ -427 people per month

39. ≈ 407 people per month

41. The average rate of change for $2f(x)$ is twice as big as the rate for change for $f(x)$.

43. The vertical distance between $(2, f(2))$ and $(2.01, f(2.01))$ is 0.04.

45.

h	Secant slope $= h^{-2/3}$
1.0	1.0
0.1	≈ 4.642
0.01	≈ 21.544
0.001	≈ 100

As h gets very small, the secant slope will get very large. As h approaches 0, the slope of the secant line appears to approach ∞.

Section 3.2 *(page 179)*

1.

h	Difference Quotient
0.1	4.1
0.01	4.01
0.001	4.001
Instantaneous rate ≈ 4.0	

3.

h	Difference Quotient
0.1	-65.6
0.01	-64.16
0.001	-64.016
Instantaneous rate ≈ -64.0	

5.

h	Difference Quotient
0.1	4
0.01	4
0.001	4
Instantaneous rate ≈ 4.0	

7.

h	Difference Quotient
0.1	0
0.01	0
0.001	0
Instantaneous rate ≈ 0	

9.

h	Difference Quotient
0.1	1,120
0.01	1,075
0.001	1,070.5
Instantaneous rate $\approx 1,070$	

11. 4   **13.** -64 **15.** 4   **17.** 0   **19.** 1070

21.

h	Difference Quotient
0.1	-0.06940
0.01	-0.07358
0.001	-0.07402
Instantaneous rate ≈ -0.074	

23.

h	Difference Quotient
0.1	-1.7305
0.01	-1.7388
0.001	-1.7396
Instantaneous rate ≈ -1.740	

25.

h	Difference Quotient
0.1	14,198.385
0.01	9,588.682
0.001	9,231.052
Instantaneous rate $\approx 9,200$	

27.

h	Difference Quotient
0.1	0.4879
0.01	0.4988
0.001	0.4999
Instantaneous rate ≈ 0.50	

29.

h	Difference Quotient
0.1	70.271
0.01	61.170
0.001	60.347

Instantaneous rate ≈ 60.0

31. $1339 per year. In 2000, teacher salaries are increasing at the rate of $1339 per year.

33. $14.3 billion per year. In 2005, the amount of money spent on prescription drugs will be increasing at the rate of $14.3 billion per year.

35. $0.871 billion per year. Video game sales in the first quarter of 2007 will be increasing at a rate of $0.871 billion per year.

37. $31.91 per year. In 2001, the cost of full-time resident tuition at Green River Community College was increasing at a rate of $31.91 per year.

39. 242.5 million discs per year. In 2002, the demand for CD-R discs was increasing at a rate of 242.5 million discs per year.

41. ≈ -51 feet per second

43. ≈ -12 feet per second

45. ≈ -15 feet per second

47. No. Consider $f(x) = x^3$. For this function, $f'(-1) = f'(1) = 3$. However, $-1 \neq 1$. Therefore, different values of x may have the same derivative.

49. Since $f'(x) = 0$, $f(x) = c$ for some constant c. If $c = 0$, there are infinitely many x-intercepts. Otherwise, there are no x-intercepts.

Section 3.3 *(page 194)*

1.

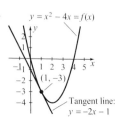

$y = x^2 - 4x = f(x)$

(1, -3)

Tangent line: $y = -2x - 1$

3.

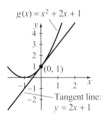

$g(x) = x^2 + 2x + 1$

(0, 1)

Tangent line: $y = 2x + 1$

5.

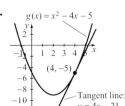

$g(x) = x^2 - 4x - 5$

(4, -5)

Tangent line: $y = 4x - 21$

7.

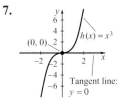

$h(x) = x^3$

(0, 0)

Tangent line: $y = 0$

9.

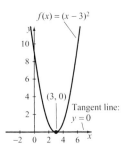

$f(x) = (x - 3)^2$

(3, 0)

Tangent line: $y = 0$

11. 2.356 billion gallons per year; 55.75 billion gallons

13. 20.27 square feet per year; 2238.5 square feet

15. Increasing at the rate of 0.128 student per teacher per year; 15.157 students per teacher

17. When the number of cassette tapes shipped is 250 million, the value of the tapes is increasing at a rate of 7.775 million dollars per million cassette tapes shipped. The estimated shipment value when 251 million cassette tapes are shipped is about 2134 million dollars.

19. The rate of change in *private* college enrollment when 12,752 *public* students are enrolled is 0.340 thousand *private* college enrollments per thousand *public* college enrollments. (In other words, 340 *private* enrollments per 1000 *public* enrollments.) We estimate that when 12,753 *public* college students are enrolled, 3879 *private* college students are enrolled.

21. $1750 per year. In 1997, the annual wage was increasing at a rate of $1750 per year.

23. -13.1 newspapers per year. In 1995, the number of different daily newspapers in the United States was decreasing at a rate of about 13 newspapers per year.

25. $26.50 per year; $34 per year. The rate of increase is itself increasing.

27. 1.1 million people per year. In 1997, the attendance at American League baseball games was increasing at a rate of 1.1 million people per year.

29. Increasing at a rate of 5.1 million people per year

31. Possible pairs: (1, 1) and (3, 9) or (0, 0) and (4, 16) (answers may vary). Each pair is horizontally equidistant from the point (2, 4).

33. $f'(-1) = 3$ and $f'(1) = 3$. The graph is increasing at the same rate at both points.

35. $p'(0) = f'(0) + g'(0)$

Section 3.4 *(page 203)*

1. $f'(x) = 2x - 4$ **3.** $g'(x) = 2x + 2$

5. $g'(x) = 2x - 4$ **7.** $h'(x) = 3x^2$

9. $f'(t) = 2t - 6$

11.

x	$g'(x)$
1	5
3	13
5	21

13.

x	$f'(x)$
1	0
3	4
5	8

15.

x	$h'(x)$
1	0
3	0
5	0

17.

x	$W'(x)$
1	-4
3	-4
5	-4

19.

x	$S'(x)$
1	4
3	16
5	28

21. $P'(x) \approx 1.099(3^x)$

23. $C'(x) \approx -4.162(4^x)$

25. $R'(x) \approx -0.102(0.98^x)$

27. 2000   **29.** Northeast

31. 1999

33. Since $t = 19$ is the end of 2009, we estimate that in mid-2010 per capita prescription drug spending will be increasing at a rate twice that of the 2000 rate.

35. 5.517 thousand students per year

37. $g'(x) = 5x$ **39.** $x = \pm 1$

Section 3.5 *(page 208)*

1. According to the model, the weight of a 10-year-old girl is 76.64 pounds. A 10-year-old girl will gain weight at a rate of 8.24 pounds per year. An 11-year-old girl will weigh about 84.88 pounds.

3. According to the model, the annual per capita income in Washington state in 1984 was $14,203. The income in 1984 was increasing at a rate of $178.40 per year. The income in 1985 was about $14,381.40.

5. According to the model, the average daily volume of the Nasdaq market in 1998 was 918 million shares. In 1998, this volume was growing at the rate of 299 million shares per year. The volume in 1999 was about 1217 million shares.

7. According to the model, the annual volume of the New York Stock Exchange in 1998 was 168,659 million shares. This volume was growing at the rate of 38,760 million shares per year. This volume in 1999 was about 207,419 million shares.

9. According to the model, the average annual wage of an employee in the radio broadcasting industry in 1998 was $34,086. The wage in 1998 was increasing at a rate of $2750 per year. The wage in 1999 was about $36,836.

11. According to the model, the amount of money spent on billboard advertising in 2000 was $1887 million. This amount was increasing at a rate of $176 million per year in 2000.

13. According to the model, when the net number of CDs shipped is 900 million, the value of the CDs is $12,261.9 million. When the net number of CDs shipped is 900 million, the value of the CDs is increasing at a rate of $17.9 million per million CDs shipped.

15. According to the model, the number of women in the Summer Olympics in 1992 was 2662 and was growing at a rate of 150 women per year.

17. According to the model, the population of the United States in the year 2000 was 284 million people and was growing at a rate of 2.70 million people per year.

19. According to the model, the number of homicides resulting from an alcohol-related brawl in the year 2000 was 185. In 2000, this number was decreasing at a rate of 14 homicides per year.

21. According to the model, the gross revenue from sales of Johnson & Johnson and its subsidiaries at the end of 2001 was $32,317 million and was increasing at a rate of $3810 million per year.

23. According to the model, the operating profit of Frito-Lay North America at the end of 2001 was $2056 million and was increasing at the rate of $93.5 million per year.

25. According to the model, the flow of the Snoqualmie River at noon on November 24, 2002, was 300 cubic feet per second and was decreasing at 45 cubic feet per second per 24-hour period. The river rafting company could predict when the flow was fast, slow, safe, unsafe, and so on.

27. According to the model, the per capita personal income of Colorado at the end of 1998 was $28,634 and was increasing at a rate of $1700.10 per year.

29.

	Per Capita Income in 1998 (dollars)	Projected Growth in Income (dollars)	Projected per Capita Income in 1999 (dollars)
California	23,712	791.7	24,503.7
Colorado	28,634	1,700.1	30,334.1
Connecticut	36,767	1,998.6	38,765.6

31. The VCR price decreased; $P'(t)$ will approach 0 dollars per year.

33. 1989; 1996   **35.** 0

CHAPTER 3 REVIEW EXERCISES

Section 3.1 *(page 212)*

1. 6   **3.** $230.2 per year   **5.** 12

Section 3.2 *(page 212)*

7.

h	Difference Quotient
0.1	4.2
0.01	4.02
0.001	4.002
Instantaneous rate ≈ 4.0	

9. 4

11.

h	Difference Quotient
0.1	0.9777
0.01	0.9767
0.001	0.9766
Instantaneous rate ≈ 0.977	

13. $T'(7) \approx 0.0687$ second per rank. In the 100-meter men's race in the 2000 Olympics, the seventh-place runner's time is increasing at the rate of 0.0687 seconds per place (rank).

Section 3.3 *(page 213)*

15. $y = 2x - 2$

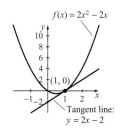

17. $\approx$ $36.33 per year; $\approx$ $996.12

19. -649.2 fatalities per year. In 1985, the number of highway fatalities was decreasing at a rate of about 649 fatalities per year.

Section 3.4 *(page 214)*

21. $f'(x) = 4x$

23. 8 at $x = 1$; 20 at $x = 3$; 32 at $x = 5$

25. $P'(x) \approx 0.693(2^x)$ **27.** 1999

Section 3.5 *(page 214)*

29. According to the model, McDonald's sales were $27,015 million in 2002 and were increasing at a rate of $1233.5 million per year.

CHAPTER 4

Section 4.1 *(page 223)*

1. 0   **3.** $15t^2 + 20t^{-3}$ **5.** $-2 + 12x^2$

7. $-4t^{-2}$ **9.** $3t^2 + 6t + 6$

11. $y = 0.5x + 0.5$ **13.** $y = -7x - 1.4$

15. $y = 551.23t - 1203.9$

17. $y = 3.8n + 102.2$ **19.** $y = 3.66x - 12.74$

21. 140.05 thousand acres per year

23. Yes   **25.** 9-year-old girl

27. No   **29.** 2000

31. No. Consider $f(x) = g(x) = x$. $h(x) = x^2$ and $h'(x) = 2x$. If we use the Constant Multiple Rule, we get

$$h'(x) = f(x)g'(x)$$
$$= x(1)$$
$$= x$$

This is a contradiction.

33.

$$s'(x) = \lim_{h \to 0} \frac{s(x + h) - s(x)}{h}$$
$$= \lim_{h \to 0} \frac{[f(x + h) + g(x + h)] - [f(x) + g(x)]}{h}$$
$$= \lim_{h \to 0} \frac{f(x + h) - f(x) + g(x + h) - g(x)}{h}$$
$$= \lim_{h \to 0} \frac{f(x + h) - f(x)}{h} + \lim_{h \to 0} \frac{g(x + h) - g(x)}{h}$$
$$= f'(x) + g'(x)$$

35. 0

Section 4.2 (page 231)

1. $2(3x + 4) + 3(2x)$

3. $2(10t + 5) + 10(2t - 6)$

5. $6n(n^3 - n) + (3n^2 - 1)(3n^2 + 8)$

7. $(2t + 2)(t^2 + 4t + 4) + (2t + 4)(t^2 + 2t + 1)$

9. $\dfrac{3x^2(x^4 - x^3) - (4x^3 - 3x^2)(x^3 - 1)}{(x^4 - x^3)^2}$

11. $-\dfrac{1}{2}$ **13.** -5.5 **15.** 56

17. 589   **19.** -4

21. ≈ 1504.35 million dollars per year

23. ≈ 178.74 million dollars per year

25. ≈ 1369.96 million dollars per year

27. $f'(40) = 1$ means that when 70 trees per acre are planted, planting an additional tree will *increase* the overall yield by one bushel of apples per acre.

$f'(45) = 0$ means that when 75 trees per acre are planted, planting an additional tree will not change the overall yield of apples per acre.

$f'(50) = -1$ means that when 80 trees per acre are planted, planting an additional tree will *decrease* the overall yield by one bushel of apples per acre.

29. $0.19 per pound

Section 4.3 (page 241)

1. $2x^2 + 11$ **3.** $7x^2 - 14x + 3$

5. $3x^6 - 11x^3$

7. $f(x) = x^2 - 6x + 9;\ g(x) = x + 1$

9. $f(x) = 2\ln(x);\ g(x) = x^2 + 4$

11. $30(3x + 1)$

13. $12(t^3 - t^2 + 1)^3(3t^2 - 2t)$

15. $20(n^2 + 2n + 1)(n + 1)$ or $20(n + 1)^3$

17. $4(-80x^2 + 42x - 5)(-80x + 21)$

19. $(3t + 2)(5t + 2)(12t + 1)$

21. $-\dfrac{13}{4}$ **23.** $\dfrac{13}{81}$ **25.** $-11{,}280$

27. $y = \dfrac{5}{8}t + \dfrac{1}{16}$ **29.** $y = 14{,}641$

31. 2.124 inches per year

33. $8x[(x^2 + 1)^2 + 1](x^2 + 1)$

35. 40 sales per year

Section 4.4 (page 249)

1. $\ln(4)4^x$ **3.** $\dfrac{4}{t}$

5. $5e^x + \dfrac{1}{x}$ **7.** $5e^x\left[\ln(x) + \dfrac{1}{x}\right]$

9. $\ln(2)(2^{x^2 + 5x})(2x + 5)$

11. $\dfrac{6}{5\ln(10)}$ **13.** $\log(2)$

15. $\dfrac{3\ln(3)}{e}$ **17.** $3\log(5)$ **19.** $15e^2$

21. $y = 3\ln(2)t - 3\ln(2) + 1$

23. $y \approx 9.862x - 4.443$ **25.** $y = \dfrac{1}{e}$

27. 23.63 dollars per year

29. ≈ 1.63 percentage points per year. In 1999, the percentage of homes with a VCR was increasing at a rate of 1.63 percentage points per year.

31. $\dfrac{\ln(4)4^{\ln(x)}}{x}$

33. $2^{2t+1}\ln(2)[\ln(2)(2t) + 1]$

35. $\dfrac{1}{t}$

Section 4.5 *(page 256)*

1. Undefined

3. -2

5. $-\dfrac{3}{2}$

7. $\dfrac{1}{3}$

9. 5

11. $\dfrac{1}{2y}$

13. $\dfrac{dy}{dx} = 0$

15. $\dfrac{y^2}{1 - xy}$

17. $-\dfrac{y}{x - y}$

19. $\dfrac{y}{xy - 2x}$

21. $(0, 1)$ and $(0, -1)$ **23.** $(0, 0)$ and $\left(0, \dfrac{1}{2}\right)$

25. $(1, -2)$ and $(-1, 2)$

27. $\dfrac{dr}{dt} = \dfrac{\dfrac{dV}{dt} - \pi r^2\dfrac{dh}{dt}}{2\pi rh}$

29. We have $\dfrac{dy}{dx} = \dfrac{e^x y - e^y}{xe^y - e^x}$.

Although the coordinate $(1, 1)$ is the only coordinate that makes the numerator of $\dfrac{dy}{dx}$ equal to zero, it also makes the denominator equal zero. So $\dfrac{dy}{dx}$ is undefined at $(1, 1)$. Thus $\dfrac{dy}{dx} \neq 0$ everywhere.

CHAPTER 4 REVIEW EXERCISES

Section 4.1 *(page 257)*

1. 0

3. $21t^2 + 12t^{-4}$

5. 1999: Decreasing at a rate of 12.773 thousand people per year

2000: Decreasing at a rate of 7.648 thousand people per year

Section 4.2 *(page 257)*

7. $18x^2 - 24x$

9. $-6t^2 + 16t + 4$

11. $\approx 24{,}422$ prisoners per year

Section 4.3 *(page 257)*

13. $-6x^2 + 3x + 4$ **15.** $54(9x + 5)^2$

17. 2,219.31 million dollars per year

Section 4.4 *(page 258)*

19. $\ln(9)9^x$

21. $\dfrac{2x}{x^2 + 3}$

23. 20.1 thousand people per year

Section 4.5 *(page 258)*

25. $-\dfrac{1}{21}$ **27.** -4 **29.** $\dfrac{-2\ln(2) + 1}{6\ln(2) - 3}$

CHAPTER 5

Section 5.1 *(page 277)*

1. $f'(x) = 0$ at $(1, -1)$

3. $f'(x) = 0$ at $(0, 1)$ **5.** $f'(x) = 0$ at $(2, 0)$

7. $f'(x)$ is undefined at $(2, 4)$

9. $f'(x) = 0$ at $(0, 200)$ and $(4, -56)$

11. $x = 2$; stationary   **13.** $t = 0, 4$; stationary

15. $x = 0$; singular
$x = \pm 1$; stationary

17. $x = \pm 1$; stationary   **19.** $p = \pm 1$; stationary

21.
$f'(x)$ $\xleftarrow{\quad - \quad | \quad + \quad}$
$\qquad\qquad$ min
$\qquad\qquad$ 2

x	$f(x)$ on $[-1, 5]$	
-1	5	Absolute maximum
2	-4	Relative and absolute minimum
5	5	Absolute maximum

23.
$h'(t)$ $\xleftarrow{\;+\;\; \max \;\; - \;\; \min \;\; +\;}$
$\qquad\qquad$ 0 $\qquad\quad$ 4

t	$h(t)$ on $[-1, 7]$	
-1	-7	
0	0	Relative maximum
4	-32	Relative and absolute minimum
7	49	Absolute maximum

25.

x	$g(x)$ on $[-3, 3]$	
−3	−1.327	
−1	−2	Relative and absolute minimum
0	0	
1	2	Relative and absolute maximum
3	1.327	

27.

x	$h(x)$ on $[-2, 2]$	
−2	−0.4	
−1	−0.5	Relative and absolute minimum
1	0.5	Relative and absolute maximum
2	0.4	

29.

p	$C(p)$ on $[-3, 3]$	
−3	−65	Absolute minimum
−1	15	Relative maximum
1	−1	Relative minimum
3	79	Absolute maximum

31.

x	$f(x)$ on $[-3, 3]$	
−3	−15	Absolute minimum
$-\sqrt{\frac{4}{3}}$	≈ 3.08	Relative maximum
$\sqrt{\frac{4}{3}}$	≈ −3.08	Relative minimum
3	15	Absolute maximum

33.

t	$h(t)$ on $[0.1, 3]$	
0.1	≈ −0.2303	
$\frac{1}{e}$	≈ −0.3679	Relative and absolute minimum
3	≈ 3.2958	Absolute maximum

35.

x	$g(x)$ on $[0.1, 3]$	
0.1	≈ 3.408	
0.567	≈ 2.330	Relative and absolute minimum
3	≈ 18.987	Absolute maximum

37. $25   **39.** $40

41. From about 1983 through 1988

43. Highest: 1999
Lowest: 1990

45. 59.457 million   **47.** $y = e^x$

49. Each time $p(x)$ changes from increasing to decreasing (or vice versa) $p'(x)$ will change sign. Since $p'(x)$ is continuous, $p'(x)$ will equal zero between sign changes. Since there are at most $n - 1$ sign changes, there are at most $n - 1$ stationary points.

Section 5.2 (page 299)

1. (a) $P(t) = 564.3t^2 - 613.7t + 3658.9$
(b) 2000

3. (a)

	Hotel Pack	Single Slice
Surface area	455.8 in²	680.5 in²
Volume	654.9 in³	869.0 in³

(b) ≈ 8.716 in × 8.716 in × 8.716 in
(c) ≈ 10.65 in × 10.65 in × 10.65 in
(d) The box is a cube, since the length, width, and height are all equal.

5. (a) $s(t) = 17.8(1.04)^{t-2}$ billion unit cases
[equivalent to $s(t) \approx 16.5(1.04)^t$ billion unit cases]

(b) $R'(s) = -0.8058s + 14.44$ dollars per case sold

(c) 17.92 billion cases

(d) $R(s(t)) = -0.4029[17.8(1.04)^{t-2}]^2$
$+ 14.44[17.8(1.04)^{t-2}] - 109.2$

(e) Early 2002

7. (a) 2002

(b) $C(t) = -120.5t^2 + 623.5t + 6553$ million dollars; t is the number of years since the end of 1999.

(c) 2002

(d) They are reasonable, since the maximum number of sales could incur additional costs.

9. 2000   **11.** 30 trees   **13.** $0.19 per pound

15. $116.92 million per year

17. ≈ 27.5 feet $\times$ 18.2 feet   **19.** 1200 books

21. 500 books   **23.** 50 vehicles

25. 43 vehicles

27. Limit the number of cars sold to the amount that maximized revenue.

29. 250 pounds

31. $\dfrac{dx}{dt} = 3t^2 - 18t + 23$ meters per second is the rate of horizontal change in position; $\dfrac{dy}{dt} = 6 - 2t$ meters per second is the rate of vertical change in position.

33.

t	$x(t)$	
0	0	Absolute minimum
1.85	18.08	Relative maximum
4.15	11.92	Relative minimum
6	30	Absolute maximum

t	$y(t)$	
0	0	Absolute minimum
3	9	Relative and absolute maximum
6	0	Absolute minimum

35. 0 meters per second (*Note:* speed is the absolute value of velocity.)

Section 5.3 *(page 322)*

1. $f(x) = x^3 - 3x^2$

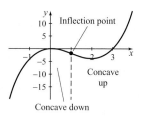

3. $f(x) = x^4 - 12x^2$

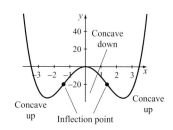

5. $y = x^4 - 8x^3 + 18x^2$

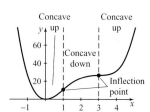

7. $y = x^4 - 2x^3 + 12x^2$

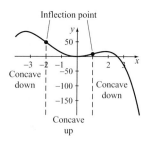

9. $f(x) = \dfrac{x^2}{x^2 + 1}$

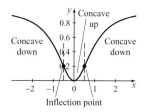

11. $(1, -2)$

13. $\left(-\sqrt{2}, -20\right); \left(\sqrt{2}, -20\right)$

15. $(1, 11); (3, 27)$ **17.** $(-2, 48); (1, 9)$

19. $\left(-\sqrt{\dfrac{1}{3}}, \dfrac{1}{4}\right); \left(\sqrt{\dfrac{1}{3}}, \dfrac{1}{4}\right)$

21. $(0, 0)$ relative maximum; $(2, -4)$ relative minimum

23. $\left(-\sqrt{6}, -36\right)$ relative minimum; $(0, 0)$ relative maximum; $\left(\sqrt{6}, -36\right)$ relative minimum

25. $(0, 0)$ relative minimum; $(3, 27)$ inflection point

27. $(0, 0)$ relative minimum; $(-3.312, 83.967)$ relative maximum; $(1.812, 16.721)$ relative maximum

29. $(0, 0)$ relative minimum

31. 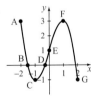 $A\left(-2, -6\dfrac{2}{3}\right)$ absolute minimum on $[-2, 4]$
$B(0, 0)$ relative maximum; x- and y-intercept
$C\left(1, -\dfrac{2}{3}\right)$ inflection point
$D\left(2, -\dfrac{4}{3}\right)$ relative minimum
$E(3, 0)$ x-intercept
$F\left(4, 5\dfrac{1}{3}\right)$ absolute maximum on $[-2, 4]$

33. $A(-2, 3)$ absolute maximum on $[-2, 2]$
$B(-1.532, 0)$ x-intercept
$C(-1, -1)$ relative and absolute minimum on $[-2, 2]$
$D(-.347, 0)$ x-intercept
$E(0, 1)$ inflection point and y-intercept
$F(1, 3)$ relative and absolute maximum on $[-2, 2]$
$G(2, -1)$ absolute minimum on $[-2, 2]$

35. $A(-4, 16)$ absolute maximum on $[-4, 4]$
$B(-3.464, 0)$ x-intercept
$C(-2, -16)$ relative and absolute minimum on $[-4, 4]$
$D(0, 0)$ x- and y-intercept; inflection point
$E(2, 16)$ relative and absolute maximum on $[-4, 4]$
$F(3.464, 0)$ x-intercept
$G(4, -16)$ absolute minimum on $[-4, 4]$

37. $A(-3, -129.6)$ absolute minimum on $[-3, 5]$
$B(0, 0)$ x- and y-intercept; relative and absolute maximum on $[-3, 5]$
$C(3, -32.4)$ inflection point
$D(4, -51.2)$ relative minimum
$E(5, 0)$ x-intercept; absolute maximum on $[-3, 5]$

39. $A(-2, 1.636)$
$B(0, 1)$ x-intercept; relative and absolute minimum on $[-2, 2]$
$C(2, 2.614)$ absolute maximum on $[-2, 2]$

41. Minimum increase: 1975, $-\$5.765$ per year
Maximum increase: 2000, $\$1.465$ per year

43. Wages should continue to increase to compensate for diminishing wages during the years directly after 1975.

45. 1995; 2000   **47.** 1995; 1998

49. 1991; 299   **51.** 1

53. $s''(t) = -32$ feet per second squared, acceleration due to the force of gravity

55. 3.9 seconds

57. $(23.16, 12.56)$; in the 23rd week, the box office sales were increasing most rapidly at about 12.56 million dollars per week.

59. $x'' = 0$ at $t = 3$; $y'' \neq 0$

61. 18 square meters per second   **63.** Yes

Section 5.4 *(page 333)*

1. $\dfrac{dA}{dt} = \dfrac{dl}{dt}w + \dfrac{dw}{dt}l$

   $\dfrac{dA}{dt}$ = change in area with respect to time in square inches per minute

   $\dfrac{dl}{dt}$ = change in length with respect to time in inches per minute

   $\dfrac{dw}{dt}$ = change in width with respect to time in inches per minute

3. $\dfrac{dV}{dt} = \dfrac{dl}{dt}wh + \dfrac{dw}{dt}lh + \dfrac{dh}{dt}lw$

   $\dfrac{dV}{dt}$ = change in volume with respect to time in cubic inches per minute

   $\dfrac{dl}{dt}, \dfrac{dw}{dt}$, same as in Exercise 1.

   $\dfrac{dh}{dt}$ = change in height with respect to time in inches per minute.

5. $\dfrac{dA}{dt} = 2\left(\dfrac{dl}{dt}w + \dfrac{dw}{dt}l + \dfrac{dw}{dt}h + \dfrac{dh}{dt}w \right.$
   $\left. + \dfrac{dl}{dt}h + \dfrac{dh}{dt}l \right)$

   Physical meanings same as in Exercises 1 and 3

7. $\dfrac{dA}{dt} = 8\pi r \dfrac{dr}{dt}$

   $\dfrac{dA}{dt}$, same as in Exercise 1

   $\dfrac{dr}{dt}$ = change in radius with respect to time in inches per minute

9. $\dfrac{dA}{dt} = 2\pi r \dfrac{dr}{dt}$

   Physical meanings same as in Exercise 7

11. 3.782 inches per hour; 12.69 hours

13. $0.2\overline{6}$ inches per second or $0.677\overline{3}$ centimeters per second

15. 0.1014 foot per hour

17. 8.83 degrees per hour

19. 4.20 degrees per hour

CHAPTER 5 REVIEW EXERCISES

Section 5.1 *(page 334)*

1. $f'(x) = 0$ at $(1.5, 2.25)$ and $(2.5, 2.25)$
   $f'(x)$ undefined at $(2, 2)$

3. $\left(-\sqrt{\dfrac{12}{5}}, \approx 15.95 \right); (0, 10); \left(\sqrt{\dfrac{12}{5}}, \approx 4.051 \right)$

5. $(0, 4)$

7.
x	$f(x)$ on $[-2, 2]$	
-2	10	
$-\sqrt{12/5}$	≈ 15.95	Relative and absolute maximum
0	10	
$\sqrt{12/5}$	≈ 4.051	Relative and absolute minimum
2	10	

9.
x	y on $[-3, 3]$	
-3	-32	Absolute minimum
0	4	Relative and absolute maximum
3	-32	Absolute minimum

11. 469.4 miles

Section 5.2 *(page 334)*

13. **(a)** 2002
    (b) $C(t) = -88.5t^2 + 410.5t + 3019$
    (c) 2002
    (d) The same year is reasonable because more sales could generate more costs.

Section 5.3 *(page 335)*

15.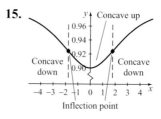

17. $\left(-\sqrt{\dfrac{10}{3}}, 0.925\right), \left(\sqrt{\dfrac{10}{3}}, 0.925\right)$

19. $(0, .9)$ stationary point and relative and absolute minimum

21. 4; 17.72 miles per day

Section 5.4 *(page 336)*

23. $\dfrac{dA}{dt} = 10s\dfrac{ds}{dt}$

$\dfrac{dA}{dt}$ = change in area with respect to time in square centimeters per second

$\dfrac{ds}{dt}$ = change in side length with respect to time in centimeters per second

25. 0.32 inch per minute   **27.** 0.20 inch per minute

29. 0.0159 foot per hour

CHAPTER 6

Section 6.1 *(page 346)*

1. $2x + C$

3. $\ln|t| - \dfrac{3t^2}{2} + C$

5. $\dfrac{t^2}{4} + 20t + C$

7. $\dfrac{t^4}{4} - t^3 + \dfrac{3t^2}{2} - t + C$

9. $2.3 \ln|x| + x + C$

11. $\dfrac{3^x}{\ln(3)} - \dfrac{x^4}{4} + C$

13. $\dfrac{-4}{t} + 2 \ln|t| + t + C$

15. $\dfrac{2(3^x)}{\ln(3)} - \dfrac{3(2^x)}{\ln(2)} + C$ **17.** $2000 \ln|p| + \dfrac{500}{p} + C$

19. $-5t^{-1} + \dfrac{2^t}{\ln(2)} + C$ **21.** $\dfrac{3x^2}{2} - 5x + C$

23. $t - 2 \ln|t| + C$

25. $-3x^{-1} - x^4 + 2x + C$

27. $-5t^{-1} - 16\dfrac{t^3}{3} - 9t + C$

29. $5u - 4 \ln|u| - u^{-1} + C$

31. $-400x^{-1} + 200 \ln|x| + 50x + C$

33. $4 \ln|x| + 5x^{-1} + C$ **35.** $2t^2 + \dfrac{4^t}{\ln(4)} + C$

37. $0.013p + 5$ dollars, p = number of pages

39. $-0.53325t^4 + 1.333t^3 - 0.07665t^2 + 0.3167t$

41. No

43. $F'(x) > 0$ for all x so, $F(x)$ is increasing

45. $x = 2$ only

Section 6.2 *(page 354)*

1. $\dfrac{(x^2 + 3)^6}{6} + C$ **3.** $2x^3 + 5x^2 + C$

5. $\dfrac{(e^x + 1)^2}{2} + C$

7. $0.4x^3 - 1.2x^2 + 0.6x + C$

9. $\dfrac{1}{2}e^{x^2 - 4x} + C$ **11.** $\dfrac{1}{2}e^{3x^2 - 6} + C$

13. $\dfrac{4}{3}e^{3t} + C$ **15.** $[\ln(x)]^2 + C$

17. $\ln|t^2 - t + 2| + C$ **19.** $2 \ln|x^3 - 9| + C$

21. $\ln|\ln(x)| + C$

23. $\dfrac{6}{5}(x - 2)^{5/2} + \dfrac{4}{3}(x - 2)^{3/2} + C$

25. $\dfrac{2}{3}(x^2 + 1)^{3/2} + C$ **27.** $\dfrac{1}{3}(t^2 - 1)^{3/2} + C$

29. $\ln|x^3 - x^2 + x + 1| + C$

31. $\dfrac{1}{4}(2x - 1 + \ln|2x - 1|) + C$

33. $\dfrac{1}{4}\ln|4x^2 - 5| + C$ **35.** $t^4 - 3t^3 + t^2 + C$

37. $\ln(2p^2 + 1) - \dfrac{2}{3}(2p^2 + 1)^{3/2} + C$

39. $2 \ln|t^2 - t| + C$ **41.** Yes; $u = x^2 - 1$

43. No   **45.** No

Section 6.3 *(page 372)*

1. $LHS = 8$ **3.** $LHS = 5.121$

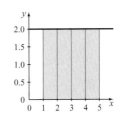

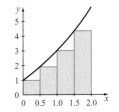

5. $LHS = 89.0$

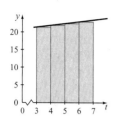

7.

n	LHS
2	0.0625
4	0.1406
10	0.2025

9.

n	LHS
2	9.333
4	8.167
10	7.564

11.

n	LHS
2	6.119
4	6.779
10	7.151

13.

n	LHS
2	50.00
4	61.00
10	67.92

15.

n	LHS
2	27.20
4	21.41
10	18.41

17.

LHS	17.75
RHS	23.75
Avg.	20.75

19.

LHS	2.39
RHS	2.14
Avg.	2.27

21.

LHS	2.00
RHS	2.00
Avg.	2.00

23.

LHS	60.00
RHS	72.00
Avg.	66.00

25.

LHS	90.00
RHS	102.00
Avg.	96.00

27.

LHS	1.40
RHS	2.15
Avg.	1.78

29.

LHS	0.339
RHS	0.686
Avg.	0.512

31.

LHS	0.776
RHS	1.205
Avg.	0.990

33.

LHS	2.67
RHS	3.36
Avg.	3.01

35.

LHS	324.69
RHS	441.86
Avg.	383.28

37. (a) $3600F(t)$ cubic feet per hour
(b)

LHS	685,115 cubic feet	$\approx 5{,}125{,}037$ gallons
RHS	597,275 cubic feet	$\approx 4{,}467{,}946$ gallons
Avg.	641,195 cubic feet	$\approx 4{,}796{,}492$ gallons

(c) More hydroelectric power is needed at 5:45 a.m. than at 7:45 a.m. (Answers may vary.)

39. About 1,399.686 million gallons. That is, roughly 1.4 billion gallons.

41. $f(x) = 3$ on $[0, 2]$ (Answers may vary.)

43. The left-hand sum will be an underestimate, and the right-hand sum will be an overestimate of the area. An average of the left- and right-hand sums will be the best estimate.

45. Yes. In the case of $f(x) = |-2x + 2|$ on the interval $[0, 2]$, the exact area is 2.0. Using $n = 2$, the left-hand sum estimate yields the exact area. However, using $n = 3$, the left-hand sum estimate yields 2.2, which is greater than the exact area. Increasing the number of rectangles worsened the area estimate.

Section 6.4 *(page 386)*

1. 21   **3.** -3 **5.** 36

7. 2.71875   **9.** 6.875

11. Area = 2   **13.** Area = 8

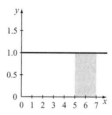
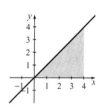

15. Area = 2   **17.** Area = 4

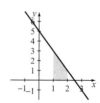
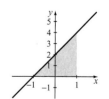

19. Area = 4

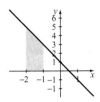

21. 10   **23.** 10   **25.** 4   **27.** 3   **29.** −23

31. $\displaystyle\int_0^3 (-x^2 + 5x - 6)\, dx$

33. $\displaystyle\int_{-2}^4 (-x^3 + 4x + 20)\, dx$

35. $\displaystyle\int_{-2}^4 (-|x - 3| + 3)\, dx$

37.   **39.**

41. 10.5 gallons per person. From the end of 1990 to the end of 1999, the consumption of bottled water increased by 10.5 gallons per person.

43. 5.06 pounds per person. From the end of 1980 to the end of 1995, the consumption of mozzarella cheese increased by 5.06 pounds per person.

45. According to the model, the number of students participating in the AB Calculus exam increased by roughly 80,690 students from the end of 1992 to the end of 2002. That is, the number of students taking the exam in 2002 is calculated to be 80,690 students higher than the number of students taking the exam in 1992.

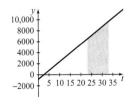

47. This inequality is equivalent to the left-hand sum $\leq \displaystyle\int_a^b f(x)\, dx \leq$ right-hand sum, which is true for an increasing function.

49. $f(x) = 0,\ f(x) = (x - 3),\ f(x) = (x - 3)^3$
(Answers may vary.)

Section 6.5 *(page 400)*

1. $\displaystyle\int_5^7 1\, dx = 2$ **3.** $\displaystyle\int_0^4 x\, dx = 8$

5. $\displaystyle\int_1^2 (-3x^2 + 5)\, dx = -2$

7. $\displaystyle\int_{-1}^1 (8x^3 + 2x)\, dx = 0$

9. $\displaystyle\int_1^3 \left(\frac{1}{x}\right) dx = \ln(3) \approx 1.099$

11. $\displaystyle\int_2^5 (2^t)\, dt \approx 40.40$

13. $\displaystyle\int_{-2}^4 (5 \cdot 2^x)\, dx \approx 113.61$

15. $\displaystyle\int_0^1 \left(\frac{2x}{x^2 + 1}\right) dx = \ln(2) \approx 0.693$

17. $\displaystyle\int_0^2 (x^2 - 1)(x^3 - 3x)^3\, dx = \frac{4}{3} \approx 1.33$

19. $\displaystyle\int_{-2}^{-1} (-2x + 1)\, dx = 4$

21. Area $= 78\frac{1}{12} \approx 78.08$

23. Area $= \frac{1}{12} \approx 0.08$ **25.** Area $= 2$

27. Area ≈ 5.28 **29.** Area $= 1\frac{5}{6} \approx 1.83$

31. $\displaystyle\int_{33}^{43} (315.70t - 770.64)\, dt \approx 112,260$
According to the model, 112,260 more people will take the exam in 2012 than in 2002.

33. $\displaystyle\int_0^{1.5} V(t)\, dt \approx 1.318$ miles

35. $\displaystyle\int_{10}^{19} (0.0009778t^3 - 0.02958t^2 + 0.2392t$
$- 0.06311)\, dt = 2.3$ pounds per person
Between the end of 1990 and the end of 1999, the per capita mozzarella cheese consumption increased by 2.3 pounds per person.

37. $F(4) = 2\ln(4) + 1 \approx 3.77$

39. Area ≈ 3.46

CHAPTER 6 REVIEW EXERCISES

Section 6.1 *(page 401)*

1. $G(x) = -x^2 + 5x + C$

3. $H(x) = 2x^2 - 3\ln|x| + C$

5. $x^4 - \dfrac{5}{2}x^2 - x + C$ **7.** $t + \dfrac{1}{2}\ln|t| + C$

9. $0.018p + 1.30$ dollars, where p is the number of pages

Section 6.2 *(page 402)*

11. $\dfrac{1}{12}(x^2 - 9)^6 + C$ **13.** $[\ln(x)]^2 + C$

15. $3\ln|t| + \dfrac{1}{2}t + C$

Section 6.3 *(page 402)*

17.

n	LHS
2	48.0
4	40.0
10	35.2

19.

n	LHS
2	6.417
4	6.954
10	7.276

21.

LHS	0.906
RHS	1.156
Avg.	1.031

23.

LHS	3.5095
RHS	3.8845
Avg.	3.697

25.

LHS	2.34375
RHS	1.59375
Avg.	1.96875

27. (a) $3600f(t)$
 (b) The total amount of water that flowed past the Priest Rapids Dam between 5:00 p.m. and 9:45 p.m. on November 22, 2002, is estimated to be 1,584,925,729 cubic feet.
 (c) 15,849 swimming pools

Section 6.4 *(page 403)*

29. 15

31. Area = 21

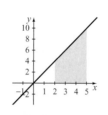

33. Area = 9

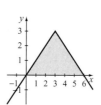

35. Area = 4

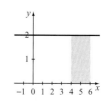

37. 12

39. 1

41.

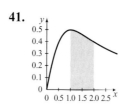

43. (a) -1.385 gallons per person. From the end of 1993 to the end of 1999, the per capita consumption of milk decreased by 1.385 gallons.
 (b) Answers may vary.

Section 6.5 *(page 404)*

45. 8.5   **47.** 0   **49.** 0.693   **51.** 132.5

53. 4.605   **55.** $\dfrac{16}{3}$ **57.** ≈ 5821.04 dollars per person

CHAPTER 7

Section 7.1 *(page 414)*

1. $-e^{-x}(x + 1) + C$

3. $\dfrac{3}{2}t^2 \ln(t) - \dfrac{3}{4}t^2 - 5t\ln(t) + 5t + C$

5. $\dfrac{1}{4}t^4 \ln(t) - \dfrac{1}{16}t^4 + C$

7. $e^{2x+1}(2x - 1) + C$ **9.** $e^{2x}(x - 1) + C$

11. $\dfrac{\ln|3t - 1|}{\ln(3)} + C$ **13.** $3xe^x - 3e^x - x^3 + C$

15. $\dfrac{1}{2}x^4 - 9x^2 + C$ **17.** $\dfrac{1}{2}t^6 - t^3 + C$

19. $\dfrac{3^x}{\ln(3)}\left(x^2 - \dfrac{2x}{\ln(3)} + \dfrac{2}{[\ln(3)]^2} - 3\right) + C$

21. $\dfrac{1}{2}e^{3x^2} + C$ **23.** $\dfrac{1}{3}e^{2t^3} + C$

25. $\dfrac{2^{\ln(x)}}{\ln(2)} + C$

27. ≈ 280.05 million dollars

29. ≈ 3773.7 million dollars

31. $e^x(4x^2 - 8x + 8) + C$

33. $4t^{3/2}\ln(t) - \dfrac{8}{3}t^{3/2} + C$

35. When the integrand is the product or quotient of two different types of functions (i.e. polynomial, exponential, etc.), the integration by parts method is often needed.

Section 7.2 *(page 424)*

1. $\dfrac{1}{6}$ **3.** $\dfrac{64}{3}$ **5.** $\dfrac{1}{6}$ **7.** $10\dfrac{2}{3}$

9. 5980.50   **11.** ≈ 7.647 **13.** 1

15. ≈ 0.0975 **17.** ≈ 3.460 **19.** ≈ 0.0858

21. $\approx 40,804.7$ million dollars

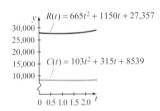

23. 3,569,382 people. According to the model, the number of adults in prison, in jail, on probation, or on parole increased by 3,569,382 more people than the number of students enrolled in private colleges from the end of 1980 through 1998.

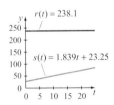

25. 773 million dollars. According to the model, the annual profit increased by 773 million dollars from the end of 1990 through 2001.

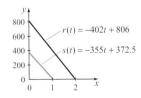

27. 13.168   **29.** 1.5

Section 7.3 *(page 434)*

1. $y = \pm\sqrt{x^2 + C}$ **3.** $y = \left(\dfrac{x}{2} + C\right)^2$

5. $y = \dfrac{x}{1 + Cx}$ **7.** $y = \pm Ce^{0.25x} + 4$

9. $y = \dfrac{4Ce^{4x}}{1 + Ce^{4x}}$ **11.** $y = \sqrt{2x^2 + 14}$

13. $y = -e^{3x}$ **15.** $y = -5e^{25x} + 5$

17. $y = \dfrac{4e^{4x}}{7 + e^{4x}}$ **19.** $y = \dfrac{100e^{2x}}{50e^{2x} - 49}$

21. (a) $T(t) = 76 - 41.7e^{-0.01344t}$
(b) $0.49°F/min$; $0.29°F/min$

23. ≈ 19.08 min

25. About 5 hours 8 minutes

27. About 5 hours 24 minutes

29. $y = 1000e^{0.1t}$ **31.** $A(t) = 2500e^{0.17293t}$

33. $\dfrac{dA}{dt} = Pe^{kt}(k) = kA$ **35.** $y = \dfrac{\pm C}{\sqrt{|x|^3}}$

Section 7.4 *(page 449)*

1. (a) 14.56 cm
(b) 16.25 cm
(c) Female

3. (a) 2.86 cm/year
(b) 4.921 cm/year
(c) Male; male is 20 cm at $t \approx 1.994$ years and female is 20 cm at $t \approx 1.555$ years

5. (a) 84.00 cm
(b) 11.438 cm/year

7. (a) 927
(b) $\dfrac{dN}{dt} = -0.000084335N(919.9 - N)$
(c) ≈ -5.86 cases per year. According to the model, the number of new pediatric AIDS cases was dropping at a rate of approximately 6 cases per year at the end of 2000.
(d) Yes. It appears that the number of new cases will continue to drop.

9. (a) $y = \dfrac{85.88}{1 + 29.65e^{-0.9249t}}$. This equation represents the percentage of public schools with access to the Internet, where t is the number of years since the end of 1994.

(b) Increasing at the rate of 7.4 percentage points per year

11. (a) $\dfrac{dN}{dt} = 0.02311N(27.80 - N)$

(b) Increasing at \$4.03 per year and \$1.98 per year, respectively

13. 57.1 minutes; 2124 people per minute

15. 22 homicides per year

17. (a) Week 1

(b) The model predicts that the cumulative gross box office sales will level out at about 242 million dollars after Week 51. Therefore, new sales are diminishing.

19. About 69.75%; about 0.15 percentage point per year

21. Proof:

$$\frac{dy}{dt} = k(M - y)$$

$$\frac{dy}{(M - y)} = k\,dt$$

$$\int \frac{dy}{M - y} = \int k\,dt$$

$$-\ln(M - y) = kt + C \qquad \text{Assuming } M > y$$
$$\ln(M - y) = -kt - C$$
$$e^{-kt-C} = M - y$$
$$y = M - e^{-kt}e^{-C}$$

Since $y(0) = 0$, we have

$$0 = M - e^{-k \cdot 0}e^{-C}$$
$$M = e^{-C}$$

So

$$y = M - e^{-kt}(M)$$
$$= M(1 - e^{-kt})$$

23. Proof:

$$D(t) = \frac{100}{1 + 200e^{-0.1t}}$$
$$= 100(1 + 200e^{-0.1t})^{-1}$$
$$D'(t) = -100[(1 + 200e^{-0.1t})^{-2}]$$
$$[200e^{-0.1t}(-0.1)] \qquad \text{Chain Rule}$$
$$= \frac{2000e^{-0.1t}}{(1 + 200e^{-0.1t})^2}$$

We will now check to see if $D'(t) = 0.001[D(t)]$ $\{100 - [D(t)]\}$ yields the same result.

$$D'(t) = 0.001[D(t)]\{100 - [D(t)]\}$$

$$= 0.001\left(\frac{100}{1 + 200e^{-0.1t}}\right)\left(100 - \frac{100}{1 + 200e^{-0.1t}}\right)$$

$$= 10\left(\frac{1}{1 + 200e^{-0.1t}}\right)\left(1 - \frac{1}{1 + 200e^{-0.1t}}\right)$$

$$= 10\left(\frac{1}{1 + 200e^{-0.1t}}\right)\left(\frac{(1 + 200e^{-0.1t}) - 1}{1 + 200e^{-0.1t}}\right)$$

$$= 10\left(\frac{1 + 200e^{-0.1t} - 1}{(1 + 200e^{-0.1t})^2}\right)$$

$$= \frac{2000e^{-0.1t}}{(1 + 200e^{-0.1t})^2}$$

Since both approaches yielded the same result, $D'(t) = 0.001[D(t)]\{100 - [D(t)]\}$.

CHAPTER 7 REVIEW EXERCISES

Section 7.1 *(page 454)*

1. $-2e^{-x}(x + 1) + C$

3. $e^x(3x^2 - 6x + 6) + C$ **5.** $\dfrac{1}{4}[\ln(x)]^4 + C$

7. $\dfrac{5}{2}x^2 \ln(x) - \dfrac{5}{4}x^2 - 3x \ln(x) + 3x + C$

9. $e^{4x}\left(x - \dfrac{1}{4}\right) + C$

Section 7.2 *(page 454)*

11. 72   **13.** ≈ 3.46 **15.** ≈ 29.264

17. According to the model, the accumulated change in profit from the end of 1999 through 2001 is 3918 million dollars.

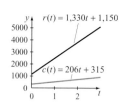

19. The gap between public- and private-school teachers is predicted to increase by 116.15 thousand teachers between the year 2000 and the year 2010.

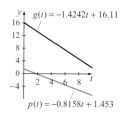

Section 7.3 *(page 455)*

21. $y = \pm k|x|$

23. $A(t) = 16 \pm ke^{-2t}$

25. $A(t) = \dfrac{4ke^{4t}}{1 + ke^{4t}}$

27. $A(t) = \dfrac{1}{5t - 2}$

29. $y = 10 + e^{0.1x}$

31. $T(t) = 70 + 28.6e^{-0.01189t}$; about 88 minutes

Section 7.4 *(page 455)*

33. (a) $\dfrac{dN}{dt} = 0.01473N(121.8 - N)$

(b) 41.91; 14.87 million units per year

35. (a) 69.07 cm

(b) 9.3 cm per year

37. 0.134 homicide per 100,000 people per year (about 1.34 homicides per million people per year)

CHAPTER 8

Section 8.1 *(page 469)*

1. $z = 3x - 2y$

x ↓	$y \to$		
	-1	0	1
-2	-4	-6	-8
0	2	0	-2
2	8	6	4

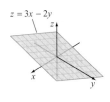

3. $z = 4xy + 4$

x ↓	$y \to$		
	-1	0	1
-2	12	4	-4
0	4	4	4
2	-4	4	12

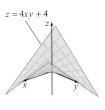

5. $f(x, y) = e^{-xy}$

x ↓	$y \to$		
	-1	0	1
-2	0.135	1.00	7.39
0	1.00	1.00	1.00
2	7.39	1.00	0.135

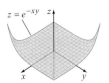

$z = e^{-xy}$

Note: In the graph, each square of the grid has dimensions $\frac{1}{4} \times \frac{1}{4}$.

7. $f(x, y) = \dfrac{x^2 + 1}{y^2 + 1}$

x ↓	$y \to$		
	-1	0	1
-2	2.5	5.0	2.5
0	0.5	1.0	0.5
2	2.5	5.0	2.5

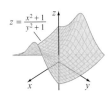

$z = \frac{x^2+1}{y^2+1}$

Note: In the graph, each square of the grid has dimensions $\frac{1}{4} \times \frac{1}{4}$.

9. $z = (x + y)^3$

x ↓	$y \to$		
	-1	0	1
-2	-27	-8	-1
0	-1	0	1
2	1	8	27

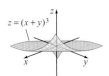

$z = (x + y)^3$

Note: In the graph, each square of the grid has dimensions $\frac{1}{4} \times \frac{1}{4}$.

11. 6889 thousand vehicles

13. \$4938

15. 504 square inches

17. A 187-pound man who is 71 inches tall has body mass index ≈ 26.1. (Answers will vary.)

19. $-15°F$

21. $\{(x, y)\,|\,|x| \geq 1 \text{ and } y \neq 0\}$

23. A number of variables affect a person's take-home pay, including any of the following: gross pay amount, federal income tax withholding, OASI deduction, Medicare deduction, medical insurance premium deduction, retirement plan deduction, etc. (Answers may vary.)

Section 8.2 *(page 485)*

1. $f_x = 3y$
$f_y = 3x$
$f_{xx} = 0$
$f_{xy} = 3$
$f_{yy} = 0$

3. $f_x = 4y$
$f_y = 4x - 2y$
$f_{xx} = 0$
$f_{xy} = 4$
$f_{yy} = -2$

5. $W_t = v^{0.5}$
$W_v = 0.5tv^{-0.5}$
$W_{tt} = 0$
$W_{tv} = 0.5v^{-0.5}$
$W_{vv} = -0.25tv^{-1.5}$

7. $V_r = 2\pi rh$
$V_h = \pi r^2$
$V_{rr} = 2\pi h$
$V_{rh} = 2\pi r$
$V_{hh} = 0$

9. $z_x = \ln(xy) + 1$
$z_y = xy^{-1}$
$z_{xx} = \dfrac{1}{x}$
$z_{xy} = \dfrac{1}{y}$
$z_{yy} = -xy^{-2}$

11. $V_r = 2\pi rh$
$V_h = \pi r^2$

$V_r(3, 2) = 12\pi$ means that when the height of the cylinder is 3 inches and the radius is 2 inches, increasing the radius by 1 inch while leaving the height unchanged will increase the volume by about 12π cubic inches.
$V_h(3, 2) = 4\pi$ means that when the height of the cylinder is 3 inches and the radius is 2 inches, increasing the height by 1 inch while leaving the radius unchanged will increase the volume by about 4π cubic inches.

13. $W_T = 0.6215 + 0.4275V^{0.16}$
$W_V = 0.16V^{-0.84}(-35.75 + 0.4275T)$

$W_T(27, 5) \approx 1.2$ means that when the temperature is 27°F and the wind speed is 5 mph, increasing the temperature by 1°F while leaving the wind speed unchanged will *increase* the wind chill temperature by about 1.2°F.
$W_V(27, 5) \approx -1.0$ means that when the temperature is 27°F and the wind speed is 5 mph, increasing the wind speed by 1 mph while leaving the temperature unchanged will *decrease* the wind chill temperature by about 1°F.

15. $B_H = -1406WH^{-3}$
$B_W = 703H^{-2}$

$B_H(67, 155) \approx -0.7$ means that when a person is 67 inches tall and weighs 155 pounds, increasing the person's height by 1 inch while leaving the weight constant will decrease the body mass index by about -0.7 index point.
$B_W(67, 155) \approx 0.2$ means that when a person is 67 inches tall and weighs 155 poounds, increasing the person's weight by 1 pound while leaving the height constant will increase the body mass index by about 0.2 index point.

17. $R_E = -PE^{-2}$
$R_P = E^{-1}$

$R_E(4, 32) = -2$ means that when the earnings are $4 per share and the price is $32 per share, increasing the earnings per share by $1 while leaving the share price constant will decrease the price-to-earnings ratio by about 2 units.
$R_P(4, 32) = 0.25$ means that when the earnings are $4 per share and the price is $32 per share, increasing the share price by $1 while leaving the earnings per share constant will increase the price-to-earnings ratio by about 0.25 unit.

19. According to the model, increasing the number of employees from 26,424 to 26,425 while leaving the amount of money spent on advertising constant would *increase* sales by about $0.17 million.
According to the model, increasing advertising spending from $519.2 million to $520.2 million while leaving the number of employees constant would *decrease* sales by about $0.37 million.

21. The tangent line in the x-direction is $z = 0x - 4$.
The tangent line in the y-direction is $z = 8y + 4$.

23. The tangent line in the x-direction is $z = x + 1$.
The tangent line in the y-direction is $z = 0y + 1$.

25. The tangent line in the x-direction is $z = 0x + e$.
The tangent line in the y-direction is
$z = -2ey - e$.

27. $f(x, y) = 2x + 3y$ (Answers will vary.)

29. $a = 2, b = 3, c = -4$. When $x = \dfrac{11}{41}$ and
$y = -\dfrac{1}{41}$, $f_x(x, y) = 1$ and $f_y(x, y) = 1$.

Section 8.3 *(page 500)*

1. $f(x, y) = x^2 + 4y^2 - 10$ has a relative minimum at $(0, 0, -10)$.

3. $f(x, y) = -x^2 - y^3 + 3y - x$ has a relative maximum at $\left(-\frac{1}{2}, 1, 2\frac{1}{4}\right)$ and a saddle point at $\left(-\frac{1}{2}, -1, -1\frac{3}{4}\right)$.

5. $f(x, y) = x^2y - y^2 + x$ has a saddle point at $\left(-1, \frac{1}{2}, -\frac{3}{4}\right)$.

7. $f(x, y) = x^4 - 4x + y^3 - y^2$ has a saddle point at $(1, 0, -3)$ and a relative minimum at $\left(1, \frac{2}{3}, -\frac{85}{27}\right) \approx (1, 0.667, -3.15)$.

9. $f(x, y) = x^3 - 6x + xy - y^2$ has a saddle point at $\left(\frac{4}{3}, \frac{2}{3}, -\frac{140}{27}\right)$ and a relative maximum at $\left(-\frac{3}{2}, -\frac{3}{4}, \frac{99}{16}\right)$.

11. $f(x, y) = \frac{x^2 + 1}{y^2 + 1}$ has a saddle point at $(0, 0, 1)$.

13. $f(x, y) = \frac{-1}{x^2 + 1} + y^4 - 4y^2$ has a saddle point at $(0, 0, -1)$ and relative minima at $\left(0, -\sqrt{2}, -5\right)$ and $\left(0, \sqrt{2}, -5\right)$.

15. $f(x, y) = x^3 - 3x + y^3 - 12y$ has a relative maximum at $(-1, -2, 18)$, a relative minimum at $(1, 2, -18)$, and saddle points at $(-1, 2, -14)$ and $(1, -2, 14)$.

17. $f(x, y) = x^2 - x - 3y^2 + 3y$ has a saddle point at $\left(\frac{1}{2}, \frac{1}{2}, \frac{1}{2}\right)$.

19. $f(x, y) = \frac{x^2 + y^2 + 1}{e^{x^2+y^2}}$ has a relative maximum at $(0, 0, 1)$.

21. $f(x, y) = \frac{x^2 - y^2}{e^{x^2+y^2}}$ has relative maxima at $\left(\pm 1, 0, \frac{1}{e}\right)$, relative minima at $\left(0, \pm 1, -\frac{1}{e}\right)$, and a saddle point at $(0, 0, 0)$.

23. $f(x, y) = e^{-x^2 + 2x - y^2 + y + 1}$ has a relative maximum at $\left(1, \frac{1}{2}, e^{9/4}\right)$.

25. $f(x, y) = x^3y + xy^3 - 4xy$ has relative maxima at $(-1, 1, 2)$ and $(1, -1, 2)$, relative minima at $(-1, -1, -2)$ and $(1, 1, -2)$, and saddle points at $(-2, 0, 0)$, $(0, -2, 0)$, $(0, 0, 0)$, $(0, 2, 0)$, and $(2, 0, 0)$.

Section 8.4 *(page 515)*

1. A relative minimum occurs at $(2, 9, -14)$.

3. A relative maximum occurs at $\left(\frac{3}{8}, \frac{1}{8}, \frac{1}{8}\right)$.

5. A relative maximum occurs at $(4, 4, 32)$, and a relative minimum occurs at $(0, 0, 0)$.

7. A relative minimum occurs at $\left(\frac{1}{5}, -1, -\frac{4}{5}\right)$.

9. A relative minimum occurs at $\left(-\frac{50}{51}, \frac{53}{51}, \approx 21.99\right)$, and a relative maximum occurs at $(-2, -1, 31)$.

11. The box whose dimensions are 30 inches $\times$ 28.24 inches $\times$ 28.24 inches has maximum volume. The volume of this box is 23,917 cubic inches.

13. According to the model, Ford should spend $18,232 million on labor and $21,768 million on capital in order to produce the maximum number of vehicles (10,680 thousand) while remaining within the $40,000 million budget.

15. According to the model, Coca-Cola should spend $44 million on labor and $873 million on capital in order to produce the maximum number of unit cases (18.9 billion) while remaining within the $917 million budget.

CHAPTER 8 REVIEW EXERCISES

Section 8.1 *(page 517)*

1. $z = 4x - 5y$

x ↓	$y \rightarrow$		
	-1	0	1
-2	-3	-8	-13
0	5	0	-5
2	13	8	3

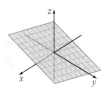

3. $z = xy + 4y$

x ↓	$y \rightarrow$		
	-1	0	1
-2	-2	0	2
0	-4	0	4
2	-6	0	6

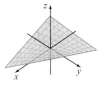

5. For Apple Computer, we have $A = \$3950$ million. For General Electric, we have $A \approx \$59,887.5$ million. Microsoft's debt/equity ratio of 0 means that all of the company's assets are financed through equity. The company doesn't have any debt.

Section 8.2 *(page 518)*

7. $f_x = 4xy - 4y^2$
$f_y = 2x^2 - 8xy$
$f_{xx} = 4y$
$f_{yy} = -8x$
$f_{xy} = 4x - 8y$

9. $f_x = y \ln(y)$
$f_y = x \ln(y) + x$
$f_{xx} = 0$
$f_{yy} = \dfrac{x}{y}$
$f_{xy} = \ln(y) + 1$

11. $V_r(h, r) = 2\pi rh$
$V_r(4, 5) = 40\pi$

When the height is 4 inches and the radius is 5 inches, increasiing the radius by 1 inch will increase the volume by *about* $40\pi \approx 126$ cubic inches.

$V_h(h, r) = \pi r^2$
$V_h(4, 5) = 25\pi$

When the height is 4 inches and the radius is 5 inches, increasing the height by 1 inch will increase the volume by *about* $25\pi \approx 79$ cubic inches.

13. The tangent line in the x-direction is $z = x + 3$.
The tangent line in the y-direction is $z = -y + 2$.

Section 8.3 *(page 519)*

15. $f(x, y) = x^2 - xy + 4y^2 + 10$ has a relative minimum at $(0, 0, 10)$.

Section 8.4 *(page 519)*

17. A relative maximum occurs at $(1, 2, 6)$.

19. A relative maximum occurs at $(6.60, 4.33, 86.6)$. The dimensions of the garden that maximize the area subject to the constraint are shown in the figure.

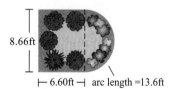

8.66ft

├─ 6.60ft ─┤ arc length =13.6ft

Index

Derivative Rules

For differentiable functions $f(x)$, $g(x)$, and $u = h(x)$ and constants n and c, the following rules apply:

Constant Rule $$\frac{d}{dx}(c) = 0$$	
Power Rule $$\frac{d}{dx}(x^n) = nx^{n-1}$$	**Generalized Power Rule** $$\frac{d}{dx}(u^n) = nu^{n-1}u'$$
Constant Multiple Rule $$\frac{d}{dx}[c \cdot f(x)] = c \cdot f'(x)$$	
Sum and Difference Rule $$\frac{d}{dx}[f(x) \pm g(x)] = f'(x) \pm g'(x)$$	
Product Rule $$\frac{d}{dx}[f(x) \cdot g(x)] = f'(x)g(x) + g'(x)f(x)$$	
Quotient Rule $$\frac{d}{dx}\left[\frac{f(x)}{g(x)}\right] = \frac{f'(x)g(x) - g'(x)f(x)}{[g(x)]^2}$$	
Exponential Rule $$\frac{d}{dx}[b^x] = \ln(b) \cdot b^x$$	**Generalized Exponential Rule** $$\frac{d}{dx}[b^u] = \ln(b) \cdot b^u u'$$
Logarithmic Rule $$\frac{d}{dx}[\log_b x] = \frac{1}{\ln(b) \cdot x}$$	**Generalized Logarithmic Rule** $$\frac{d}{dx}[\log_b u] = \frac{1}{\ln(b) \cdot u}u'$$
Chain Rule $$\frac{d}{dx}\{f[g(x)]\} = f'[g(x)]g'(x)$$	**Chain Rule (Alternate Form)** $$\frac{dy}{dx} = \frac{dy}{dt} \cdot \frac{dt}{dx}$$

Integral Rules

For integrable functions $f(x)$, $g(x)$, $u(x)$, and $v(x)$ and constants n, b, c, and C, the following rules apply:

Power Rule
$$\int x^n \, dx = \frac{x^{n+1}}{n+1} + C \text{ for } n \neq -1$$

Rule for $\dfrac{1}{x}$
$$\int \frac{1}{x} \, dx = \ln|x| + C$$

Constant Multiple Rule
$$\int [c \cdot f(x)] \, dx = c \cdot \int f(x) \, dx$$

Sum and Difference Rule
$$\int [f(x) \pm g(x)] \, dx = \int f(x) \, dx \pm \int g(x) \, dx$$

Exponential Rule
$$\int b^x \, dx = \frac{b^x}{\ln(b)} + C$$

Integration by Parts
$$\int u \, dv = uv - \int v \, du$$

Common Functions	Equation
Constant	$y = c$
Linear	$y = mx + b$
Quadratic	$y = ax^2 + bx + c$
Cubic	$y = ax^3 + bx^2 + cx + d$
Quartic	$y = ax^4 + bx^3 + cx^2 + dx + k$
Power	$y = x^n$
Exponential	$y = ab^x$
Logarithmic	$y = \log_b(x)$

Index of Businesses, Products, and Associations